GW00771931

BRITAIN'S INSECTS

A field guide to the insects of Great Britain and Ireland

Paul D. Brock

Published by Princeton University Press,
41 William Street, Princeton, New Jersey 08540
99 Banbury Road, Oxford OX2 6JX
press.princeton.edu

First published 2021

British Library Cataloging-in-Publication Data is available

Library of Congress Control Number 2020930983
ISBN 978-0-691-17927-8
Ebook ISBN 978-0-691-20499-4

Production and design by **WILD***Guides* Ltd., Old Basing, Hampshire UK.
Printed in Italy

10 9 8 7 6 5 4 3 2

Contents

Introduction

Insects are the most abundant and successful animals on Earth, with almost 25,000 species recorded in Great Britain and Ireland alone. At first glance, this seems an impressive number, but when compared to the world's 1·07 million insect species (80% of all animals), it is actually a relatively small figure. The main reason for this disparity is that the majority of insects favour warmer climates. Nonetheless, the British and Irish insect fauna includes an amazing diversity of shapes and colours and fascinating behaviours – more than enough variety to delight the field naturalist, whether just by watching them, photographing or filming them, or undertaking serious research. And with insects there is always the possibility of finding a new species for the region, or even the ultimate discovery – one completely new to science!

Like many plants and animals around the world, insects are undoubtedly being affected by the combined impacts of human overpopulation and overconsumption, as well as climate change. Recent scientific studies have found that many insects are declining at an alarming rate, and Great Britain and Ireland are no exception. Increasingly, there are press headlines about the threats posed to insect populations, and the question that entomologists (scientists who study insects) are often asked is "*But what use are they?*" For starters, insects are crucial pollinators of many plants (including the crops upon which we rely), break down wastes (including dead animals), help control pests and provide food for other animals and birds. But that is not all. Insects also serve many other important functions, such as producing honey (from honey bees) and silk (from cocoons of some Emperor Moths), although some of these are not yet fully understood. Conservation action is crucial, and incumbent on us all, if insects are to be safeguarded, not only for their intrinsic value, but to ensure they are able to continue to fulfil their vital role in maintaining a healthy environment. Indeed, to quote Sir David Attenborough, "*If we and the rest of the back-boned animals were to disappear overnight, the rest of the world would get on pretty well. But if the invertebrates were to disappear, the world's ecosystems would collapse.*" This is a sobering thought.

But to understand the roles that insects play and to inform any conservation action it is essential to be able to identify the species or types of species concerned – and that was the motivation for producing this book.

Insects can be divided into 25 broad 'types', or orders, reflecting their overall structural and genetic differences. Although many of the orders are rather poorly understood, covering what is known about all the species that have been recorded in Britain and Ireland in a comprehensive way would still require many volumes. Indeed, quite a number of the orders comprise numerous tiny insects, and identifying the species often requires detailed examination of highly technical and often subjective characters. The aim of this book is to provide an up-to-date overview of all the insect orders found in Great Britain and Ireland, to help you identify many of the commoner species, as well as a selection of those that are particularly distinctive, although sometimes rare.

Using this book

This is a photographic guide designed to help you put a name to many of the adult insects you encounter without having to understand overly technical terms. It is divided into separate sections for each of the 25 orders, the introduction to which covers the key characteristics and provides an overview of its general biology and ecology.

For the larger orders, there is a guide to its 'subdivisions' (families and in some cases genera), with an explanation of the constraints in identifying the species correctly. It is important to recognize that in many instances it is not possible to identify species without microscopic study, and that identification to a 'higher' level, such as to genus, is otherwise as far as you can go. Some insect recording groups are taking advantage of photographs on-line, indicating those species that can be identified in the field and what features need to be seen or photographed to ensure a confident identification. The level of detail provided in this book therefore varies depending on the degree of complexity and ease of identification, with some of the more 'popular' and regularly encountered groups being covered in full.

Where to start if you know little or nothing about insects

Identify your adult insect using a step-by-step approach.

– Do you have an insect at all? If in doubt, check the features (see *page 6*).

– The *guide to insect orders* (*page 10*) should help to get you to the right part of the book.

– Further visual keys aim to help narrow down the options to the appropriate **family**, and where covered, to **genus** and/or **species**, predominantly using characters that can be seen in the field with the naked eye or using a hand lens.

Want to learn more?

For each order, suggested sources of further information are provided, including relevant publications, websites and organizations from which more detail can be obtained.

DEFINITIONS

SPECIES' NAMES Each species has an **English name**, in common use, and a ***scientific name*** (in *italics*)

English names generally match those in the National Biodiversity Network (NBN) Atlas https://nbnatlas.org, if up-to-date. For some species, different names can be found in use across a range of websites and publications – which can be confusing! Popular groups, including butterflies, larger macro moths and dragonflies have long-standing English names that are in widespread usage. For those species, families and orders that do not have English names, these have been created for the purposes of this book, using already published names, with some edits, if these are sensible.

Scientific names are stable and used worldwide, only being amended as a result of peer-reviewed research (*e.g.* molecular studies) that changes the view of a species classification, such as reassignment to a different genus. The scientific name of a species consists of two Latin or Latinized words that together form a unique reference to that species. The first word refers to the genus (a group of allied species), the second to the species within that genus.

Some entomologists take the view that scientific names must be learned. However, these are often difficult to pronounce or remember, and many people therefore prefer to use English names. But if discussing species with others it is generally easier to be aware of the scientific name to reduce the chance of confusion. Having said that, Great Britain and Ireland has sometimes been several years behind European or worldwide research in accepting changes to scientific names, although this is gradually being reflected as updated checklists of our insect fauna are published.

GEOGRAPHICAL AREA COVERED BY THE BOOK

Great Britain (GB) – refers to England, Wales and Scotland, including the entomologically rich offshore Scottish islands, including the Hebrides, Orkney and Shetland.

Ireland – refers to Northern Ireland and the Republic of Ireland.

Biogeographically, the Channel Islands are not part of Great Britain or Ireland and are therefore omitted.

Insects and their relatives

The chart below summarizes the wide range of living invertebrates (animals without backbones) that are classed within the phylum **Arthropoda**. Arthropods (name derived from the Greek for 'jointed foot') are invertebrates with a hard external skeleton, segmented body and jointed limbs. The class **Insecta** (insects) is included in the subphylum **Hexapoda** (from the Greek for 'six feet'), together with the class **Entognatha**, which contains the three primitive non-insect orders: **Collembola** (springtails), **Diplura** (two-tailed bristletails) and **Protura** (proturans) (formerly regarded as insects). Entognathans are wingless, have mouthparts enclosed within the head and show very little or no metamorphosis through their life-cycle. Members of the orders Diplura and Protura are eyeless.

Phylum	ARTHROPODA			
Subphylum	**Hexapoda**	**Crustacea**	**Chelicerata**	**Myriapoda**
Class	**INSECTA Insects** ENTOGNATHA [Non-insect hexapods] (Springtails, two-tailed bristletails and proturans)	Crabs, lobsters, shrimps, isopods, crayfish, water fleas, barnacles, woodlice, etc.	ARACHNIDA (spiders and their relatives – which include, harvestmen, mites and ticks) XIPHOSURA (horseshoe crabs) PYCNOGONIDA (sea spiders)	PAUROPODA (pauropods) DIPLOPODA (millipedes) CHILOPODA (true centipedes) SYMPHYLA (pseudocentipedes)

Is it an insect?

Hexapods are unlikely to be confused with **crustaceans** or any **myriapods** – the simplest way of distinguishing between a hexapod and all other arthropods is that hexapods have six legs (three pairs): the thorax having three consolidated segments, each bearing a single pair of legs. In other arthropods, there are at least four pairs of legs, sometimes many more. Arachnids (particularly spiders) are often mistaken for insects, but can be distinguished by several easily recognizable features:

Feature	Insects	Arachnids
Body parts	**3** (head, thorax and abdomen)	**2** (cephalothorax and abdomen), although the distinction is not obvious in all arachnids
Legs	**3 pairs of legs**	**4 pairs of legs**
Eyes	Compound eyes (each comprising a large array of simple photoreception elements)	Simple eyes
Wings	Often have wings	Never have wings
Antennae	1 pair of antennae	No antennae, but do have pedipalps
Metamorphosis	Undergo some sort of metamorphosis: egg → larva → pupa → adult or egg → nymph → adult	Emerge from the egg with all the adult body parts (apart from reproductive organs)

SOME 'INSECT-LIKE' ARTHROPODS

Hexapoda
ENTOGNATHA
6 legs

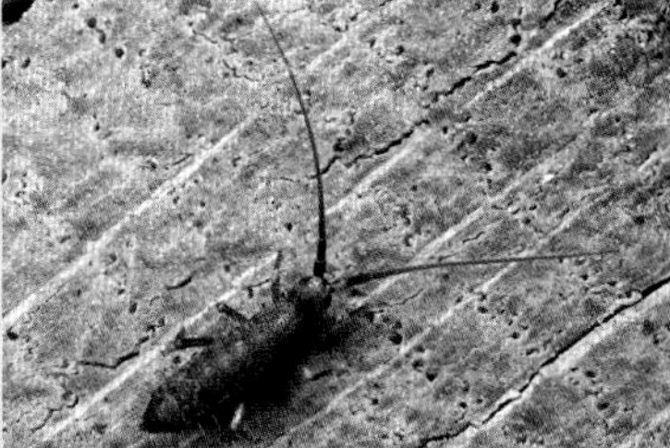

Order Collembola Springtails

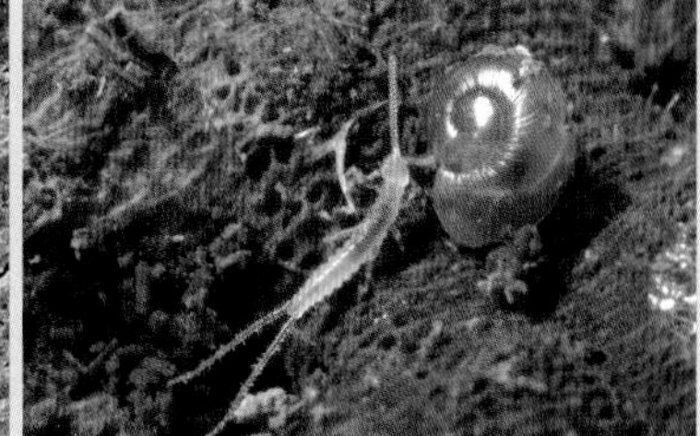

Order Diplura Two-tailed bristletails

Chelicerata
ARACHNIDA
8 legs

Order Araneae Spiders

Order Opiliones Harvestmen

Order Pseudoscorpionida Pseudoscorpions

Order Acari Ticks & mites

Myriapoda
many legs, number varies between species

Class Diplopoda Millipedes

Class Chilopoda Centipedes

Crustacea
8–10 pairs of legs, depending on the species

Class Isopoda Woodlice & relatives

CLASS INSECTA Insects

BI | 540 families [**approx. 25,000 species**] **W** | 1,089 families [**approx. 1,067,000 species**]

The class Insecta can be broadly classified into two 'high-level' groupings: **APTERYGOTA** (wingless, primitive insects) and **PTERYGOTA** (winged or secondarily wingless insects). The subclass **PTERYGOTA** is split into two divisions, **Palaeoptera** and **Neoptera**, the latter having three subdivisions – **Polyneoptera**, **Paraneoptera** and **Holometabola** (or **Endopterygota**).

Each of the divisions and subdivisions is further separated into **orders**, of which there are 25 in total. Representatives of all these orders occur in Britain and Ireland and are presented in this book broadly in the conventional sequence of classification, based on their structure (body features) and means of reproduction (*i.e.* whether metamorphosis is regarded as incomplete (**egg → nymph → adult**) or complete (**egg → larva → pupa → adult**), as illustrated *below*). The sequence followed is as shown *opposite*, with brief explanatory notes. However, for the benefit of those who may not be familiar with this sequence, to help you to reach the correct insect order, the illustrated guides that follow – to adults (*pages 10–18*) and immature stages (*pages 19–25*) – show orders with similar features as close together as possible, cross-referenced to the relevant section of the book.

Complete metamorphosis – Emperor Moth (*p. 410*)

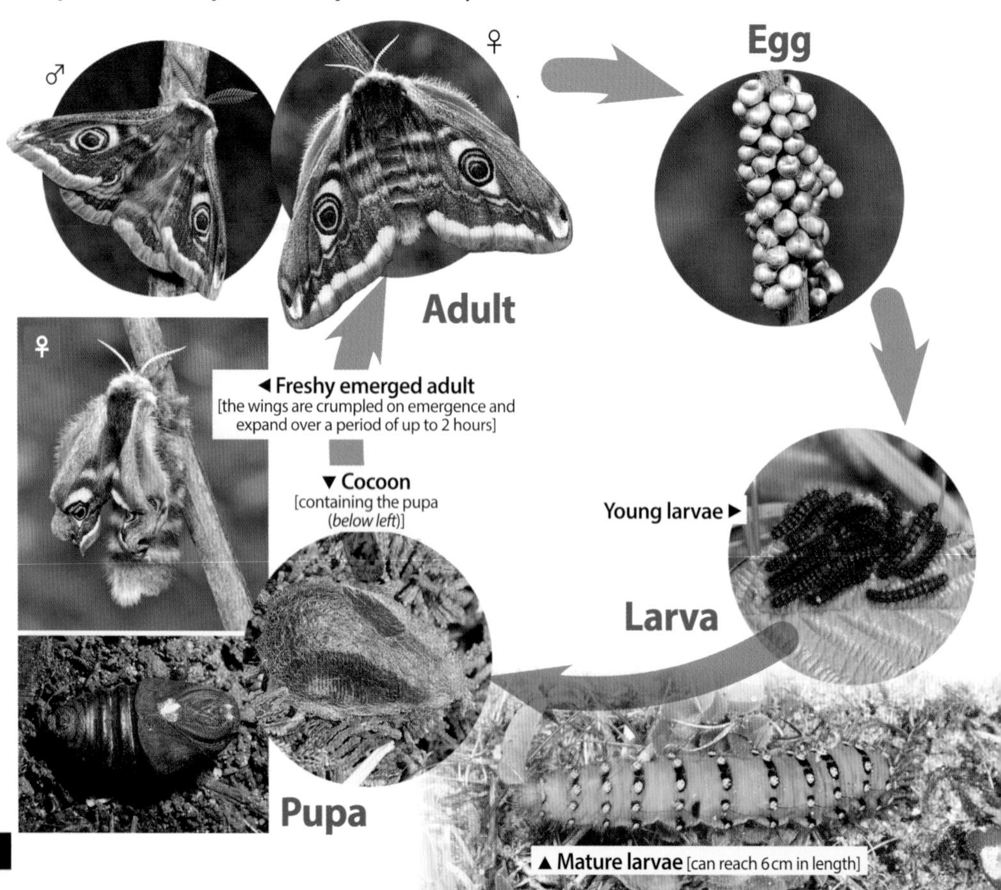

Classification of insects

SUBCLASS APTERYGOTA

Wingless primitive insects	ARCHAEOGNATHA Bristletails	p. 34	ZYGENTOMA Silverfish	p. 38

SUBCLASS PTERYGOTA

Winged or secondarily wingless insects; thorax usually large, the mesothorax and metathorax (see *Glossary, page 28*) **bearing the wings.**

a) **Wings develop externally; the larval forms are known as nymphs, which resemble small wingless adults**

DIVISION PALAEOPTERA

Wings not folded over the abdomen	EPHEMEROPTERA Mayflies	p. 40	ODONATA Dragonflies & damselflies	p. 46

DIVISION NEOPTERA – *see subdivisions for details*

SUBDIVISION POLYNEOPTERA

Wings (if present) folded flat over abdomen, or hidden beneath toughened forewings

PLECOPTERA Stoneflies	p. 74	EMBIOPTERA Webspinners	p. 115	PHASMIDA Stick-insects	p. 117
DERMAPTERA Earwigs	p. 80	MANTODEA Mantids	p. 116	BLATTODEA Cockroaches & termites	p. 125
ORTHOPTERA Grasshoppers & crickets	p. 84				

SUBDIVISION PARANEOPTERA

Often winged, variable in form, mouthparts evolved for piercing and sucking

PSOCODEA Lice, barklice & booklice	p. 132	HEMIPTERA Bugs	p. 137	THYSANOPTERA Thrips	p. 228

b) Wings **develop internally during a pupal stage.**
Larval stages include the well-known 'caterpillars' of butterflies & moths

SUBDIVISION HOLOMETABOLA (= Endopterygota)

mainly winged insects

NEUROPTERA Lacewings & antlions	p.230	DIPTERA Flies	p.330	SIPHONAPTERA Fleas	p.378
MEGALOPTERA Alderflies	p.237	MECOPTERA Scorpionflies	p.365	LEPIDOPTERA Butterflies & moths	p.380
COLEOPTERA Beetles	p.240	TRICHOPTERA Caddisflies	p.369	HYMENOPTERA Ants, bees, wasps & relatives	p.464
RAPHIDIOPTERA Snakeflies	p.327	STREPSIPTERA Stylops	p.377		

Insect orders of Great Britain and Ireland – a simplified guide to adults

The following is a guide to the general characters of each **order** to help in the identification process of an unfamiliar insect. Whilst not foolproof (as some winged orders have species/sexes/forms that are wingless), orders are grouped primarily by wing characteristics, supplemented with a brief list of distinguishing features, notes on variation and taxonomy, and potentially confusing orders.

Reference codes are given to enable quick cross-reference between adult and larva; a **blue box** (number) indicates information about the adult (*pages 11–18*); a **green box** (letter) the immature stage (*pages 19–25*) of that order.

DEFINING WINGLESS and WINGED INSECTS

WINGLESS *p. 11*

Those insect orders found in the region that consist predominantly of species that lack wings entirely, or have wings that are vestigial or unable to be used in sustained flight.

WINGS LESS OBVIOUS *pp. 12–13*

Those insects that, when at rest, have wings that are not immediately obvious unless a closer look is taken. This group consists predominantly of those species with hardened forewings that close over the body to protect the membranous hindwings.

WINGS OBVIOUS *pp. 14–18*

Those insects that, when at rest, have wings that are obvious. This group consists predominantly of those species with membranous wings, but also includes those with scaly, hairy and feather-like wings.

WINGLESS

1 Bristletails | Archaeognatha *p.34*

BL (BODY LENGTH) **RANGE** 8–15 mm; **BODY** scaly; cylindrical; **CERCI 3, medium–long,** held near central structure (epiproct); **HEAD** compound eyes large, with 3 ocelli; elongated 7-segmented maxillary palps.

2 Silverfish | Zygentoma *p.38*

BL RANGE 11–20 mm; **BODY** scaly; less cylindrical, tapered towards rear; **CERCI 3, medium–long,** held pointing outwards; **HEAD** compound eyes reduced or absent, 1–3 ocelli; 5-segmented maxillary palps.

3 Webspinners | Embioptera *p.115*

BL RANGE 6–7 mm; **BODY** elongated; legs short; **WINGS** ♀ wingless; ♂ some winged (as in single known British species). **NOTES:** seldom seen; live inside webs.

4 B Fleas | Siphonaptera *p.378*

BL RANGE 1–6 mm; **BODY** laterally flattened; usually heavily 'armoured'; **LEGS** long, able to jump well. **NOTES:** usually on mammals or birds.

Wingless members of winged groups: some of the most commonly encountered groups are shown below. Some other insect orders have a few species with wingless forms, such as moths (Lepidoptera (*e.g.* Vapourer moth female)) and termites (Blattodea).

12 Aphid HEMIPTERA *p.223*

BL RANGE 1–6 mm; **BODY** soft and often broad, many species with a distinctive pair of organs (siphunculi) on abdomen S5 or S6; **MOUTHPARTS** tube-like rostrum (for sucking).

see also *p. 13*, *p. 15*

13 Lice & booklice PSOCODEA *p.132*

BL RANGE 1–6 mm; **BODY** flattened, broad or elongated; eyes small or absent; **MOUTHPARTS** mandibles (for biting). **NOTES:** some ectoparasites of mammals or birds, others in houses and food stores on moulds.

see also *p. 15*

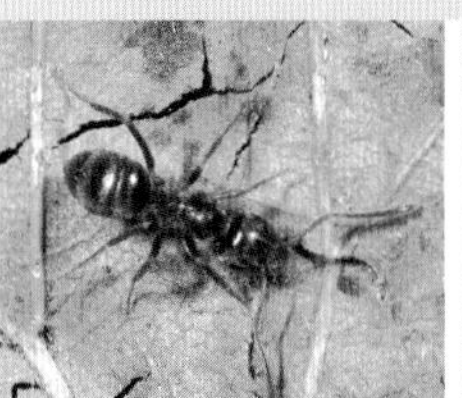

ABDOMEN with waist-like constriction.

23 Ant HYMENOPTERA *p.490*

BL RANGE 1–6 mm (workers; wingless queens much larger). Antennae distinctly 'elbowed'.

23 Wasp HYMENOPTERA *p.466*

BL RANGE 1–5 mm (note presence of ovipositor in ant-like females). Antennae not 'elbowed'.

see also *p. 16*

WINGS LESS OBVIOUS

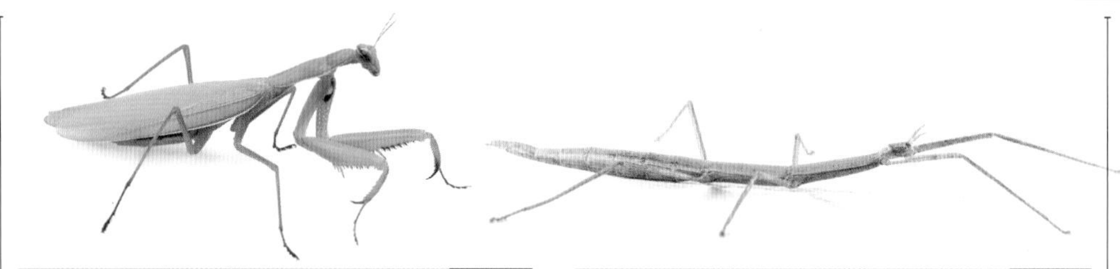

5 Mantids | Mantodea p.116

BL RANGE 43–88 mm; **BODY** large, elongated stick-like; **HEAD** large, triangular eyes; **FORELEGS** with spines; held in praying pose. **NOTE**: very rare in the region.

6 N Stick-insects | Phasmida p.117

BL RANGE 64–125 mm; **BODY** large, very elongated and resembling a stick; **LEGS** long, plain. **NOTE**: UK species wingless and mostly ♀.

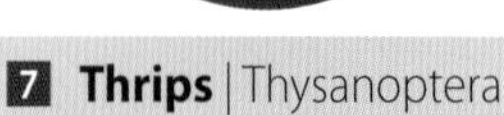

7 Thrips | Thysanoptera p.228

BL RANGE 1–7 mm (usually ≤2 mm); **BODY** small; **WINGS** 2 pairs **feathery** (where present).

8 Earwigs | Dermaptera p.80

BL RANGE 4–26 mm (usually ≤15 mm); **BODY** tip of abdomen with 'forceps' (modified cerci); **WINGS** short forewings, membranous hindwings folded beneath.

GRASSHOPPER

♀

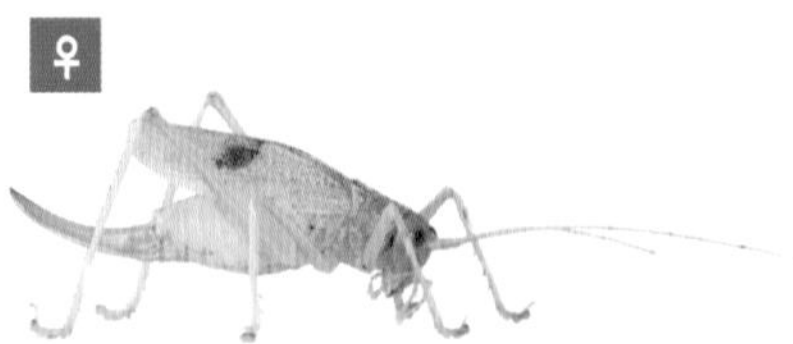

BUSH-CRICKET

long ovipositor in females

TERMITES

COCKROACH

9 M Grasshoppers & crickets | Orthoptera p.84

BL RANGE 8–55 mm; **BODY** broad, often robust; antennae short to very long; **LEGS** most species with enlarged long hindlegs used for jumping; **WINGS** 2 pairs; partly toughened forewings used in flight which cover delicate fan-folded hindwings [rarely wingless]. **NOTE**: many species 'sing' (stridulate).

10 Cockroaches & termites | Blattodea p.125

BL RANGE 4–43 mm; **BODY** rather flattened and broad; head hidden beneath pronotum. **Cockroach** (scavenger) soft-bodied and pale; workers and soldiers wingless. **Termite** (in mounds/underground).

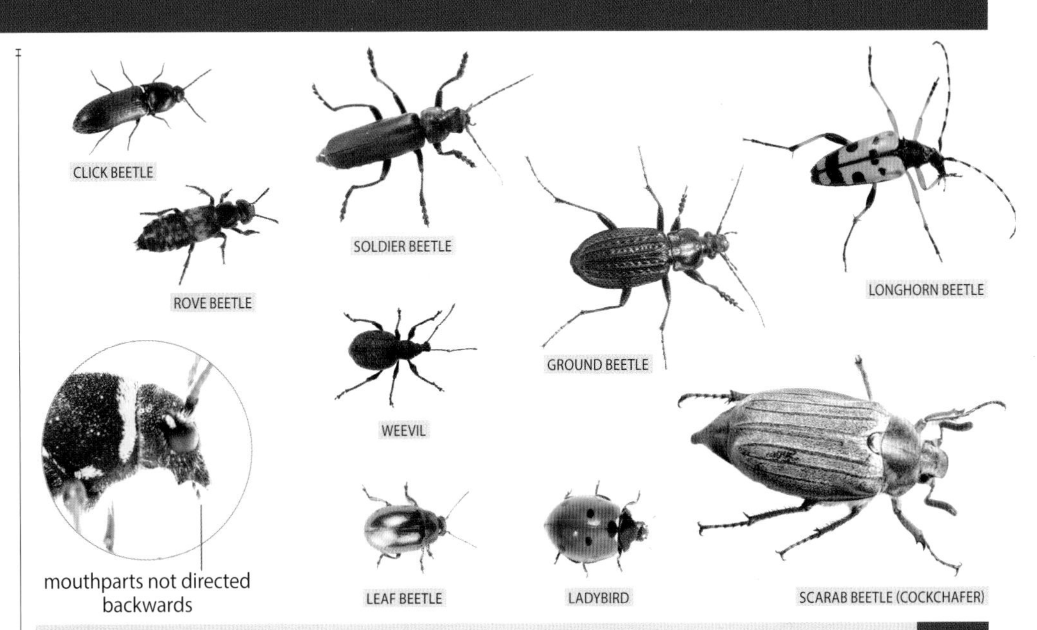

11 C **Beetles** | Coleoptera p. 240

BL RANGE 1–80 mm; **BODY** often robust and compact, some heavily 'armoured'; **WINGS** 2 pairs; toughened forewings (elytra) not used in flight and hardened to protect membranous hindwings; **NOTE:** elytra may be shortened in some species, rarely absent (where long, they meet in a straight line down the back).

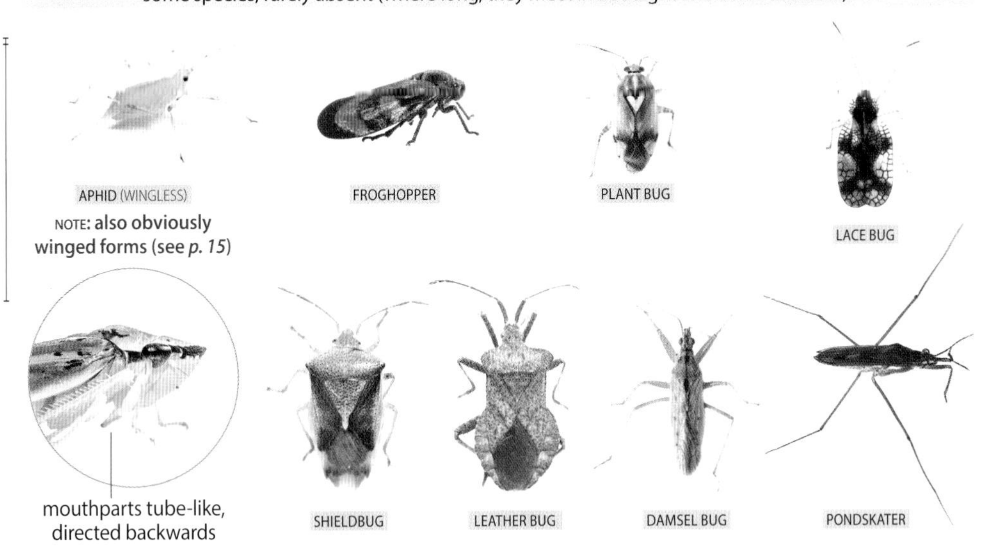

12 L1 & L2 **Bugs** | Hemiptera – **see also** ***p. 11, p. 15*** p. 137

BL RANGE 1–35 mm; **BODY** varied body forms (some families wingless); **HEAD** mouthparts tube-like (for sucking) include an elongated rostrum, directed backwards beneath the head; **WINGS** 2 pairs; forewings (where present) hardened to protect membranous hindwings (wings reduced or absent in some species; also softer in aphids and others).

WINGS OBVIOUS 1/3

19 F Caddisflies | Trichoptera p.369

FWL RANGE 3–28 mm; **BODY** small, broad and soft; **HEAD** mouthparts lack a proboscis; **WINGS** 2 pairs (rarely wingless) held 'tent-like'; **covered in hairs.**
NOTES: found near water.

18 J Mayflies | Ephemeroptera p.40

WS RANGE 7–38 mm; **BODY** not scaly; with 2 or 3 long 'tails' (cerci); **HEAD** antennae small; **WINGS** 2 pairs; held in a vertical plane above the dorsal surface.
NOTES: found near water.

17 Alderflies | Megaloptera p.237

BL approx. 10 mm; **WS RANGE** 22–34 mm; **BODY** broad; **WINGS** 2 pairs; held 'tent-like'; blackish-brown; few, if any, forewing veins forked at hind margin.
NOTES: found near water.

20 K Stoneflies | Plecoptera p.74

BL RANGE 4–25 mm; **BODY** not scaly; with 2 short or long 'tails' (cerci) of equal length; **HEAD** antennae long; **WINGS** 2 pairs; wrapped around abdomen.
NOTES: found near water.

21 E Snakeflies | Raphidioptera p.327

BL RANGE 7–15 mm; **WS RANGE** 15–30 mm; **BODY** 'snake-like' elongated pronotum; ♀ with ovipositor; **WINGS** 2 pairs.

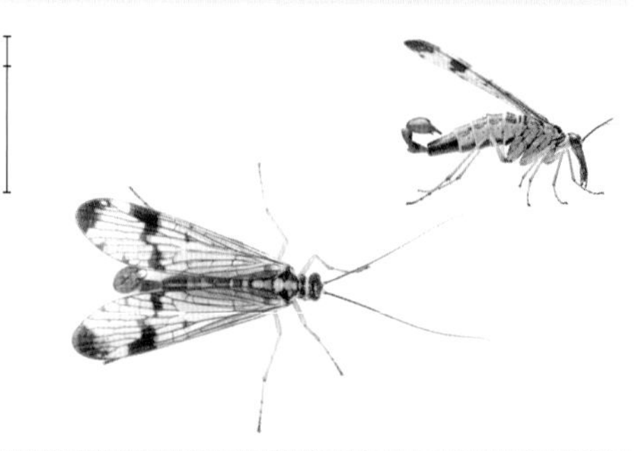

22 Scorpionflies | Mecoptera p.365

BL RANGE 3–15 mm; **WS RANGE** 32–33 mm; **BODY** ♂ 'scorpion-like' raised genital capsule; **HEAD** down-pointing, beak-like mouthparts; **WINGS** 2 pairs; often mottled with black patches; 1 species with forewings reduced.

NOTE: Wax-flies are superficially similar to whiteflies and some aphids (both Hemiptera); wax-flies have mandibles; whiteflies a tube-like rostrum (see Bugs inset on *p. 13*).

WAXFLY

LACEWING

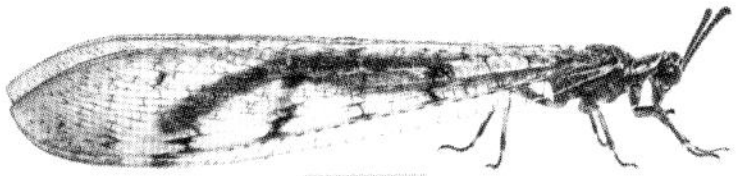
ANTLION

15 D **Lacewings** and **antlions** | Neuroptera p. 230

BL RANGE 2–35 mm; **WS RANGE** 3–50 mm; **BODY** elongated; **WINGS** 2 pairs; held 'tent-like'; brown or green but transparent with several cross-veins.

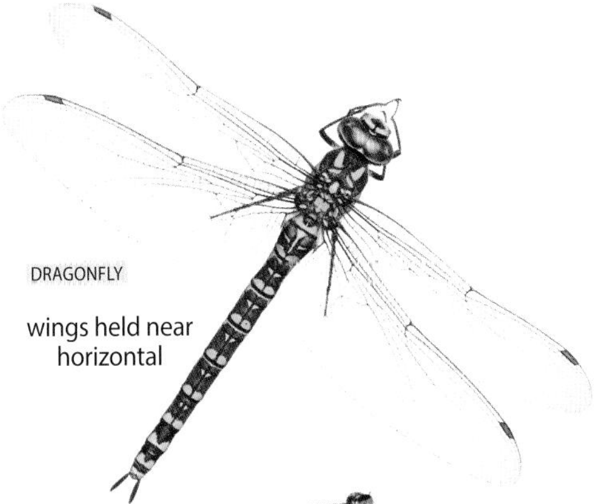
DRAGONFLY

wings held near horizontal

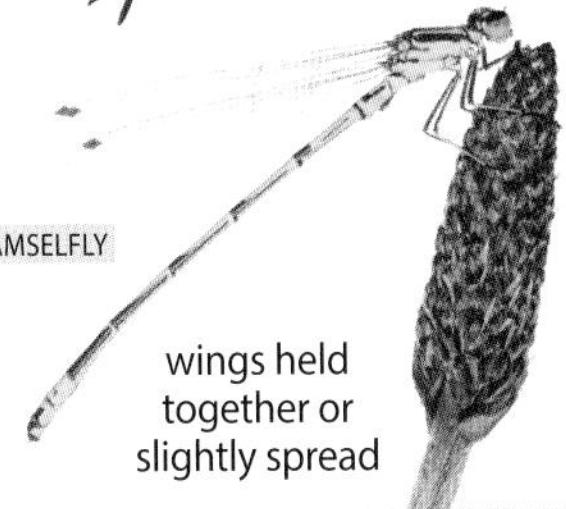
DAMSELFLY

wings held together or slightly spread

16 I **Dragonflies** & **damselflies** | Odonata p. 46

BL RANGE 25–84 mm; **BODY** elongated; **HEAD** eyes large; antennae short; **WINGS** 2 pairs; narrow; folded in vertical plane or slightly spread at rest = **damselfly**; held near horizontally at rest = **dragonfly**.

Obviously winged members of groups that include species/forms/sexes that have less obvious wings or are wingless.

Winged barklice could be confused with aphids and psyllids but barklice differ in their broader heads and biting (not sucking) mouthparts.

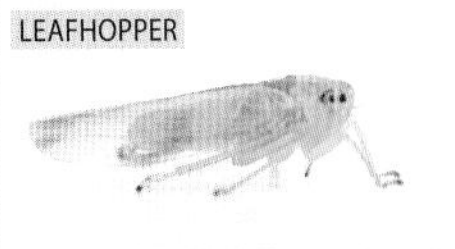
LEAFHOPPER

APHID

JUMPING PLANT LOUSE

12 Aphids, leafhoppers and other winged hemipterans | HEMIPTERA – see also *p. 11*, *p. 13* p. 210 p. 222

BOOKLOUSE

BARKLOUSE

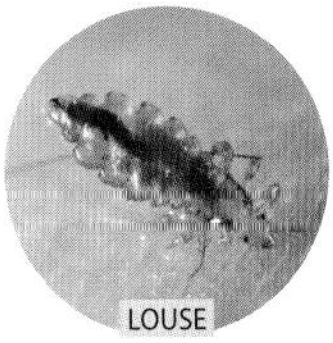
LOUSE

NOTE: Lice are wingless; booklice typically have no wings or wings so small as to appear wingless.

13 **Lice, barklice** and **booklice** | Psocodea – see also *p. 11* p. 132

BL RANGE 1–10 mm; **BODY** small, broad and soft; **MOUTHPARTS** with mandibles (for biting); **WINGS** 2 pairs; held 'tent-like' where present; some species and individuals wingless.

♂

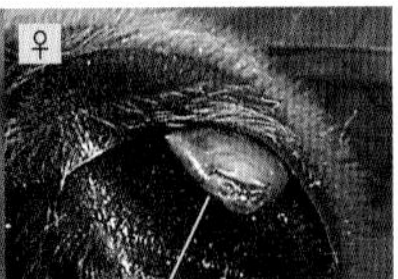
♀

14 **Stylops** | Strepsiptera p. 377

BL RANGE 2–4 mm; **BODY** ♀ larva-like (lives in host); **WINGS** ♂ 1 pair; forewings reduced to 'knobs'; hindwings large; ♀ wingless.

WINGS OBVIOUS 2/3

Mimicry

Certain species resemble those from other orders, some superficially, whereas others are very similar. This is most prevalent in the Diptera (flies), in particular the hoverflies (Syrphidae *p. 353*), and can cause potential confusion when attempting to identify an insect. However, there are structural differences that, although involving small details, provide an easy way to avoid misidentification.

Three moths that mimic hymenopterans – left to right Hornet Moth, Yellow-legged Clearwing and Broad-bordered Bee Hawk-moth.

23 B & G **Ants, wasps, bees & relatives** | Hymenoptera *p. 464*

BL RANGE 1–50 mm; **BODY** diverse range of shapes, many with narrow 'waist'; ovipositor present in ♀ of some species; **HEAD** eyes large, antennae short; **WINGS** **2 pairs**; **hindwings much shorter than forewings**; some species and forms of species wingless (*e.g.* worker ants).

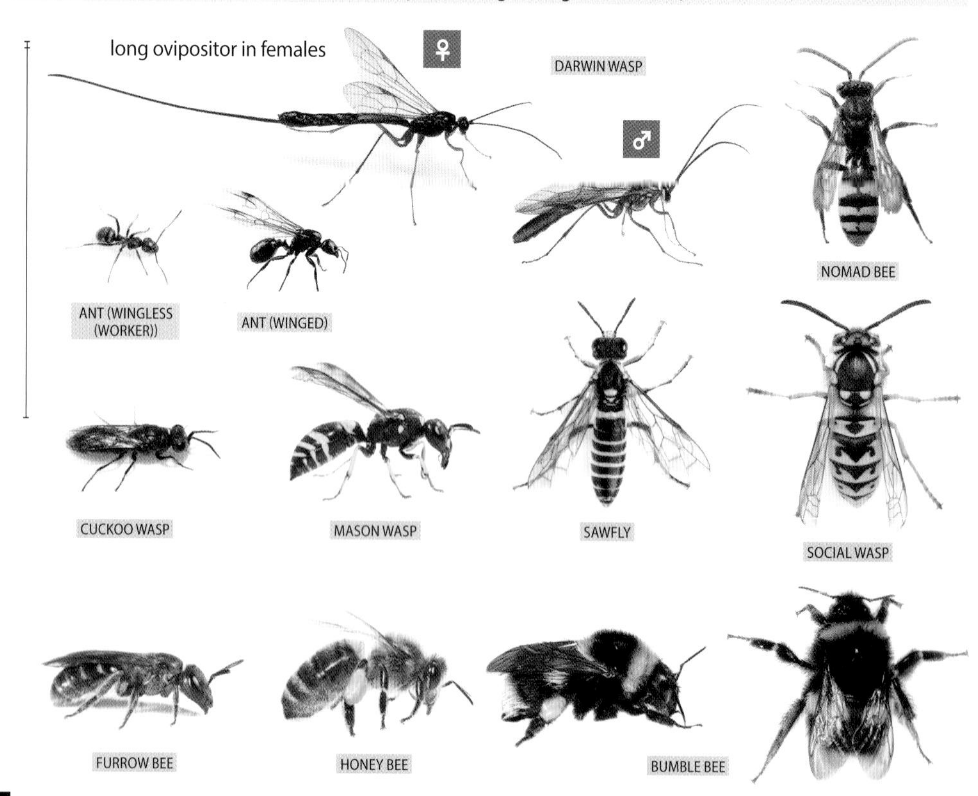

Other bees (*e.g.* mason bees, mining bees, flower bees) are similar to one of the examples shown.

IS IT A FLY (Dipteran) OR A WASP (Hymenopteran)?

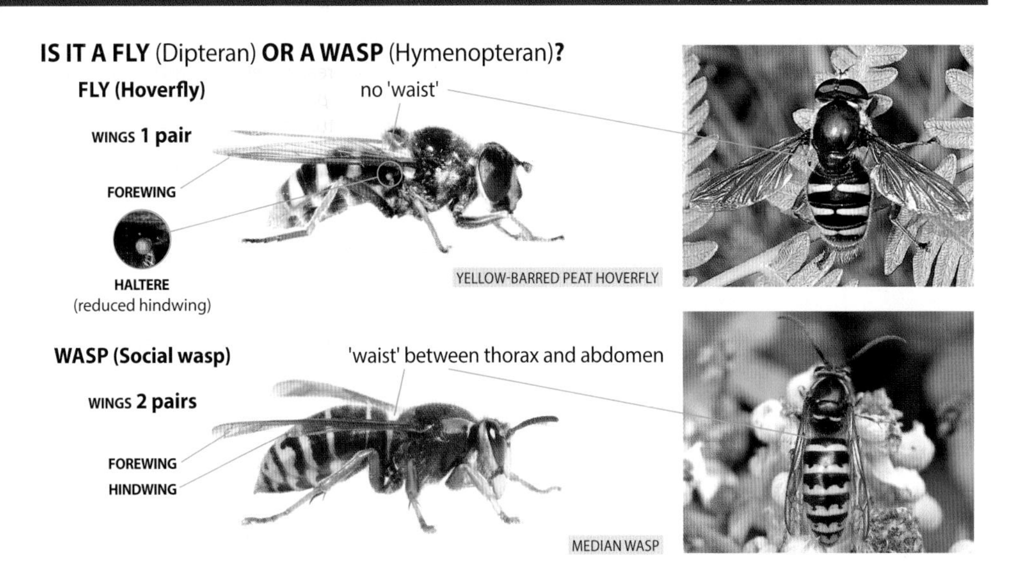

24 A **Flies** | Diptera p.330

FWL RANGE 1–30 mm; BODY diverse range of shapes; HEAD compound eyes large, mouthparts designed to suck or pierce to take liquids, antennae often very short; WINGS **1 pair**; **hindwings reduced to 'knobs' (halteres)**.

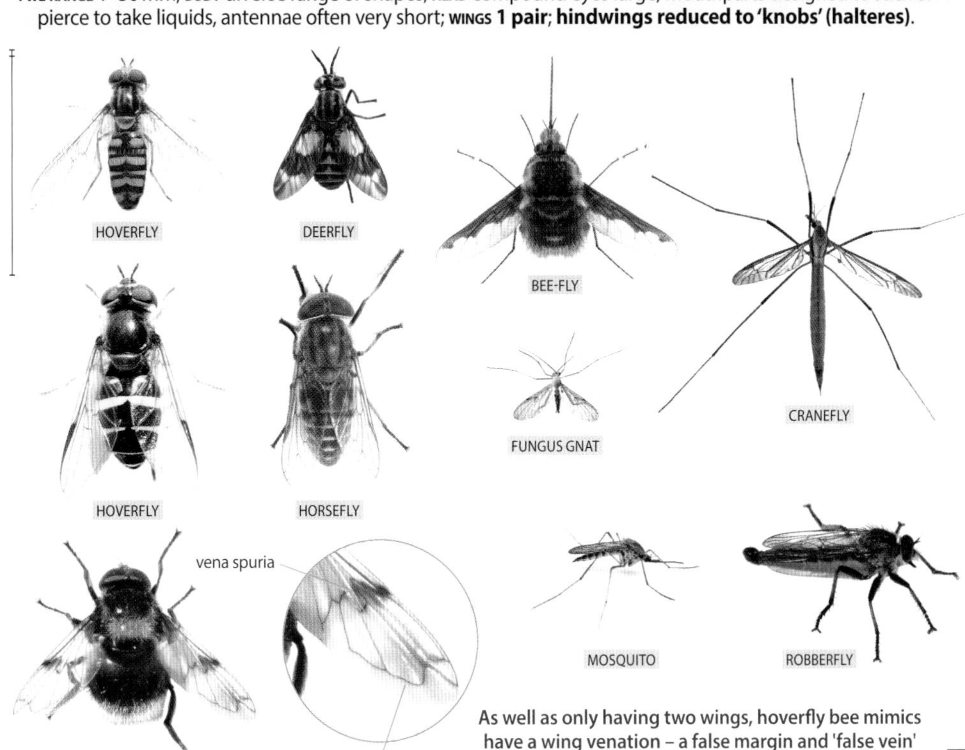

As well as only having two wings, hoverfly bee mimics have a wing venation – a false margin and 'false vein' (vena spuria) – not found in bees.

WINGS OBVIOUS 3/3

25 H **Butterflies** & **moths** | Lepidoptera *p.380*

WS RANGE 3–135 mm (a few moths wingless); BODY elongated; HEAD **Butterfly** mouthparts a coiled tube (proboscis), antennae clubbed; **Moth** antennae thin or feathery; WINGS 2 pairs, **covered in coloured scales**; patterns distinguish larger species and sex in some species.

NOTE: **some moth species have wingless ♀s**

♀ VAPOURER (MOTH)

see also *p. 16* for more moth examples

Insect orders of Great Britain and Ireland – selected larvae and nymphs

This book deals with identification of adult insects. However, it is useful to know what immature stages are like. Larvae (known as nymphs in some orders) vary considerably between species and families. This section shows the larval types commonly encountered. There are some specialist works on identification of the larvae of certain groups, although it is an understudied area and many life-histories are unknown.

WITHOUT LEGS

Larvae that go on to form pupae (mainly terrestrial but includes aquatic Diptera)

A 24 **Flies** | Diptera p.330

Aquatic and terrestrial, larvae commonly known as 'maggots'; fang-like or hook-like mouthparts.

1 **Beautiful Tachinid** *Sturmia bella* (TACHINIDAE) larva (*right*), pupa (*left*) – a parasitoid on nymphalid butterfly (including Small Tortoiseshell) larvae; 2 **Long-palped cranefly** *Tipula* sp. (TIPULIDAE) larva – known as a 'leather-jacket'; 3 **Hoverfly** *Syrphus* sp. (SYRPHIDAE) larva; 4 **Drone Fly** *Eristalis tenax* (SYRPHIDAE) known as a 'rat-tailed maggot'. The tail functions as a breathing tube when the larva is submerged.

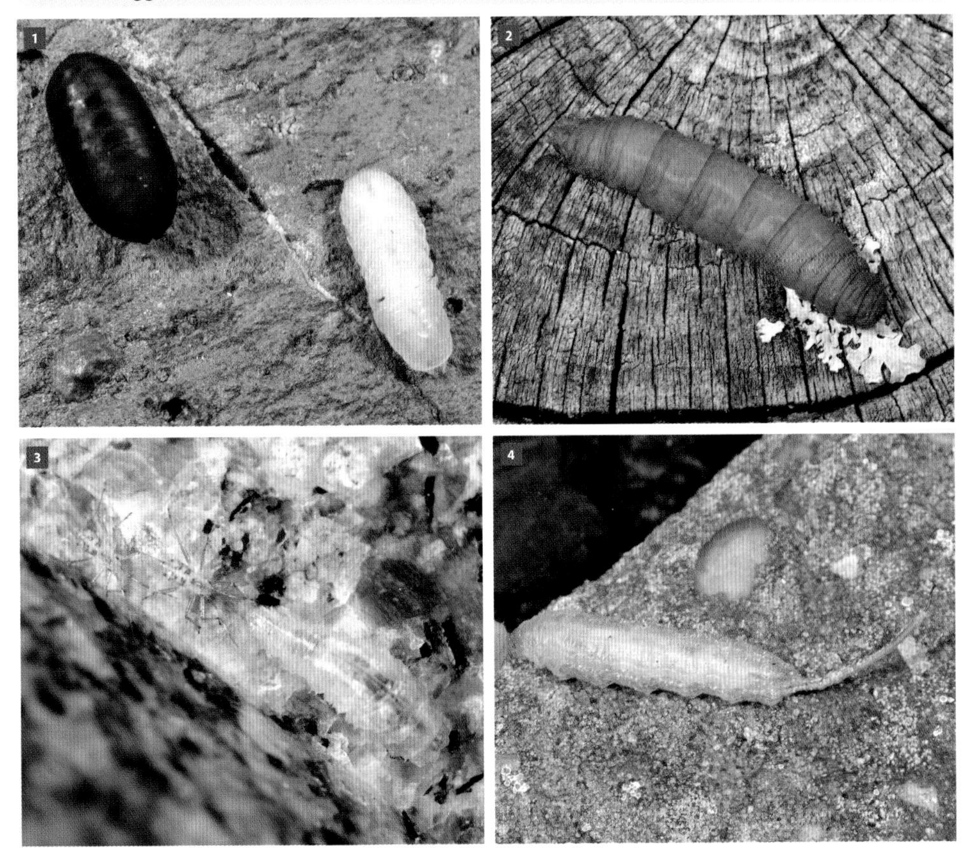

B 23 **Ants, wasps, bees** & **relatives** | Hymenoptera [part] p. 464

Often 'C'-shaped and pointed one end; mouthparts not hook-like.

1 **Large-headed Resin Bee** *Heriades truncorum* (APIDAE) nest cells exposed with larvae; 2 **Hornet** *Vespa crabro* (VESPIDAE) larvae – note also cocoons in the Hornet nest cells; 3 **Small Black Ant** *Lasius niger* (FORMICIDAE), ants tending to larvae; 4 **Small Black Ant** *Lasius niger* (FORMICIDAE) pupae.

NESTS: Larvae hidden in nests, and often tended by 'workers'; two examples shown here: 5 **Yellow Meadow Ant** *Lasius flavus* mound; 6 **Hornet** *Vespa crabro*, with 'workers' guarding entrance to nest, located in a tree.

ALSO IN THIS CATEGORY: some 'micro-moths', mainly small leaf-miners (LEPIDOPTERA); some beetles (COLEOPTERA) and fleas (SIPHONAPTERA).

WITH LEGS | 3 pairs of legs on thoracic segments

Larvae that go on to form pupae (mainly terrestrial)

C 11 Beetles | Coleoptera

p.240

Aquatic and terrestrial, larvae commonly known as 'maggots'; fang-like or hook-like mouthparts. NB some also without legs (see B).

1 **7-spot Ladybird** *Coccinella septempunctata* (COCCINELLIDAE); 2 **Strand-line Burrower** *Broscus cephalotes* (CARABIDAE); 3 **Diving beetle** *Dytiscus* sp. (DYTISCIDAE) – aquatic (NB see also Odonata I); 4 **Cockchafer** *Melolontha melolontha* (SCARABAEIDAE); 5 **Gloomy Carrion Beetle** *Silpha tristis* (SILPHIDAE); 6 **Bloody-nosed Beetle** *Timarcha tenebricosa* (CHRYSOMELIDAE); 7 **Fleabane Tortoise Beetle** *Cassida murraea* (CHRYSOMELIDAE) – with old skins and droppings forming a fecal shield, used in defence against predators; 8 **Glow-worm** *Lampyris noctiluca* (LAMPYRIDAE); 9 **Click beetle** *Melanotus* sp. (ELATERIDAE) – known as 'wire worm'.

D 15 **Lacewings** and **antlions** | Neuroptera

p.230

1, 2 **Suffolk Antlion** *Euroleon nostras* (MYRMELEONTIDAE); 3, 4 **Lacewing** *Chrysoperla* sp. (CHRYSOPIDAE) – larvae sometimes cover themselves in debris (see image 4)

E 21 **Snakeflies** | Raphidioptera

p.327

1 **Snakefly** sp. (RAPHIDIIDAE)

F 19 **Caddisflies** | Trichoptera

p.369

Often case-builders, one terrestrial species (shown), otherwise aquatic.

1 **Land Caddis** *Enoicyla pusilla* (LIMNEPHILIDAE)

G 23 **Ants, wasps, bees** and **relatives** | Hymenoptera [part]

p.464

3 pairs of legs on thoracic segment with 7 or 8 pairs of fleshy 'false' legs on abdomen segments.

1 **Sawfly** sp. (SYMPHYTA)

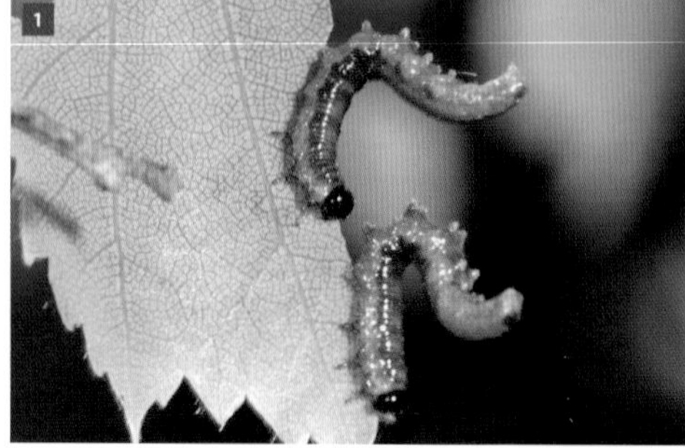

WITH LEGS | 3 pairs of legs on thoracic segments with up to 5 pairs of 'false' legs (prolegs) on abdomen segments

Larvae that go on to form pupae (terrestrial)

H 25 **Butterflies** and **moths** | Lepidoptera p.380

The 'false' legs are equipped with small hooks to cling onto leaves or surfaces.

NB some also without legs (see B).

BUTTERFLIES 1 **Brimstone** *Gonepteryx rhamni* (PIERIDAE); 2 **Swallowtail** *Papilio machaon* (PAPILIONIIDAE); 3 **Small Tortoiseshell** *Aglais urtica* (NYMPHALIDAE); 4 **White-letter Hairstreak** *Satyrium w-album* (LYCAENIDAE);

MOTHS 5 **Privet Hawk-moth** *Sphinx ligustri* (SPHINGIIDAE); 6 **Goat Moth** *Cossus cossus* (COSSIDAE); 7 **Lappet** *Gastropacha quercifolia* (LASIOCAMPIDAE); 8 **Dark Crimson Underwing** *Catocala sponsa* (EREBIDAE).

WITH LEGS

Aquatic nymphs that omit the pupal stage

I 16 **Dragonflies** and **damselflies** | Odonata

p. 46

1 Broad-bodied Chaser *Libellula depressa* (LIBELLULIDAE); **2 Emperor Dragonfly** *Anax imperator* (AESHNIDAE); **3 Large Red Damselfly** *Pyrrhosoma nymphula* (COENAGRIONIDAE) – well camouflaged on stream bed.

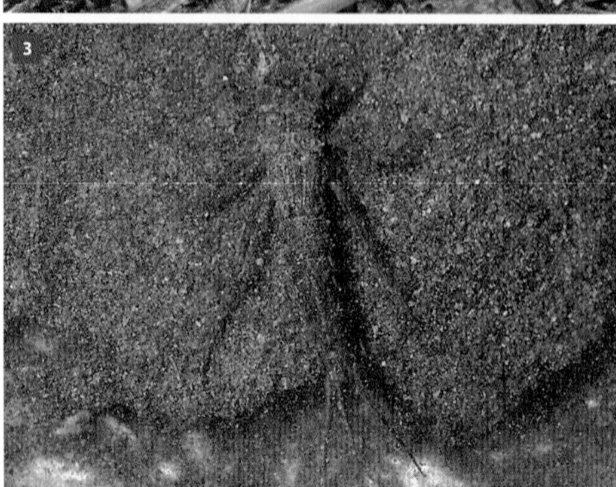

J 18 **Mayflies** | Ephemeroptera

p. 40

1 Green Drake Mayfly *Ephemera danica* (EPHEMERIDAE)

K 20 **Stoneflies** | Plecoptera

p. 74

1 Needle fly *Leuctra* sp. (LEUCTRIDAE)

L1 12 **[Water] Bugs** | Hemiptera [part]

p. 137

1 Common Backswimmer *Notonecta glauca* (NOTONECTIDAE)

WITH LEGS

Terrestrial nymphs that omit the pupal stage

L2 12 [Land] Bugs | Hemiptera [part] **p.137**

1 **Green Shieldbug** *Palomena prasina* (PENTATOMIDAE); 2 **Dock Bug** *Coreus marginatus* (COREIDAE); 3 **Common Barkbug** *Aneurus laevis* (with adult) (ARADIDAE); 4 **Nettle Groundbug** *Heterogaster urticae* (with adults) (LYGAEIDAE); 5, 6 **Common Froghopper** *Philaenus spumarius* (LYGAEIDAE) – protected in froth on stem.

M 9 Grasshoppers and crickets | Orthoptera **p.84**

1 **Mole cricket** *Gryllotalpa gryllotalpa* (GRYLLOTALPIDAE); 2 **Bog Bush-cricket** *Metrioptera brachyptera* (TETTIGONIDAE).

N 6 Stick-insects | Phasmida **p.117**

1 **Prickly Stick-insect** *Acanthoxyla geisovii* – small nymph; 2 **Unarmed Stick-insect** *Acanthoxyla inermis* – large nymph (both PHASMATIDAE).

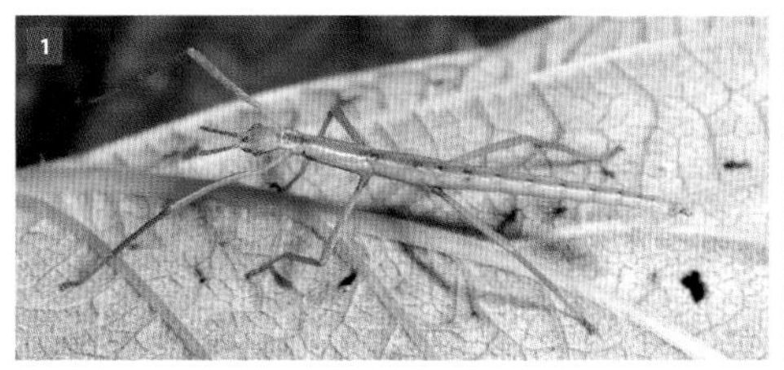

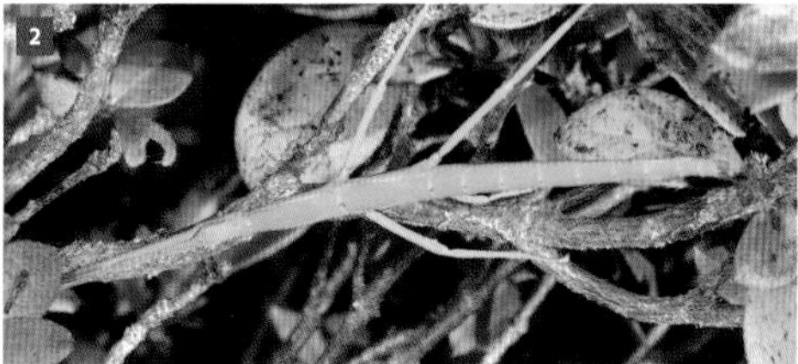

Glossary

As far as possible, the use of technical terms has been avoided in this book. However, some are in frequent use and these are defined here. Terms that are specific to a particular order are included in the annotated photograph at the beginning of the relevant section. **Emboldened** text is used to highlighted terms that are further defined in this glossary.

abdomen the third, rearmost part of an insect's body

abdomen segments subdivisions of the insect's **abdomen** (segment numbers referred to as S1, S2 *etc.*, or a range as *e.g.* S1–4)

aculeate possessing a sting (some bees and wasps (HYMENOPTERA))

adult the final (mature) stage of an insect, during which reproduction occurs

adventive not native and not necessarily established

antenna (plural: antennae) the paired sensory appendages of the head, sometimes called 'feelers' (segment numbers referred to as S1, S2 *etc.*, or a range as *e.g.* S1–4)

arista a specialized bristle or hair-like outgrowth near the **antenna** tip in some flies (DIPTERA) (see *p. 331*)

arolium pad between bases of tarsal claws

body length (BL) the measurement taken from the front of the head to the end of the abdomen, excluding **cerci** and female **ovipositor** (if present)

cell (of wing) a part of the wing membrane closed or partly closed by veins

central appendage central, thread-like attachment to end of **abdomen**, between the **cerci** (*e.g.* in mayflies (EPHEMEROPTERA))

cercus (plural: cerci) the paired, segmented appendages at the end of the **abdomen**, 'tail'-like in some families, including stoneflies (PLECOPTERA)

chrysalis alternative name for **pupa**

clavus the sharply pointed hind part of the forewing (**hemelytra**) found in some bugs (HEMIPTERA)

cleptoparasite a female (♀) seeking prey or stored food from another female, often of a different species (mainly some bees and wasps (HYMENOPTERA))

clypeus the broad lower facial plate above the **labrum** and below the **eyes**, in *e.g.* ants, bees, wasps and relatives (HYMENOPTERA)

cocoon a silk case made by some pupating larvae

complete metamorphosis the **egg**–**larva**–**pupa**–**adult** transition process found in *e.g.* beetles (COLEOPTERA), ants, bees, wasps and relatives (HYMENOPTERA), butterflies and moths (LEPIDOPTERA)

connexivum the flattened, often rounded abdominal border of some bugs (HEMIPTERA)

costa longitudinal wing vein, running along the front edge of the wing, ending near the tip

coxa (plural: coxae) the basal (first) segment of a leg that connects the **thorax** to the **trochanter** (second segment) and the rest of the leg

cross-vein a short vein joining any two neighbouring longitudinal veins

cuneus a wedge-shaped area in some bugs (HEMIPTERA), located at the tip of the thicker, leathery part of the **forewings**

dorsal the upper surface

egg first stage of an insect, which hatches into a **larva** in butterflies and moths (LEPIDOPTERA) [often called a caterpillar], a **nymph** in grasshoppers and crickets (ORTHOPTERA)

elongate used to describe long, slender structures or whole insects, the classic example being a stick-insect (PHASMIDA)

elytra (forewings) a term often used to describe the rigid wing covers (which are not used in flight) of many beetles (COLEOPTERA)

elytral suture the midline between the two wings, often used in beetles (COLEOPTERA)

entomology the study of insects [a person studying insects is an entomologist]

exuvia larval or nymphal skin that remains after moulting; term often used for dragonflies and damselflies (ODONATA)

eye a large eye (compound eye) made up of many separate units (see also **ocellus**)

family in zoological classification, a rank below **order** and above **genus**; name always ends in '–idae'

femur (plural femora) the third segment, and longest part of an insect leg, between the **trochanter** (second segment) and the **tibia** (fourth segment)

forceps moveable pincer-like appendages at the tip of the **abdomen**; modified **cerci** in earwigs (DERMAPTERA)

forewings paired wings of the second thoracic segment, also known as **elytra** and **tegmina**

frons the upper area of the face between the **clypeus** and top of head

gall abnormal plant growth caused by a bacterium, virus, fungus, mite or insect

gaster see **metasoma**

genus an assemblage of species that share one or more character(s) (the first part of the two-part scientific name used to describe a **species** (*e.g. Lucanus* in the Stag Beetle *Lucanus cervus*)

gonapophyses four hardened structures forming the **ovipositor**

gynandromorph an insect with secondary male (♂) or female (♀) characters in the same individual (*e.g.* stick-insects (PHASMIDA))

haltere pin-shaped knob, which is actually a modified **hindwing** in flies (DIPTERA) that assist with balance when the insect is flying (see *p. 331*)

head the first division of the insect body, containing the mouth, **eyes** and **antennae**

hemelytra the forewing of some bugs (HEMIPTERA) characterized by a thickened base and a membranous apex

hindwings the paired wings of the third thoracic segment (sometimes known as alae)

immigrant an insect that has reached another region by natural flight

incomplete metamorphosis the **egg nymph adult** transition process found in *e.g.* bugs (HEMIPTERA), dragonflies and damselflies (ODONATA), grasshoppers and crickets (ORTHOPTERA)

inquiline an insect living as a 'guest' of another species, perhaps in the nest

keel a ridge, for example, on the **pronotum** of some grasshoppers and crickets (ORTHOPTERA)

key a tabulation of identification characters of **species**, **genera**

labium (plural: labia) the second **maxillae** – the lower 'lip' of an insect's mouth

labrum a hinged plate on the face that protects the mouthparts

larva (plural: larvae) the immature stage of an insect that develops by **complete metamorphosis** (*i.e.* those hatched from an egg) – includes caterpillars (butterflies and moths (LEPIDOPTERA))

lateral at or from the side

life-cycle the time between the fertilization of the **egg** and the death of the **adult**; stages of reproduction, growth and development

malar space distance between the **mandible** and nearest part of the **eye** margin

mandible jaws that are adapted for biting or cutting

maxilla (pl. maxillae) accessory jaws beneath the **mandibles**

maxillary palps **antennae**-like sensory appendages from the **maxillae**

median segment basal (first) **abdomen segment** (S1), fused to a varying extent in stick-insects (PHASMIDA) and appearing to form part of the **metanotum**

membranous thin, semi-transparent tissue

mesonotum **dorsal** (upper) surface of the **mesothorax** (see also **prothorax** and **metathorax**)

mesosoma	part of **thorax** between **head** and **metasoma**, in ants, bees, wasps and relatives (HYMENOPTERA)
mesothorax	the second thoracic segment (see also **prothorax** and **metathorax**)
metamorphosis	the changes that occur during successive stages of development (see **incomplete metamorphosis**)
metanotum	**dorsal** (upper) surface of the **metathorax** (3rd thoracic segment)
metasoma	part of **abdomen** behind a constricted 'waist', (**petiole**) in ants, bees, wasps and relatives (HYMENOPTERA) [= **gaster** in some works on Hymenoptera]
metathorax	the third thoracic segment (see also **prothorax** and **mesothorax**)
micropylar plate	the often longitudinal, scar-like area surrounding the **micropyle** in stick-insects (PHASMIDA)
micropyle	a small pore in eggshells of some insects, including stick-insects (PHASMIDA)
moult (ecdysis)	to shed or cast the skin or outer covering of the body
nocturnal	active at night
nymph	the immature stage of an insect (between **egg** and **adult**) that develops by **incomplete metamorphosis** (*e.g.* bugs (HEMIPTERA))
ocellus (plural: ocelli)	generally three simple, often small and circular **eyes**, situated on top of the **head**, between the much larger compound eyes
ootheca	**egg** pod in cockroaches (BLATTODEA) and mantids (MANTODEA)
operculum	i) the last **ventral abdomen segment** (subgenital plate) of **adult** females (♀) in stick-insects (PHASMIDA); ii) the 'lid' of an **egg**, in stick-insects (PHASMIDA), which is pushed off by an emerging **nymph**
order	a major insect classification based on structural differences (*e.g.* butterflies and moths (LEPIDOPTERA))
ovipositor	the **egg**-laying apparatus of a female (♀), concealed in some insects but large in others (*e.g.* bush-crickets (ORTHOPTERA))
oviscape	the non-retractile basal sheath of an **ovipositor**
palps	segmented, paired sensory structures from the **maxilla** and **labium** that are used in the tasting of food
parasite	an organism that lives in or on another (the host), from which it obtains food, shelter or other requirements
parasitoid	an insect spending much of its development **in** (endoparasitoid) or **attached to** (ectoparasitoid) a host, which it often eventually sterilizes or kills (some wasps (HYMENOPTERA), and flies (DIPTERA))
parthenogenesis	**egg** development without fertilization (*e.g.* stick-insects (PHASMIDA))
petiole	the stalk-like constricted waist found in some HYMENOPTERA, connecting the **propodeum** to the first segment of the **metasoma**
pollen basket	pollen-carrying part of the hindleg (hairs or bristles) of some bees (*e.g.* see Honey Bee on *p. 16*)
postpetiole	an additional segment in some ants (HYMENOPTERA), between the **petiole** and the **metasoma**
proboscis	extended mouth structure (*e.g.* butterflies and moths (LEPIDOPTERA) have a coiled 'tongue' adapted for feeding on nectar)
pronotum	**dorsal** (upper) surface of the **prothorax** (first thoracic segment)
propodeum	the hind part of the central **mesosoma**, technically part of the **abdomen**, in some HYMENOPTERA
prothorax	the first thoracic segment (see also **mesothorax** and **metathorax**)
pterostigma	the darkened area near the front margin of the **forewing** and sometimes **hindwing** (in HYMENOPTERA, PSCOPTERA, MECOPTERA, ODONATA, RAPHIDIOPTERA)
pupa (plural: pupae)	final immature stage (between **larva** and **adult**) of an insect that develops by **complete metamorphosis** (known as **chrysalis** in butterflies and moths (LEPIDOPTERA)). Does not feed or move about (but can wriggle)

pygidium raised area on hind tip of last abdomen segment, mainly in female (♀)

rostrum tubular, sucking mouthparts of some insects (*e.g.* bugs (HEMIPTERA)); term often used for extended part of the **head** of scorpionflies (MECOPTERA) and some weevils (COLEOPTERA)

scape basal (first) segment of the **antenna**, attached to the **head**

scutellum **dorsal** (upper) cover of the hind part of mid- or hind **thorax**; term often used in some bugs (HEMIPTERA)

soldier 'worker' responsible for colony defence in some social insects, notably some ants (HYMENOPTERA) and termites (BLATTODEA); often large-headed, with powerful **mandibles**

species (form; kind) individual organisms that are alike in appearance and structure, mating freely, with fertile offspring resembling each other and their parents (the second part of the two-part scientific name used to describe a species (*e.g. cervus*) in the Stag Beetle *Lucanus cervus*)

spermatophore a 'sperm package' produced by a male (♂) in order to fertilize a female (♀) (*e.g.* in some grasshoppers and crickets (ORTHOPTERA))

sting modified **ovipositor** in some bees and wasps (Aculeate HYMENOPTERA) used for injecting venom

striae fine parallel lines or ridges, *e.g.* often used when referring to **elytra** in beetles (COLEOPTERA)

stridulate producing sound from body parts, including 'song' from some grasshoppers and crickets (ORTHOPTERA)

subadult (or subimago) non-feeding, pre-adult winged form, not yet sexually mature in mayflies (EPHEMEROPTERA); these quickly moult to become **adult**

subgenital plate the last ventral **abdomen segment** (*e.g.* grasshoppers and crickets (ORTHOPTERA)), although referred to as **operculum** in some insects, including females (♀) of stick-insects (PHASMIDA)

suborder a classification division of an **order**, higher than a **family**; based on character(s) common to usually a large series of species

tarsus (plural: tarsi) the segmented foot of an insect connected to the **tibia** (fourth segment)

taxonomy a scheme of classification, arranging **species** and groups according to their relationship to each other

tegmen (plural: tegmina) **forewing**, sometimes used for the leathery wings covering the **hindwings** found in *e.g.* some grasshoppers and crickets (ORTHOPTERA) and cockroaches (BLATTODEA)

tegula a shield-like scale covering the base of the wings on the **thorax**, in some bees and wasps (HYMENOPTERA)

teneral a recently emerged **adult**, in which the colour is not fully developed (term often used when referring to dragonflies and damselflies (ODONATA))

thorax second major division of the insect body, to which the legs and, where present, the wings are attached

tibia (plural: tibiae) the fourth segment (and second long portion of the leg) between the **femur** (third segment) and the **tarsus** (terminal segment)

tubercles small knobs, present on the body of some insects

ventral the underside

vertex the top of the **head** above the **eyes**, including **ocelli**, If present

vesicle (plural: vesicles) a small sac, bladder or cyst

vestigial small, poorly developed or non-functional; term generally used for wings

wingspan (WS) twice the distance from the centre of the **thorax** to the tip of the **forewing**. NB **wing length** (**WL**) is often used in flies (DIPTERA) and is measured as the distance from the base (at the thorax) to the tip of the forewing

Watching and photographing insects

Many naturalists just take pleasure in enjoying the country air and watching or photographing insects. No special equipment is needed: the author usually goes out with just camera gear in a camera bag, but some might use good quality binoculars with close focus down to 50 cm, which are suitable for butterflies, dragonflies, *etc.*; others prefer a hand lens with 20× magnification. A notebook is also useful to record finds and retain details that can later be submitted to recording schemes. Entomologists conducting surveys use a range of equipment, including nets and traps; they use established methods to preserve specimens, the identification of which can then be confirmed using a microscope and with reference to specialist keys and texts. Nowadays there are fewer people making private collections than, for example, during the Victorian era; there is plenty of information available on collecting and preserving insects (equipment for which is readily available from entomological dealers), and the subject is therefore not covered in this book.

Finding insects

Methodical searching of flowers and vegetation should produce interesting insects, even in a garden, park or waterside location – but nature reserves can be of course be particularly prolific. Even searching bare ground can be productive, and it is well worth turning over logs or watching for insects on fallen logs in ancient woodland on sunny days. The more you know about the habits of these insects, including their season, habitat and food sources, the more likely you are to find them.

Classic sand dune habitat at Ballyteige Burrow Nature Reserve, Co. Wexford, Ireland. A search on Wild Asparagus here in 2017 resulted in the **Asparagus Beetle** [*inset* adult and larva] being found, the first record for Ireland.

Photographing insects

A lightweight, digital single-lens reflex model is shown *below*, with a dedicated image-stabilized macro lens that focuses from infinity to life-size (1:1) – so important for macro work. This, combined with a dedicated macro flash system, comprising adjustable wireless remote speedlights on a lens attachment ring, all controlled by a wireless commander on the hot shoe connection, has produced the majority of the images featured in this book. Although this equipment comes at a cost, it is possible to obtain reasonable results with compact and bridge cameras with macro capability. However, even opting for the more expensive equipment, there are other options to consider, such as whether to shoot hand-held, as the author does, or use a tripod – which takes time to set up at the right angle. Photographs can be taken in various modes, with adjustments made to the various settings to obtain the desired result. Many of the images in this book were shot in aperture priority mode at *f*16 or *f*22 in order to ensure the maximum depth of field (range of sharp focus). However, some photographers prefer images with a more blurred background and shoot at *f*11 or so. For small insects ≤10 mm long, a screw-on close-up lens filter is ideally needed, although this means you need to be very close (a few centimetres) from the subject.

Some insects are just too quick in the wild to photograph but with patience you can normally obtain good results. Assuming the species or site is not protected, as a last resort, the insect can be boxed in a small plastic container and placed for a short time in a cool bag containing an ice block. This may still only result in a second or less of time to obtain a photograph before the insect warms up and flies away, so preparation is important.

Whether just watching insects or taking time to photograph them, there is always the possibility that you might end up with a new discovery or a once-in-a-lifetime opportunity – so it is definitely worth persevering!

Photographing Smooth Stick-insects in Tresco Abbey Gardens, Tresco, Isles of Scilly.

Insect behaviour

Insects have a wide range of unexpected behaviours, including building elaborate structures, having a wide selection of hosts and prey, and elaborate defence behaviour when threatened. Welcome to the fascinating world of insects, which certainly warrants close observation! A few examples are shown here:

Southern Wood Ant (*p. 496*), tending to **Rose Chafer** (*p. 283*), permitting it entry to the ants' nest; the chafer lays eggs and its resulting larvae help keep the nest clean, in a symbiotic (mutually beneficial) relationship.

Sap run from a Goat Moth-infected oak (*p. 422*) caused by holes the larvae make at the trunk's surface. This often attracts other insects by day and night, including moths, beetles and others (*e.g.* woodlice).

Ladybird Braconid *Dinocampus coccinellae* (BL approx. 3 mm) parasitoid of **7-spot Ladybird** (*p. 310*) – note the wasp cocoon beneath the ladybird.

A **Common Green Grasshopper** (*p. 111*) becomes prey to the only slightly larger **Hornet Robberfly** (*p. 361*).

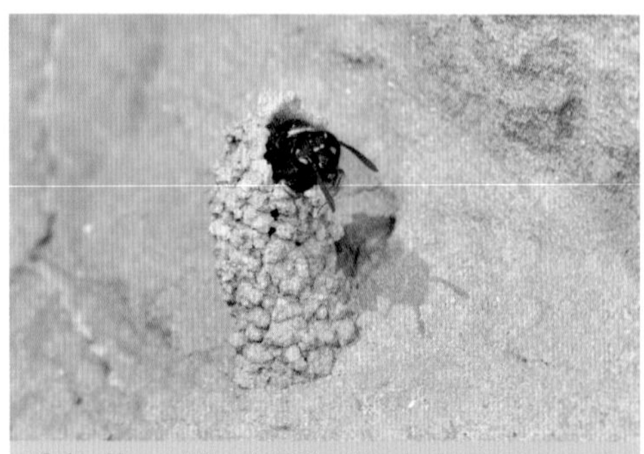

Spiny Mason Wasp (*p. 519*) exiting burrow entrance in the form of a chimney.

Noble Cuckoo Wasp (*p. 485*) investigating **digger wasp** *Cerceris* sp. (*p. 532*) nesting sites.

The Species Accounts

The **species accounts** that follow vary in approach between **orders**, covering every species in some (*e.g.* dragonflies and damselflies (**Odonata**), grasshoppers and crickets (**Orthoptera**), butterflies (but not moths) (**Lepidoptera**) and some of the smaller orders), and giving an overview in the case of others (*e.g.* larger orders such as ants, wasps, bees and relatives (**Hymenoptera**), flies (**Diptera**) and beetles (**Coleoptera**). The individual sections and accounts are fully cross-referenced throughout. Where species are featured elsewhere in the book, only the English name is given. The text is presented in a consistent sequence and provides concise information covering the following, as appropriate:

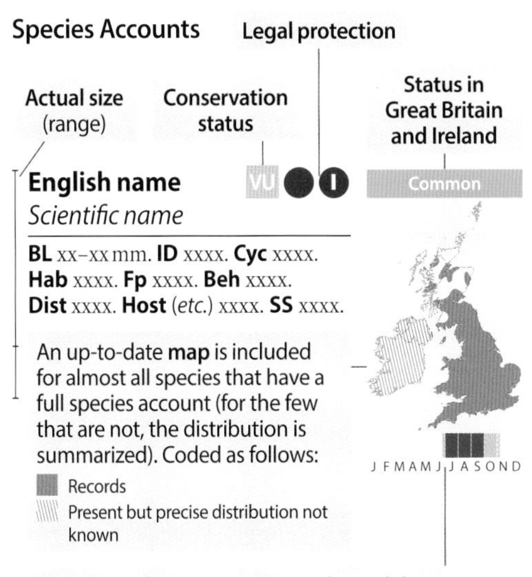

Phenology chart summarizing when adults may be seen: dark green shading is used to highlight the peak period(s); buff indicates usually not present or unlikely to be found.

Measurements: body length (**BL**) and/or wingspan (**WS**) or forewing length (**FWL**) (also shown as a bar either to the left of the species text or to the right of the map to indicate actual size).

Key identification features (ID): such as colour and physical characteristics.

Life-cycle (Cyc): indicating years to maturity.

Habitat (Hab): a summary of the type(s) of habitat in which the species is usually found.

Foodplant(s) (Fp): specific or types of plant on which the insect is most likely to be found.

Behaviour (Beh): where this is likely to be helpful for identification.

Distribution (Dist): where information additional to that shown in the map is considered helpful.

Host(s): Parasitoid(s) or **Cleptoparasite(s)** for species involved in parasitic relationships.

Similar species (SS): with an indication of the key identification features.

Status in Great Britain and Ireland (see *page 570*)

A colour-coded **status box** is included above the map for species that are the subject of a full species account. This provides a concise summary of the species' overall status. Six broad categories are used:

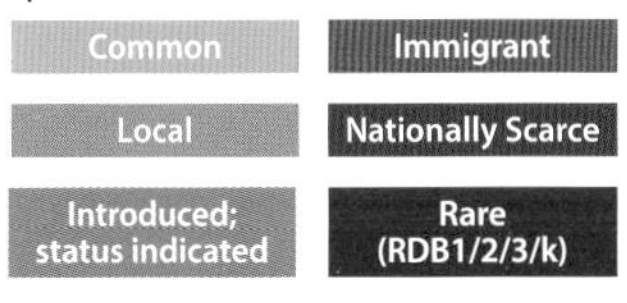

Legal protection (see *page 571*)

Species afforded legal protection are indicated as follows:

● – Protected in Great Britain

Ⓘ – Protected in Ireland

Conservation status (see *page 570*)

The red "Rare" status box is further coded to indicate the conservation (or Red List) status of the species.

For those that have been assessed and are considered to be of conservation concern, an icon is shown. For species that are on the Irish Red List, the icon has a green border, as shown *below*:

RE	RE	Regionally Extinct
CR	CR	Critically Endangered
EN	EN	Endangered
VU	VU	Vulnerable
NT	NT	Near Threatened
DD	DD	Data Deficient

Other symbols and codes used

♂ male	BI – Britain and Ireland
♀ female	W – Worldwide
† extinct	N/I – Not illustrated

ORDER ARCHAEOGNATHA Bristletails

[Greek: *archaeo* = old; *gnatha* = jaw]

BI | 1 family [3 genera, **7 species**]

W | 2 families [**approx. 500 species**]

Bristletails are elongated, medium-sized, poorly studied, primitive, wingless insects, 6–25 mm long (up to ±15 mm in Great Britain and Ireland), with a humped thorax and long appendage. Combined with the elongated pair of shorter cerci, it looks like they have three tails. The head has a pair of touching or near-touching, large, compound eyes, the shape of which helps to distinguish some genera; they also have ocelli. The mouthparts are directed downwards. A glossy appearance results from scales on the body, which may offer some protection against predators. They can also run fast and are able to jump several times their own body length, by flexing the abdomen backwards. British species belong to the family Machilidae; some species are rarely seen and have a restricted distribution. Males of most species attach sperm droplets to silken lines, or deposit these on the ground, attached to stalks, which the females need to locate. All stages feed on leaf-litter, algae, lichens and mosses, laying eggs that overwinter. Nymphs resemble miniature adults.

FINDING BRISTLETAILS The well-camouflaged bristletails are occasionally seen in leaf-litter or on vegetation. However, they are most likely to be seen under stones or on rocks above the high-water mark along the coast. Being mainly nocturnal, they are particularly obvious in torch-light. One rarity, Cave Bristletail (*p. 37*), is known to occur in caves, where it appears to at least overwinter.

Sea Bristletail on a rock at dusk.

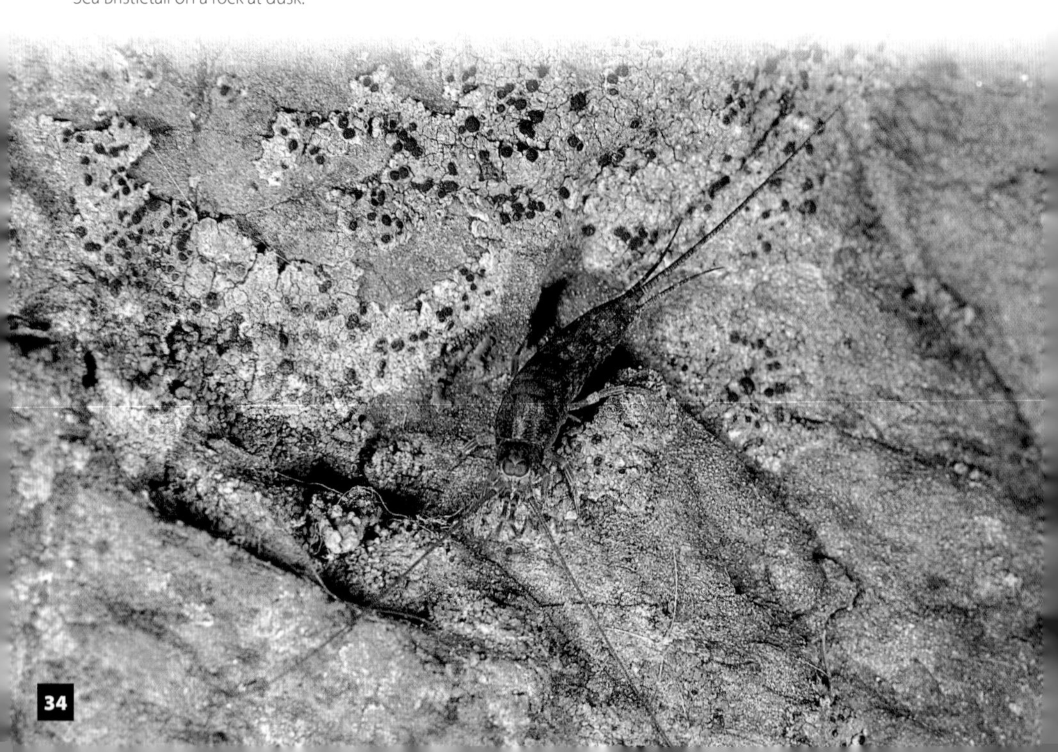

Structure of a bristletail

DILTA SP.

FAMILY **Machilidae** (Machilids) GUIDE TO GENERA | 3 BI

The largest of the two families in the Archaeognatha. These elongate, wingless insects with large eyes are cosmopolitan. Some species are restricted to rocky shorelines. The other family in the order (Meinertellidae) lacks scales at the base of hindlegs and antennae.

ANTENNAE short

ABDOMEN [VENTRAL]
all segments with a pair of coxal vesicles (yellow)

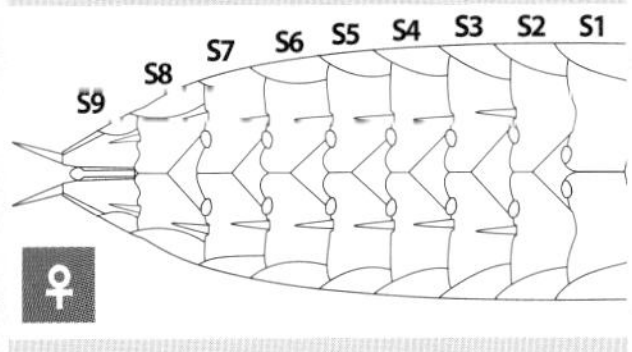

EYES rectangular

FRONT OF HEAD

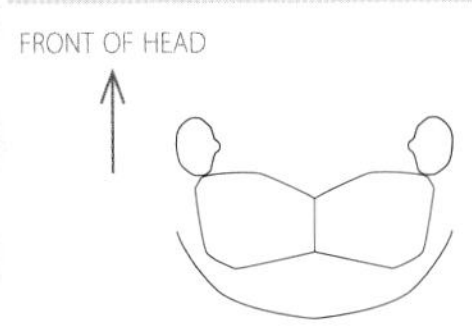

OCELLI (YELLOW)
at outer edge of eyes

ANTENNAE
≤ ½ body length

Dilta
[4 species | 3 ILL.] **p. 36**

ANTENNAE long [*beware, these may be partly broken off*]

ABDOMEN [VENTRAL]
segments 2–5 with 2 pairs of coxal vesicles (yellow)

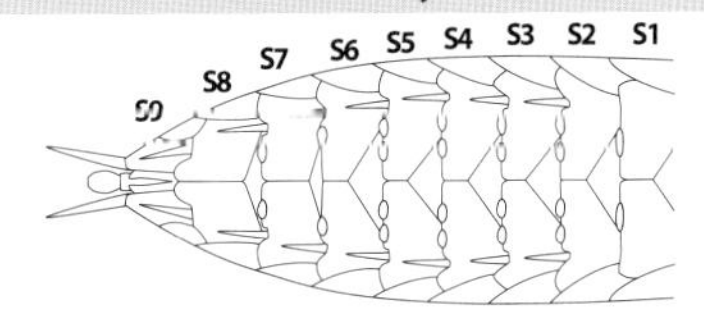

EYES rounded

OCELLI (YELLOW)
slit-like and small

ANTENNAE
≈ body length; scaled

Trigoniophthalmus
[Cave Bristletail] **p. 37**

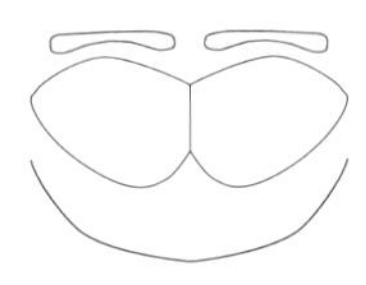

OCELLI (YELLOW)
large; almost length of eyes

ANTENNAE
> body length; scaled only on two basal segments

Petrobius
[2 species | 2 ILL.] **p. 37**

BRISTLETAILS WITH SHORT ANTENNAE

Dilta 4 spp. | 3 ILL.

Easily recognized by short antennae, British species are brown with glossy scales, some having dark mottling, including the antennae and appendage. **Cyc** Like all bristletails, post-hatching to adult takes a year or more; adults overwinter. **All *Dilta* species are very similar; ♀s difficult to identify, only possible by minor technical differences in the genitalia; ♂s may be identified by hairs on the lateral surface of the 2nd segment of the labial palp** (examination at high magnification needed). The three species most likely to be encountered are shown here; the fourth, **Irish Bristletail** *Dilta saxicola* [N/I] (BL 7–11 mm) is little known (initially reported from Howth Head, Dublin): ♂ lacks spine-like hairs on 2nd segment of labial palp.

Chater's Bristletail
Dilta chateri

BL ±10 mm. **ID** Brown. **Hab** Inland and coastal grassland, heathland and woodland with plenty of leaf-litter. **Dist** Little-known, described in 1995; at present only known from parts of South Wales. **SS** Other *Dilta* species.

Heathland Bristletail
Dilta littoralis

BL 10–11 mm. **ID** Greyish-brown, with some darker markings. **Hab** Various inland and coastal heathland and grassland habitats. **SS** Other *Dilta* species.

Southern Bristletail
Dilta hibernica

BL ±11 mm. **ID** Brown, rather plainer than other *Dilta* species. **Hab** Various inland and coastal habitats. **Dist** Also recorded in Ireland. **SS** Other *Dilta* species.

♂ LABIAL PALP 2ND SEGMENTS COMPARED

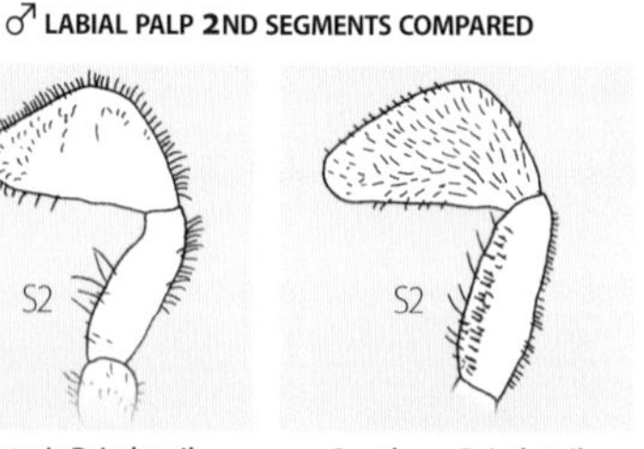

Chater's Bristletail with few spine-like hairs

Southern Bristletail with irregular pattern of spine-like hairs

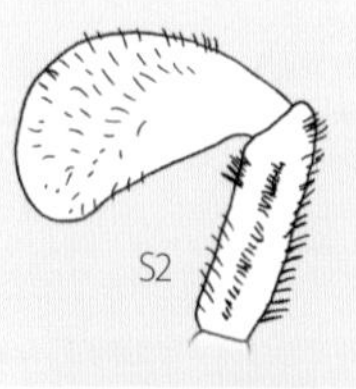

Heathland Bristletail with group of ±10 central spine-like hairs

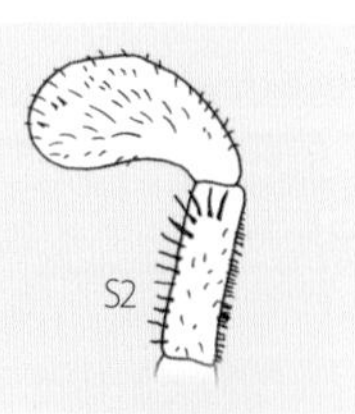

Irish Bristletail lacks spine-like hairs

BRISTLETAILS WITH LONG ANTENNAE

Petrobius 2 spp.

Easily recognized by the long antennae, well exceeding body length. Usually grey and variably mottled, including the antennae and appendage. **Cyc** Like all bristletails, post-hatching to adult takes a year or more; adults overwinter. The Sea Bristletail is the most widespread bristletail species in Great Britain & Ireland. **Both *Petrobius* species are very similar and sometimes occur together**; ♂s distinguished by details of the underside of abdomen S8; ♀s by details of the antennae.

Sea Bristletail
Petrobius maritimus

BL 13–15 mm. **ID** Mottled grey. **Hab** Mainly rocky coastal areas above high-water mark; inland in some western sites. **SS** Western Sea Bristletail, Cave Bristletail.

♀ **ANTENNAE** clearly ringed
♂ **SUBCOXAE** lacking rounded lobes

♀ **ANTENNAE**

Common

J F M A M J J A S O N D

Western Sea Bristletail
Petrobius brevistylis

BL 13–14 mm. **ID** Mottled grey. **Hab** Rocky coastal areas above high-water mark. **SS** Sea Bristletail, Cave Bristletail.

♀ **ANTENNAE** plain
♂ **SUBCOXAE** with rounded lobes

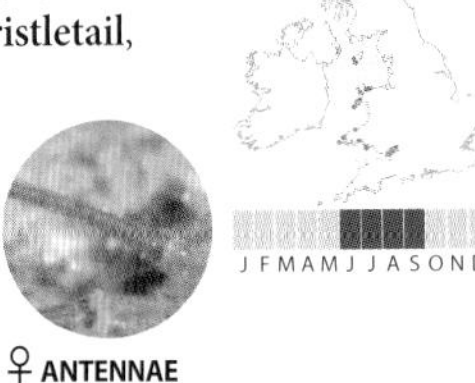

♀ **ANTENNAE**

Local

J F M A M J J A S O N D

Trigoniophthalmus 1 sp.

Easily recognized by antennae about matching body length. The only British species is little known but otherwise looks similar to other bristletails. **Cyc** Like all bristletails, post-hatching to adult takes a year or more; adults overwinter.

Cave Bristletail
Trigoniophthalmus alternatus

BL ±10 mm. **ID** Mottled brown with irregular paler areas; can be dark. **Hab** Known to hibernate in caves, including inland sites and presumably lives in rocky surrounds in warmer months. **SS** Sea bristletails.

Rare (not Red Listed)

J F M A M J J A S O N D

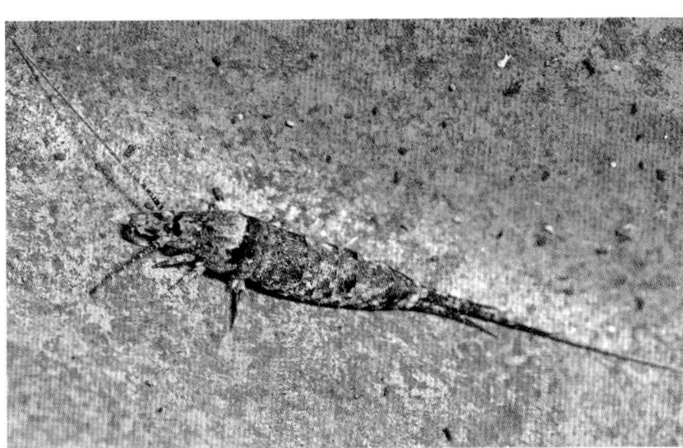

ORDER ZYGENTOMA — Silverfish

[Greek: *zygon* = bridge; *entoma* = cut into, relating to anatomy and perceived link with other orders]

BI | 1 family [3 genera, **4 species**]

W | 5 families [**approx. 574 species**]

This order was formerly known as Thysanura [Greek: *thysanos* = fringe and *oura* = tail]. These primitive insects are small to medium-sized, compressed and wingless, often covered in scales. Mouthparts are directed downwards to forwards. Compound eyes are small or absent in many non-British species although one to three ocelli may be present. Some species, including those in Great Britain and Ireland, have three 'tails', the paired cerci nearly as long as the central appendage. They can run fast and the scales may afford some protection from predators such as spiders; their flattened body enabling them to hide in crevices. Reproduction is by spermatophores deposited on the substrate by males, which are picked up by females. All stages feed on litter such as plant debris, and some species reside in caves. Several species, including the pest species featured, live in buildings (often in kitchens and bathrooms), where they are mainly nocturnal. Nymphs resemble miniature adults.

FINDING SILVERFISH Likely to be seen in houses.

Structure of a silverfish

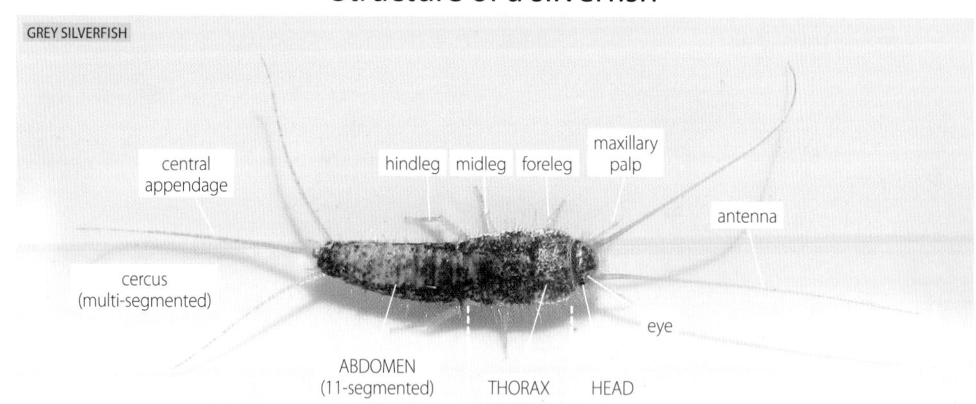

FIREBRAT

FAMILY Lepismatidae 3 GEN. | 4 spp.

The largest of the families in the Zygentoma, with two well-known cosmopolitan scavengers. A third species, Grey Silverfish *Ctenolepisma longicaudata*, was added to the list of British insects in 2016.

Lepisma 1 sp.

Silverfish *Lepisma saccharina* Common

BL 11–15 mm. **ID** Body covered in silvery scales; antennae about ⅔ body length. **Cyc** Multi-brooded but growth depends on temperature and humidity. Adults can be long-lived. **Hab** Cool, damp parts of houses (including outbuildings) with available food, such as cereals, paper and paper products – including wallpaper and books. Can live for several months without feeding. **SS** Firebrat.

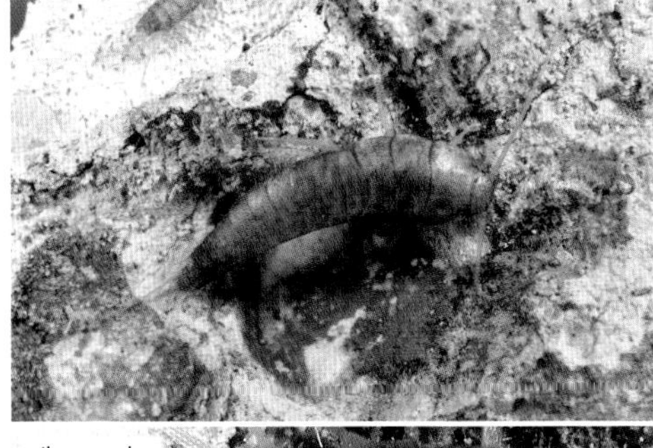

Thermobia 1 sp.

Firebrat *Thermobia domestica* Common

BL 11–15 mm. **ID** Body brown with dark mottled patches and long hairs on thorax and abdomen segments; antennae ≥ body length. **Cyc** Multi-brooded. ♀s can live 3–5 years, laying approx. 6,000 eggs. The eggs hatch in approx. 14 days, the resulting nymphs maturing in as little as three months. **Hab** Near ovens and hot pipes such as in bakeries, where they eat paper and starchy foods. **Dist** Could be spread by escapees from reptile food cultures. In warmer countries, also found outdoors. **SS** Silverfish.

Ctenolepisma 2 spp.

Grey Silverfish *Ctenolepisma longicaudata*

SHOWN IN ANNOTATED PHOTO *OPPOSITE*. **BL** up to 20 mm. **ID** Body grey-brown with bristles at the side of the body; long cerci; antennae = body length. **Dist** First recorded in Reading, Berkshire in 2016 and is understood to be spreading. **SS** **Four-lined Silverfish** *Ctenolepisma lineata* [N/I] (**BL** up to 14 mm) a possible introduced pest, which is brown, abdomen marked with four longitudinal dark brown lines.

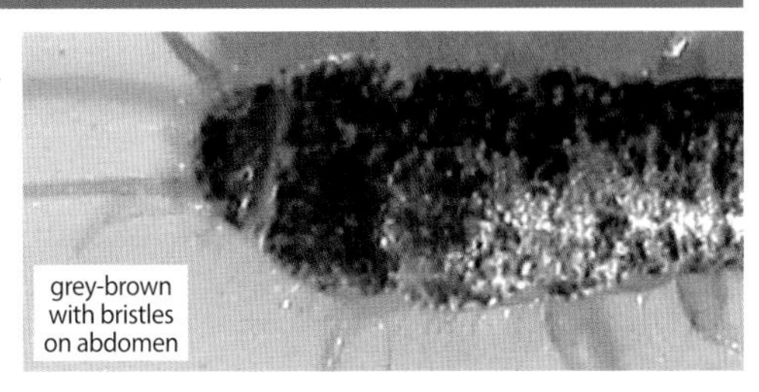

 Further reading p. 45

ORDER EPHEMEROPTERA — Mayflies

[Greek: *ephemeros* = lasting a day; *pteron* = wing]

BI | 10 families [21 genera, **51 species**] **W** | 42 families [**approx. 3,340 species**]

Primitive small to large winged insects with hindwings much smaller than the forewings, rarely absent; they rest in a characteristic manner with wings closed, rather similar to butterflies. The mouthparts are reduced, compound eyes large and antennae short. Abdomen more slender than thorax, tip of abdomen with two or three 'tails', the paired cerci often as long as the central appendage (where present). There are at least two generations a year. The plump nymphs are aquatic, spend up to two years in the water, and moult 11–45 times depending on the species and conditions. The nymphs possess three 'tails' (cerci) and the abdomen has up to seven pairs of lateral gills. Before reaching adulthood there is a non-feeding winged stage (known as the subimago). Mature adults also do not feed and are short-lived, weak fliers, that can be easily seen in mating swarms, often over water. In some species mating occurs at the subimago stage. Identification of some species requires careful, detailed examination of features including the number of 'tails' (two or three), size of hindwings (absent in some species) and forewing markings/colour. The sexes usually look very similar but males have a pair of claspers at the end of the abdomen and usually larger eyes and longer forelegs than females.

FINDING MAYFLIES Most likely to be seen flying over waterways (lakes, ponds, rivers and streams) or resting on nearby vegetation. Scotland is particularly rich in mayflies, with 38 species recorded. Trout fishermen actively fish in the mayfly season, which provides a feeding frenzy for trout; on average the peak is during the 3rd week of May. Artificial flies are used by the fishermen, as these imitate mayflies.

Structure of a mayfly

ORDER **Ephemeroptera** (Mayflies) — GUIDE TO FAMILIES | 10 BI

Mayflies are best identified to family group by a combination of how many 'tails' (cerci and central appendage, if present) there are at the tip of the abdomen and the size/shape of the hindwings. Further differentiation is by the features outlined in the table below:

hindwing large

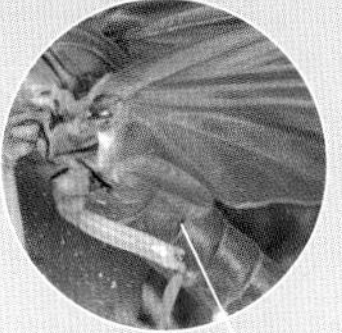
hindwing small

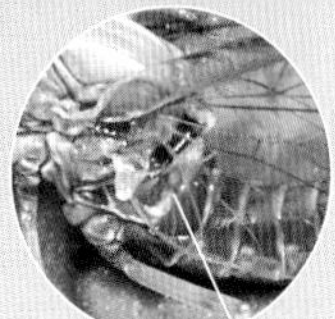
hindwing absent

2 'tails'		
HINDWINGS small/absent		
Small–medium-sized [ws 11–23 mm]	**Baetidae** [4 gen. | 14 spp.]	*p. 42*
HINDWINGS large		
Medium-sized–large [ws 20–38 mm]; **TARSUS** 4-segmented; hind tarsus 1½ × as long as tibia	**Siphlonuridae** [1 gen. | 3 spp.]	n/c
Small–large [ws 13–38 mm]; **TARSUS** 5-segmented	**Heptageniidae** [5 gen. | 11 spp.]	*p. 43*
Medium-sized–large [ws approx. 20 mm]; **TARSUS** 4-segmented; hind tarsus shorter than tibia	**Ameletidae** [1 species]	n/c
Medium-sized [ws 24 mm]; only one record (Middlesex, 1920)	**Arthropleidae** [1 species]	n/c

HEPTAGENIIDAE
Yellow May Dun

3 'tails'		
HINDWINGS small/absent		
Small [ws 7–11 mm]	**Caenidae** [2 gen. | 9 spp.]	n/c
HINDWINGS large		
Large [ws 30–51 mm]; **FOREWINGS** with black marks	**Ephemeridae** [1 gen. | 3 spp.]	*p. 45*
Medium-sized–large [ws 25–32 mm]; yellow, including wings	**Potamanthidae** [1 species]	n/c
Small–medium-sized [ws 15–23 mm]; **FOREWINGS** with short veins between each long vein	**Ephemerellidae** [2 gen. | 2 spp.]	*p. 44*
Small–medium-sized [ws 13–27 mm]; **FOREWINGS** lacking short veins between each long vein	**Leptophlebiidae** [3 gen. | 6 spp.]	*p. 44*

LEPTOPHLEBIIDAE
Sepia Dun

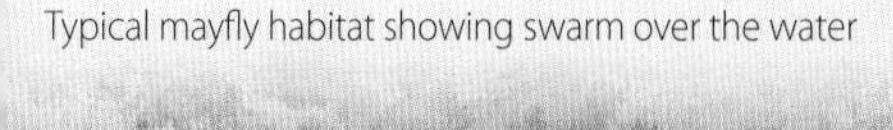
Typical mayfly habitat showing swarm over the water

Artificial flies, which are imitations of mayflies, are used by trout fishermen.

MAYFLIES WITH 2 'TAILS' + HINDWINGS SMALL/ABSENT

FAMILY Baetidae — 4 GEN. | 14 spp. | 3 ILL.

Small to medium-sized species with two tails. Hindwings are absent in three species, but are present (small) in all others. Genera/species can be separated as follows:

HINDWINGS small, oval	***Baetis*** [9 species| 1 ILL.]
HINDWINGS small, spur-shaped, pointed tip	***Centroptilum*** [1 species| ILL.]
HINDWINGS **Tiny Sulphur Dun** *Procloeon pennulatum*: small spur-shaped, rounded tip. **Pale Evening Dun** *Procloeon bifidum*: absent; TARSI basal segment 3× length of 2nd segment	***Procloeon*** [2 species| N/I]
HINDWINGS absent; TARSI basal segment 2× length of 2nd segment	***Cloeon*** [2 species| 1 ILL.]

Baetis — 9 spp.

Large Dark Olive
Baetis rhodani

BL 7–9 mm. **WS** 12–25 mm.
ID Two-tailed; hindwings small, oval. Body and legs olive-green; wings pale grey. **Hab** Rivers and small streams. **SS** Other *Baetis* species (although this species is the largest in Britain and Ireland).

Centroptilum — 2 spp.

Small Spurwing
Centroptilum luteolum

BL 6–7 mm. **WS** 13–16 mm.
ID ♂ with large orange-red, turret-like eyes and translucent body, the final three abdomen segments pale orange. ♀ eyes dark brown, top half of body pale orange, underside paler cream. Hindwings small, spur-shaped, tip pointed. **Hab** Pools, river margins and streams. **SS** Other species in Baetidae, but only **Tiny Sulphur Dun** *Procloeon pennulatum* [N/I] (**WS** 16–19 mm) (absent from Ireland) also has spur-shaped hindwings, but with rounded tip.

MAYFLIES WITH 2 'TAILS' + HINDWINGS LARGE

Cloeon 2 spp.

Pond Olive *Cloeon dipterum*

Common

BL 6–10 mm. **WS** 13–25 mm.
ID ♂ cream, the final three abdomen segments darker; wings transparent with distinctive brown veins. Tails ringed with brown. ♀ similar except orange or brown with yellow edge to forewings. Hindwings absent in both sexes.
Hab Pools, river margins and streams, ponds and some larger lakes. **SS** Lake Olive *Cloeon simile* [N/I] (**WS** 17–23 mm), which has 9–11 cross-veins at tip of forewing (Pond Olive has 3–5) and ♀ lacks yellow edge to forewing. Other Baetidae [see table *opposite*].

♀

FAMILY Heptageniidae 5 GEN. | 11 spp. | 2 ILL.

Small to large species with two tails. The genera are difficult to distinguish without an adult male to study, when eyes and legs can help. The eyes touch, or nearly touch in *Ecdyonurus* but are farther apart in *Heptagenia*.

Heptagenia 2 spp.

Yellow May Dun *Heptagenia sulphurea*

Common

BL 8–11 mm, **WS** 17–27 mm. **ID** ♀ body and wings (partly) bright sulphur-yellow, eyes bluish; ♂ brownish, with dark brown wing venation. **Hab** Large rivers, in riffle sections. **SS** Could be confused with Scarce Yellow May Dun *Heptagenia longicauda* [N/I], which has two paler rings on fore femur and a single black dot above hindleg, and possibly Yellow Mayfly *Potamanthus luteus* [N/I] [Potomanthidae] (**WS** 25–32 mm), which has three 'tails'.

Ecdyonurus 4 spp.

Large Brook Dun *Ecdyonurus torrentis*

Common

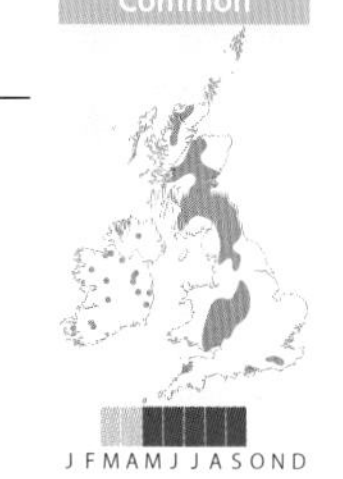

BL approx. 15 mm, **WS** 28–32 mm.
ID Subimago unmistakable: brown with reddish sides, underside purple; wings pale brownish, mottled with black patches and with yellowish margins. Adult body olive-brown, wings transparent and dark-veined.
Hab Small stony streams and rivers.
SS Subimago of Large Brook Dun is distinctive; other *Ecdyonurus* species can only be distinguished by forewing colour and patterning.

♀ [subimago]

♂ [subimago]

MAYFLIES WITH 3 'TAILS'

FAMILY Ephemerellidae — 2 GEN. | 2 spp. | 1 ILL.

Two small to medium-sized species with large hindwings; forewing with **short veins between each long vein**. Species can be separated as follows:

ABDOMEN segments with pair of small spines	**Blue-winged Olive**
ABDOMEN spines absent; an uncommon mainly northern species, also in Ireland	**Yellow Hawk** *Ephemerella notata* [N/I]

Serratella — 1 sp.

Blue-winged Olive
Serratella ignita

Common

J F M A M J J A S O N D

BL 7–10 mm, **WS** 15–23 mm. **ID** ♂ reddish-brown to olive; ♀ greenish-olive; tails pale, ringed with brown. Wings dark bluish-grey. **Hab** Fast-flowing rivers and streams. **SS** Yellow Hawk *Ephemerella notata* EN [N/I] (**WS** 17–23 mm) [see table *above*]; Ditch Dun *Habrophlebia fusca* [N/I] [Leptophlebiidae] (**WS** 13–15 mm), which lacks detached veins at edge of forewings (present in Blue-winged Olive).

♀ [subimago]

FAMILY Leptophlebiidae — 3 GEN. | 6 spp. | 1 ILL.

Small to medium-sized species with three tails; forewings with **no regular short veins between each long vein**; hindwings large.

Sepia Dun VU
Leptophlebia marginata

Common

J F M A M J J A S O N D

BL 7–11 mm, **WS** 13–23 mm. **ID** Subimago body reddish-brown; wings brownish-grey with brown veins. Adult body brown with wings brownish, notably forewing tips. **Hab** Streams, ponds and lakes. **SS** Claret Dun *Leptophlebia vespertina* [N/I] (**WS** 15–21 mm), the subimago of which has pale grey forewings and much paler hindwings.

♀ [subimago]

FAMILY Ephemeridae — 1 GEN. | 3 spp. | 2 ILL.

Three large species with large hindwings; **forewing with black marks**. Species distinguishable by distinctive pattern of dark markings on abdomen as follows:

ABDOMEN dark markings only distinct on S7–9 (no triangles or lines)	**Green Drake Mayfly**
ABDOMEN dark triangular markings; [mainly R. Thames and R. Wye]	**Striped Mayfly** *Ephemera lineata* [N/I]
ABDOMEN dark triangular markings; [mainly SE England]	**Drake Mackerel Mayfly**

Ephemera — 3 spp.

Drake Mackerel Mayfly
Ephemera vulgata

Common

BL 13–25 mm, **WS** 30–51 mm.
ID Subimago body creamy-yellow; wings grey. Legs pale with black marks. Abdomen segments have distinctive brown, triangular marks. Adult body similar, wings transparent with brown veins and forewing with dark patches.
Hab Rivers with muddy bases.
SS **Green Drake Mayfly**, **Striped Mayfly** *Ephemera lineata* [N/I] (**WS** 30–44 mm) [see table *above*].

J F M A M J J A S O N D

♀

Green Drake Mayfly
Ephemera danica

Common

BL 15–20 mm, **WS** 30–46 mm.
ID Forewing with several dark patches. Subimago creamy-yellow. Abdomen with dark markings on final segments.
Cyc Completed in a year in warmer parts of S England, otherwise two years. ♀ lays up to 8,300 eggs during several visits to suitable waters. Nymphs feed on organic detritus.
Hab Fast-flowing lakes and rivers. **SS** **Striped Mayfly** *Ephemera lineata* [N/I] (**WS** 30–44 mm) and **Drake Mackerel Mayfly** have different abdominal markings and more limited distributions [see table *above*].

J F M A M J J A S O N D

♀

RECOMMENDED FURTHER READING AND USEFUL WEBSITES

Macadam, C. & Bennett, C. 2010. ***A pictorial guide to British Ephemeroptera***. Field Studies Council (AIDGAP).

www.riverflies.org The Riverfly Partnership

ORDER ODONATA Dragonflies & damselflies

[Greek: *odontos* = tooth]

BI | 9 families [25 genera, **58 species** (incl. 2 extinct spp. (†) and 15 rare or breeding immigrants)]

W | 38 families [**6,650 species**]

Medium-sized to large winged insects, often with elongated body. The mobile head has large, compound eyes, separated in the generally slender damselflies (suborder Zygoptera), but generally touching in the more robust-looking dragonflies (suborder Epiprocta (which includes intraorder Anisoptera)). The antennae are short and bristle-like. Mouthparts are with mandibles. The wings are large and heavily veined, with hindwings wider than forewings only in dragonflies. The thorax has large muscles, enabling larger hawker dragonflies to reach speeds of up to 36 km/h, but they also sometimes just hover. At rest, the wings are spread flat. By comparison, damselflies have a weak flight and rest with wings held together above the body, or slightly angled. The 10-segmented abdomen possesses claspers in both sexes, and females often have an ovipositor for egg-laying. The aquatic nymphs are robust or elongate, with a labial 'mask'; they can escape from possible predators at speed by ejecting water from the anus. Nymphs feed on aquatic life such as insect nymphs (including the same species); when large they sometimes prey on tadpoles. Adults, however, catch prey in flight. The shed skins of nymphs (exuviae) are a common sight on waterside vegetation; freshly emerged adults harden and in the first few days are termed 'teneral', with colours gradually developing as the insect matures (the lack of pigmentation can cause confusion when trying to identify such individuals). The male transfers sperm from the abdomen to accessory genitalia under the front of the abdomen; during mating, claspers are used to attach to the back of the 'neck' of a female and a 'wheel' position is adopted to enable the sperm to be transferred. This process can take a few minutes to several hours and when they separate, the female soon starts laying eggs (in some species with the male still attached).

FINDING DRAGONFLIES & DAMSELFLIES Most likely to be seen flying over or near waterways (canals, ditches, lakes, ponds, rivers and streams), resting or hunting nearby or along woodland edges.

♂ EMPEROR DRAGONFLY

Structure of a dragonfly and a damselfly

RECOMMENDED FURTHER READING AND USEFUL WEBSITES

Smallshire, D. & Swash, A. 2018. ***Britain's Dragonflies: A field guide to the damselflies and dragonflies of Great Britain and Ireland***. 4th Edition. Princeton University Press (Princeton WILD*Guides*).
Particularly user-friendly: covers adults and nymphs.

Smallshire, D. & Swash, A. 2020. ***Europe's Dragonflies: A field guide to the damselflies and dragonflies***. Princeton University Press (Princeton WILD*Guides*).

Brooks, S., Cham, S. & Lewington, R. 2018. ***Field Guide to the dragonflies and damselflies of Great Britain and Ireland***. 5th Edition. Bloomsbury Publishing.

Dijkstra, K-D.B., Schröter, A. & Lewington, R. 2020. ***Field Guide to the Dragonflies of Britain and Europe***. 2nd Edition. Bloomsbury Publishing.

www.dragonflysoc.org.uk British Dragonfly Society

ORDER **Odonata** (Dragonflies & damselflies)

FORM **small** (<50 mm); **slender-bodied**. EYES **small**; well separated. WINGS **forewings same shape as hindwings**; closed vertically together along abdomen or held open (usually <60° to body). FLIGHT **weak**.

DAMSELFLIES (Zygoptera)

BODY small (<40 mm); pale whitish or blue; LEGS white with a black line; **hind tibiae 'flattened'**

White-legged Damselfly

White-legged damselflies
Platycnemididae
[White-legged Damselfly]

p. 53

BODY small (<40 mm); black, red or blue with blue or black markings; LEGS hind tibiae not 'flattened'

Variable (top) and Large Red Damselflies (bottom)

Blue/red damselflies
Coenagrionidae
[6 genera | 14 species (1†)]

p. 54

BODY medium-sized (37–55 mm); **metallic green or brown** with or without some bluish areas; WINGS **translucent**; usually held open at <60° to body

Emerald Damselfly

Emerald damselflies
Lestidae
[3 genera | 5 species]

p. 51

BODY medium-sized (45–50 mm) metallic (green or blue); WINGS **fully or partly coloured**; usually held closed

Banded Demoiselle

Demoiselles
Calopterygidae
[1 genus | 2 species]

p. 50

FORM **large** (>50 mm); **robust body**. EYES **large**; touching (NB separated in Gomphidae) WINGS **forewings narrower than hindwings**; wings **held open** (±90° to body). FLIGHT **strong**.

DRAGONFLIES (Anisoptera)

Further details and fuller descriptions of both adult and larvae can be found in the companion WILD*Guides* volume *Britain's Dragonflies* (see *page 47* for details).

BODY medium-sized (45–55 mm); **metallic dark green**; EYES **bright green**

Downy Emerald

Emerald dragonflies
Corduliidae
[3 genera | 5 species (1†)]

p. 66

BODY medium-sized (45–50 mm) black and yellow; **abdomen S8–9 clubbed**; **EYES** distinctly separated

Club-tailed Damselfly

Club-tailed dragonflies
Gomphidae
[Club-tailed Dragonfly]

p. 65

BODY medium-sized (<50 mm) red, blue, brown or yellow (**not green**); if yellow, abdomen S8–9 not clubbed; **EYES** touching

Ruddy Darter

Darters, skimmers and chasers
Libellulidae
[6 genera | 16 species]

p. 68

BODY **large (>75 mm), robust**; black with bright yellow stripes on on abdomen; appendages short; ♀ with long ovipositor

Golden-ringed Dragonfly

Golden-ringed dragonflies
Cordulegastridae
[Golden-ringed Dragonfly]

p. 65

BODY large (>50 mm), particularly robust; abdomen black or brown (never metallic or red), spotted blue and/or green and/or yellowish, or brown; appendages long; ♀ with short ovipositor

Common Hawker

Hawker dragonflies
Aeshnidae
[3 genera | 12 species]

p. 60

Guide to families *pp. 48–49*

■ FAMILY **Calopterygidae** (Demoiselles) **1 GEN. | 2 spp.**

Large, metallic damselflies found by flowing water. Flight quite 'butterfly'-like. ♂s have all or partly blue wings; ♀s have coloured wings – see accounts for differences.

Banded Demoiselle *Calopteryx splendens*

BL 45–48 mm. **ID** ♂ metallic bluish green; wings translucent, except for broad dark blue band on forewings and hindwings. ♀ with a metallic green abdomen; wings greenish-yellow, with a distinctive white spot near tip of forewings. **Cyc** 1–2 years. **Hab** Slow flowing streams and rivers, on waterside vegetation, occasionally in company with Beautiful Demoiselle. Can be found in large numbers at some sites, alighting on emergent vegetation. **SS** Beautiful Demoiselle.

♂ **WINGS** translucent with large blue band
♀ **WINGS** greenish

Beautiful Demoiselle *Calopteryx virgo*

BL 45–49 mm. **ID** ♂ metallic bluish green; wings completely dark blue. ♀ with a metallic green abdomen; wings brown, with an indistinct white spot near tip of forewings. **Cyc** 2 years. **Hab** Fast-flowing streams and rivers, with a sandy or gravel base. Often seen perched on waterside vegetation. **SS** Banded Demoiselle.

♂ **WINGS** all dark blue
♀ **WINGS** brown

FAMILY Lestidae (Emerald damselflies) 3 GEN. | 5 spp.

Elongate, mainly metallic green damselflies, most of which rest with wings partly open. Species can be separated as follows:

ABDOMEN **brown**; [late-flying immigrant, extremely rare]	Winter Damselfly	*p.51*
ABDOMEN **very elongated**; WING-SPOTS **pale brown** [rare]	Willow Emerald Damselfly	*p.51*
ABDOMEN **elongated; WING-SPOTS black or brown-and-white**	*Lestes* spp. [3 spp.]	
WING-SPOTS brown; ♂ ABDOMEN blue pruinescence on S1, **S2** (**all**) and S9–10; ♀ ABDOMEN green (no blue); **near-triangular spots on pale S2**; OVIPOSITOR **=S10**; dark below	Emerald Damselfly	*p.52*
WING-SPOTS brown; ♂ ABDOMEN blue pruinescence on S1, **S2** (± ½) and S9–10; ♀ ABDOMEN green (no blue); **squarish spots on pale S2**; OVIPOSITOR **>S10**; dark below	Scarce Emerald Damselfly	*p.52*
WING-SPOTS **½ brown, ½ white**; ♂ + ♀ ABDOMEN uniform green white on sides of S9–10; ♀ OVIPOSITOR **whitish**	Southern Emerald Damselfly	*p.53*

Sympecma 1 sp.

Winter Damselfly *Sympecma fusca*

BL 34–39 mm. **ID** Brown. **Dist** Very rare immigrant (1 record, December 2009). **SS** None.

Chalcolestes 1 sp.

Willow Emerald Damselfly *Chalcolestes viridis*

Rare (not Red Listed) (recent colonist)

BL 39–48 mm. **ID** Metallic green, no blue pruinescence. Abdomen very elongated. Wing-spots pale brown. ♂ appendages pale, dark-tipped; ♀ ovipositor whitish with dark patches [see illustration on *p.* 53). **Cyc** 1 year. **Hab** Still waters such as ponds and canals with overhanging trees such as willows, which these inconspicuous insects frequent. **Beh** Females oviposit in living twigs. **Dist** Has spread after becoming established in SE Suffolk in 2009. **SS** Southern Emerald Damselfly (*p.* 53), which has two-tone wing-spots.

WING-SPOTS pale brown

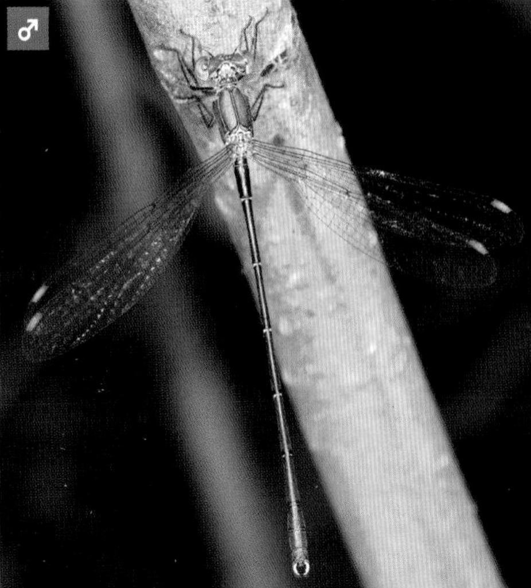

Guide to families *pp. 48–49*

Lestes 3 spp.

Emerald Damselfly
Lestes sponsa

Common

BL 35–39 mm. **ID** Metallic green. Wing-spots brown. ♂ with blue pruinescence, including on abdomen S1–2 and S9–10. ♀ with two almost triangular-shaped spots on abdomen S1; ovipositor does not extend beyond S10 [see illustration *opposite*]. **Cyc** 1 year. **Hab** Pools with emergent plants, such as rushes.
SS Scarce Emerald Damselfly.

***CHALCOLESTES* and *LESTES* ABDOMENS S1 AND S2**

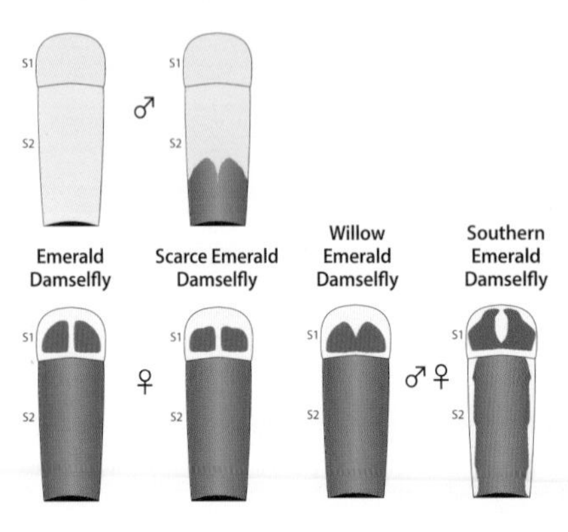

Scarce Emerald Damselfly NT
Lestes dryas NT

Rare

BL 34–37 mm. **ID** Metallic green, rather robust appearance. Wing-spots brown. ♂ with blue pruinescence, including on abdomen S1, the first half only of S2 and S9–10. ♀ duller, with two square spots on abdomen S1; ovipositor extends beyond S10 [see illustration *opposite*]. **Cyc** 1 year. **Hab** Ditches and pools with emergent plants, such as rushes or sedges (sometimes together with Emerald Damselfly).
SS Emerald Damselfly.

J F M A M J J A S O N D

WING-SPOTS
pale brown (as Emerald Damselfly *above*)

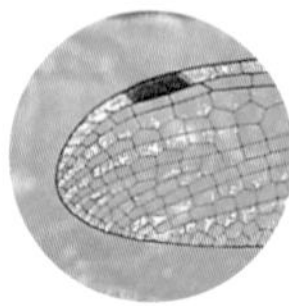

Southern Emerald Damselfly
Lestes barbarus

BL 40–45 mm. **ID** Metallic green, with no blue pruinescence. Abdomen very elongated. Wing-spots half black, half white. ♂ with whitish abdomen S10 and dark-tipped, whitish appendages. ♀ ovipositor whitish [see illustration *below*]. **Cyc** 1 year. **Hab** Pools, such as dune slacks. **SS** Willow Emerald Damselfly (*p. 51*), which has uniformly pale brown wing-spots.

Rare (recent colonist)

J F M A M J J A S O N D

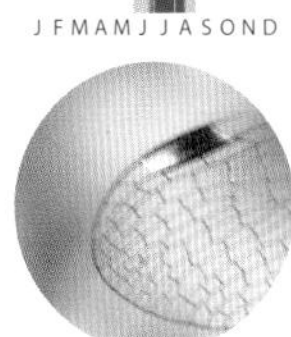

WING-SPOTS
½ brown, ½ white

♀ *CHALCOLESTES* and *LESTES* OVIPOSITORS

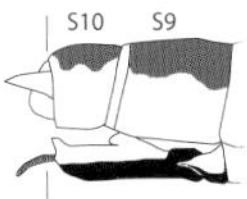

Emerald Damselfly

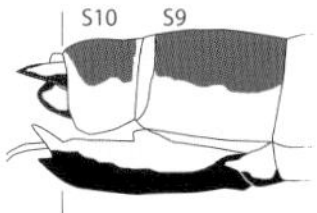

Scarce Emerald Damselfly

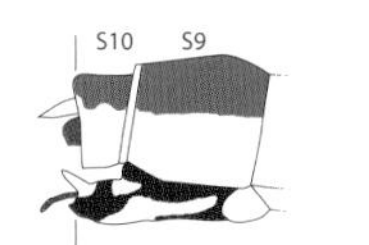

Willow Emerald Damselfly (*p. 51*)

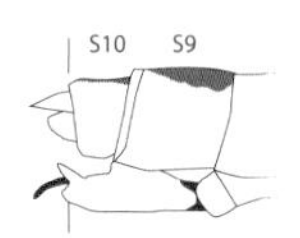

Southern Emerald Damselfly

FAMILY Platycnemidae (White-legged damselflies) 1 sp.

Small damselflies mainly associated with rivers and characterised by the broad, 'flattened', feather-like hind tibiae and narrow head.

White-legged Damselfly
Platycnemis pennipes

BL 35–37 mm. **ID** Pale, with black markings: ♂ pale blue; ♀ cream. Notably, the hind tibiae are 'expanded' (feather-like), white, with a black line. **Cyc** Usually 2 years. **Hab** Slow-flowing rivers and stream; sometimes well away from water. **SS** Other blue damselflies [Coenagrionidae] (*p. 54*).

Local

J F M A M J J A S O N D

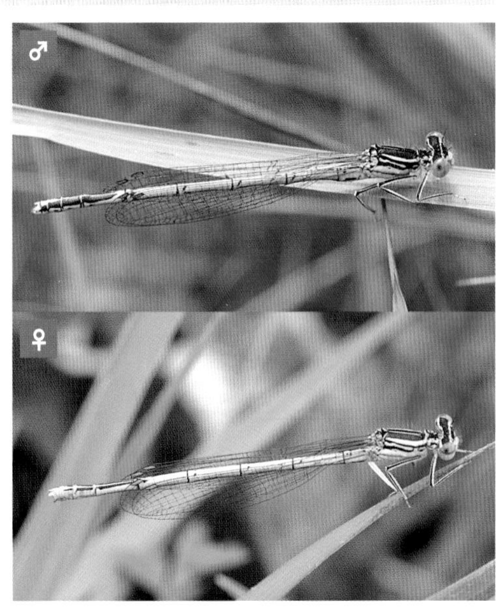

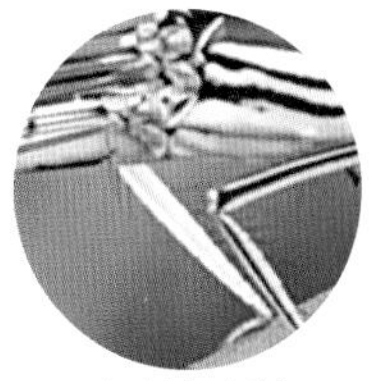

expanded hind tibiae are diagnostic of the species

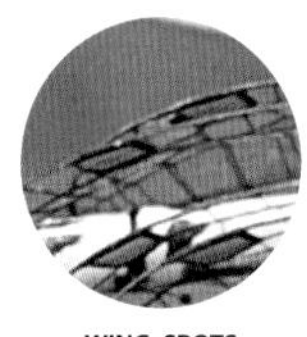

WING-SPOTS
distinctive, chestnut-brown

 Guide to families *pp. 48–49*

■ FAMILY **Coenagrionidae** (Red and blue/black damselflies)

6 GEN. | 14 spp. (1†) | 12 ILL.

Elongated, red or blue-and-black damselflies. Identified by abdomen colour and pattern, and shape of hind margin of pronotum, as illustrated in the table below. (**Norfolk Damselfly** *Coenagrion armatum* is extinct in Great Britain, last seen in Norfolk in 1957.)

ABDOMEN red

LEGS AND WING-SPOTS **black**; ABDOMEN ♂ red-and-black; ♀ with some red	Large Red Damselfly	*p.55*
LEGS AND WING-SPOTS **reddish**; ABDOMEN ♂ red; ♀ red-and-black, red, or black	Small Red Damselfly	*p.55*

ABDOMEN blue-and-black

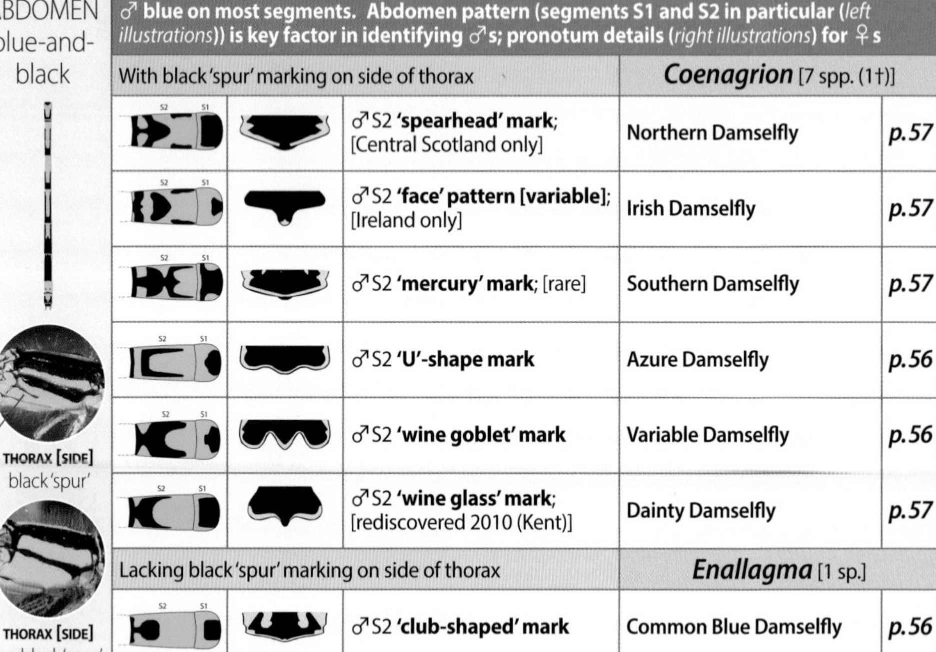

♂ blue on most segments. Abdomen pattern (segments S1 and S2 in particular (*left illustrations*)) is key factor in identifying ♂s; pronotum details (*right illustrations*) for ♀s

With black 'spur' marking on side of thorax	***Coenagrion*** [7 spp. (1†)]	
♂ S2 **'spearhead' mark**; [Central Scotland only]	Northern Damselfly	*p.57*
♂ S2 **'face' pattern [variable]**; [Ireland only]	Irish Damselfly	*p.57*
♂ S2 **'mercury' mark**; [rare]	Southern Damselfly	*p.57*
♂ S2 **'U'-shape mark**	Azure Damselfly	*p.56*
♂ S2 **'wine goblet' mark**	Variable Damselfly	*p.56*
♂ S2 **'wine glass' mark**; [rediscovered 2010 (Kent)]	Dainty Damselfly	*p.57*
Lacking black 'spur' marking on side of thorax	***Enallagma*** [1 sp.]	
♂ S2 **'club-shaped' mark**	Common Blue Damselfly	*p.56*

THORAX [SIDE] black 'spur'

THORAX [SIDE] no black 'spur'

ABDOMEN mainly black

♂ blue only on S1 (base) and S9–10 (posterior) segments

Eyes red | Usually on ponds with water-lilies or other floating vegetation	***Erythromma*** [2 spp.]	
ABDOMEN ♂ blue S1, S9 and S10; ♀ black with blue division to final abdomen segments	Red-eyed Damselfly	*p.58*
ABDOMEN ♂ blue S1, S9 and S10 and notably the sides of S2–4 and S8; ♀ black with blue on S10 and also on part on S9	Small Red-eyed Damselfly	*p.58*
Eyes bluish	***Ischnura*** [2 spp.]	
ABDOMEN S8 blue (pale brown in some ♀)	Blue-tailed Damselfly	*p.59*
ABDOMEN ♂ with tip of S8 and S9 blue (usually with tiny black spots); ♀ black (orange in immature form *aurantiaca*)	Scarce Blue-tailed Damselfly	*p.59*

DAMSELFLIES WITH RED ABDOMEN

Pyrrhosoma 1 sp.

Large Red Damselfly
Pyrrhosoma nymphula

BL 33–36 mm. **ID** Legs and wing-spots black; thorax black with red sides (or yellow in ♀ forms). ♂ abdomen red, with bronze-black on S7–9, the black more extensive in ♀ (occurs on all abdomen segments), particularly in the mainly black form *melanotum*. **Cyc** Normally takes 2 years, but can be 1–3 years. **Hab** Wetlands, often those with plentiful waterside vegetation. **SS** Small Red Damselfly.

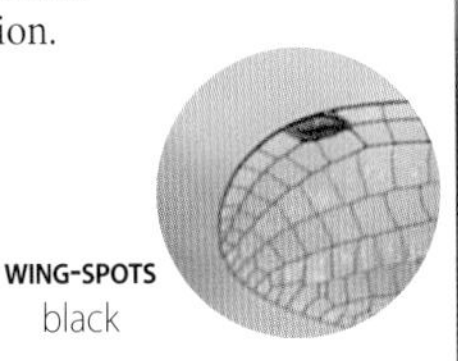

Ceriagrion 1 sp.

Small Red Damselfly
Ceriagrion tenellum

BL 25–35 mm. **ID** Generally small, with a weak flight. Legs and wing-spots reddish; thorax bronze. ♂ abdomen completely red. ♀ usually with abdomen S1–3 and S9–10 red, although in the form *melanogastrum* reddish only appears on the divisions of S8–10. **Cyc** 2 years. **Hab** Boggy heathland. **SS** Large Red Damselfly.

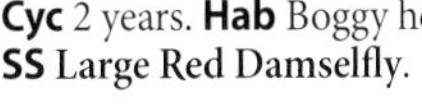

♀ form *melanogastrum*

Guide to families *pp. 48–49*

DAMSELFLIES WITH BLUE-AND-BLACK ABDOMEN

Enallagma — 1 sp.

No black 'spur' marking on side of thorax – see *p. 54* for comparison of blue-and-black damselflies.

Common Blue Damselfly
Enallagma cyathigerum

BL 29–36 mm. **ID** ♂ abdomen S8–9 blue (often two small black spots on S9). Side of thorax lacks black 'spur' marking seen on all other blue-and-black species. ♀ blue or green, but with a mainly black abdomen. **Cyc** 1 year, 2 or more in the north. **Hab** Various still and flowing waters; may occur around the margins, resting on vegetation, or be seen flying over water to reach water-lily pads. **SS** Other blue-and-black damselfly species.

Coenagrion — 7 spp. | 6 ILL.

Black 'spur' marking on side of thorax – see *p. 54* for comparison of blue-and-black damselflies.

Azure Damselfly
Coenagrion puella

BL 33–35 mm. **ID** ♂ black 'U'-shape on abdomen S2. ♀ green or blue, with a mainly black abdomen. **Cyc** 1 year, 2 in the north. **Hab** Various standing waters, including garden ponds; often found around the margins. **SS** Other blue-and-black damselfly species.

Variable Damselfly — NT
Coenagrion pulchellum

BL 33–38 mm. **ID** ♂ black 'wine goblet' marking on abdomen S2 and S9 with much black. Some ♀ extensive blue, others with mainly black abdomen. Markings vary between individuals. **Cyc** 1–2 years. **Hab** Various well-vegetated waters, including ditches, fens and ponds, as well as bogs. Often found resting on vegetation. **SS** Other blue-and-black damselfly species, particularly **Common Blue Damselfly**, **Azure Damselfly** and, in Ireland, **Irish Damselfly**.

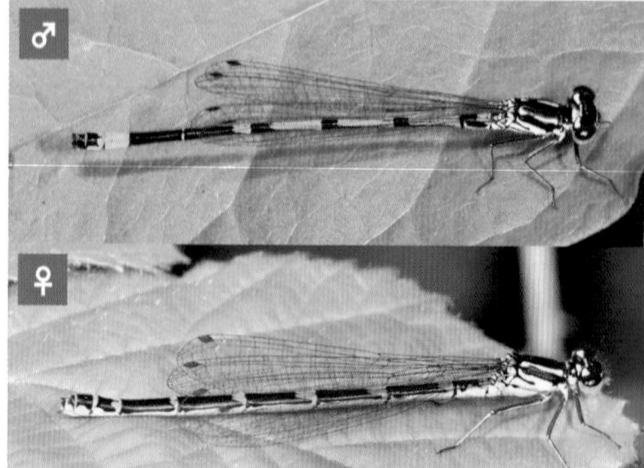

Northern Damselfly EN

Coenagrion hastulatum

BL 31–33 mm. **ID** ♂ abdomen S8–9 entirely blue, except for two black dots on S9. ♀ green-and-black; underside green. **Cyc** 1–2 years. **Hab** Pools with emergent vegetation. **SS** In its restricted range, **Common Blue Damselfly**.

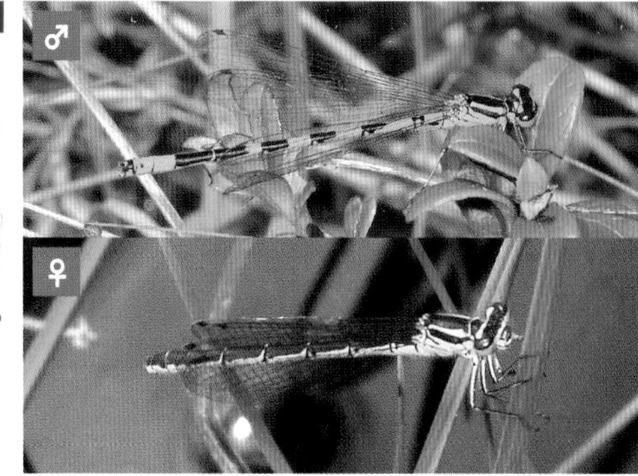

Irish Damselfly VU I

Coenagrion lunulatum

BL 30–33 mm. **ID** ♂ rather dark, as most of the abdomen is black; however, abdomen S8–9 are blue, with only a pair of small black dots on each segment. ♀ with even less blue, mainly at the front of S8. **Cyc** Little known, believed to be about 2 years. **Hab** Small lakes and large ponds, often resting on pondweeds and water-lilies. **SS** Variable Damselfly.

Southern Damselfly EN

Coenagrion mercuriale

BL 27–31 mm. **ID** ♂ distinctive black 'mercury' mark on abdomen S2. Other segments have spear-shaped black markings, particularly S3–4. ♀ usually green but some individuals blue, with a mainly black abdomen. **Cyc** 2 years. **Hab** Heathland flushes, also fens and chalk streams. **Dist** Stronghold is in the New Forest, Hampshire; also occurs in the Preseli mountains, Pembrokeshire; otherwise only a few small populations. **SS** Azure Damselfly.

Dainty Damselfly *Coenagrion scitulum*

BL 30–33 mm. **ID** ♂ black 'wine glass' mark on abdomen S2. ♀ with 'rocket-shaped' markings on S3–4 (similar to Common Blue Damselfly). Both sexes have lemon yellow undersides and wing-spots longer than other *Coenagrion* (2× as long as broad). **Cyc** 2 years. **Hab** Well-vegetated waters, both brackish and fresh. **Dist** Rare: rediscovered in 2010 on the Isle of Sheppey, Kent; also Sandwich Bay, Kent since 2019. **SS** Other blue-and-black damselfly species.

Guide to families *pp. 48–49*

DAMSELFLIES WITH MAINLY BLACK ABDOMEN (reddish/brownish eyes)

Erythromma 2 spp.

Two similar species best separated by details of the thorax and abdomen S10.

Red-eyed Damselfly

Erythromma najas

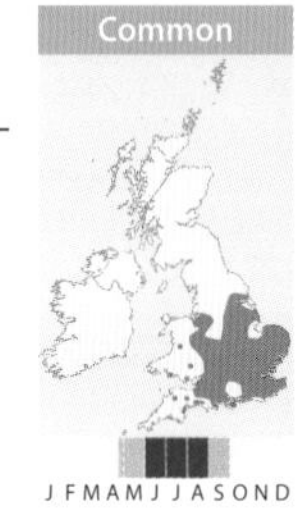

BL 30–36 mm. **ID** ♂ eyes red; thorax bronze-black with blue sides; abdomen black, S1 and S9–10 blue. ♀ mainly black: thorax with greenish sides; abdomen with bluish divisions to final segments. **Cyc** 1 or usually 2 years. **Hab** Large ponds, lakes and other waters with floating vegetation; often seen on floating leaves, notably water-lilies, along with other damselflies, where the red eyes are obvious, particularly in sunny weather. **SS** Small Red-eyed Damselfly.

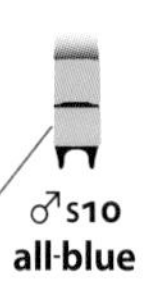

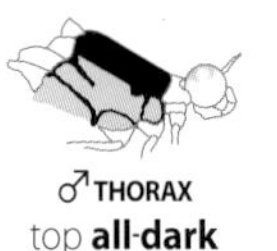

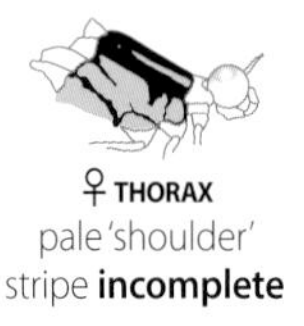

Small Red-eyed Damselfly

Erythromma viridulum

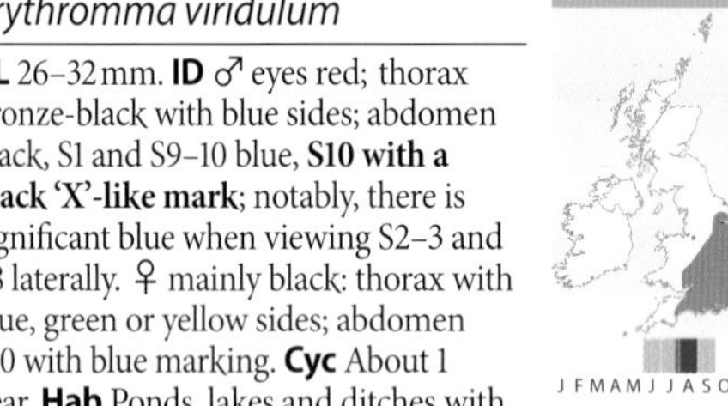

BL 26–32 mm. **ID** ♂ eyes red; thorax bronze-black with blue sides; abdomen black, S1 and S9–10 blue, **S10 with a black 'X'-like mark**; notably, there is significant blue when viewing S2–3 and S8 laterally. ♀ mainly black: thorax with blue, green or yellow sides; abdomen S10 with blue marking. **Cyc** About 1 year. **Hab** Ponds, lakes and ditches with floating algae, pondweeds or water-lilies; often seen on the floating vegetation. **Dist** Range has expanded rapidly since first colonizing Great Britain in 1999. **SS** Red-eyed Damselfly.

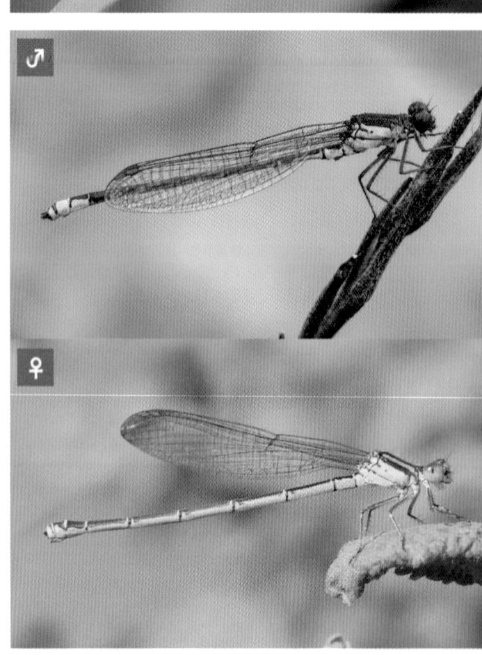

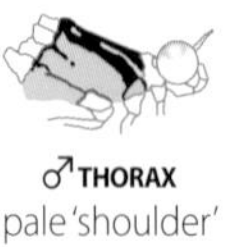

DAMSELFLIES WITH MAINLY BLACK ABDOMEN (bluish eyes)

Ischnura 2 spp.

Two similar species best separated by details of abdomen S8.

Blue-tailed Damselfly
Ischnura elegans

BL 30–34 mm. **ID** ♂ mainly black, except for blue on much of thorax and abdomen S8. ♀ similar, except thorax colour and pattern varies between individuals (in two of the forms abdomen S8 is pale brown). **Cyc** 1 year in the south, 2 years in the north. **Hab** Various waters; usually seen in marginal vegetation. **SS** Scarce Blue-tailed Damselfly.

♂ s8
all-blue

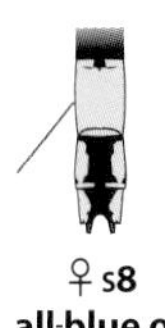
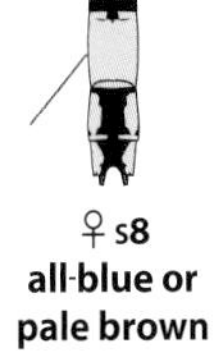

♀ s8
all-blue or pale brown

Scarce Blue-tailed Damselfly NT VU
Ischnura pumilio

BL 26–31 mm. **ID** ♂ mainly black, except for blue on much of thorax and abdomen S9 (which has two black marks), tip of S8 and part of S10. Seen from the side, the blue on S8 is more extensive. ♀ black-and-green, but the immature form *aurantiaca* is distinctive orange-and-black. **Cyc** 1–2 years. **Hab** Shallow wetland sites, such as heathland streams and flushes. Generally on low vegetation; easily overlooked. **SS** Blue-tailed Damselfly.

♂ s8
blue only at tip

♀ s8
no blue

Guide to families *pp. 48–49*

FAMILY **Aeshnidae** (Hawker dragonflies) **3 GEN. | 12 spp.**

The largest, most powerful dragonflies in flight. Known as hawkers, they often fly for long periods and will occasionally rest on branches. Some are long-distance immigrants. Identified by thorax pattern initially along with details of the abdomen, wings and eyes.

THORAX plain green or brown

THORAX plain

ABDOMEN **blue, green or brown with broad black central stripe; WINGS clear**	***Anax*** [4 spp. incl. 2 vagrants]	
THORAX **green**		
THORAX ♂ with bold blue patches above forewing bases; ABDOMEN mainly blue or green; EYES bluish-green	**Emperor Dragonfly**	***p.* 61**
Similar to **Emperor Dragonfly** but; EYES **brownish** [extremely rare immigrant from North America]	**Common Green Darner**	***p.* 61**
THORAX **brown**		
EYES **green**; ABDOMEN dark with a distinctive yellow ring ahead of a blue 'saddle' on S2 and part of S3 [rare breeder]	**Lesser Emperor**	***p.* 61**
EYES **brown**; ABDOMEN dark with blue 'saddle' restricted to S2 (absent in some ♀s) [rare immigrant]	**Vagrant Emperor**	***p.* 61**

THORAX dark brown with coloured bands on side

THORAX [SIDE] coloured bands

FORM small; noticeably hairy body; ABDOMEN blackish with blue (♂) or yellow-green (♀) oval spots; WINGS clear with long, thin, brown wing-spots	**Hairy Dragonfly**	***p.* 62**
FORM **large; body at most only slightly hairy;** ABDOMEN **black or brown; either no strong pattern or with a 'mosaic' of blue/green and/or yellow markings;** WINGS **clear or brown, with rather 'thick', black wing-spots**	***Aeshna*** [7 spp. incl. 1 rare immigrant]	
ABDOMEN **predominantly brown with no strong pattern** WINGS **clear or brown**		
ABDOMEN **brown**; WINGS **brown**; EYES yellowish-brown	**Brown Hawker**	***p.* 64**
ABDOMEN **brown, S2 with bold yellow triangle**; WINGS **clear**; EYES green	**Norfolk Hawker**	***p.* 64**
ABDOMEN **dark with contrasting 'mosaic' pattern;** WINGS **clear**		
THORAX [TOP] with **pair of large yellow patches**; ABDOMEN **band across S9–10**; ♂ with mainly **green** spots	**Southern Hawker**	***p.* 62**
THORAX [SIDE] **narrow, wavy pale streaks** (blue in ♂); [TOP] ♂ with short blue streaks; ABDOMEN blue or yellow spots on S9–10; ♂ all spots **blue**	**Azure Hawker**	***p.* 62**
THORAX [TOP] distinctive streaks; COSTA **yellow** ABDOMEN no yellow triangle on S2	**Common Hawker**	***p.* 63**
THORAX [TOP] faint streaks at most; COSTA **brown** ABDOMEN **yellow triangle** on S2	**Migrant Hawker**	***p.* 63**
As Migrant Hawker but **no dark areas on thorax sides**; ♂ much bluer with blue on thorax sides and ABDOMEN **blue triangle** on S2 [mainly rare immigrant]	**Southern Migrant Hawker**	***p.* 63**

THORAX [TOP] short streaks

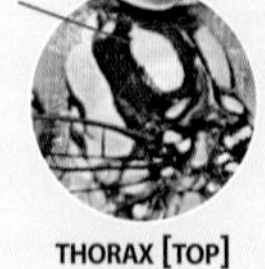
THORAX [TOP] patches

HAWKER DRAGONFLIES WITH PLAIN THORAX

Anax 4 spp.

Emperor Dragonfly *Anax imperator*

BL 66–84 mm. **ID** Thorax greenish. Abdomen with thick black central line. ♂ abdomen vivid blue; ♀ abdomen green. **Cyc** 1–2 years. **Hab** Various ponds, lakes standing waters. **Beh** Usually seen on the wing, including when eating prey, but sometimes found resting on vegetation. **SS** **Common Green Darner** [extremely rare immigrant], which has brownish eyes, and both **Vagrant** and **Lesser Emperors** [both rare immigrants that have bred], which both have a brown thorax (Lesser Emperor has green eyes; Vagrant Emperor brown eyes).

A female Emperor Dragonfly egglaying.

RARE IMMIGRANT *ANAX* DRAGONFLIES

From Europe or Africa (North America in the case of the Common Green Darner)

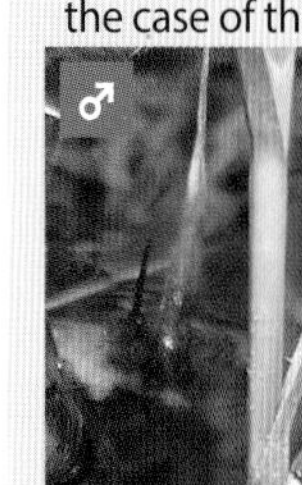

Lesser Emperor *Anax parthenope*

Vagrant Emperor *Anax ephippiger*

Common Green Darner *Anax junius*

Guide to families *pp. 48–49*

HAWKER DRAGONFLIES WITH PATTERNED THORAX 1/2

Brachytron 1 sp.

Hairy Dragonfly
Brachytron pratense

BL 54–63 mm. **ID** Smaller than other hawkers and the only one with a thickly hairy thorax; long, thin brown wing-spots; appendages long. Abdomen with spots, blue in ♂, yellow in ♀. **Cyc** 1–3 years. **Hab** Ponds, lakes and ditches with emergent vegetation. A spring-flying small hawker is most likely to be Hairy Dragonfly. **SS** Other hawkers (superficially similar).

Aeshna 7 spp.

Azure Hawker VU
Aeshna caerulea

BL 54–64 mm. **ID** Thorax with narrow, wavy, blue stripes **on side**. ♂ abdomen dark with large, paired blue spots (appears blue). ♀ blue colour form similar, but blue spots smaller; brown form has buff or yellowish spots. **Cyc** 3–4 years. **Hab** Bog pools with bog mosses. **Beh** Often settles on rocks, tree trunks or vegetation. **SS Common Hawker.**

Southern Hawker
Aeshna cyanea

BL 67–76 mm. **ID** Large yellow patches on top of thorax (often visible in flight); abdomen S9–10 banded (paired spots in other hawkers). ♂ black with green markings, or blue on S8–10 and abdomen side. ♀ similar but dark brown with green markings. **Cyc** Eggs overwinter; nymphs take 2–3 years to develop. **Hab** Various waters, including garden ponds. **Beh** Often hunts in woodland clearings and rides, landing frequently; can be inquisitive. **SS** Other hawkers.

Common Hawker
Aeshna juncea

BL 65–80 mm. **ID** Dark brown or black, with paired spots on abdomen; costa yellow. ♂ has blue spots and yellow flecks on abdomen segments. ♀ similar, but spots usually yellow or green. **Cyc** 3 or more years. **Hab** Acidic standing waters, including heathland and moorlands. **Beh** Often hunts away from water. **SS** Other hawkers, particularly **Migrant Hawker**.

Migrant Hawker
Aeshna mixta

BL 56–64 mm. **ID** Dark brown. Thorax usually has two short yellow stripes on top, bold yellow stripes on side. Abdomen with paired spots; S2 has a yellow triangle; costa brown. ♂ has blue spots and yellow flecks on abdomen segments. ♀ similar, but smaller spots usually yellow; appendages long. **Cyc** Less than 1 year. **Hab** Various waters. **Beh** Often rests (hangs) on vegetation. **Dist** Numbers bolstered by immigrants. **SS** Other hawkers, such as **Common Hawker**, but particularly **Southern Migrant Hawker** [rare immigrant and now breeding at sites such as Sandwich Bay, Kent], which has no dark markings on sides of thorax and a blue triangle on abdomen S2.

Southern Migrant Hawker *Aeshna affinis*

HAWKER ♂ THORAX SIDES
(♀s similar – green replaces blue)

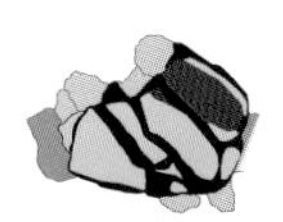
Hairy Dragonfly

Common Hawker

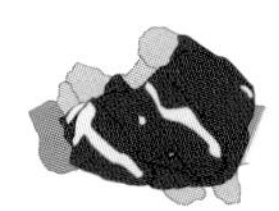
Azure Hawker

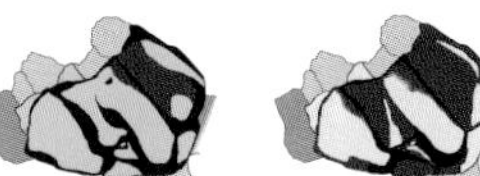
Southern Hawker

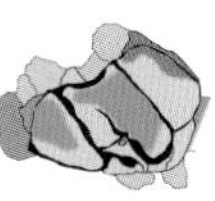
Migrant Hawker

Southern Migrant Hawker

Guide to families *pp. 48–49*

HAWKER DRAGONFLIES WITH PATTERNED THORAX 2/2

Brown Hawker
Aeshna grandis

BL 70–77 mm. **ID** Largely brown, including wings. Thorax with bold yellow stripes on side. ♂ has blue spots on abdomen S8–10, including sides. ♀ similar, with yellow and/or blue markings. **Cyc** Eggs overwinter; nymphs take 2–4 years to develop. **Hab** Various standing or slow-flowing waters, including ponds and lakes. **Beh** Often hunts in woodland clearings, along rides and hedgerows, and over meadows. **SS** Norfolk Hawker.

Brown Hawker
EYES brown
S2 no triangle

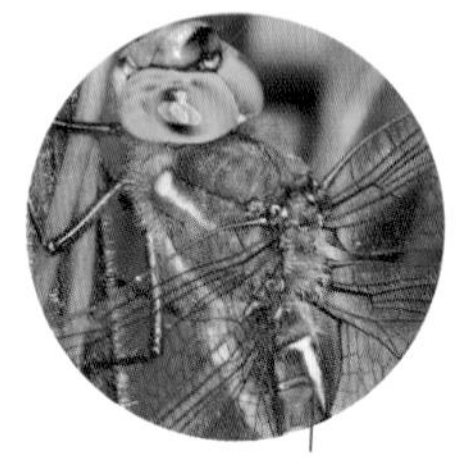

Norfolk Hawker
EYES green
S2 yellow triangle

Norfolk Hawker
EN

Aeshna isosceles

BL 62–67 mm. **ID** Brown, lacking coloured spots or markings of other hawkers. Eyes green. Thorax with pair of pale yellow stripes on side. Abdomen S2 with distinctive yellow triangle. **Cyc** 2 years. **Hab** Fen ditches and grazing marsh, with good Water-soldier cover. **SS** Brown Hawker.

FAMILY Gomphidae (Club-tailed dragonflies)

2 GEN. | 2 spp. | 1 ILL.

Broad, medium-sized, greenish-yellow and black dragonflies with abdomen S8–9 distinctly clubbed. (**River Clubtail** *Stylurus flavipes* was recorded in East Sussex in 1818.)

Common Clubtail *Gomphus vulgatissimus*

NT | Rare

BL 45–50 mm. **ID** Black-and-yellow or black-and-green. Eyes green, well separated. Abdomen S8–9 broadened (abdomen tip appears club-like), particularly in ♂. **Cyc** 3–5 years. **Hab** Slow-flowing rivers. Newly emerged individuals are found regularly at key sites; mature adults encountered much less frequently. **SS** None.

Club-tailed Dragonfly favours slow-flowing rivers where silt accumulates.

FAMILY Cordulegastridae (Golden-ringed dragonflies)

1 sp.

Large; black with bright yellow stripes. Generally observed patrolling streams. The only British species sometimes settles (hangs) on vegetation. ♀ has a long ovipositor.

Golden-ringed Dragonfly *Cordulegaster boltonii*

Common

BL 74–84 mm. **ID** Eyes green; thorax and abdomen banded black and yellow. ♀ with long ovipositor. **Cyc** 2–5 years. **Hab** Acidic running waters, often in heathland and woodland. **Beh** ♀s are sometimes seen stabbing eggs into the substrate. ♂s patrol sections of stream, engaging in aerial battles with rival males. Both sexes periodically use low perches. **SS** None.

Guide to families *pp.* 48–49

FAMILY **Corduliidae** (Emerald dragonflies) **3 GEN. | 5 spp. (1†)**

Medium-sized, metallic dark green dragonflies with bright green eyes. Rarely settle and are usually found in habitats with trees close by. **Orange-spotted Emerald** *Oxygastra curtisii* is extinct in England (last recorded in 1963). The three resident species are best identified by details of the frons; the extinct and rare immigrant species both have paler markings on the abdomen.

FRONS green	
FRONS dark, lacking yellow markings (yellow only present below eyes); ♂ ABDOMEN bulbous towards tip	**Downy Emerald**
FRONS **dark, with yellow markings; ♂ ABDOMEN less bulbous towards tip**	
THORAX + ABDOMEN **metallic green**; FRONS with broad yellow 'U' between eyes	**Brilliant Emerald**
THORAX **metallic green;** ABDOMEN **much darker, almost black**; FRONS with yellow spot next to each eye	**Northern Emerald**

Downy Emerald
FRONS plain

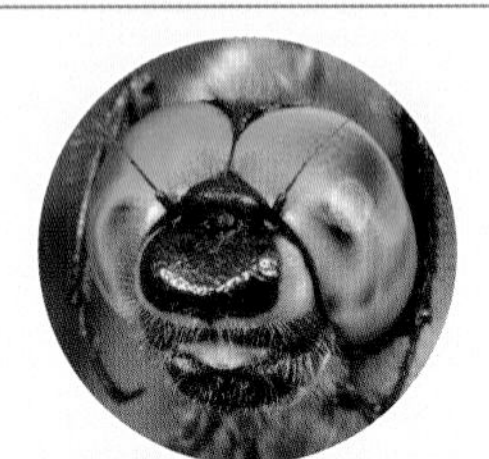

Brilliant Emerald
FRONS with yellow 'U'

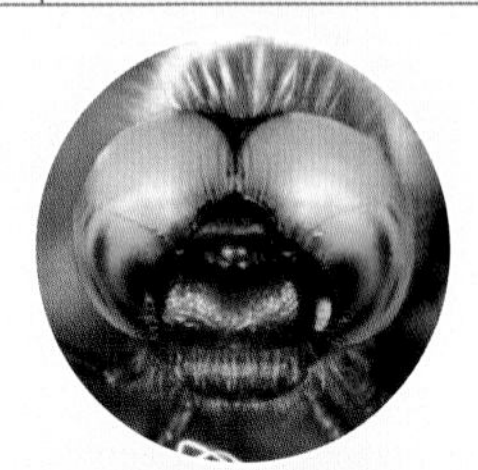

Northern Emerald
FRONS yellow spot next to eyes

ABDOMEN **with orange or yellow markings on most segments**	
ABDOMEN **slender with yellow markings** [1 record: 2018]	**Yellow-spotted Emerald**
ABDOMEN **slender with orange markings S1–7** [last recorded 1963]	**Orange-spotted Emerald**

VERY RARE EMERALD DRAGONFLIES WITH ABDOMINAL MARKINGS

Yellow-spotted Emerald
Somatochlora flavomaculata

Orange-spotted Emerald
Oxygastra curtisii †

Cordulia 1 sp.

Downy Emerald EN Local
Cordulia aenea

BL 47–55 mm. **ID** Eyes bright green, no yellow on frons; thorax downy, metallic greenish-bronze; abdomen darker. ♂ abdomen bulbous towards tip; 3rd segment distinctively narrow (giving 'waisted' appearance); ♀ abdomen broad. **Cyc** 2–3 years. **Hab** Ponds, lakes and canals. **Beh** Males patrol territories and rarely settle. Once a female is located near the water, the pair flies high up into the canopy of nearby trees to mate. Also feeds in the tree canopy. **SS** Other (rarer) emeralds.

Somatochlora 3 spp.

Brilliant Emerald VU Rare
Somatochlora metallica

BL 50–55 mm. **ID** Eyes, thorax and abdomen metallic green; frons with broad, yellow 'U'-shape between eyes. ♂ with narrower abdomen S3 (giving slightly 'waisted' appearance). **Cyc** 3–4 years. **Hab** Ponds, lakes and canals near wooded areas. **SS** Other emeralds.

Northern Emerald NT EN Rare
Somatochlora arctica

BL 45–51 mm. **ID** Eyes bright green when mature (brown in immatures), frons with yellow spots near each eye. Thorax metallic greenish-bronze. Abdomen black with paired yellow spots and marks at base. ♂ appendages calliper-like. **Cyc** 2 or more years. **Hab** Boggy areas, including peat bogs, near wooded areas. **SS** Other emeralds.

Guide to families *pp. 48–49*

FAMILY **Libellulidae** (Chasers, skimmers and darters)

6 GEN. | 16 spp. (including **6** rare immigrants)

Small to medium-sized dragonflies with a range of abdomen colours (often bright) – blue, yellow, red, and brown, but not green. Often settle on, and dart from, regularly used perches, such as marginal vegetation. ♂s are fairly straightforward to identify; ♀s are typically a rather nondescript yellowish-brown and less easy, but can be distinguished using abdomen, thorax, leg and wing features.

WINGS dark patches at base of forewings and/or hindwings

ABDOMEN **broad or tapered**; 'FACE' **various colours, but not 'bright' white**; COSTA dark (compare with *Orthetrum*)

Libellula
[3 species]
p. 73

ABDOMEN **'waisted' or parallel-sided**; 'FACE' **'bright' white**

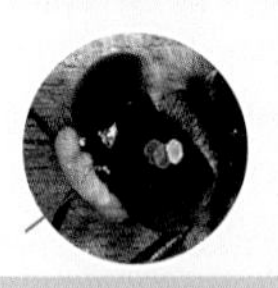

Leucorrhinia
[2 species, incl. 1 very rare immigrant]
p. 72

WINGS without dark patches (clear or yellow at base)

ABDOMEN **rather broad** (about as wide as thorax); ♂ at least partly blue; COSTA **yellow** (compare with *Libellula*)

Orthetrum
[2 species]
p. 72

ABDOMEN **rather narrow and slender**; ♂ reddish (black in one species)

Sympetrum
[7 species, incl. 3 rare immigrants]
p. 69

ABDOMEN **orange, with red markings;** HINDWINGS **broad**

Wandering Glider *Pantala flavescens*
[very rare immigrant or accidental introduction]

ABDOMEN **noticeably broadened; ♂ reddish**

Scarlet Darter *Crocothemis erythraea*
[very rare immigrant]

VERY RARE IMMIGRANT CHASERS, SKIMMERS and DARTERS

♂ **Yellow-spotted Whiteface**
Leucorrhinia pectoralis

♂ **Wandering Glider**
Pantala flavescens

♂ **Scarlet Darter**
Crocothemis erythraea

DARTERS WITHOUT DARK PATCH AT BASE OF WINGS 1/2

Sympetrum 7 spp.

Black Darter and mature ♂s relatively easy to distinguish; ♀s more challenging but possible with close examination of the features summarized in the table *below*:

♂	♀		
ABDOMEN **black**	ABDOMEN **black with yellow markings**	Black Darter	*p.71*
ABDOMEN red	ABDOMEN yellowish/brownish		
EYES **bottom half blue;** WINGS **distinctive red veins**	EYES **bottom half blue;** WINGS **distinctive yellow veins**	Red-veined Darter STATUS local immigrant/ occasional breeder	*p.71*
WINGS **suffused orange with yellow basal veins**		Yellow-winged Darter STATUS rare immigrant	*p.69*
WINGS **clear with black bands**		Banded Darter STATUS one record (1995)	*p.69*
ABDOMEN blood-red, distinctly 'waisted'; THORAX **plain reddish-brown**; FRONS with black 'eye' line; LEGS **all-black**	THORAX **yellowish with dark 'T' marking on top**; FRONS with black 'eye' line; LEGS **all-black**	Ruddy Darter STATUS common GB (SE) and Ireland **'FACE'** frons with dark 'eye' line	*p.70*
ABDOMEN orange-red, slightly 'waisted'; THORAX **brown with yellowish patches on side**; FRONS **no black 'eye' line**; LEGS typically **with pale stripe**	THORAX **brown with two yellow lines (can be obscure) on top**; FRONS typically no black 'eye' line; LEGS typically **with pale stripe**	Common Darter STATUS common **'FACE'** frons without 'eye' line	*p.70*
As Common Darter but THORAX poorly marked; FRONS WITH black 'eye' line	As Common Darter but FRONS WITH black 'eye' line; ABDOMEN WITH prominent vulvar scale on underside of S9	Vagrant Darter STATUS rare immigrant **'FACE'** frons with dark 'eye' line	*p.69*

VERY RARE IMMIGRANT *SYMPETRUM* DARTERS

Yellow-winged Darter
Sympetrum flaveolum

Banded Darter
Sympetrum pedemontanum

Vagrant Darter
Sympetrum vulgatum

DARTERS WITHOUT DARK PATCH AT BASE OF WINGS 2/2

Common Darter
Sympetrum striolatum

BL 33–44 mm. **ID** Legs black with white stripe. No black line down side of frons. ♂ abdomen orange-red, slender, only slightly 'waisted'; thorax brown with yellow sides, split by a brown panel. ♀ abdomen yellowish, darkening and becoming redder with age; thorax brown with two yellow lines/patches on top (obscure in some individuals). A darker form occurs in the Scottish Highlands. **Cyc** 1 year. **Hab** Various waters. **Beh** Frequently returns to the same perch, including bare ground. **SS** Other red darters.

NOTE: ♀s from the Scottish Highlands closely resemble ♀ **Black Darter**, which differs in having a black triangle on the top of the thorax.

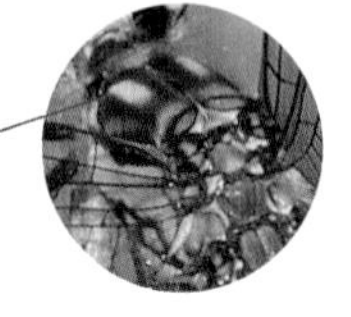

♀ **THORAX** yellow patches on top

Ruddy Darter
Sympetrum sanguineum

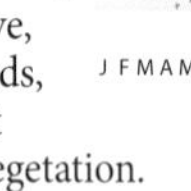

BL 34–39 mm. **ID** Legs black. Black line down side of frons. ♂ abdomen blood-red, rather short and robust, 'waisted'; thorax reddish brown; eyes reddish brown above, dark green below. ♀ abdomen yellowish; thorax with dark 'T' marking on top; eyes brown above, yellow beneath. **Cyc** 1 year. **Hab** Ponds, ditches, lakes, *etc.* often frequent just away from water, perching on low vegetation. **SS** Other red darters.

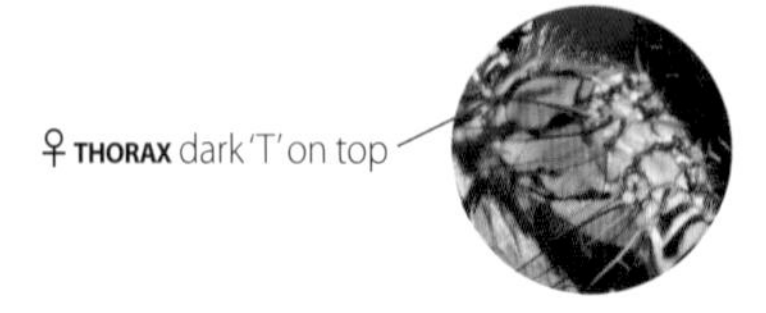

♀ **THORAX** dark 'T' on top

Black Darter
Sympetrum danae

J F M A M J J A S O N D

BL 29–34 mm. **ID** Small; mainly black. Side of thorax has yellow bands separated by black band with three yellow spots. Legs black. ♂ abdomen black with yellow marks on side. ♀ abdomen with extensive yellow markings, turning brown with age; thorax with dark triangle on top. **Cyc** 1 year. **Hab** Heathland pools and lake margins, including boggy areas. **Beh** Adults often perch on low vegetation or stones. **SS** **White-faced Darter** (*p. 72*), which has a white face; ♀/immature ♂s of other darters.

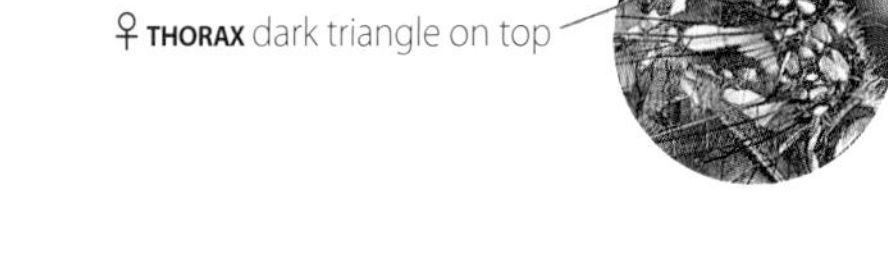
♀ **THORAX** dark triangle on top

Red-veined Darter
Sympetrum fonscolombii

J F M A M J J A S O N D

BL 33–40 mm. **ID** Bottom half of eyes blue. Base of wings yellow; wing-spots pale, outlined in black. Legs black with white stripe. ♂ eyes reddish brown above; front of head and thorax reddish. Side of thorax with pale yellow stripe. Abdomen deep red, with black marks on S8–9. Veins towards front of forewings and hindwings red. ♀ thorax and abdomen yellowish. Wing veins as ♂ but yellow. **Cyc** 1 year (sometimes has two generations in a year). **Hab** Warm, still waters; perches on vegetation or bare ground. **SS** Other red darters.

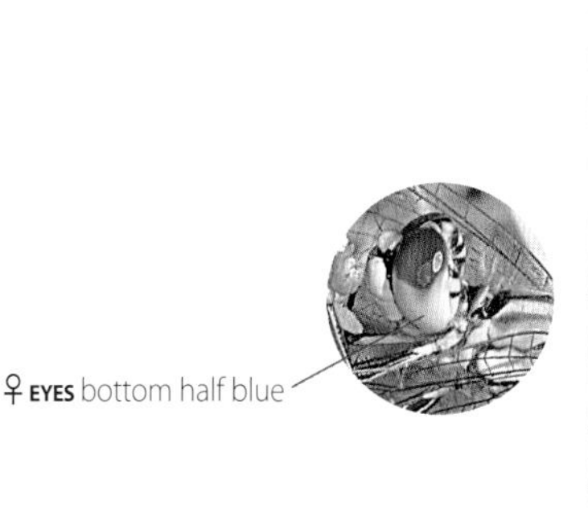
♀ **EYES** bottom half blue

Guide to families *pp. 48–49*

SKIMMERS

Orthetrum 2 spp.

Black-tailed Skimmer *Orthetrum cancellatum*

BL 45–50 mm. **ID** Wings clear; costa yellow; wing-spots black; thorax olive-brown. Abdomen tapered: ♂ blue with a black tip; ♀ yellow with black longitudinal stripes. **Cyc** 1–3 years. **Hab** Ponds, lakes and canals. **Beh** Sometimes skimming the water, but often resting on rocks or soil, as well as perches. **SS** **Scarce Chaser, Keeled Skimmer.**

Keeled Skimmer *Orthetrum coerulescens*

BL 36–45 mm. **ID** Wings clear; costa yellow; wing-spots brown. Thorax dark brown with buff stripes (fading with age in males). Abdomen slender, tapered: ♂ wholly blue; ♀ brown with black central stripe. **Cyc** About 2 years. **Hab** Pools, streams and flushes on bogs and wet heathland. **SS** **Black-tailed Skimmer.** NB ♀/imm resemble some ♀ darters, but differ in having pale thoracic stripes and a row of yellow cross-veins between 2nd and 3rd main veins from front of each wing.

DARTERS WITH DARK PATCH AT BASE OF WINGS + CHASERS

Leucorrhinia 2 spp. (see also *p. 68*)

White-faced Darter EN
Leucorrhinia dubia

BL 31–37 mm. **ID** Small. Wings with black base. **'Face' white.** ♂ black with red markings on thorax and abdomen S1–7 (mainly S1–2 and S6–7). ♀ with more extensive markings, but yellow. **Cyc** About 2 years. **Hab** Bog pools. **Beh** Often perches on vegetation or bare ground. **SS** ♂ **Black Darter** (*p. 71*), which has a black face.

Libellula 3 spp.

Broad-bodied Chaser *Libellula depressa*

BL 39–48 mm. **ID** Extensive dark mark at base of all wings. Eyes brown. Top of thorax with two pale stripes. Abdomen very broad (blue in ♂, yellowish-brown in ♀, with yellow sides). **Cyc** 1–3 years. **Hab** Ponds, including in gardens, and small lakes. **Beh** Periodically uses low perches. **Dist** Mainly C, E and S Britain, more local farther north (absent from Ireland). **SS** Other chasers (but no other has such a broad body); ♂ skimmers.

Four-spotted Chaser *Libellula quadrimaculata*

BL 40–48 mm. **ID** Base of hindwings dark; all wings with yellow patch at base and black patch halfway along the front edge (form *praenubila* has additional (often ill-defined) black patch extending from wing-spot). Eyes and thorax brown; abdomen brown, with a black tip and yellow sides. **Cyc** 2 years or more. **Hab** Various standing waters, ♂ often frequenting a favourite perch. **SS** None.

Scarce Chaser *Libellula fulva* NT

BL 42–45 mm. **ID** All wing bases with dark patch, most extensive on hindwings; no patch halfway along wing. Eyes bluish-grey, ♂ thorax black; abdomen blue with a black tip (often has dark marks in centre of abdomen caused by pigmentation being rubbed off by the female's legs during mating). ♀ thorax brown; abdomen brown or orange, with black marks down centre of each segment. Tip of wings often black. **Cyc** About 2 years. **Hab** Small rivers and dykes, some ponds with plentiful emergent vegetation. **SS** **Black-tailed Skimmer**, particularly ♂, although that species has clear wing bases.

Further reading *p. 79*

ORDER PLECOPTERA Stoneflies

[Greek: *plekein* = to fold; *pteron* = wing]

BI | 7 families [19 genera, **34 species**] **W** | 16 families [**3,930 species**]

Slender brown or yellowish, medium-sized flattened insects. The compound eyes are bulging and there are two to three ocelli. The mouthparts are developed for chewing and they have long, thread-like antennae. They are weak fliers, the wings membranous, at rest partly wrapped around the abdomen and often extending beyond end of abdomen (but shortened in some species). The soft-bodied abdomen has 10 segments, the paired cerci (tails) conspicuous when extended beyond wings. The aquatic nymphs often have abdominal gills and generally resemble wingless adults. Adults rarely feed and are sometimes active by day, but mainly nocturnal. Some males attract females by drumming the substrate with their abdomens. Egg masses are released underwater or laid under stones in damp areas. Nymphs feed on plants and rotten wood, sometimes other insects and may have slow development. However, all but the two longest British species (family Perlidae) complete the life-cycle in a year; this also applies to species covered in this book, unless otherwise stated. Fishermen use artificial stonefly flies as an effective lure when fly fishing. Selected representative species from the families are illustrated.

FINDING STONEFLIES Most likely to be seen on vegetation, fences or rocks near cool running streams and rivers; perhaps fluttering around. Nymphs often develop in stony or gravelly streams and cast skins are often seen after emergence.

Structure of a stonefly

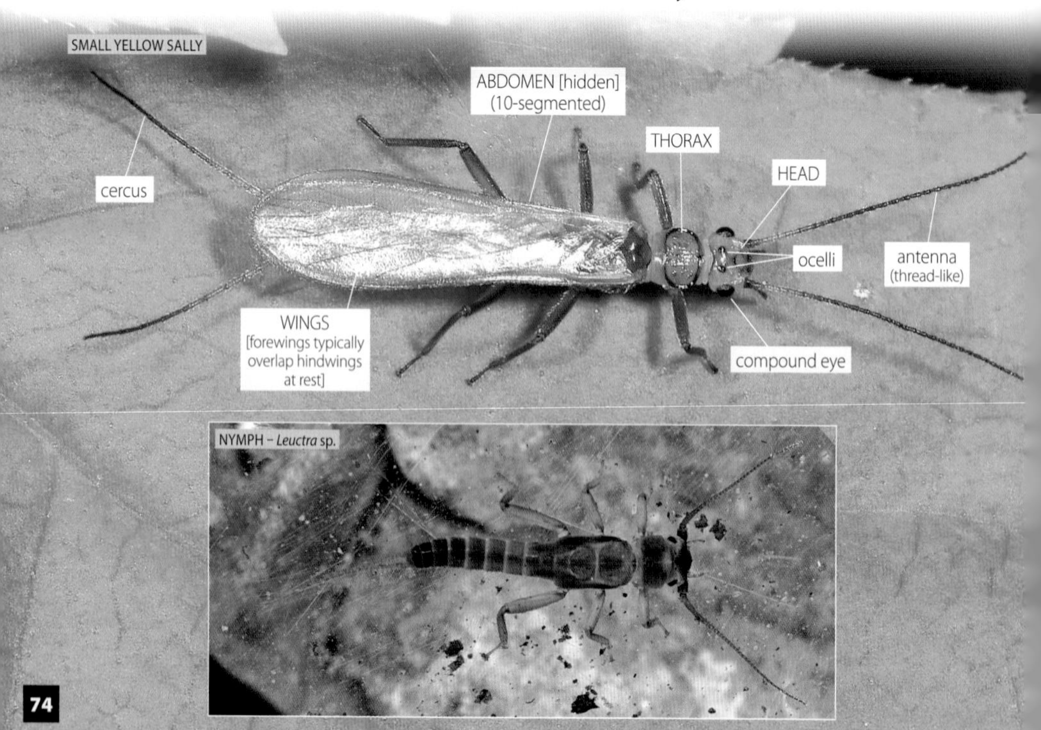

ORDER Plecoptera (Stoneflies) GUIDE TO FAMILIES | 7 BI

Stoneflies are best identified to family group by a combination of the length of the cerci ('tails'), wing pattern, tarsal segment (S) lengths and colour/size as outlined in the table below.

'TAILS' short
not visible beyond end of forewings

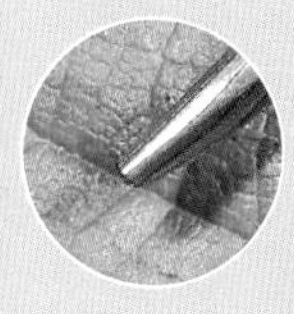

TARSUS **3 segments of similar length**

TARSUS **S2 much shorter than S3**

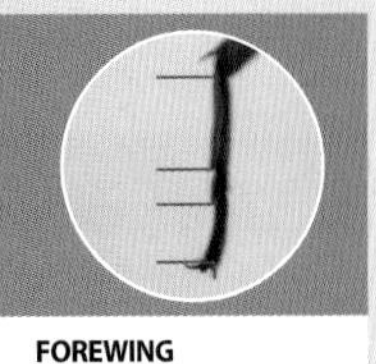

FOREWING with **at least one darker band**

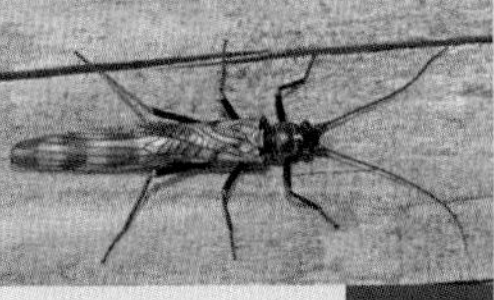

February reds
Taeniopterygidae
[3 genera | 4 species]
p. 77

FOREWING **with** 'X'-like figure

Small brown stoneflies
Nemouridae
[4 genera | 11 species]
p. 76

FOREWING **lacking** 'X'-like figure

Needle flies
Leuctridae
[1 genus | 6 species]
p. 76

'TAILS' long
extending well beyond end of forewings

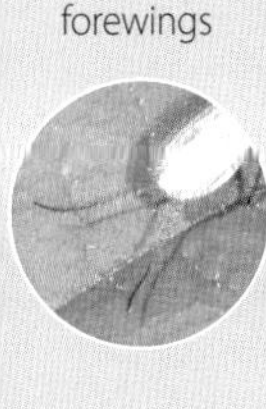

FORM often larger 15–23 mm; HEAD **orange**

Large stoneflies
Perlidae
[2 genera | 2 species]
p. 79

FORM 8–23 mm; HEAD AND PRONOTUM **top with yellow or orange stripe**

Perlodid stoneflies
Perlodidae
[4 genera | 5 species]
p. 78

FORM **small 4–8 mm; dark**

Black stoneflies
Capniidae
[2 genera | 3 species]
p. 79

FORM **small 5–9 mm; yellow**

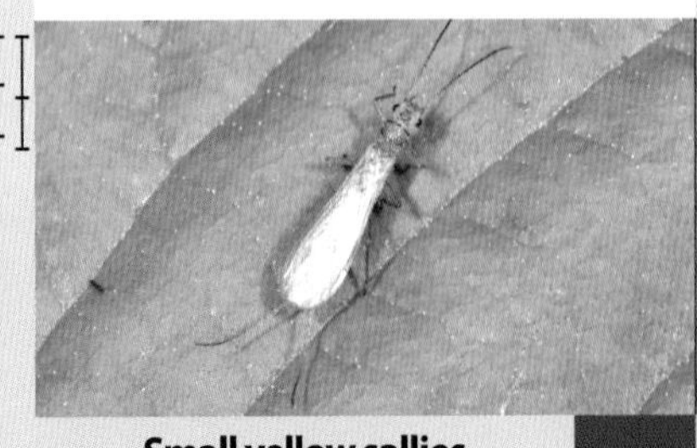

Small yellow sallies
Chloroperlidae
[3 genera | 3 species]
p. 77

Guide to families *p. 75*

STONEFLIES WITH SHORT 'TAILS'

FAMILY **Leuctridae** (Needle flies) — 1 GEN. | 6 spp. | 1 ILL.

Six similar-looking species (4–11 mm), with tarsus S2 shorter than S3 (see *p. 75*). Species identified by microscopic examination of genitalia.

Willow Fly *Leuctra geniculata*

Common

BL 7–11 mm, **WS** 20–25 mm. **ID** Brown, usually with darker thorax, including pronotum and in some individuals head; tips of antenna segments hairy. Legs brown or yellowish, femora with dark subbasal mark or band. **Hab** Rivers and large streams with stony substrates. **SS** Other similar-sized *Leuctra* species [N/I], but these fly earlier in the season and the tips of the antenna segments lack hairs.

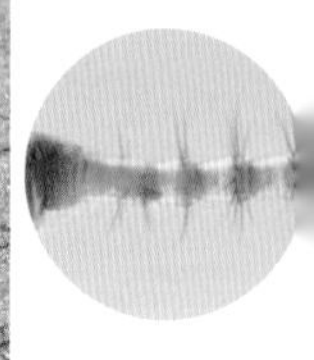

FAMILY **Nemouridae** (Small brown stoneflies) — 4 GEN. | 11 spp. | 1 ILL.

The largest family of stoneflies in Great Britain and Ireland, with 11 mostly widespread and common species (4–9 mm), with tarsus S2 shorter than S3 (see *p. 75*). Species identified by microscopic examination of genitalia and shape of male cerci.

Nemoura — 5 spp. | 1 ILL.

Small Dull Brown *Nemoura cinerea*

BL 6–9 mm, **WS** 15–20 mm. **ID** Small, brown stonefly. **Hab** Flowing and still waters with emergent vegetation. **SS** Other Nemouridae [N/I], which require microscopic examination of genitalia for identification.

Common

FAMILY Taeniopterygidae (February reds) 3 GEN. | 4 spp. | 1 ILL.

Four small, dark, mostly early year species (peak numbers March) with at least one darker band on the forewings and with tarsal segments of a similar length (see *p. 75*).
Species separated by slight differences in forewing venation and antenna structure.

Brachyptera 2 spp. | 1 ILL.

Northern February Red *Brachyptera putata*

BL 7–10 mm, **WS** approx. 17 mm. **ID** Brown; three dark bands on forewings with darkened tip. The short-winged male is unable to fly. **Hab** Slower parts of rivers. **Dist** Sronghold is in NE Scotland and the Highlands, rare in Wales (river Usk) and Herefordshire (river Wye). This species is found nowhere else in the world. **SS** The **Common February Red** *Brachyptera risi* [N/I] (**BL** 7–11 mm) is more widespread and found mainly from March to May; its forewing tip is not darkened and the antennal basal segments are wider than long.

Nationally Scarce

STONEFLIES WITH LONG 'TAILS' 1/2

FAMILY Chloroperlidae (Small yellow sallies) 3 GEN. | 3 spp. | 1 ILL.

Includes two small (5–9 mm), widespread to common, yellow-bodied species.

Siphonoperla 1 sp.

Small Yellow Sally *Siphonoperla torrentium*

BL 5–8 mm, **WS** 13–18 mm. **ID** Small, conspicuous and fragile-looking, yellow insect. **Hab** Flowing and still waters with stony substrates. **SS** **Western Small Yellow Sally** *Chloroperla tripunctata* [N/I] (**BL** 6–9 mm); distinction requires examination of genitalia. **SS** Superficially similar to **Common Yellow Sally** (*p. 78*) which has a yellow central stripe on the thorax and femora with longitudinal black stripes.

Common

Guide to families *p. 75*

STONEFLIES WITH LONG 'TAILS' 2/2

FAMILY **Perlodidae** (Perlodid stoneflies) 4 GEN. | 5 spp. | 2 ILL.

Small–large species (8–23 mm), some rather different in appearance but **all with a yellow or orange line on the head and thorax.**

Perlodes 1 sp.

Orange-striped Stonefly *Perlodes mortoni* RE (Ireland)

[Listed as *Perlodes microcephalus* in most other publications, but *mortoni* is the valid species name]

BL 13–23 mm, **WS** 12–35 mm. **ID** Large, brown and easily recognized by the bold orange-striped thorax, also with orange central spot(s) on head. Short and long-winged forms occur. **Hab** Flowing rivers and streams with stony substrates, sometimes lake shores in Scotland. **SS** None; larger and often darker than other Perlodidae.

Isoperla 2 spp. | 1 ILL.

Common Yellow Sally *Isoperla grammatica*

BL 8–13 mm, **WS** 16–27 mm. **ID** Robust-looking yellow species with yellow central stripe on thorax. All femora with longitudinal black stripes. **Hab** Flowing and still rivers and streams with stony substrates; lake shores in Scotland. **SS** Superficially similar to **Small Yellow Sally**, (*p. 77*), which has a plain thorax and femora. **Scarce Yellow Sally** *Isoperla obscura* [N/I] (**BL** 8–12 mm) is presumed extinct (last recorded 1920) and would require detailed examination of genitalia.

FAMILY Perlidae (Large stoneflies) 2 GEN. | 2 spp. | 1 ILL.

Great Britain's largest stoneflies (15–25 mm). Genera/species can be separated as follows:

PRONOTUM yellowish with two spots	**Large Two-spotted Stonefly**
FORM some darker than Large Two-spot. Stonefly; PRONOTUM plain	**Large Dark Stonefly** *Dinocras cephalotes* [N/I]

Perla 1 sp.

Large Two-spotted Stonefly *Perla bipunctata*

BL 16–24 mm, **WS** 25–52 mm. **ID** Large, brown with orange head. Pronotum yellowish with black margin; either side of the black central line is a dark patch (the scientific name bipunctata = 'two spots'). **Cyc** Research in Ireland indicates a three year life-cycle, with two periods of egg hatching; the larvae feed on mayflies and flies. **Hab** Flowing rivers and streams with unstable stony substrates. **Dist** N and W GB, also part of Ireland. **SS** Large Dark Stonefly *Dinocras cephalotes* [N/I] (**BL** 15–24 mm) [see table *above*].

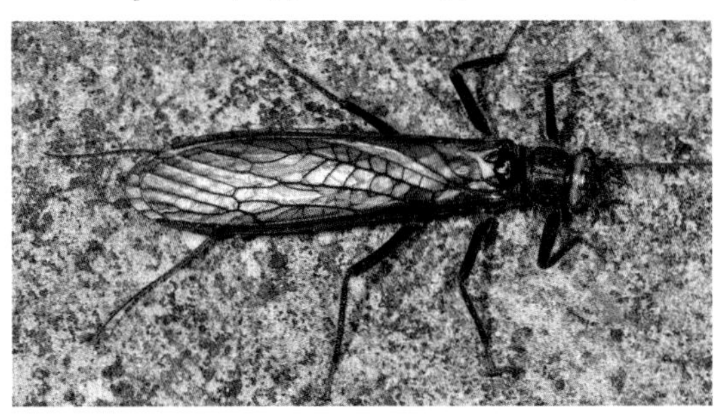

FAMILY Capniidae (Black stoneflies) 2 GEN. | 3 spp. | 1 ILL.

Small (4–8 mm), black species with long cerci. Only the Common Black Stonefly is common (recently transferred from the genus *Capnia* to *Zwicknia*). The two similar-looking *Capnia* species are mainly found in Scotland, but with few modern records, hence both have a Red List conservation status.

Zwicknia 1 sp.

Common Black Stonefly *Zwicknia bifrons*

BL 4–8 mm, **WS** approx 15 mm. **ID** Black with dark brown wings, the male short-winged. **Hab** Deeper parts of river and stream beds (flowing and still). **SS** *Capnia* species [N/I] (**BL** 4–8 mm), which require microscopic examination of genitalia.

RECOMMENDED FURTHER READING AND USEFUL WEBSITES

Hynes, H.B.N. 1977. ***A key to the adults and nymphs of British stoneflies (Plecoptera) with notes on their ecology and distribution.*** *Scientific Publication No. 17, 3rd Edition.* Freshwater Biological Association. *Out of print; photocopy available via* www.fba.org.uk/scientific-publications-sps.

Pryce, D., Macadam, C. & Brooks, S. 2007. ***Guide to the British stonefly (Plecoptera) families: adults and larvae.*** Field Studies Council (AIDGAP). *Useful chart, but Hynes needed to identify to species level.*

www.riverflies.org The Riverfly Partnership

ORDER DERMAPTERA — Earwigs

[Greek: *derma* = skin; *pteron* = wing]

BI | 4 families [6 genera, **7 species** (incl. 3 introductions, 2 of these now extinct)]

W | 12 families [**approx. 2,000 species**]

Small to medium-sized, elongate, flattened insects with short legs. The antennae are of short to moderate length. Where present, hindwings are folded beneath short, leathery forewings. Conspicuous forceps (modified cerci) usually present, which are used for various purposes, including in defence. Nymphs resemble small adults, but are wingless in early stages.

FINDING EARWIGS A wide range of habitats are suitable, including gardens, where these mainly nocturnal insects feed on plant or animal material. In a rare case of maternal care in the insect world, females guard their eggs and young until the nymphs can fend for themselves.

Structure of an earwig

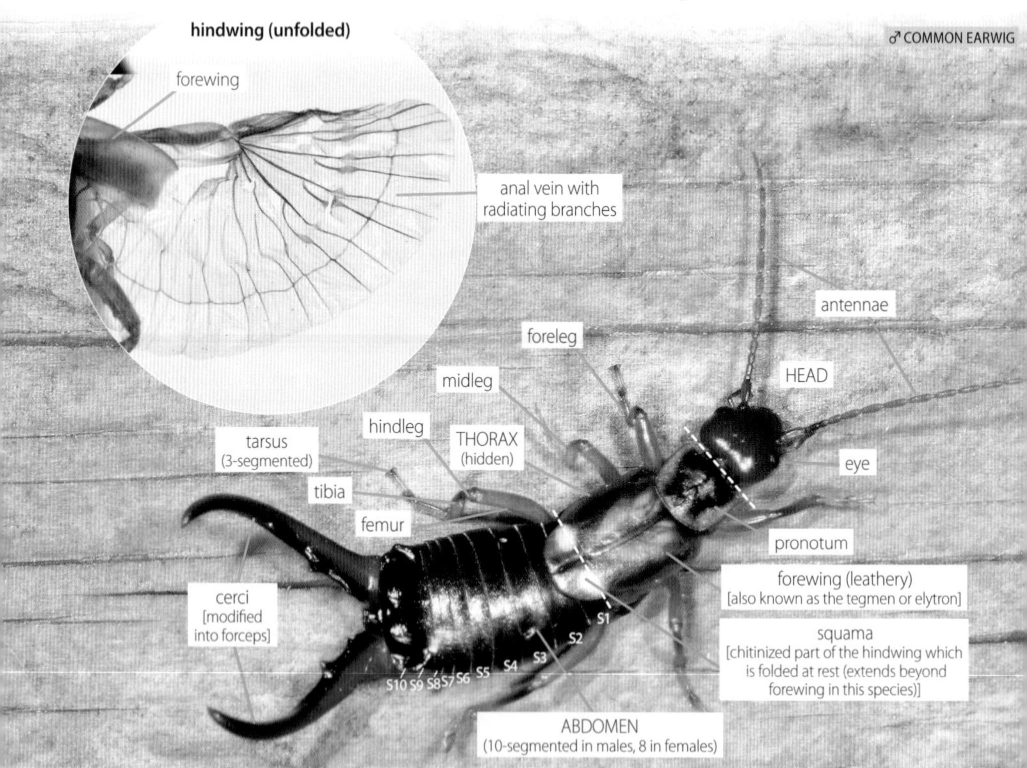

FURTHER READING AND USEFUL WEBSITES

Marshall, J.A. & Haes, E.C.M. 1990. ***Grasshoppers and allied insects of Great Britain and Ireland*** *(revised edition, original 1988)*. Colchester: Harley Books. Out of print, but not difficult to check in a library or obtain a used copy.

Marshall, J. & Ovenden, D. 1999. ***Guide to British grasshoppers and allied insects***. Field Studies Council (AIDGAP). *Fold-out illustrated identification chart to adult Orthoptera and allies, ideal for field trips.*

DermapteraSpeciesFilehttp://dermaptera.speciesfile.org/

Orthoptera & allied insects Recording Scheme www.orthoptera.org.uk

ORDER Dermaptera (Earwigs) GUIDE TO FAMILIES | 4 BI

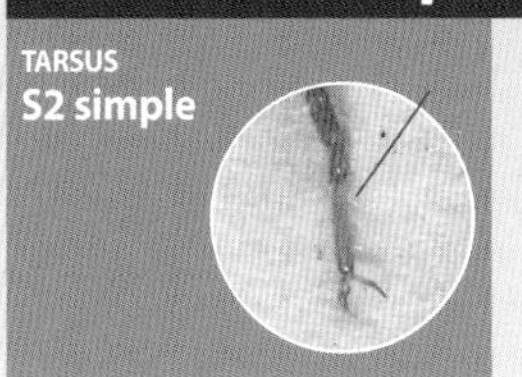

Anisolabididae [Ring-legged Earwig]
Spongiphoridae [Lesser Earwig]
Labiduridae
[3 genera | 4 species]

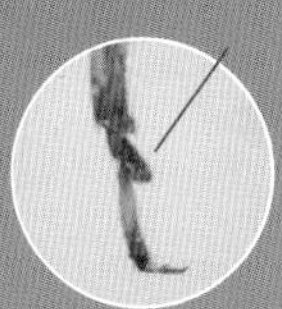

WINGED | ANTENNAE 11–14 segments

Forficulidae [2 genera | 3 species] p. 82

WINGLESS | ANTENNAE 16–17 segments; LEGS pale with darker bands

FAMILY Anisolabididae 1 SP. (introduced)

Ring-legged Earwig *Euborellia annulipes* Introduced; rare

BL 9–11 mm. **ID** Wingless, robust-looking, blackish-brown; legs yellow; femora with dark bands (indistinct in some). **Cyc** Potentially multi-brooded, with food stored in or near the underground nest. **Hab** Places with a warm micro-climate (*e.g.* greenhouses) where they are known to survive for some years. Formerly imported in ship's ballast, in 1991 recorded from Warwickshire via imported sea shells. **Dist** Few recent records (*e.g.* greenhouses at the Eden Project, Cornwall and Tropiquaria, Somerset). A cosmopolitan species, widespread in the Mediterranean. **SS** None.

J F M A M J J A S O N D

WINGED | small (4–6 mm); ANTENNAE 11–12 segments; LEGS uniform

FAMILY Spongiphoridae 2 GEN. | 2 spp. (incl. 1 introduced, now †)

Lesser Earwig *Labia minor* Local

BL 4–6 mm. **ID** Small, yellowish-brown with dark head and pubescent body; antennae with 11–12 segments; legs uniform. Has forewings and hindwings and can fly well. **Cyc** Takes about three months in the heat of a dung heap, so when found all stages may be noted. **Hab** Active by day in sunshine and at night; readily attracted to lights. Breeds in dung and rubbish heaps. **SS** None.

Bone-house Earwig *Marava arachidis* [N/I] (**BL** 4–6 mm), not recorded in GB since 1950s, is dark reddish-brown and lacks hindwings.

WINGED | large (12–26 mm); ANTENNAE 20+ (often 27–30) segments; LEGS uniform

FAMILY Labiduridae Giant Earwig *Labidura riparia*

The only British species [probably extinct here] is readily recognized by its large size. Possibly originally an accidental introduction (although native status cannot be ruled out), resident in Christchurch and Boscombe, Dorset from 1808 to possibly the 1930s (also Folkestone, Kent, in the 1870s). There is an unconfirmed 1985 sighting from Seaton, Devon and it may still survive in a warm part of the south coast.

EARWIGS WITH TARSUS S2 FLATTENED

FAMILY Forficulidae 2 GEN. | 3 spp. | 3 ILL.

The well-known Common Earwig has hindwings and occasionally flies; the other two species found in Great Britain (one in Ireland) have shortened hindwings and are unable to fly. Genera/species can be separated as follows:

WINGS folded hindwings visible beyond forewings	**Common Earwig**
WINGS only forewings visible; ♂ **FORCEPS** broadened basally	**Lesne's Earwig**
WINGS only forewings visible; ♂ **FORCEPS** not broadened basally	**Short-winged Earwig**

Forficula 2 spp.

This genus of 75 species worldwide includes **the Common Earwig, one of our most familiar insects.**

Common Earwig
Forficula dentata

BL 10–15 mm. **ID** Abdomen fairly broad; chestnut-brown with yellowish legs. A chitinized part of the hindwings, known as the squama, is visible beneath the forewings. ♀ forceps straight, in ♂ curved, much larger (up to 10 mm long) in form *forcipata*. White earwigs are newly moulted and yet to harden and darken. **Cyc** Each female lays 30–50 eggs in an underground cavity; eggs overwinter and hatch in spring, with maternal care provided during these stages (including eggs licked clean). Nymphs moult three times and adults mature by July. **Hab** Many, in most gardens and waste scrub; in flowers, on vegetation, under bark, on or beneath the soil. **SS** Lesne's Earwig.

NOTE: preliminary molecular study (2020) shows that the British/Irish species is *Forficula dentata*, not *F. auricularia* as previously believed; further research is needed on these cryptic species.

The cerci of ♂s are noticeably larger than those of ♀s.

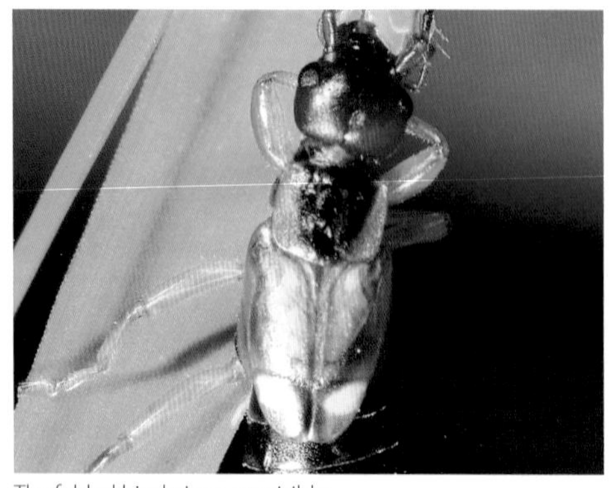
The folded hindwings are visible.

Lesne's Earwig
Forficula lesnei

Nationally Scarce

BL 6–7 mm. **ID** A robust-looking, small, brown species with short forewings; hindwings absent or very small, concealed by forewings; darker head and abdomen. ♂ forceps broadened at base, then incurved. ♀ forceps straight. **Cyc** Presumed similar to the Common Earwig. **Hab** Hedgerows and scrub, including coastal sites. Checking hedgerows with plentiful Ivy at night might be productive. **SS** Common Earwig, Short-winged Earwig (♀). *Guanchia pubescens* [N/I] (**BL** approx. 10 mm) was found in a nursery in Sussex, 2006, but appears to have died out and is not formally included on the British List.

Only the forewings are visible (as in Short-winged Earwig).

Lesne's Earwig
♂ FORCEPS
broadened at base

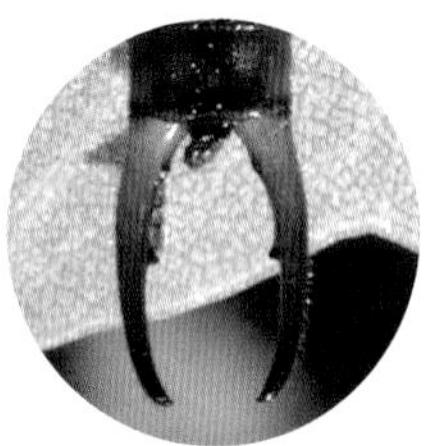

Short-winged Earwig
♂ FORCEPS
not broadened at base

Apterygida 1 sp.

Short-winged Earwig
Apterygida media

Nationally Scarce

BL 6–10 mm. **ID** Elongate, reddish brown, with yellow legs. Forewings light brown, hindwings just tiny lobes, hidden by forewings. ♂ forceps long, slightly incurved. ♀ forceps straight. **Cyc** Presumed similar to the Common Earwig, known from Europe all year. **Hab** Woodland, hedgerows and gardens, on trees and shrubs (they chew petals and are most likely to be seen in flowers). Formerly frequent in hop gardens. **SS** Lesne's Earwig (♀).

Further reading *p. 86*

ORDER ORTHOPTERA Grasshoppers & crickets

[Greek: *orthos* = straight; *pteron* = wing]

BI | 8 families [30 genera, **39 species** (incl. 5 immigrants)] **W** | 42 families [**approx. 28,000 spp.**]

The Orthoptera are mostly medium-sized to large, winged insects (although a few species lack wings). There are two suborders: Caelifera (grasshoppers and locusts) and Ensifera (bush-crickets and crickets), which often have large hindlegs adapted for jumping. Head with large compound eyes, some with three ocelli between the eyes; the antennae are short to long. At rest, the toughened forewings cover the rather delicate fan-folded hindwings. Females have an ovipositor for egglaying. Besides having considerable colour variation, short and long-winged forms can occur in some species, which may confuse even experienced observers, but checking certain features, as shown, will help ensure accurate identification. Nymphs moult 4 to 10 times, depending on the species, many completing the life-cycle within a year. Adults feed on vegetation, often grasses in the case of grasshoppers, although some crickets are predators. Males make distinctive sounds through stridulation by scraping one part of the body against another (this can be via forewings (in Tettigoniidae, Gryllidae) or hindlegs (in Acrididae)). The sound is distinctive to each species and is referred to as 'song', although some might regard it as a 'chirp'! Females of some species can also stridulate, but are quieter. Sound and courtship are very important in Orthoptera, and grasshoppers lacking both hindlegs (sometimes encountered) cannot 'sing'. Even males that have lost one hindleg are at a disadvantage, as the female may not recognize its 'song'.

NOTE Species recorded in the UK very rarely, such as those imported in consignments of plants, are omitted if there is no evidence they can survive here and breed. Details of these mainly historic finds may be found on the Orthoptera & Allied Insects website (see *p. 86*). Various immigrants are, however, at least illustrated, including locusts, well known for their ability to fly long distances. An example of an omission is the **Italian Locust** *Calliptamus italicus* (one specimen recorded from Dorset in 1933), quite likely to be an accidental one-off import.

FINDING GRASSHOPPERS & CRICKETS Many grassland or scrubby sites can be productive for Orthoptera, including downlands, cliffs, embankments, unspoilt verges and damp areas. Dungeness, Kent, is noted for the most recent addition to the British list, the European Tree Cricket (2015), as well as a population of Sickle-bearing Bush-cricket and Large Cone-head; arrangements can be made in advance to visit the main site for these species. As Orthoptera are sometimes hidden, some recorders trace them using a bat detector, set to a frequency of 35–40 kHz, or use an app.

♀ LONG-WINGED CONE-HEAD

Structure of Orthopterans

Side view of ♀ bush-cricket (Tettigoniidae)

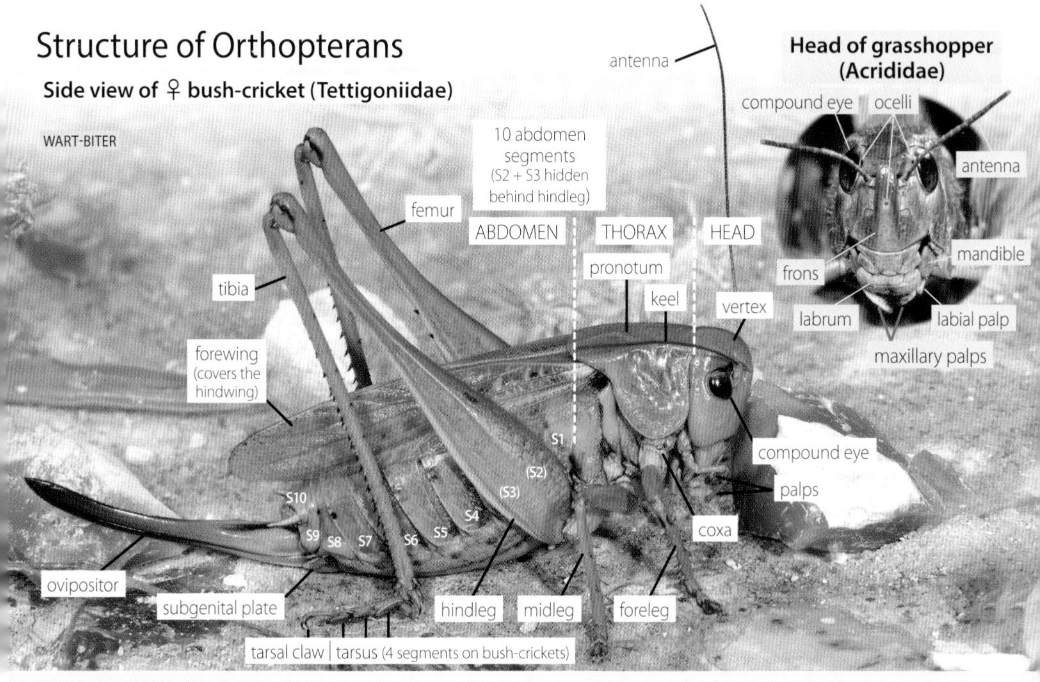

Sexing Orthopterans

BUSH-CRICKETS AND CRICKETS

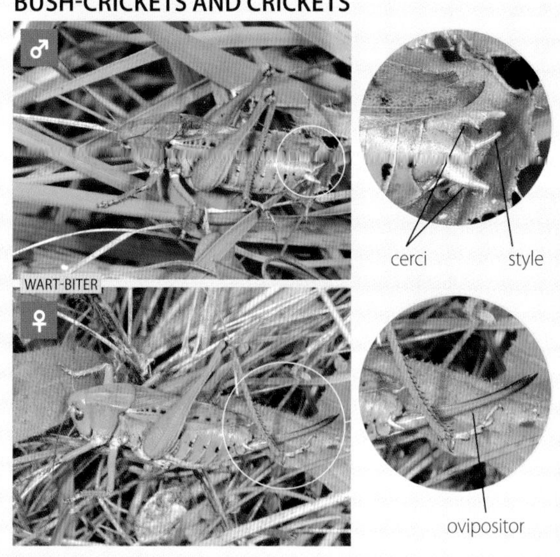

GRASSHOPPERS

Bush-crickets and crickets

Females have an **ovipositor** (absent in the male).
Males have antennae that are longer than those of females.

Grasshoppers

Females have **genital valves** (absent in the male).
Males have antennae that are longer than those of females.

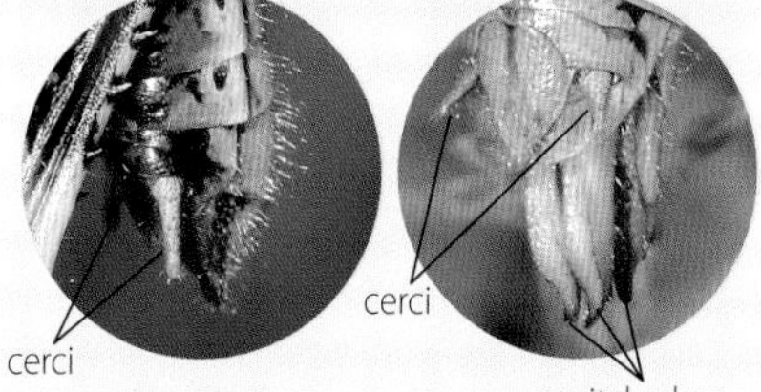

ORDER Orthoptera (Grasshoppers & crickets)

A NOTE ABOUT 'SONGS'

All 31 species of singing grasshoppers and crickets featured in this book can be identified by their 'song'. These songs are presented here as oscillograms, which show the pattern and length of the sound fragments; these can be useful when using a bat detector or making a recording. The QR codes provided enable immediate access to recordings using a readable bar code app on a mobile phone. The iRecord Grasshoppers mobile app http://www.brc.ac.uk/article/irecord-grasshoppers-mobile-app also includes 'songs'.

ANTENNAE **short and thick**; ♀ OVIPOSITOR **short**.

GRASSHOPPERS & GROUNDHOPPERS

PRONOTUM saddle-like, **does not extend backwards** over top of abdomen; SIZE mainly larger 12–36 mm

Meadow Grasshopper

Grasshoppers
Acrididae
[10 genera | 14 species]

p. 104

PRONOTUM **extends backwards** over top of abdomen; SIZE small 8–14 mm

Common Groundhopper

Groundhoppers
Tetrigidae
[1 genus | 3 species]

p. 114

FURTHER READING AND USEFUL WEBSITES

Orthoptera & allied insects Recording Scheme www.orthoptera.org.uk

Marshall, J.A. & Haes, E.C.M. 1990. ***Grasshoppers and allied insects of Great Britain and Ireland*** *(revised edition, original 1988)*. Colchester: Harley Books. Out of print, but not difficult to check in a library or obtain a used copy.

Sutton, P., Beckmann, B. & Lewington, R. in press. ***Field Guide to the Grasshoppers and Crickets of Great Britain and Ireland***. Bloomsbury Wildlife Guides.

Benton, T. 2012. ***Grasshoppers & Crickets***. London: Collins, New Naturalist. Includes a DVD featuring most species.

Evans, M. & Edmondson, R. 2007. ***A photographic guide to the grasshoppers & crickets of Britain & Ireland***. King's Lynn: WGUK. Useful when examining colour forms of species, although colour reproduction is of varying quality.

ANTENNAE long and thin; ♀ OVIPOSITOR long.

WINGED

♂ CERCI short; TARSI 4-segmented

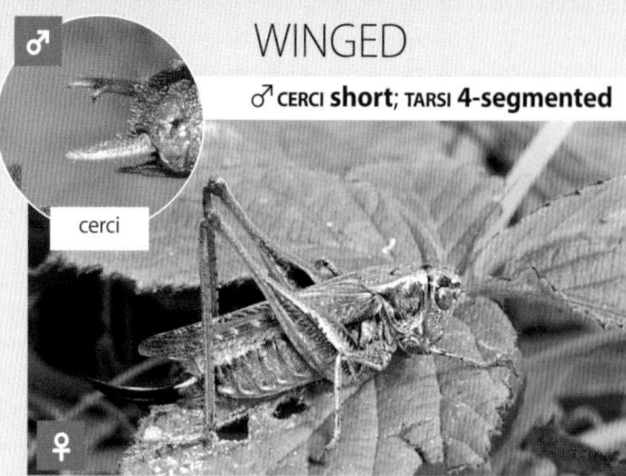

Grey Bush-cricket

Bush-crickets
Tettigoniidae
[11 genera | 13 species]

p. 88

CRICKETS

WINGLESS

FORM pale brown, mottled;
HABITAT in greenhouses

Wingless Camel-crickets
Rhaphidophoridae
[Greenhouse Camel-cricket]

p. 102

FORM greyish brown, covered in small scales;
HABITAT on shingle beaches

Scaly-crickets
Mogoplistidae
[Scaly-cricket]

p. 102

♂ CERCI long; TARSI 3-segmented

Field-cricket

Crickets + Wood-crickets
Gryllidae + Trigonidiidae
[4 genera | 5 species + 1 genus | 1 species]

p. 98

FORELEGS broadened, modified for digging
♀ **OVIPOSITOR** short

Mole-crickets
Gryllotalpidae
[Mole-cricket]

p. 103

Guide to families *pp. 86–87*

■ FAMILY **Tettigoniidae** (Bush-crickets) — **11 GEN. | 13 spp.**

A large family of about 7,850 species worldwide, mostly well camouflaged and often with loud 'songs'. Many species are nocturnal, others are active day and night, feeding on vegetation and small insects. Some species complete the life-cycle within a year; others take much longer, the eggs overwintering twice. (In North America and some other countries bush-crickets are known as katydids.) Genera/species best identified initially by size (body length not including wings), and then by wing length, body pattern and shape of the genitalia as follows:

LARGE SPECIES | body length usually ≥30 mm; usually green

KEY FEATURES **black marks on pronotum and wings**; ♀ OVIPOSITOR long, **upcurved**

Decticus
[Wart-biter]
p. 91

KEY FEATURES **brown stripe along back**; ♀ OVIPOSITOR long, **downcurved**

Tettigonia
[Great Green Bush-cricket]
p. 90

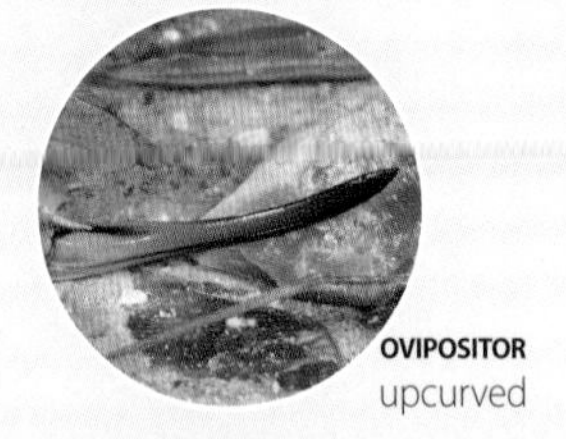

OVIPOSITOR upcurved

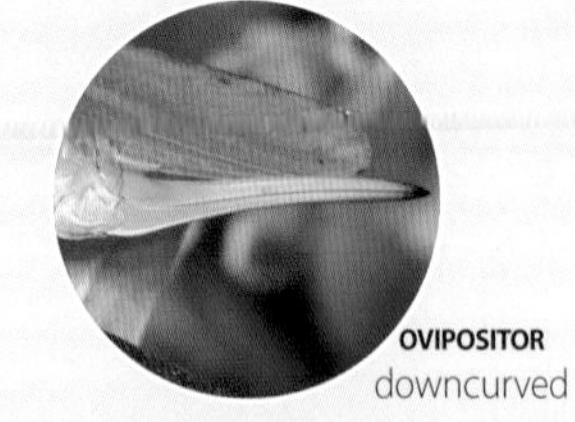

OVIPOSITOR downcurved

SMALLER SPECIES 1/2 | body length ≤30 mm; long-winged

KEY FEATURES **greyish-brown**; **a coastal species**

Platycleis
[Grey Bush-cricket]
p. 97

KEY FEATURES slender; green, **body with copious black spots**; ♀ OVIPOSITOR **sickle-shaped**

Phaneroptera
[Sickle-bearing Bush-cricket]
p. 94

KEY FEATURES slender; green, **white line across tip of pointed head**; ♀ OVIPOSITOR **straight**

Ruspolia
[Large Cone-head]
p. 94

SMALLER SPECIES 2/2 | body length often ≤30 mm; most typically short-winged; two species long-winged

A NOTE ABOUT WING LENGTH
Occasionally, some species that are normally short-winged have longer-winged (macropterous) forms; this most often occurs in large populations or under favourable breeding conditions, and is mentioned within the text for the species concerned.

KEY FEATURES as Southern Oak Bush-cricket but long-winged

♂

Oak Bush-cricket

KEY FEATURES as Short-winged Cone-head but long-winged; ♀ **OVIPOSITOR almost straight**

♀

Long-winged Cone-head

LONG-WINGED

SHORT-WINGED

KEY FEATURES robust; green, **body with tiny black speckles**; ♀ **OVIPOSITOR** broad, upcurved

♀

Leptophyes
[Speckled Bush-cricket] p. 92

KEY FEATURES green with **yellowish stripe along back**

♀

Southern Oak Bush-cricket

Meconema
[2 species] p. 93

KEY FEATURES green with **brown stripe along back**; **wings brown**; ♀ **OVIPOSITOR upcurved**

Long-winged form rare

♀

Short-winged Cone-head

Conocephalus
[2 species] p. 95

KEY FEATURES brown; **wings very short**

♂

Pholidoptera
[Dark Bush-cricket] p. 97

KEY FEATURES dark brown, some individuals partly green; **PRONOTUM** margin cream or green; **THORAX three yellow or green spots on side**

Long-winged form rare

♀

Roeseliana
[Roesel's Bush-cricket] p. 96

KEY FEATURES brown, often with green on top of head, pronotum and part of wings

Long-winged form rare

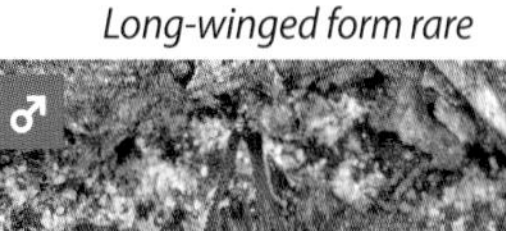

♂

Metrioptera
[Bog Bush-cricket] p. 96

Guide to Tettigoniidae *pp. 88–89*

BUSH-CRICKETS ≥ 30 mm long; green

Tettigonia 1 sp.

Great Green Bush-cricket *Tettigonia viridissima*

BL 40–55 mm. **ID** GB's largest breeding Orthopteran; wings long. Green, with a brown dorsal stripe. ♀ ovipositor long, slightly downcurved. **Cyc** 1 or sometimes 2 years. **Hab** Tall grassland, particularly on or near the coast, gardens and hedgerows with plentiful Bramble. Bare ground is needed for egglaying. Adults feed on various plants as well as insects, including their own kind. **Dist** Also one record from Ireland (imported). **SS** None.

'Song' Loud and rasping, produced in continuous bursts; far-carrying; can be heard from up to 200 m away.

Decticus 1 sp.

Wart-biter *Decticus verrucivorus*

J F M A M J J A S O N D

BL 30–38 mm. **ID** Bulky and flightless, even though wings extend to about tip of abdomen. Green, with black speckling on pronotum and wings on most individuals. ♀ ovipositor long, upcurved. Grey and purple forms also occur, but are rare. Well camouflaged and easily overlooked. **Cyc** Usually 2 years, but can up to 7 years. **Hab** Chalk grassland with a mosaic of clumps of long grass, short turf and bare ground. Adults feed on invertebrates, including grasshoppers. **Dist** Known from only five sites that are specifically managed for this species, including two reintroduction sites. **SS** None.

'Song' Lengthy, repeated high-pitched clicks; usually from mid-morning to early afternoon.

Guide to Tettigoniidae *pp. 88–89*

BUSH-CRICKETS ≤ 30 mm long; green with black speckles; forewings only

Leptophyes 1 sp.

Speckled Bush-cricket
Leptophyes punctatissima

BL 9–18 mm. **ID** Robust and 'round-backed', with tiny forewings and no hindwings. Green, heavily speckled black. **Cyc** Eggs are laid into crevices in bark, where they overwinter (some lying dormant for an extra year before they hatch). Nymphs emerge in May and mature to adults from late July. **Hab** Scrub, gardens, hedgerows, parks, woodland edge; often associated with Bramble. **SS** None; one of the most frequently recorded species of bush-cricket.

'Song' A bat detector is needed to hear the clicks made by the insects, given every few seconds. For all Orthoptera, set the detector frequency control knob to around 35–40kHz, with the volume control at about half. If desired, once singing is picked up, the knob can then be adjusted to the exact frequency for the species.

A Speckled Bush-cricket ovipositing.

BUSH-CRICKETS ≤ 30 mm long; green with yellow dorsal stripe

Meconema	2 spp.
WINGS **longer** than body	Oak Bush-cricket
WINGS **shorter** than body	Southern Oak Bush-cricket

Oak Bush-cricket

Meconema thalassinum

Common

J F M A M J J A S O N D

BL 13–17 mm. **ID** Rather delicate and fully winged. ♂ cerci long, incurved. ♀ ovipositor slightly upcurved towards tip. Pale green with yellowish dorsal stripe. **Cyc** 1 or sometimes 2 years. **Hab** Woodland, hedgerows, parks and gardens; found on trees and shrubs, where eats some vegetation, but feeds mainly on other insects and their larvae, frequently aphids. **Beh** May be found on fences or tree trunks at night, and is also attracted to lights. **SS** Southern Oak Bush-cricket.

'Song' Does not stridulate; instead, males drum on leaves.

Southern Oak Bush-cricket

Meconema meridionale

Introduced; spreading

J F M A M J J A S O N D

BL 11–17 mm. **ID** Rather delicate; wings reduced to small stubs in both sexes. ♂ cerci long, incurved. ♀ ovipositor upcurved towards tip. Pale green with yellowish dorsal stripe. **Cyc** 1 year. **Hab** Woodland, parks and gardens; found on various trees and shrubs, including oaks. **Dist** First recorded in Surrey as recently as 2001 but has since spread to various parts of England (clearly, in at least some cases, transported on vehicles). **SS** Oak Bush-cricket.

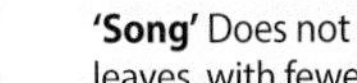

'Song' Does not stridulate; instead, males drum on leaves, with fewer taps than the Oak Bush-cricket.

♂

♀ egglaying in bark

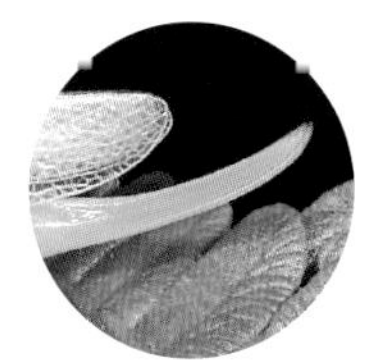

Oak Bush-cricket
OVIPOSITOR slightly upcurved

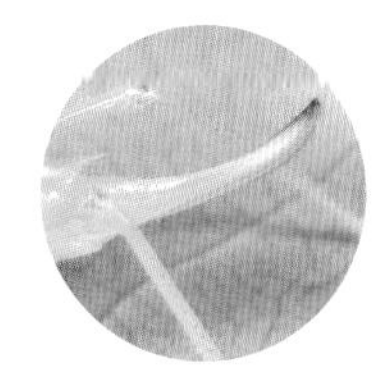

Southern Oak Bush-cricket
OVIPOSITOR upcurved

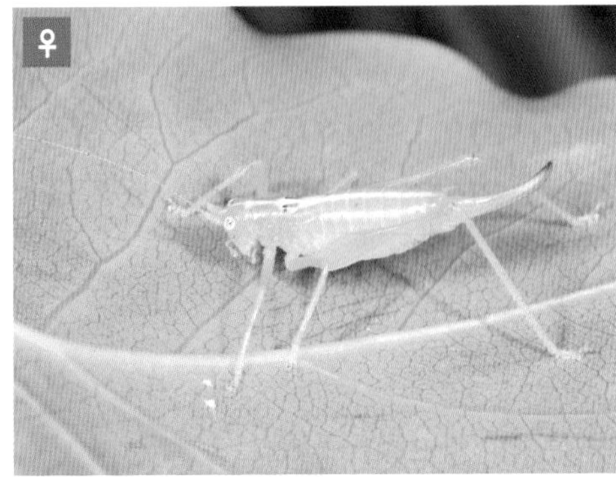

♀

Guide to Tettigoniidae *pp. 88–89*

BUSH-CRICKETS ≤ 30 mm long; green, wings well in excess of body length

Phaneroptera 1 sp.

Sickle-bearing Bush-cricket

Phaneroptera falcata

BL 12–18 mm. **ID** Slender, with wings extending well beyond tip of abdomen (some individuals are over 30mm long); legs very long. Green, with fine black speckles; forewings with a brown dorsal line, ♂ with two darker patches near basal area (used during stridulation). **Cyc** Eggs overwinter. **Hab** Scrub. **Dist** An occasional introduction and immigrant. Became established at a site in Hastings, East Sussex in 2006 but apparently died out a few years later; has bred at Dungeness, Kent since at least 2015 (first recorded there in 2009) and is also breeding elsewhere in Kent.

SS Southern Sickle-bearing Bush-cricket *Phaneroptera nana* [N/I] (**BL** 12–18 mm) was accidentally imported to a plant nursery in Nottinghamshire in 2007 and is a potential immigrant; it is very similar to Sickle-bearing Bush-cricket but cerci of ♂ much more incurved.

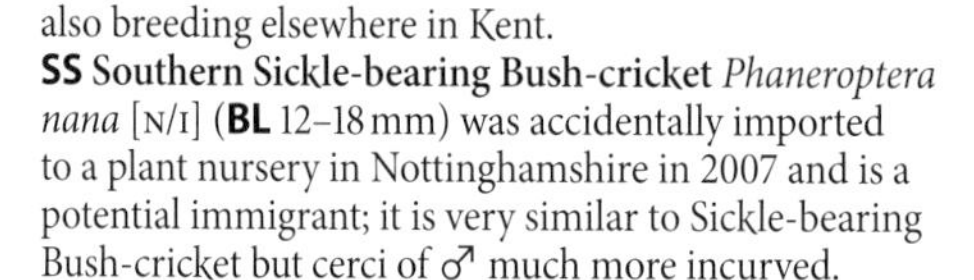

'Song' A faint *"tss tss tss."*

Ruspolia 1 sp.

Large Cone-head

Ruspolia nitidula

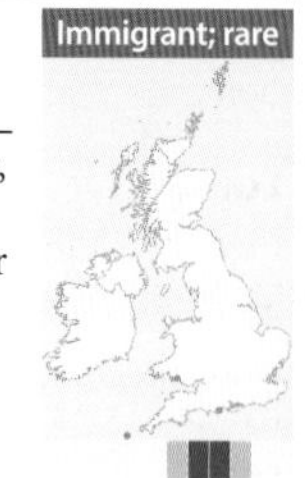

BL 20–33 mm. **ID** Rather large, slender, with wings extending well beyond tip of abdomen (some individuals are over 50 mm long, including wings); head is a distinctive pointed shape. Green or occasionally brown, with a white line at the front of the head. **Cyc** 1 year. **Hab** Dry and damp grassland and scrub. **Dist** First recorded on imported plants in 2001. Subsequent records from St. Mary's, Isles of Scilly in 2003, Canford Cliffs, Dorset in 2005/6 and Dungeness in 2020, indicate that this species is continuing to reach Britain, and suggests that it could become established. **SS** None (head shape diagnostic).

'Song' A distinctive buzzing, including squeaking sounds, which can last continuously for several minutes.

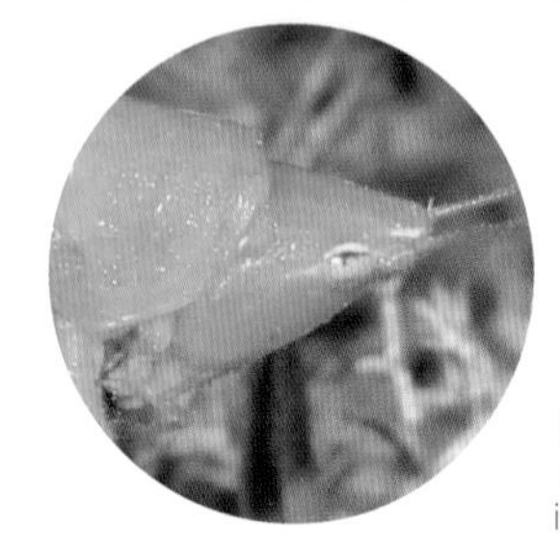

the pointed head shape is diagnostic

BUSH-CRICKETS ≤ 30 mm long; green (rarely brown) with brown dorsal stripe

Conocephalus 2 spp.

WINGS **longer** than body; ♀ OVIPOSITOR **almost straight**	Long-winged Cone-head
WINGS **shorter** than body; ♀ OVIPOSITOR **upcurved**	Short-winged Cone-head

Long-winged Cone-head
Conocephalus fuscus

BL 16–22 mm. **ID** Slender; wings extend beyond tip of abdomen (very long in some forms). ♀ ovipositor **almost straight**. Green, with brown dorsal stripe (some individuals are all-brown). **Cyc** 1 year. **Hab** Dry and damp grassland. **Dist** Has spread rapidly since 2000. **SS** Short-winged Cone-head.

Head of a ♂ Long-winged Cone-head showing the prominent dorsal stripe typical of the two *Conocephalus* species.

'Song' Long bursts of high-pitched hissing sounds.

NOTE: this image shows an extra-long-winged form (see *p. 89* for image of a 'normal' ♀).

Short-winged Cone-head
Conocephalus dorsalis

BL 11–18 mm. **ID** Small; short winged (although there is a rare long-winged form, *burri*). ♀ ovipositor upcurved. Green, with brown dorsal stripe (rarely all-brown). **Cyc** 1 year. **Hab** Saltmarshes and sand dunes; found mainly on maritime rushes and grasses, Also occurs in various wetland habitats, but usually in low numbers. **Dist** Has spread from S GB since the 1990s. **SS** Long-winged Cone-head.

'Song' Long bursts interrupted by clicking sounds.

Guide to Tettigoniidae *pp.* 88–89

BUSH-CRICKETS ≤ 30 mm long; dark brown, with some green/yellow

Roeseliana 1 sp.

Roesel's Bush-cricket
Roeseliana roeselii

BL 13–26 mm. **ID** Robust; wings short (although a fully winged form, *diluta*, is not uncommon in some years). **Thorax has green or yellow spots on side**; pronotum has prominent pale margin below. **Cyc** 1 year (although some eggs laid late in the season lie dormant for more than a year and overwinter twice before they hatch). **Hab** Coarse grassland; coastal reed beds in Ireland. **Dist** Has spread considerably since 2000. **SS** Bog Bush-cricket.

'Song' Long, continuous, high-pitched bursts, rather like a 'buzz'.

Metrioptera 1 sp.

Bog Bush-cricket
Metrioptera brachyptera

BL 11–21 mm. **ID** Robust; wings short (although a rare, fully winged form, *marginata*, occurs). Head and pronotum green or almost uniform dark brown, with blackish markings; pronotum margin with pale band on hind part only. Wings usually partly green. Abdomen underside often bright green. Hind femora have a conspicuous black streak down side. **Cyc** 2 years. **Hab** Heathland and woodland clearings, often in damp areas such as bogs, but occasionally in drier habitats. **Dist** Has spread since 1970. **SS** Roesel's Bush-cricket.

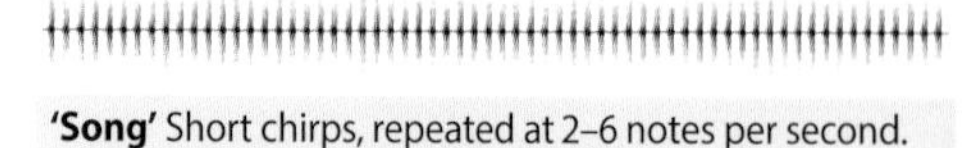

'Song' Short chirps, repeated at 2–6 notes per second.

BUSH-CRICKETS ≤ 30 mm long; greyish or brown

Pholidoptera 1 sp.

Dark Bush-cricket *Pholidoptera griseoaptera*

Common

BL 11–21 mm. **ID** Robust; forewings short on ♂, only tiny flaps on ♀; hindwings absent. Various shades of brown or greyish. **Cyc** 2 years. **Hab** Grassland, hedgerows, woodland edges and rides, and scrub; feeds on bramble leaves and small insects. **Beh** Often well hidden in bushes and, although mainly nocturnal, both sexes may be seen sunbathing on vegetation by day. **SS** None.

'Song' Brief high-pitched chirps, repeated at intervals of a few seconds.

Platycleis 1 sp.

Grey Bush-cricket *Platycleis albopunctata*

Nationally Scarce

BL 18–28 mm. **ID** Robust; fully winged. Greyish-brown with blackish flecks, merging into patches, notably on pronotum sides. Hind femora often black-streaked. ♀ ovipositor upturned, black. **Cyc** 1 year. **Hab Coastal grassland and scrub** (rarely inland), typically in bramble bushes or taller dune vegetation. **Beh** Sometimes seen sunbathing on vegetation or sandy banks. **SS** None (other 'dark' bush-crickets have much shorter wings and are usually not coastal).

'Song' Brief chirps, 2–4 times a second for several minutes.

Guide to families *pp. 86–87*

■ FAMILY Gryllidae (Crickets) — 4 GEN. | 5 spp.

A fairly large family of more than 3,250 species worldwide. Although the hind femora are thickened, they are rarely used for jumping; these insects are built to run and burrow (although tree-crickets frequent trees and bushes). They have conspicuously long cerci. The antennae are generally shorter than those of bush-crickets. Males have longer antennae than those of the females, which also have a conspicuous ovipositor. Following recent molecular work, the Wood-cricket and allies were transferred from Grylliidae to Trigonidiidae (see *below*). The genera/species from these two families, collectively referred to simply as 'crickets' can be separated as indicated in the guide *below*:

■ FAMILY Trigonidiidae (Wood-crickets) — 1 GEN. | 1 sp.

A family of almost 1,000 species worldwide, mostly sword-tail crickets. Following recent molecular work, the Wood-cricket (*Nemobius*) and allies were transferred from Grylliidae to Trigonidiidae.

LARGER SPECIES; BL usually ≥ 13 mm

shiny black; robust; head large

WINGS **short of** abdomen tip

[Field-cricket]

WINGS **extend past** abdomen tip

[Southern Field-cricket]

Gryllus [2 species] *p. 100*

brown, black bars across head

FOREWINGS **long**; HINDWINGS **long**

Acheta [House-cricket] *p. 101*

FOREWINGS **short**; HINDWINGS **short**

Gryllodes [Tropical House-cricket] *p. 101*

SMALLER SPECIES; BL 7–15 mm

brown (black streaks in *Nemobius*)

FOREWINGS **long**; HINDWINGS **long**

Oecanthus [European Tree-cricket] *p. 99*

FOREWINGS **short**; HINDWINGS **short**

Nemobius [Wood-cricket] *p. 99*

CRICKETS usually ≤ 15 mm long

Nemobius 1 sp.

Wood-cricket
Nemobius sylvestris

Nationally Scarce

BL 7–12 mm. **ID** Small; forewings short (plain in ♂; streaked in ♀), hindwings absent. Dark brown, with lighter brown areas, including pronotum. **Cyc** 2 years. **Hab** Woodland; occasionally more open areas, such as roadside banks; among leaf-litter. Often seen close to and under logs and fallen bark; easy to find at night by torch-light. Feeds on dead leaves and fungi. **Dist** Stronghold is in the New Forest, Hampshire. **SS** None.

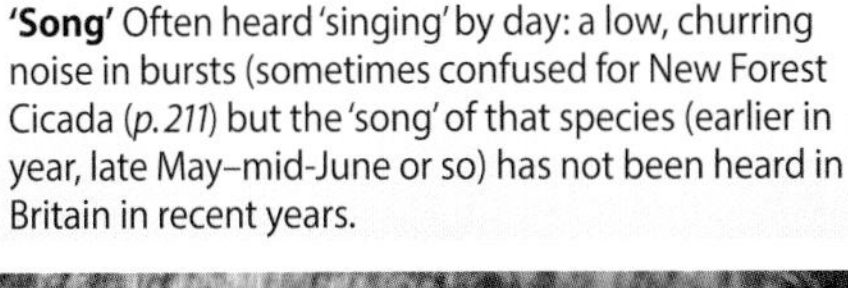

'Song' Often heard 'singing' by day: a low, churring noise in bursts (sometimes confused for New Forest Cicada (*p. 211*) but the 'song' of that species (earlier in year, late May–mid-June or so) has not been heard in Britain in recent years.

Oecanthus 1 sp.

European Tree-cricket
Oecanthus pellucens

Introduced; rare

BL 9–15 mm. **ID** Rather slender, with a yellowish to pale brown ground colour. ♀ cerci almost as long as ovipositor. **Cyc** 1 year. **Hab** Scrub; found on trees and bushes. **Dist** First reported from Dungeness, Kent in 2015, where it has subsequently become well established. **SS** None.

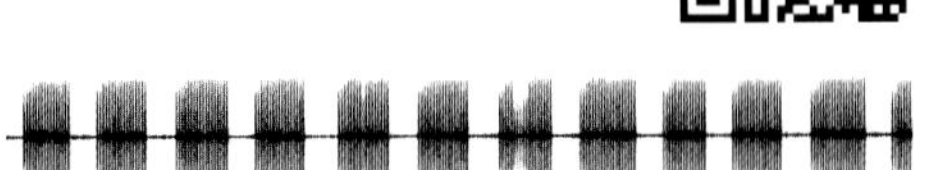

'Song' A very loud "*tsrruu*," around dusk, audible from up to about 50 m away.

Guide to Gryllidae *p. 98*

CRICKETS usually ≥ 13 mm long

Gryllus	2 spp.
Field-cricket	WINGS **just short of, or at most reaching**, abdomen tip
Southern Field-cricket	WINGS **extend beyond** abdomen tip

Field-cricket

VU

Gryllus campestris

Rare

BL 17–23 mm. **ID** Robust, with large head; forewings may reach tip of abdomen on ♂, not quite on ♀; hindwings vestigial. Both sexes shiny black with brown forewings that have a yellow band across the base. **Cyc** Eggs are laid from May and hatch within a few weeks; the resulting nymphs overwinter and reach adult stage about mid-May. **Hab** South-facing slopes with some areas of short grass and bare ground, into which overwintering nymphs and adults burrow. **Dist** Only known from a few populations, including reintroduction sites, which are monitored closely. **SS** Southern Field-cricket.

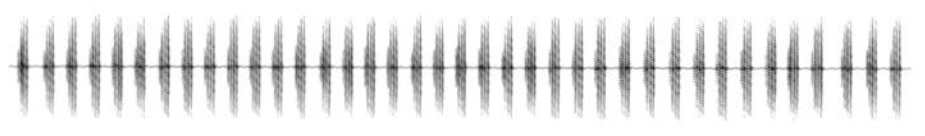

'Song' Males sing from a platform in front of the underground burrow: a series of loud, ringing chirps.

Southern Field-cricket

Gryllus bimaculatus

Introduced; local

BL 17–23 mm. **ID** Robust, with large head; forewings extend beyond tip of abdomen. Shiny black, ♂ with brownish-orange legs at least in part, entirely in some, and dark brown forewings; ♀ with front of forewings yellow. **Cyc** Breeds continuously in captivity. **Hab** Ground-dwelling, in areas of scrub or grassland. **Dist** Presumably mainly short-lived escapees or releases from culture (this is a Mediterranean species and the British climate is likely to be unsuitable for colonization). **SS** Field-cricket.

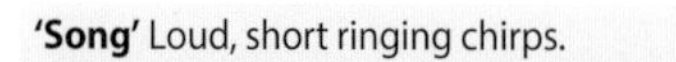

'Song' Loud, short ringing chirps.

Acheta 1 sp.

House-cricket
Acheta domesticus

Introduced; local

BL 14–20 mm. **ID** Rather slim; forewings reach or extend beyond tip of abdomen. Brown, with dark bars across head. **Cyc** 3–8 months; continuously brooded. **Hab** Heated bakeries, factories, commercial kitchens, rubbish tips, manure heaps, although may survive in hedgerows in warmer weather. **Dist** A cosmopolitan species, turning up almost anywhere, but less frequently recorded since 2000. Some records may be escapees from culture. **SS** Tropical House-cricket.

'Song' A high-pitched, repeated chirp.

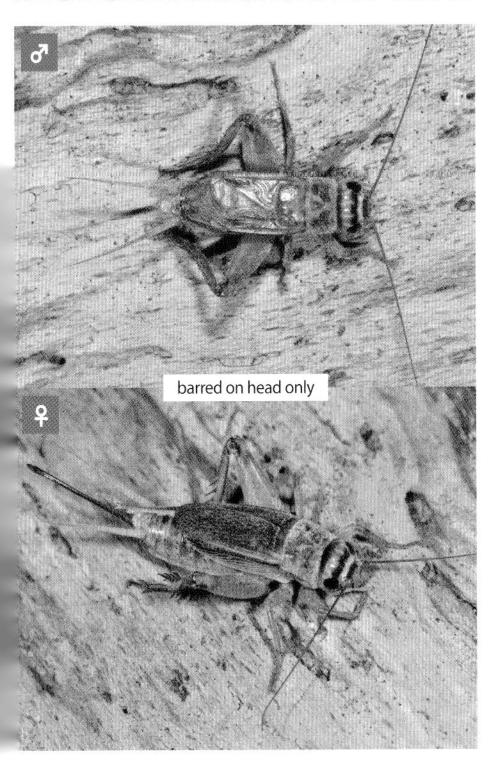

Gryllodes 1 sp.

Tropical House-cricket
Gryllodes sigillatus

Introduced; rare

BL 13–18 mm. **ID** Robust; short-winged. Brown, with dark bars across the head, thorax and abdomen. **Cyc** Continuously brooded. **Hab** Hothouses and waste ground. **Dist** Formerly bred at Kew Gardens, Surrey and may persist around London Zoo. Most probably a native of SW Asia. **SS** House-cricket.

NOTE: This is one of a few species of crickets supplied as pet food for reptiles and therefore likely to escape and live for a short while; reported more often in Great Britain since 2000 than before. Previously listed as a synonym of *Gryllodes supplicans* from Sri Lanka, but has been found to be a distinct species.

'Song' High-pitched repetitive chirps.

Guide to families pp. 86–87

FAMILY **Mogoplistidae** (Scaly-crickets) 1 sp.

Scaly-cricket
Pseudomogoplistes vicentae

VU

Rare

BL 8–13 mm. **ID** Robust; wingless and covered in small scales. Greyish-brown. **Cyc** 2–3 years. **'Song'** None (not able to stridulate). **Hab** Shingle beaches, under pebbles, just above high-tide mark. **Beh** Nocturnal, feeding on decaying animal and plant matter. May be found during the day by lifting stranded seaweed, but moves away under pebbles rapidly. **Dist** Only three known populations: Chesil Beach, Dorset; Branscombe, Devon and Marloes/Dale Peninsula, Pembrokeshire. **SS** None.

Scaly-cricket habitat – Branscombe, Devon

FAMILY **Rhaphidophoridae** (Wingless camel-crickets) 1 sp.

Greenhouse Camel-cricket
Tachycines asynamorus

Introduced; rare

BL 11–18 mm. **ID** Conspicuous hump-backed shape (typical for camel-crickets and cave-crickets of this family that live in dark places); wingless. Pale and rather mottled. **'Song'** No record of stridulation. **Cyc** Continuously brooded in glasshouses. **Hab** Greenhouses and warehouses among potted plants. **Beh** Nocturnal; scavenges on plant material and dead insects and other animals. **Dist** A cosmopolitan species, which originated from Asia, reported occasionally from GB and from Dublin in Ireland. Appears to have died out from the last known nurseries in Leicester in 2007, and more recently Clowne in Derbyshire, but probably persists elsewhere. **SS** None.

■ FAMILY **Gryllotalpidae** (Mole-crickets) 1 sp.

Mole-cricket *Gryllotalpa gryllotalpa*

BL 35–50 mm. **ID** Large, distinctive; forewings short, hindwings longer, reaching about tip of abdomen; fore femora flattened (adapted for digging). Brown with golden-brown velvet hair on thorax. **Cyc** 2–3 years. **Hab** Damp grassland or well-drained meadows, typically near streams. **Beh** Nocturnal, feeding on grasses and other plant roots, larvae and earthworms underground. Females guard small nymphs. **Dist** Only presently known in GB from a small breeding population in the New Forest, Hampshire, but this enigmatic insect may well be breeding elsewhere, undetected. **SS** Other *Gryllotalpa* species that sometimes reach Britain accidentally via imported plants: for example, the Italian **Fifteen-chromosome Mole-cricket** *Gryllotalpa quindecim* [N/I] (**BL** approx. 45 mm), which has short hindwings, was found in Oxfordshire in 2005 and also imported as a nymph in Southport, Merseyside, 2019.

'Song' Males 'sing' from underground burrows to attract a mate, but can be elusive, living up to a metre underground and rarely appearing on the surface. The 'song' can be heard from 100 m or more away and is a loud, continuous 'churr', likened to Nightjar's song. They 'sing' from dusk (when Nightjars stop) and continue all night.

Typical mole-cricket habitat; males 'sing' from beneath tussocks

Male peering from burrow.

Guide to families *pp. 86–87*

FAMILY **Acrididae** (Grasshoppers) — 10 GEN. | 14 spp.

A large family of about 6,730 species worldwide. These familiar jumping, grass-eating insects are abundant in grassland and are well known for their 'singing' ability. The antennae are often shorter than those of bush-crickets. Males lack genital valves and have longer antennae than females (see *p. 85*). Grasshoppers usually complete the life-cycle within a year, overwintering in the egg stage. Genera/species can be separated as follows:

ANTENNAE THICKENED AT TIP ('clubbed')

ANTENNAE club tip white; **BL** 14–22 mm

Gomphocerippus
[Rufous Grasshopper]
p. 108

ANTENNAE club tip dark; **BL** 12–19 mm

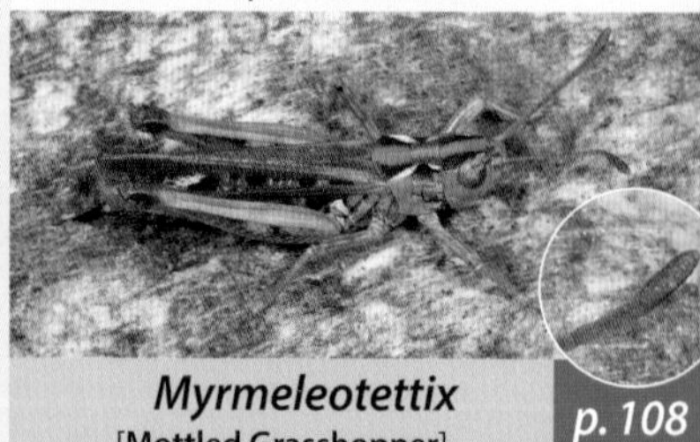

Myrmeleotettix
[Mottled Grasshopper]
p. 108

ANTENNAE NOT THICKENED AT TIP

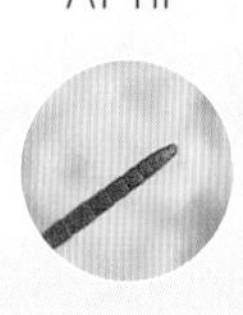

Larger species (body length usually ≥ 25mm)

WINGS strongly chequered [**Locusts** – rare immigrants; economic pests]

EYES single bar across; **BL** 32–80 mm

Locusta migratoria
[Migratory Locust]

EYES several lines down; **BL** 60–90 mm

Schistocerca gregaria
[Desert Locust]

WINGS not chequered/weakly mottled

EYES single bar across; **BL** 32–80 mm

Stethophyma
[Large Marsh Grasshopper]
p. 105

KEY FEATURES grey, brown, yellowish or olive (**BL** 32–66 mm) [irregular immigrant/import]

Anacridium aegyptium
[Egyptian Grasshopper]

Smaller species (body length usually ≤ 25mm)

Stenobothrus [2 species]
Omocestus [2 species]
Chorthippus [3 species]
Pseudochorthippus [1 species]

» KEY *p. 106*

GRASSHOPPERS ≥ 25 mm long; wings not mottled

Stethophyma 1 sp.

Large Marsh Grasshopper NT

Stethophyma grossum

BL 21–36 mm. **ID** GB's largest breeding grasshopper. Yellowish-green to olive-brown, with red on lower margin of hind femora. An uncommon purple form of the female occurs in some larger populations (usually those with plentiful heather). **Hab** Wet quaking bogs and marshes on heathland, particularly where *Sphagnum* mosses, and usually heathers, Purple Moor-grass, White-beaked Sedge and Broad-leaved Cottongrass, are present, sometimes with Bog-myrtle. Favours wetter areas that are avoided by other Orthoptera (except for Bog Bush-cricket, which is usually found near the drier periphery of bogs). **Beh** When disturbed, can fly several metres. **Dist** Restricted to its strongholds in the New Forest, Hampshire, Dorset and parts of Ireland. Recently recorded in Somerset and reintroduced to Norfolk. **SS** None.

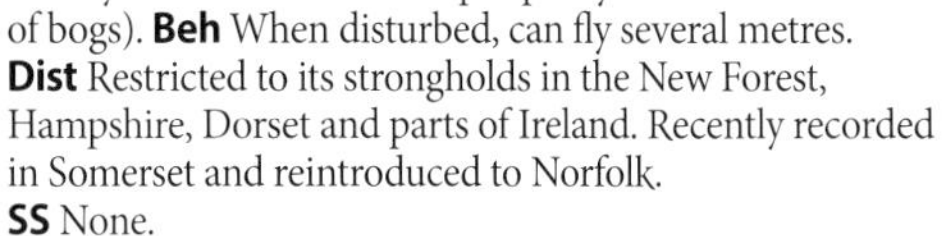

'Song' A series of brief ticking noises, quite unlike other grasshoppers.

purple form (rare)

From *p. 104* – Grasshoppers with antennae not thickened at the tip

Smaller species (body length < 25mm)

The following species are best identified by the shape of the keels and markings on the pronotum. *Pseudochorthippus* is usually short-winged; the others are all long-winged.

FORM **Nearly always short-winged**; usually green (some are brown/purple); KEEL **slightly incurved**; PRONOTUM **usually plain** (rarely with dark areas)

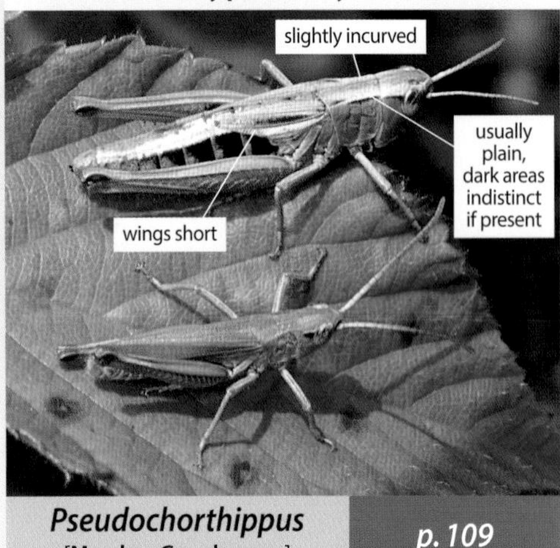

Pseudochorthippus
[Meadow Grasshopper]
p. 109

COLOUR Usually green (some with brown/purple sides); KEEL **moderately incurved**; PALPS pale (but never chalk-white); ABDOMEN TIP **never red**

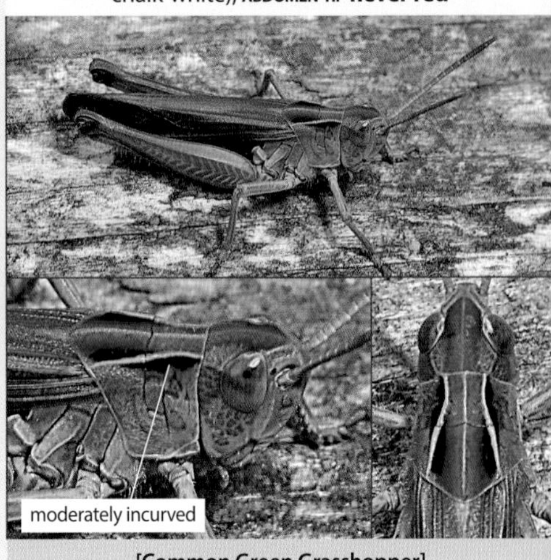

[Common Green Grasshopper]

FORM Often 'bright' green. KEEL slightly incurved, with small black marks; FOREWING **with white line and mark**; PRONOTUM distinctively marked/coloured.

Stripe-winged Grasshopper
Stenobothrus
[2 species]
p. 110

COLOUR Dark greyish-brown (some partly green); KEEL incurved; PALPS **white-tipped**; ABDOMEN TIP **usually red**

[Woodland Grasshopper]
Omocestus
[2 species]
p. 111

KEEL indented or almost straight; THORAX UNDERSIDE usually plain (rarely with indistinct dark patches)

KEEL **almost straight**; PRONOTUM **no dark 'wedges'**; THORAX underside **sparsely hairy**; ABDOMEN uniform colour; usually plain green or brown; HABITAT wetlands

[Lesser Marsh Grasshopper]

KEEL **sharply angled**; PRONOTUM **dark 'wedges' do not reach pale rear edge**; THORAX underside **very hairy**; ABDOMEN orange-red on tip (on mature ♂ only)

[Field Grasshopper]

KEEL indented; PRONOTUM **dark 'wedges' reach rear edge**; THORAX underside **sparsely hairy**. ABDOMEN red on tip (on mature adults)

[Heath Grasshopper]

Chorthippus
[3 species]

p. 112

Guide to Acrididae *pp. 104–107*

GRASSHOPPERS < 25 mm long; antennae 'clubbed'

Myrmeleotettix 1 sp.

Mottled Grasshopper
Myrmeleotettix maculatus

BL 12–19 mm. **ID** Small, with many colour forms ranging from green, brown and purple to black; usually an irregular pattern of various colours. Antennae distinctively clubbed (♂) or thickened (♀). **Hab** Grassland with some bare ground and short turf. **SS** Heath Grasshopper (*p. 113*).

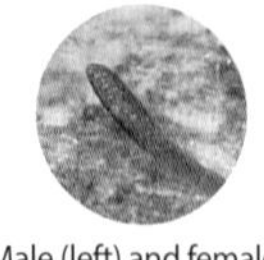

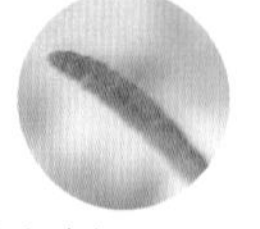

Male (left) and female (right) antenna tips

'Song' Consists of 10–30 soft 'buzzes', becoming louder.

Gomphocerippus 1 sp.

Rufous Grasshopper
Gomphocerippus rufus

BL 14–22 mm. **ID** Brown. Distinctive white-tipped, clubbed antennae. **Hab** Mainly rough calcareous grassland. **SS** None.

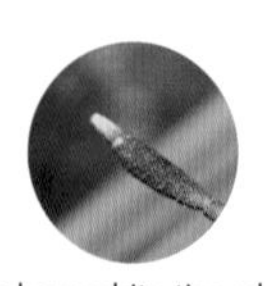

Both sexes have white-tipped antennae

'Song' A subdued trill, lasting 5 or more seconds; repeated irregularly.

GRASSHOPPERS < 25 mm long; antennae not 'clubbed' 1/3

Pseudochorthippus 1 sp.

Meadow Grasshopper
Pseudochorthippus parallelus

BL 10–23 mm. **ID** Keels on pronotum slightly incurved; forewings short (although a rare long-winged form, *explicatus*, occurs in some large populations). Mainly green, but some individuals are brown or purple. **Hab** Most grassland. **SS** None with short wings.

'Song' A 'churring' sound, in 1–3-second bursts, gradually becoming louder; repeated at irregular intervals.

Mating pair; the much smaller male is in the foreground.

Forms with a dorsal stripe have pink or purple on top of the head and pronotum and are particularly striking. A pink/purple form (left) also occurs but is uncommon (and only in ♀).

Guide to Acrididae *pp. 104–107*

GRASSHOPPERS < 25 mm long

Stenobothrus 2 spp.

Stenobothrus grasshoppers can be told from other grasshoppers by the presence of a 'tooth' on each ovipositor valve.

Stripe-winged Grasshopper

Stenobothrus lineatus

Local

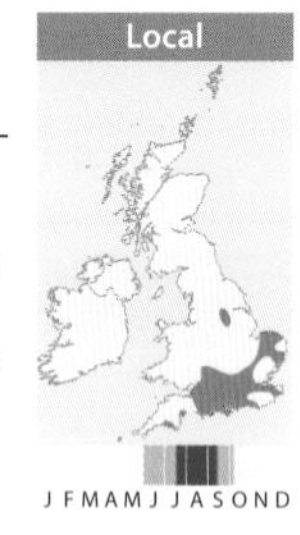

BL 15–23 mm. **ID** Green and brown, with some red on the abdomen. Forewings have a distinctive white line along top edge and elongated white mark at centre. **Hab** Mainly calcareous grassland, but also heathland and sand dunes, with some short turf. **SS** Common Green Grasshopper.

'Song' Males stridulate often but quietly, producing a 'wheezing' sound that lasts for 10–20 seconds.

Lesser Mottled Grasshopper

Stenobothrus stigmaticus

Rare (not Red Listed)

BL 10–15 mm. **ID** Very small; usually green with brownish wings with mottled sides (as shown in the image *below*), forewings of ♀ usually with white line; ♂, and sometimes ♀, orange-red on tip of abdomen. **Hab** Coastal cliffs on thin turf. **Dist** Confined to Langness Peninsula on the Isle of Man, where the only other grasshoppers that occur are Mottled Grasshopper (*p. 108*) and Field Grasshopper (*p. 112*). **SS** None in range.

'Song' A series of irregular chirps for 2–4 seconds, each becoming louder.

Omocestus 2 spp.

COLOUR mainly green; ABDOMEN **tip never red**; PALPS pale	Common Green Grasshopper
COLOUR dark greyish-brown; ABDOMEN **tip usually red**; PALPS **white**	Woodland Grasshopper

Common Green Grasshopper
Omocestus viridulus

Common

BL 14–23 mm. **ID** Pronotum keels moderately incurved. ♂ mainly green or olive-brown; ♀ green dorsally, but with green, brown or purple elsewhere on the body. The abdomen tip is never red; palps pale. **Hab** Grassland. **SS** Woodland Grasshopper, Stripe-winged Grasshopper.

J F M A M J J A S O N D

'Song' A series of loud 'ticks' in bursts of 10–30 seconds, that are within the human audible range.

♂

♀

Woodland Grasshopper
Omocestus rufipes

Nationally Scarce

BL 12–20 mm. **ID** Dark greyish-brown, some ♀s partly greenish; abdomen tip usually red. Tips of palps white. **Hab** Woodland edges, with some tall grasses; also grassland. **SS** Common Green Grasshopper.

J F M A M J J A S O N D

'Song' A series of 'ticks' lasting 5–20 seconds that starts softly and becomes louder, before stopping suddenly.

♂

♀

♀ partly green individual

Guide to Acrididae *pp. 104–107*

GRASSHOPPERS < 25 mm long 3/3

Chorthippus	3 spp.
KEEL **sharply angled**; THORAX UNDERSIDE **very hairy**	Field Grasshopper
KEEL **indented**; THORAX UNDERSIDE sparsely hairy	Heath Grasshopper
KEEL **almost straight**; THORAX UNDERSIDE sparsely hairy	Lesser Marsh Grasshopper

Field Grasshopper
Chorthippus brunneus

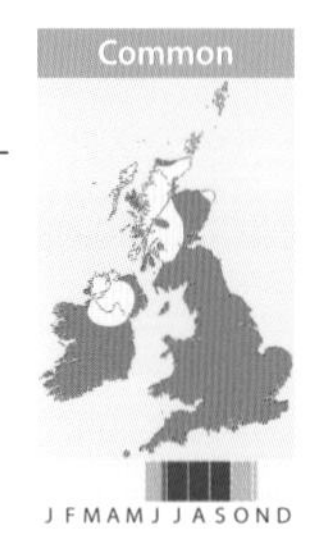

BL 15–25 mm. **ID** Keels on pronotum sharply angled; thorax underside very hairy. Several colour forms: often brown/brownish; some described as 'striped', 'mottled' or 'semi-mottled'. Orange to purple or even black forms are less common and a green form very rare, although a green dorsal surface with brown elsewhere on the body is not uncommon. Only a limited range of forms occur in even large populations. ♂ has orange-red abdomen. Dark, wedge-shaped marks on pronotum do not reach the pale hind margin. **Hab** Grassland. **SS** Heath Grasshopper.

Dark marks on the pronotum do not reach the back edge (compare with Heath Grasshopper)

'Song' A series of 6–10 short chirps, each lasting 0·5 seconds, with pauses of 1–2 seconds between each chirp; repeated regularly. ♀s can also stridulate, but are quieter.

Field Grasshopper nymph (rarely seen pink colour form)

Lesser Marsh Grasshopper
Chorthippus albomarginatus

BL 13–23 mm. **ID** Body elongated; keels on pronotum almost straight; underside of thorax sparsely hairy; wings often well short of tip of abdomen, although occasionally longer (as in the photo of ♂ shown). Plain light green or straw-coloured, rarely pinkish. No dark wedge-shaped marks on the pronotum. **Hab** Wetlands; also some drier areas. **SS** None.

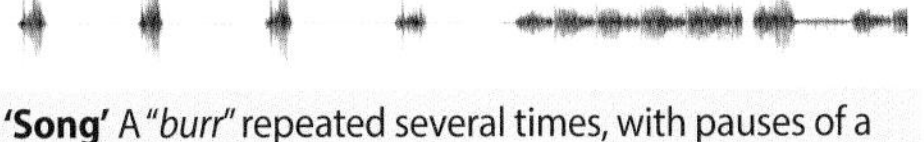

'Song' A "*burr*" repeated several times, with pauses of a second in between.

Heath Grasshopper
NT

Chorthippus vagans

BL 13–21 mm. **ID** Keels on pronotum indented; underside of thorax sparsely hairy. Always dark, greyish-brown with red abdomen; the hind femora normally have two dark bands. Dark wedge-shaped marks on the pronotum reach the hind margin. **Hab** Dry heathland; feeds on heathers and Dwarf Gorse. **Dist** Restricted to parts of Dorset and the New Forest, Hampshire. **SS** Mottled Grasshopper (*p. 108*), Field Grasshopper.

Dark marks on the pronotum reach the back edge (compare with Field Grasshopper)

'Song' Soft, in bursts of 5-seconds or more; repeated irregularly.

♀ egglaying into sandy soil

Guide to families *pp.* 86–8

FAMILY **Tetrigidae** (Groundhoppers) — 1 GEN. | 3 spp.

A family of almost 2,000 species worldwide. Small and 'thickened', groundhoppers resemble grasshoppers but do not 'sing'. They prefer damp conditions with some bare ground, such as seepages on undercliffs, where they feed on mosses, algae or plant material; they are able to swim. The life-cycle is about a year, with adults overwintering. **The three British species require close examination to separate them** (see *below*), although the most distinctive species is the only one that is widespread.

Tetrix — 3 spp.

HINDWINGS ± **abdomen tip**		Common Groundhoppe
HINDWINGS well beyond abdomen tip	VERTEX (CROWN) **narrow**, ≤1·5× eye width FEMUR OF MIDLEG **almost straight**	Slender Groundhopper
	VERTEX (CROWN) **wide**, 1·5–1·8× eye width FEMUR OF MIDLEG **wavy**	Cepero's Groundhopper

Common Groundhopper
Tetrix undulata

BL 8–11 mm. **ID** Conspicuous ridge along top of pronotum. Wings short, although a rare long-winged form occurs. **Hab** Various damp areas, with some bare ground. **SS** Other groundhoppers.

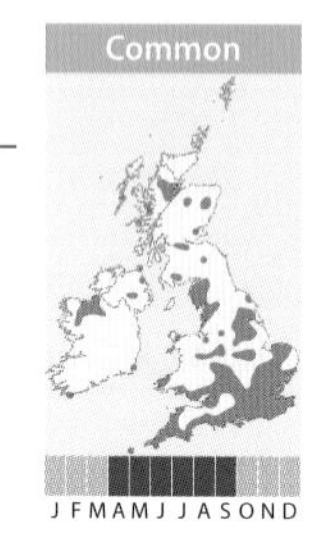

PRONOTUM with conspicuous ridge

Slender Groundhopper
Tetrix subulata

BL 9–14 mm. **ID** See *above* for key features. Wings long but short-winged form *bifasciata* sometimes occurs. Individually variable in colour. **Hab** Damp areas, with some bare ground. **SS** Other groundhoppers.

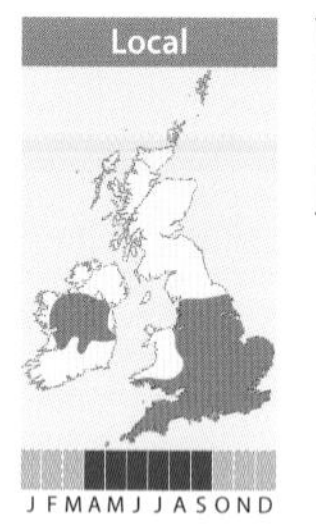

Cepero's Groundhopper
Tetrix ceperoi

BL 8–13 mm. **ID** See *above* for key features. Wings long. Individually variable in colour and often mottled. **Hab** Seepages on coastal cliffs, sand dunes, bare muddy soil in damp areas. **SS** Other groundhoppers.

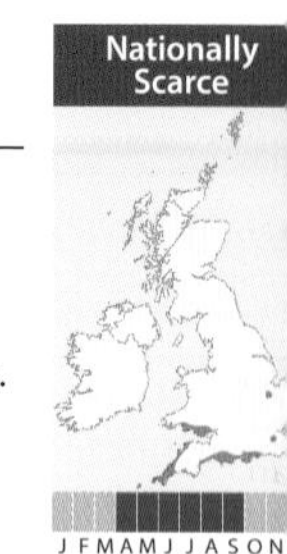

VERTEX (CROWN) **narrow**, ≤1·5× eye width
FEMUR OF MIDLEG **almost straight**

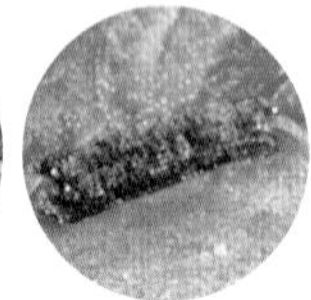

VERTEX (CROWN) **wide**, 1·5–1·8× eye width
FEMUR OF MIDLEG **wavy**

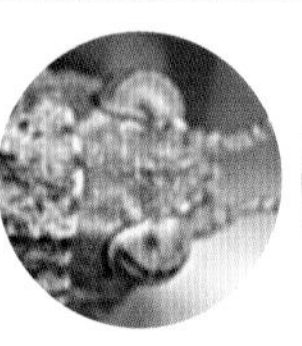
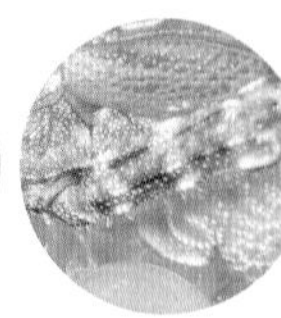

ORDER EMBIOPTERA — Webspinners

[Greek: *embio* = lively; *pteron* = wing]

BI | 1 family [1 genus, **1 species** (introduced)] W | 13 families [**464 species**]

Small to medium-sized, body elongated and cylindrical; cerci two-segmented. The compound eyes are small (larger in males) and kidney-shaped, with ocelli absent. Antennae multisegmented. Pronotum quadrate, larger than mesonotum or metanotum; abdomen 10-segmented (with rudiments of 11th segment only). Females are wingless, only males of some species are winged, with large forewings covering similar-sized hindwings. When present, wings are soft. The legs are short, tarsi three-segmented. Notably the fore tarsi are swollen basally, which contain silk glands. Hind femora also swollen. This is a poorly known, mostly tropical or subtropical order (although 13 species are found in Mediterranean Europe). Webspinners are known to live together in silken galleries in leaf-litter, in bark, rocks, tree trunks or vegetation. Within the galleries, they are afforded protection from predators and the elements and feed on algae, lichens, moss or leaf-litter; they extend galleries as necessary and tend to the offspring. Females only occasionally move from the structure and sometimes eat the short-lived males. There have been reports and interceptions of several webspinner species in Great Britain via imported plants (mainly orchids), but only one case of any being established under glass (2018).

Webspinner web

FAMILY Oligotomidae — 1 sp.

A family of 16 species worldwide, mainly from Asia.

Wisley Webspinner
Aposthonia ceylonica

BL 6–7 mm. **ID** Slender; dark brown with paler head; ♂ has wings, which are greyish-brown (♀ wingless). **Hab** Heated greenhouses, in tropical orchid roots. **Dist** First observed in 2018 in Royal Horticultural Society Garden, Wisley, Surrey. Probably native to Sri Lanka and possibly India, also recorded in Malaysia and Thailand, and farther afield in Madagascar and Mauritius. **SS** None.

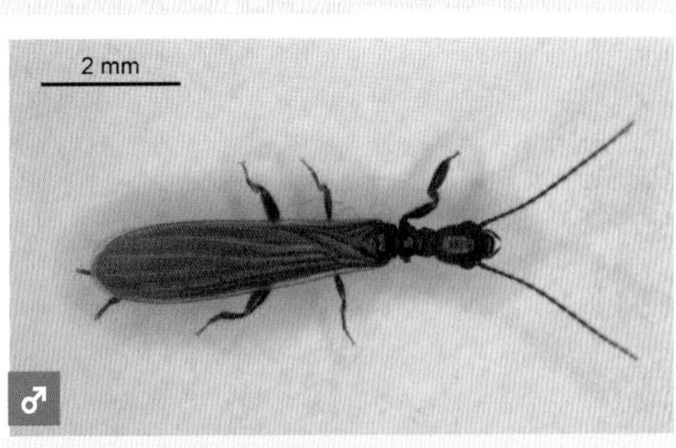

Webspinner in its web

ORDER MANTODEA — Mantids

[Greek: *mantis* = soothsayer]

BI | 1 family [1 genus, **1 species** (introduced)] **W** | 29 families [**approx 2,500 species**]

Medium-sized to large predators, usually winged. Rather like winged stick-insects, they are well camouflaged but the forelegs are designed for grabbing prey, so they are readily recognizable. With their large eyes and triangular head, they look rather alien-like and sometimes feature in science fiction or horror films. Included here briefly as they are rarely recorded in Great Britain (East Sussex in 1959, Hampshire in 2015, Kent and Oxfordshire in 2020), but in theory any continental species could be imported as an ootheca (eggcase) on vegetation, may be found as a discarded pet or escapee from culture stock, or be transported inadvertently (as with **White-spotted Mantis** *Sphodromantis viridis* [N/I] in Cornwall, in January 2018, discovered in a container from Israel; the Mediterranean **Conehead Mantis** *Empusa pennata* [N/I] was found in a school kitchen on St. Mary's, Isles of Scilly in December 2013). They might survive in favourable weather, as is the case with stick-insects from southern Europe. Some authors regard Mantodea as the sister group to Blattodea in a group called Dictyoptera, treating the present orders (Blattodea and Mantodea) as suborders; others regard them as orders within a superorder.

FAMILY Mantidae — 1 sp.

The largest mantid family, including the most widespread European species, the Praying Mantis. The species' name derives from the apparent praying pose, with forelegs clasped together.

Praying Mantis
Mantis religiosa

Introduced; rare

BL 43–88 mm. **ID** Green, brown or yellowish. Bold black or black-ringed spot at base of fore coxae, seen when walking. **Cyc** ♀ lays an ootheca (eggcase) that overwinters. Nymphs resemble tiny adults and gradually develop wings; they hatch together and prey on insects, but may be cannibalistic when food is in short supply. **Hab** Grassland or gardens with small bushes. **Dist** Sometimes reared by pet keepers, and is almost cosmopolitan; three widespread subspecies occur in Europe. **SS** None in Great Britain.

♀ showing black spot at base of fore coxa

ORDER PHASMIDA

Stick-insects

[Greek: *phasma* = ghost]

BI | 3 families [5 genera, **9 species** (all introduced)] W | 14 families [**approx. 3,400 species**]

Stick-insects are elongate, nocturnal, plant-feeding insects found mainly in the tropics and rely upon camouflage to evade predation. Although many species are winged, those occurring in Great Britain and Ireland are all wingless, established aliens. Four of the nine 'British' species are introductions from New Zealand that have colonized sheltered areas, reproducing parthenogenetically (*i.e.* females lay unfertilized eggs which hatch into females only). However, in 2016 there was a surprising find in Cornwall solving a long-standing mystery: a male Unarmed Stick-insect *Acanthoxyla inermis* was found, the first male in the genus *Acanthoxyla* ever discovered. Stick-insects have the ability to shed legs in order to escape predation and are often seen with fewer than six legs; if a leg is lost by a young nymph, it can be re-grown, but may be shorter than the original. Most of the 'British' species occur in various shades of green or brown and blend in well with their background; some individuals can even change colour overnight. Tropical stick-insects are popular as pets and there is a thriving UK-based Phasmid Study Group. Although it is illegal to release stick-insects into the wild, pets are sometimes discarded and other unlikely alien species reported. However, these are generally short-lived, unable to survive outdoors in the British climate.

Some authors prefer to use Phasmatodea or other names for the order, as there is no consensus of opinion, or rules for the construction of names. Phasmida is the oldest and simplest name, first used by Leach in 1815 and is in wide usage, with the insects often referred to as phasmids.

FINDING STICK-INSECTS Although stick-insects can be found by day resting on their foodplant or nearby vegetation, they are so well camouflaged that it is easier to find them at night by torchlight (permission is, of course, needed if searching on private land). These insects are easy to photograph, as they often remain perfectly motionless.

Smooth Stick-insects at rest by day, one female on top of another

Structure of a stick-insect

USEFUL WEBSITES

http://phasmida.speciesfile.org
Phasmida Species File Online
[**Go to search and type the species name in to see numerous references**.]

http://phasmidstudygroup.org/index.php/phasmids/uk-phasmid-sighting

Phasmid Study Group. *The Naturalized British Stick-insects*.

ORDER **Phasmida** GUIDE TO FAMILIES | 3 BI

NB species from Great Britain & Ireland belong to the suborder Euphasmatodea

MID- OR HIND TIBIAE with sunken areola (triangular region) on underside of apices

Winged or wingless; **ANTENNAE** of adults distinctly segmented; **FEMORA** in females distinctively and evenly serrated in Phasmatidae, uneven in Lonchodidae.

Species found in Great Britain and Ireland are wingless, with medium-length antennae (longer than femur of foreleg [except Thailand Stick-insect, which also has short antennae]).

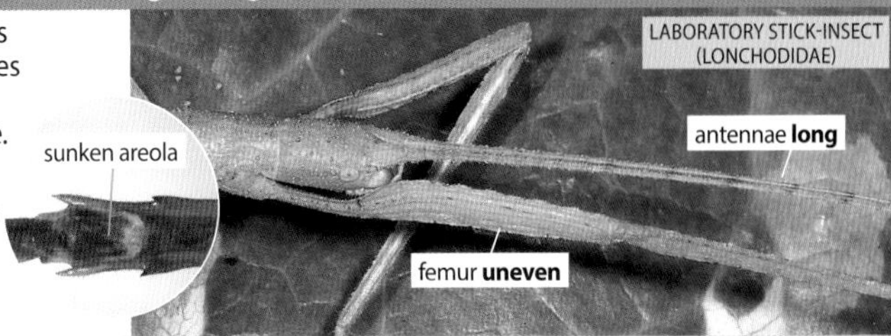

Phasmatidae and Lonchodidae

MID- OR HIND TIBIAE without sunken areola

FEMORA in females not serrated. *The two species found in Britain have short antennae (much less than ½ length of foreleg femur).*

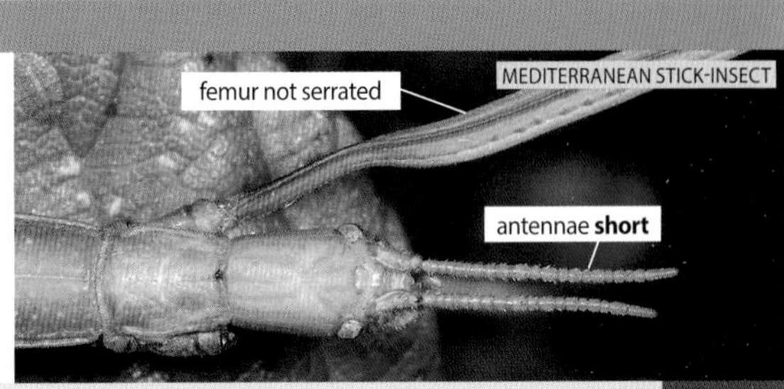

Bacillidae [2 genera | 3 species] p. 124

FAMILY **Phasmatidae** (Phasmatids) 3 GEN. | 5 spp. | 5 ILL.

A fairly large family of fewer than 700 winged and wingless species worldwide, mainly from the tropics and subtropics. Antennae rather robust-looking and distinctly segmented. Species found in Great Britain are wingless with medium-length antennae, longer than fore femora (except in *Ramulus*). Females often with serrated femora. **Females can be identified using the table below:**

FAMILY **Lonchodidae** (Lonchodids) 1 GEN. | 1 sp. | 1 ILL.

A large family of 1,170 species worldwide that usually has longer antennae than Phasmatidae.

Lonchodidae		CERCI – DORSAL VIEW	CERCI – LATERAL VIEW		
CERCI very short, hidden beneath anal segment (S10) when viewed dorsally				***Carausius*** [1 sp.]	p. 123
Phasmatidae		**CERCI – DORSAL VIEW**	**CERCI – LATERAL VIEW**		
CERCI short				***Ramulus*** [1 sp.]	p. 123
CERCI moderately long, broadly leaf-like	**OPERCULUM** basal spine or broad tubercle absent			***Clitarchus*** [1 sp.]	p. 122
	OPERCULUM basal spine or broad tubercle present			***Acanthoxyla*** [3 spp.]	p. 120

Guide to Phasmatidae *p. 119*

STICK-INSECTS with moderately long cerci **1/2**

Acanthoxyla **3 spp.**

Native to New Zealand. Elongate (up to 125 mm). Usually shades of green or brown; body with spines or tubercles, but individually variable; abdomen of some with leaf-like lobes on S2–8. Legs with apical spines on all femora and tibiae, and other serrations on femora. Base of operculum with a large spine or tubercle. Cerci long, hairy and leaf-like. Egg shape differs between species. The usual method of reproduction is by parthenogenesis; a ♀ will drop around 300 eggs to the ground. A ♂ was finally discovered in 2016 in Great Britain, despite years of fruitless searching in New Zealand.

BODY usually very spiny; **OPERCULUM** with curved, sharply pointed basal spine	**Prickly Stick-insect**
BODY moderately spiny; **OPERCULUM** with large basal tubercle	**Black-spined Stick-insect**
BODY lacking spines; **OPERCULUM** with basal tubercle	**Unarmed Stick-insect**

Prickly Stick-insect
Acanthoxyla geisovii

BL ♀ 80–110 mm. **ID** ♀ Green, brown or occasionally mottled, body often extensively black-spined. Operculum with curved basal spine. **Cyc** Eggs overwinter, hatch in spring. **Hab** Gardens, hedgerows and waste ground. **Fp** Mainly Bramble, roses, *Cupressus*. **Dist** Originally introduced to Great Britain in 1908 (Paignton, Devon) from its native New Zealand, via imported plants. **SS Black-spined Stick-insect.**

DORSAL VIEW LATERAL VIEW

EGG: 5·2 × 2·3 mm; CAPSULE dark, mottled; KEEL strongly bulging, CAPITULUM elongated

♀ operculum with basal spine

Black-spined Stick-insect

Acanthoxyla prasina

Introduced; rare

J F M A M J J A S O N D

BL ♀ 78–108 mm. **ID** ♀ Green or brown, with fewer spines on head and thorax than Prickly Stick-insect (colour forms in parts of New Zealand include attractive black-and-white-mottled specimens). Operculum with basal tubercle. **Cyc** Eggs overwinter and hatch in spring. **Hab** Gardens. **Fp** Brambles, roses and *Cupressus*. **Dist** The least common of the *Acanthoxyla* species in Great Britain; recorded rarely in Cornwall and occasionally elsewhere. The sources of this species, which is native to New Zealand (recorded in Great Britain since 1991) are not known with certainty, but possibly from released culture stock at one site, otherwise via imported plants.
SS Prickly Stick-insect.

EGG: 4·2 × 1·8 mm; CAPSULE dark, mottled; KEEL bulging outwards

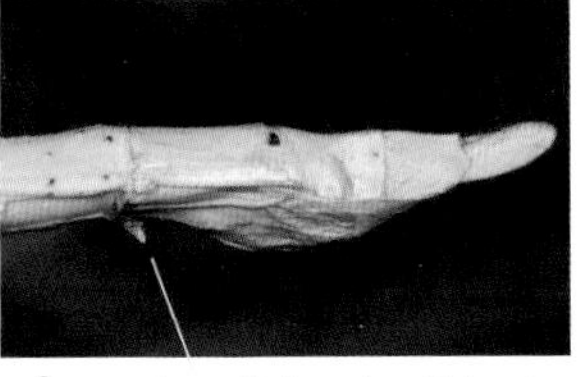
♀ operculum with large basal tubercle

Unarmed Stick-insect

Acanthoxyla inermis

Introduced; local

J F M A M J J A S O N D

BL ♂ 75 mm; ♀ 94–125 mm. **ID** The longest insect found in Britain or Ireland; green or brown, with central black longitudinal line on pronotum. Thorax usually with few tubercles; ♀ operculum with basal tubercle. Only one ♂ ever recorded: brown, with central black longitudinal line on pronotum. **Cyc** Eggs overwinter. Nymphs are most likely during May–August. Adults can survive longer in mild weather, until severe frosts. **Hab** Mainly gardens, but also hedgerows and waste ground. **Fp** Brambles, roses and *Cupressus*. **Dist** The most widespread of the stick-insects, but still local. Mainly in Cornwall, since being introduced from its native New Zealand via imported plants in the 1920s. Also found in several other counties, Wales and SW Ireland. **SS** Smooth Stick-insect (*p. 122*).

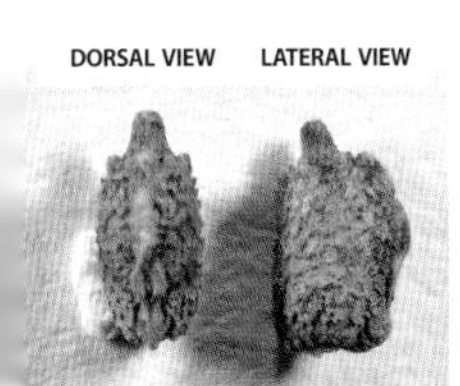

EGG: 3·9 × 1·9 mm; CAPSULE often pale; KEEL reduced

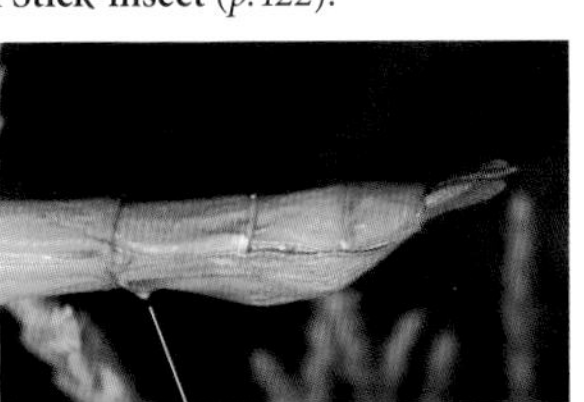
♀ operculum with basal tubercle

♀ showing black line on pronotum

Guide to Phasmatidae *p. 119*

STICK-INSECTS with moderately long cerci 2/2

Clitarchus 1 sp.

Smooth Stick-insect
Clitarchus hookeri

BL ♀ 64–91 mm. **ID** ♀ Green or brown, body smooth, thorax with an interrupted black line starting from the back of the head; operculum lacks basal tubercle; cerci long and hairy. More mottled examples are occasionally seen in New Zealand, some with tubercles on thorax. **Cyc** Each ♀ drops a few hundred eggs to the ground; eggs overwinter and hatch in spring or summer. **Hab** Gardens and waste ground. **Fp** Mainly Bramble, gum trees *Eucalyptus* spp. and roses. **Dist** Isles of Scilly only (Tresco, Bryher and St. Mary's): introduced from its native New Zealand to Tresco via imported plants in 1949, and has since spread to other islands. **SS** Unarmed Stick-insect (*p. 121*).

Native to New Zealand; there are three similar-looking species described, of which the Smooth Stick-insect is the commonest New Zealand phasmid, although numbers are being drastically reduced [along with other native wildlife] following the introduction there of the German Wasp *Vespula germanica* (*p. 513*) a predator from Europe. Only females have been recorded in Great Britain, where the method of reproduction is by parthenogenesis; males are frequent in some New Zealand populations of Smooth Stick-insect and both other *Clitarchus* species.

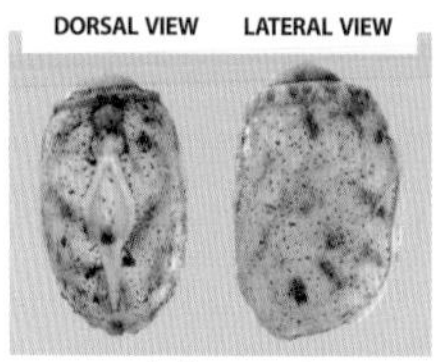

EGG: lacks capitulum

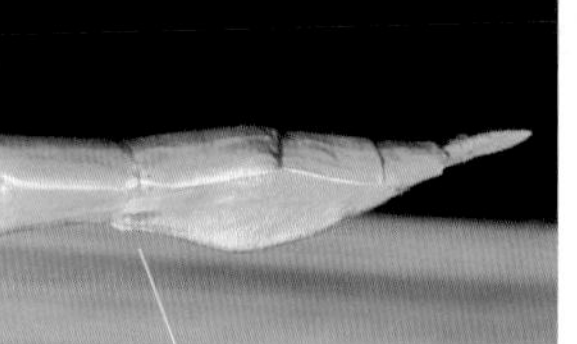
♀ operculum lacks basal tubercle

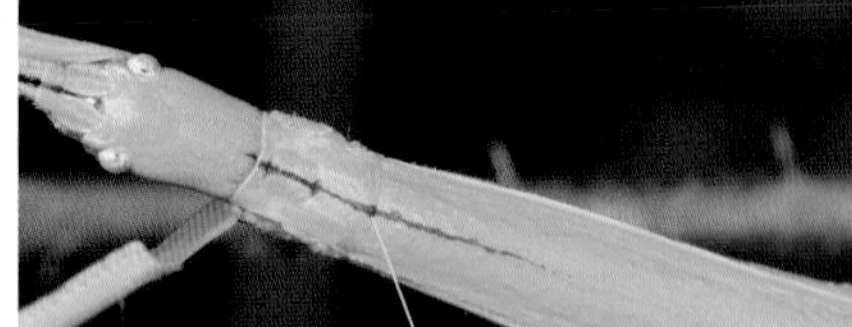
diagnostic broken black line on thorax

Tresco Abbey Gardens, Isles of Scilly, a reliable site for Smooth Stick-insect *Clitarchus hookeri*.

STICK-INSECTS – cerci short

Ramulus 1 sp.

Thailand Stick-insect *Ramulus thaii*

BL ♂ approx. 80 mm; ♀ approx. 120 mm. **ID** ♀ dark green, ♂ dark brown; individually variable, some with foliose expansions on midlegs and hindlegs. **Cyc** At least one generation a year. **Hab** Under cover in glasshouses; can probably survive outdoors in warmer weather. **Fp** Brambles, but likely to accept a wide range of plants. **Dist** First reported from a glasshouse in Cardiff in 2001, this species is still established, at least under glass. **SS** None.

Ramulus spp. are mainly native to Asia. Elongate, medium-sized, plain (usually brown), with both sexes present. This is the largest genus of phasmids, with almost 170 species worldwide

♀ head often has tufts as shown

STICK-INSECTS – cerci very short

Carausius 1 sp.

Laboratory Stick-insect *Carausius morosus*

Introduced; rare

BL ♀ 70–84 mm. **ID** Fairly stockily built and with relatively shorter legs than other 'British' stick-insects; has a number of small tubercles on thorax. Dull green or brown, some individuals with dark mottling. ♀ operculum boat-like. Very rarely, a thin, brown, ♂-like individual (50–60 mm) (believed to be a gynandromorph) occurs in culture stocks. **Cyc** Each ♀ drops several hundred eggs to the ground, which hatch in 4–6 months in warmer weather. ♀s may live several months as adults. **Hab** Gardens and hedgerows. **Fp** Ivy, but usually Privet in captivity and escapees often switch to houseplants. **Dist** Only established on Isles of Scilly (St. Agnes and St. Mary's); may turn up elsewhere but populations usually short-lived (although does persist in glasshouses, such as in Glasgow and Torquay). **SS** None.

S10 LATERAL VIEW

operculum reaches tip of abdomen

MID- OR HIND TIBIAE without sunken areola

FAMILY **Bacillidae** (Bacillids) — 2 GEN. | 3 spp. | 2 ILL.

A small family of 54 species worldwide, mainly from southern Europe and northern Africa. Wingless, with very short antennae. Some are smooth, others have tubercles (number varies between species). **There are several very similar species, which may require molecular studies to confirm identification.**

ANTENNAE short	♀ OPERCULUM about ½ length of abdomen S9	**Mediterranean Stick-insect**
	♀ OPERCULUM reaches tip of abdomen S9	**White's Stick-insect**
ANTENNAE very short (no longer than head); THORAX heavily granulated		**French Stick-insect**

Bacillus — 2 spp. | 1 ILL.

Elongate, medium-sized, plain stick-insects. ♀ operculum length varies between species (but short of anal segment), which can help distinguish between closely related species. There are six species worldwide, of which Mediterranean Stick-insect is the most widespread and occurs in both sexes in some populations, although only females are known in Great Britain, which **reproduce parthenogenetically.**

Mediterranean Stick-insect
Bacillus rossius

Introduced; local

J F M A M J J A S O N D

BL ♀ 64–105 mm. **ID** Smooth-bodied; green or brown, with cream or red side-stripe: has the ability to change colour to blend in with the background (for example, on Hayling Island the colour of these insects closely matches the colour of the beach hut they live beside!). ♀ operculum reaching about ½ length of abdomen S9. **Cyc** Each ♀ drops several hundred eggs to the ground, which hatch in 4–6 months. ♀s may live several months as adults. **Hab** Scrub, often near the coast. **Fp** Brambles. **Dist** Mainly Isles of Scilly (Tresco) and Hayling Island, Hampshire. **SS** White's Stick-insect *Bacillus whitei* [N/I] (**BL** 76–90 mm), a virtually identical-looking native of Sicily, has been recorded from Slough, Berkshire as escapees from culture (operculum reaches tip of abdomen S9). French Stick-insect *Clonopsis gallica* [BELOW, ON RIGHT] (**BL** 59–79 mm), has been reported from S England; this is easily distinguished by even shorter antennae (no longer than head) and heavily granulated thorax.

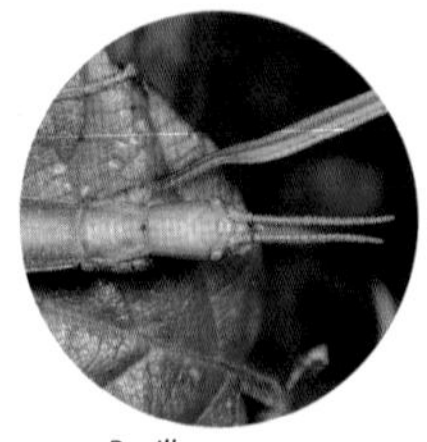

Bacillus: ANTENNAE short

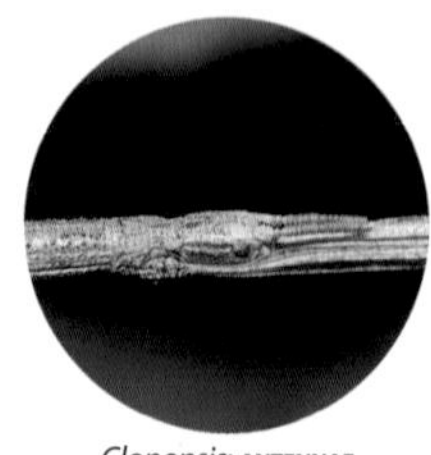

Clonopsis: ANTENNAE very short

ORDER BLATTODEA Cockroaches & termites

[Latin: *blatta* = an insect that shuns the light]

BI | 4 families [8 genera, **14 species** (incl. 11 introduced)]   **W** | 19 families [**approx. 7,750 species**]

Cockroaches are flat-bodied, have long antennae and can move fast, with a peculiar gait. The front of the head looks directly downwards; the pronotum is shield-shaped, often covering the head. The leathery, overlapping forewings [tegmina] cover membranous hindwings and some species are good fliers. Eggs are laid in a case-like ootheca. Small, native species are most likely to be encountered, and perhaps, larger, mainly nocturnal introduced pest species that are reliant upon artificial heating. These are far less common than they once were. For example, Victorian nature books refer to species such as the German Cockroach and Oriental Cockroach being numerous in hotels, restaurants, cellars and kitchens. The smaller termites are large-headed, often specialized wood or grass-eating species living in colonies; formerly placed in a separate order (Isoptera). Although popular with some pet-keepers, cockroaches and termites generally tend to be among the least popular insects. Cockroaches are hardy and fast-breeding, feeding on human and pet food; they are proven or suspected carriers of various diseases. Termites can cause major damage to wooden structures. The Blattodea are part of the superorder Dictyoptera, along with mantids [Mantodea]. Species included are generally recognized to be on the British list, as they have been found breeding at least indoors. However, other cockroaches that are sometimes imported accidentally have been omitted.

FINDING COCKROACHES Native cockroaches are most likely to be seen on hot days, resting on flowers or vegetation, but many inhabit leaf-litter and are harder to find. Cockroaches in kitchens are difficult to find as they tend to hide from light, seeking crevices in which to hide.

Structure of a cockroach

ORDER Blattodea

GUIDE TO FAMILIES | 4 BI

FORM Resemble white, wingless ants; **antennae about as long or not much longer than head**; large-headed (particularly in soldier caste); **FOOD** wood or grass; **EGGS** laid deep inside nest (not in ootheca)

TERMITES

Rhinotermitidae
[Iberian Termite]

p. 127

FORM Broad, flat-bodied with **long antennae**; head looks directly downwards
FOOD decaying material; **EGGS** usually laid in purse-shaped ootheca.

COCKROACHES

NB images below show underside of abdomen tip.

♂ **SUBGENITAL PLATE STYLES** rather simple symmetrical.
♀ **ABDOMEN TIP** 2-valved

SIZE medium-sized to large (17–43 mm)

♀

2-valved tip

♂

Blattidae
[2 genera | 4 species]

p. 127

♂ **SUBGENITAL PLATE STYLES** symmetrical, asymmetrical or one/both absent.
♀ **ABDOMEN TIP** smooth, not 2-valved

SIZE small to medium-sized (5–15 mm); **ANTENNAE** > ½ **body length**; **LEGS elongated**; **CERCI long**, extending beyond body

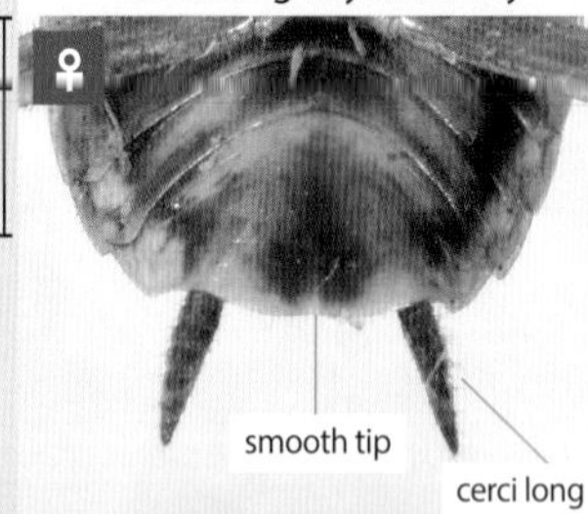

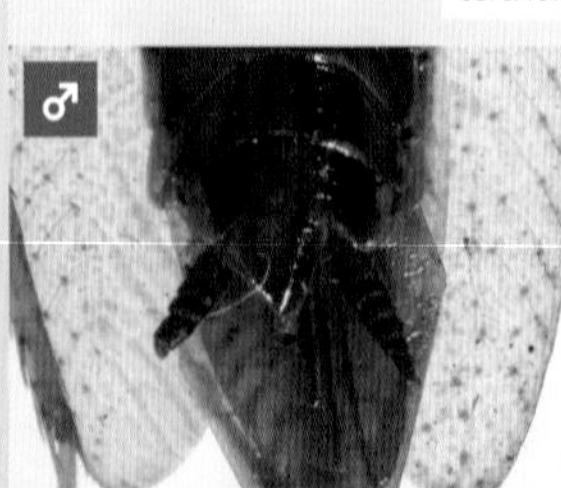

Ectobiidae
[5 genera | 8 species]

p. 129

SIZE medium-sized (20–25 mm); **ANTENNAE** < ½ **body length**; **LEGS short, stout**; **CERCI short**, not extending beyond end of body

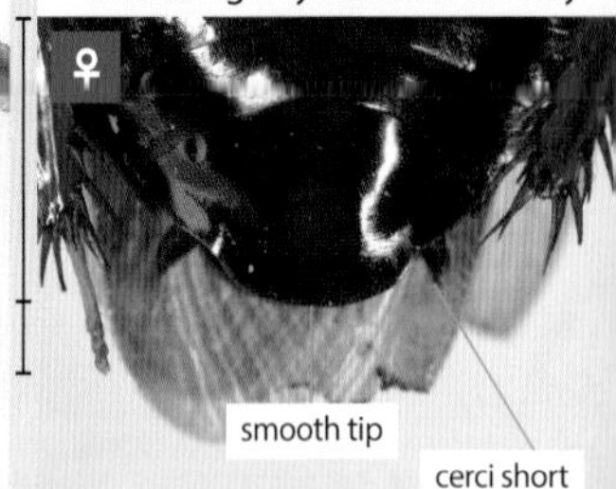

Blaberidae
[Surinam Cockroach]

p. 131

FURTHER READING AND USEFUL WEBSITES

Marshall, J.A. & Haes, E.C.M. 1990. ***Grasshoppers and allied insects of Great Britain and Ireland*** *(revised edition, original 1988)*. Harley Books.

Orthoptera & allied insects Recording Scheme
www.orthoptera.org.uk

TERMITES

FAMILY Rhinotermitidae (Termites) — 1 sp.

There are 314 species of termites in this family worldwide, with some well-known pests.

Iberian Termite

Introduced; rare

Reticulitermes grassei

J F M A M J J A S O N D

BL 4–6 mm (winged adults approx. 9 mm). **ID** Whitish, head of soldiers larger than workers. Winged form black. **Cyc** Completed well within a year. **Hab** Nests in the ground, not in wood, but return to attack wood, hence a threat to structural timbers. **Dist** Accidentally introduced to a property in Saunton, North Devon (1994) via imported plants and possibly still present, last sighted in 2010 following eradication work. A major pest in Europe. Although a £190,000 eradication programme was put in place in Saunton, only time will tell whether this has been successful or whether these evasive insects have spread elsewhere. **SS** None.

Workers and soldier caste with large head

COCKROACHES (♀ WITH 2-VALVED ABDOMEN) 1/2

FAMILY Blattidae (Blattids) — 2 GEN. | 4 spp.

Medium-sized to large species (more than 650 species worldwide), including some of the commonest and most readily recognized cosmopolitan household blattids. Genera can be separated as follows:

WINGS vestigial (lobes) or < abdomen; PRONOTUM black	*Blatta* [1 sp.]	p. 127
WINGS ≥ abdomen; PRONOTUM patterned brownish and yellow	*Periplaneta* [3 spp.]	p. 128

Blatta — 1 sp.

Oriental Cockroach

Introduced; local

Blatta orientalis

J F M A M J J A S O N D

BL 17–30 mm. **ID** Black ♀ with wings reduced to lobes. ♂ can be brown with short wings. Also known as the Common Cockroach. **Cyc** Completed well within a year, females laying several oothecae. Adults live several months. **Hab** Heated buildings such as greenhouses, hospitals, hotels, restaurants and warehouses, sometimes private houses. These hardy, nocturnal scavengers eat almost anything and visit sewers, so can carry harmful bacteria. It is possible they can survive outdoors, for example in rubbish tips. **Dist** Few recent records, but historically well recorded in almost all GB and Ireland. A cosmopolitan pest. **SS** None.

Guide to families *p. 126*

COCKROACHES (♀ WITH 2-VALVED ABDOMEN) 2/2

Periplaneta	3 spp.
PRONOTUM with broad yellowish areas	American Cockroach
PRONOTUM with yellow submarginal ring	Australian Cockroach
PRONOTUM dark brown with small upper and large lower yellow patches	Brown Cockroach

American Cockroach
Periplaneta americana

BL 27–43 mm. **ID** Reddish-brown with broad paler areas on pronotum. **Cyc** Completed in a year or more, then adults live several months. **Hab** Heated buildings including bakeries, nurseries with glasshouses, hotels, sea ports, warehouses and zoos, more rarely private houses. Hardy but keep to heated premises. **Dist** Few recent records, but historically well recorded in almost all GB and Ireland. A cosmopolitan pest. **SS** Australian Cockroach, Brown Cockroach.

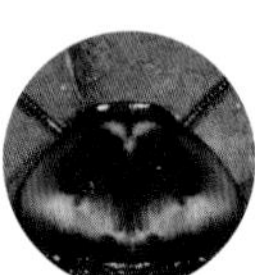

Australian Cockroach
Periplaneta australasiae

BL 27–34 mm. **ID** Reddish-brown with yellow submarginal ring on pronotum. Forewings with short yellow basal stripe. **Cyc** Completed in a year or more, then adults live several months. **Hab** Heated nurseries with glasshouses, warehouses and other heated buildings. Likely to be imported with produce, including bananas. **Dist** Has more recent records, but less extensively recorded in GB and Ireland than American Cockroach, although common in greenhouses at the Eden Project, Cornwall. A cosmopolitan pest. **SS** American Cockroach, Brown Cockroach.

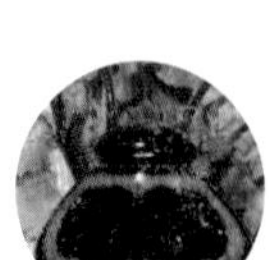

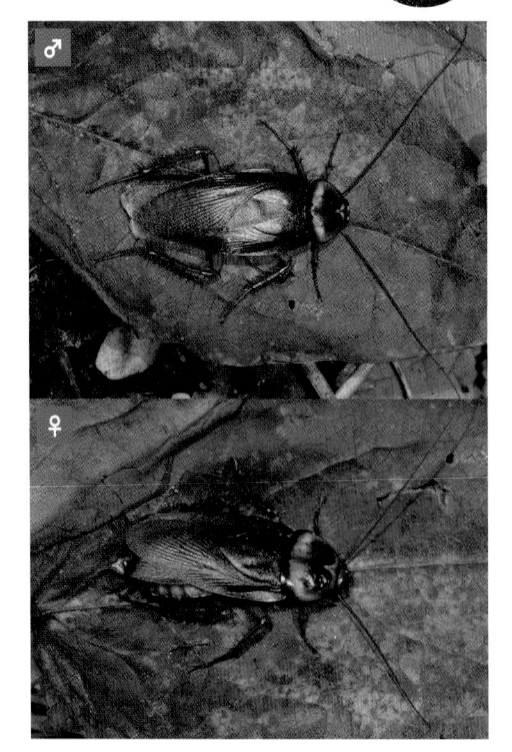

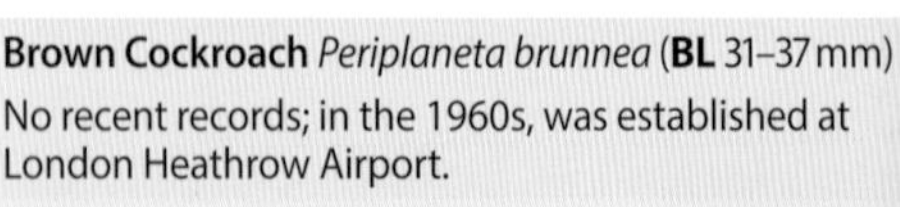

Brown Cockroach *Periplaneta brunnea* (**BL** 31–37 mm)
No recent records; in the 1960s, was established at London Heathrow Airport.

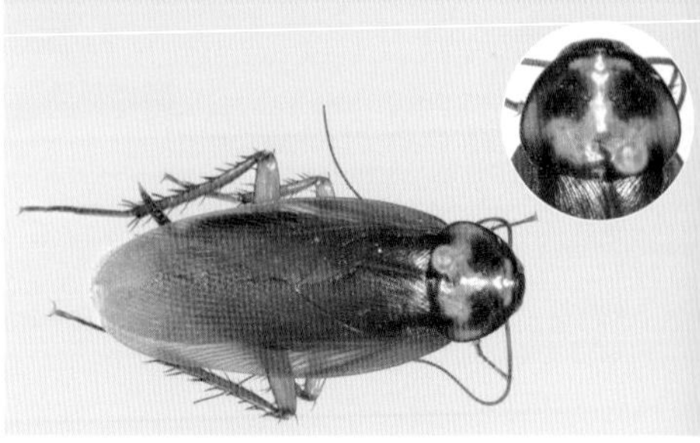

COCKROACHES (♀ WITH SMOOTH-TIPPED ABDOMEN) 1/3

FAMILY Ectobiidae (Wood cockroaches) 5 GEN. | 8 spp. | 6 ILL.

Small to medium-sized, with long cerci. One of the largest families, with almost 2,400 species worldwide.

INTRODUCED SPECIES; SURVIVING INDOORS

Blattella 1 sp.

German Cockroach
Blattella germanica

Introduced; local

BL 10–15 mm. **ID** Yellowish-brown, pair of dark stripes on pronotum. **Cyc** Completed well within a year. **Hab** Mainly heated buildings such as hotels, bakeries and warehouses. **Dist** The main pest cockroach in Great Britain, historically well recorded in much of GB and Ireland and numbers can build up without detection. A major cosmopolitan pest, readily available as reptile food. **SS** None.

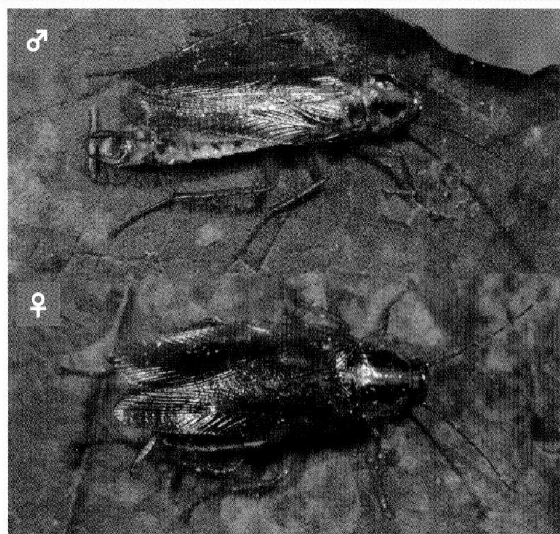

Brown-banded Cockroach *Supella longipalpa*

BL 10–14 mm. **ID** Yellowish-brown, reddish-brown or dark; pronotum with clear lateral margins. Forewings with pale basal and other bands; wings full length in ♂ (can fly well), not reaching tip of abdomen in darker ♀. **Cyc** Completed well within a year. **Hab** Houses and offices. **Dist** Apparently not recorded since 1998. **SS** None.

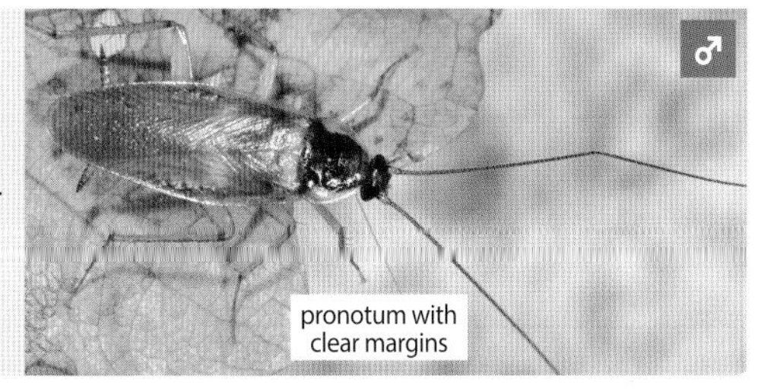

INTRODUCED SPECIES; SURVIVING OUTDOORS

Variable Cockroach *Planuncus tingitanus*

BL approx. 10 mm. **ID** Abdomen striped, with dark patches, or wholly black. Wings nearly reaching tip of abdomen in ♂, but shorter in ♀. **Cyc** Probably completed within a year. **Hab** Gardens. **Dist** Known from Hythe, Kent since at least 2011. **SS** None.

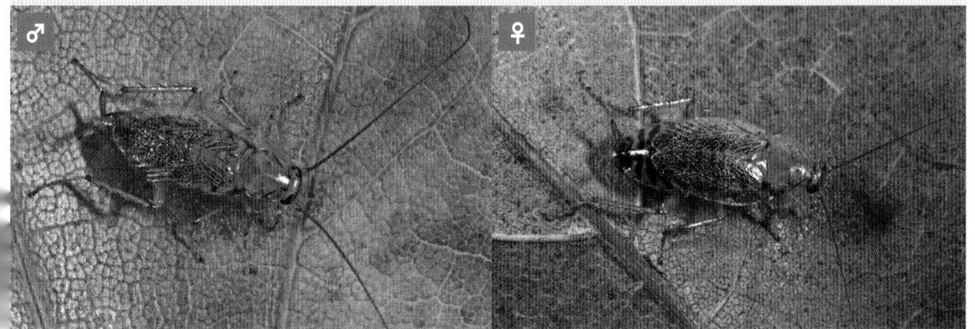

Guide to families p. 126

COCKROACHES (♀ WITH SMOOTH-TIPPED ABDOMEN) 2/3

NATIVE SPECIES; SURVIVING OUTDOORS

Ectobius 4 spp. | 2 ILL.

FOREWINGS with dark mottled pattern; ♂ PRONOTUM black; ♀ WINGS ± **abdomen length**	Dusky Cockroach
FOREWINGS plain; ♂ PRONOTUM tawny; ♀ WINGS > **abdomen length**	Tawny Cockroach

Dusky Cockroach
Ectobius lapponicus

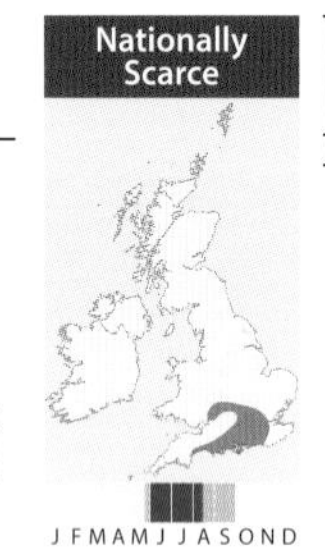

BL 7–11 mm. **ID** Pale brown or greyish-brown with some speckling on forewings; ♂ with dark brown or black pronotum; ♀ wings not quite reaching tip of abdomen. **Cyc** Two years, eggs overwintering and resulting nymphs overwintering before reaching adult stage. **Hab** Various scrubby sites, including heathland. Active both day and night, sometimes seen resting on tops of stems. **Dist** Stronghold is the New Forest, Hampshire. **SS** ♂s of Lesser Cockroach (*p. 131*) and Tawny Cockroach.

Tawny Cockroach
Ectobius pallidus

BL 8–10 mm. **ID** Tawny, fairly broad. ♀ wings extend beyond tip of abdomen. **Cyc** Two years. **Hab** Woodland rides and clearings, coastal cliffs, heathland, also chalk downlands. Often seen on flowers or vegetation. **SS** ♂s of Lesser Cockroach (*p. 131*) and Dusky Cockroach.

Ootheca visible at tip of abdomen

SS At time of going to press, at least two more *Ectobius* species have been found: **Italian Cockroach** *Ectobius montanus* [N/I] (**BL** 7–9 mm) at Dungeness, Kent (♂ has more black on pronotum than Dusky Cockroach) and **Garden Cockroach** *Ectobius vittiventris* [N/I] (**BL** 9–11 mm) in Hertfordshire (brown, rather elongated).

COCKROACHES (♀ WITH SMOOTH-TIPPED ABDOMEN) 3/3

NATIVE SPECIES; SURVIVING OUTDOORS

Capraiellus 1 sp.

Lesser Cockroach
Capraiellus panzeri

Nationally Scarce

J F M A M J J A S O N D

BL 5–8 mm. **ID** Dark brown to black, some marks and speckling on pronotum, and also abdomen of ♀. ♂ fully-winged, flies well. ♀ has short, truncate forewings; hindwings vestigial; cannot fly. **Cyc** One year, eggs overwintering until late April or May. **Hab** Coastal cliffs, sand dunes and shingle beaches, also various inland sites from grassland to heathland. Often seen on flowers or vegetation. **Dist** Inland stronghold is the New Forest, Hampshire. **SS** ♂s of **Dusky Cockroach** (*p. 130*) and **Tawny Cockroach** (*p. 130*).

FAMILY **Blaberidae** (Giant cockroaches) 1 SP.

A moderately large family, with about 1,260 species.

Surinam Cockroach
Pycnoscelus surinamensis

Introduced; rare

BL 20–25 mm. **ID** Reddish-brown, with pale band along front edge of pronotum. **Cyc** Completed well within a year. **Hab** Nurseries and botanical gardens in greenhouses, also a few reports in the past from offices and hospitals. Unlikely to survive long outdoors. **Dist** Few recent records, most in the London area and in greenhouses at the Eden Project, Cornwall. **SS** None.

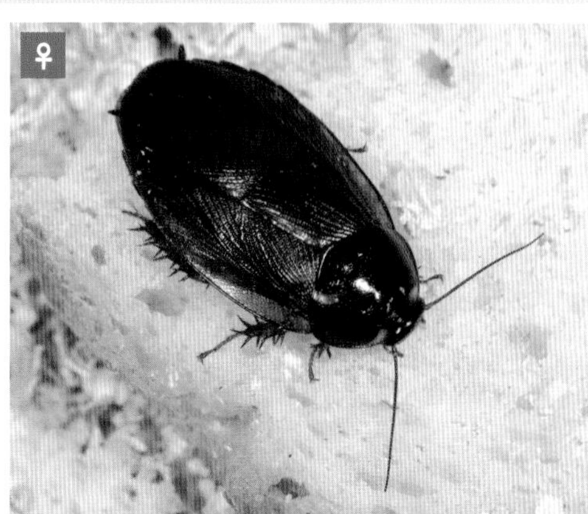

ORDER PSOCODEA — Lice, booklice & barklice

[Greek: *psocus* = biting; *eidos* = form]

BI | 36 families [approx. 140 genera, **650 species**] **W** | 65 families [**approx. 11,100 species**]

Little known, tiny to small-sized, soft-bodied insects (1–10 mm), in Great Britain and Ireland represented by 19 families comprising 107 barklice and booklice species ('Psocoptera') with a large head, large eyes and long, often 13-segmented antennae; winged species have three ocelli, wingless species have none. **Forewing veins (simple, with few crossveins) are useful characters to help place to family and species level.** There are also approximately 17 families comprising 550 flattened, wingless crab-like ectoparasitic lice ('Phthiraptera') with tiny, or no, eyes and short antennae with few segments. These are known as chewing and sucking lice, being attached to birds and some mammals, feeding on hair or blood. Where wings are present in barklice [confusingly sometimes referred to as barkflies, to widen their appeal], these are membranous. Many species known from a single import may never be encountered again and new lice are sometimes found on migrant birds and zoo animals. The life-cycle is usually completed within a year.

FINDING LICE Barklice are easiest to find, often crawling on tree trunks or leaves, but an unpopular species is probably best known, due to its association with humans; the Head Louse.

This group of species is rather specialized, little known and most are unlikely to be seen without specifically searching. For this reason, only a few representative species are included here (specialist keys are available for those wishing to learn more – see *below*).

Chewing Louse ('Phthiraptera': Mallophaga) attached to the fly *Ornithomya avicularia* (Diptera: Hippoboscidae)

FURTHER READING AND USEFUL WEBSITES

New, T.R. 2005. ***Psocids: Psocoptera (Booklice and Barklice)***. *2nd Edition. Handbooks for the Identification of British insects*, Vol. 1, Part 7. Royal Entomological Society.

www.brc.ac.uk/schemes/barkfly/homepage.htm National Barkfly Recording Scheme (Great Britain and Ireland)

Structure of a louse

HEAD LOUSE
ABDOMEN (11-segmented)
THORAX
femur
tibia
tarsus
eye
antenna
HEAD
claw

Structure of a barklouse

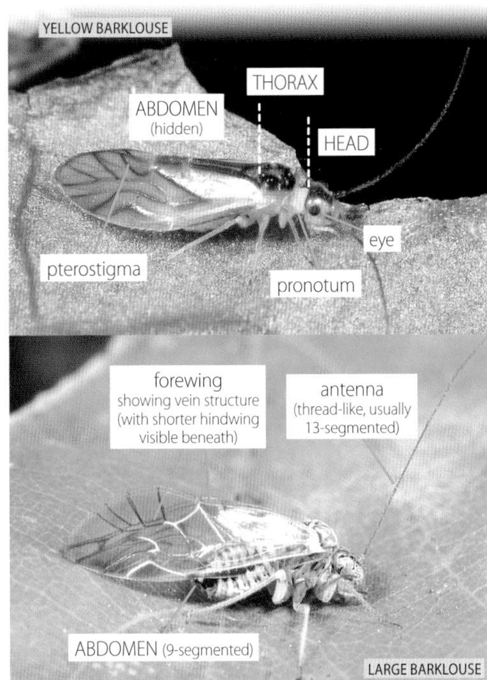

'PHTHIRAPTERA' (Chewing and sucking lice) approx. **550 spp.**

FAMILY **Pediculidae** (Primate lice) **1 sp.**

One species, which is by far the best known of the 'Phthiraptera', occurs in Great Britain and Ireland. It is represented by two subspecies: the **Body Louse** *Pediculus humanus humanus* [now uncommon] and the **Head Louse** *Pediculus humanus capitis*, which are sometimes regarded as distinct species.

Head Louse
Pediculus humanus capitis

BL approx. 3 mm. **ID** Grey to pale brown, hairy; rather broad-bodied. Reddish areas seen after feeding on blood. **Cyc** As little as one month. **Hab** All stages live on the human scalp, particularly children (the majority of whom experience these insects at least once before leaving primary school). Nymphs and adults feed on blood, transferring to other humans by direct head to head contact. **SS** The disease-carrying (including typhus) **Body Louse** *Pediculus humanus humanus* is the normal form; this is virtually identical to Head Louse but slightly larger, lives on the base of the neck or torso and lays eggs on clothing or bedding, rather than on the scalp.

'PSOCOPTERA' (Booklice and barklice) **107 spp.**

Tarsi of illustrated families

TARSI 2-SEGMENTED

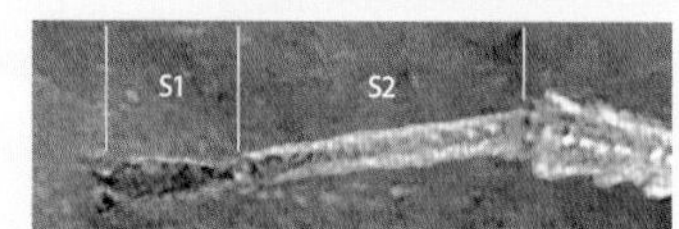

Psocidae Ectopsocidae Stenopsocidae Caeciliusidae

TARSI 3-SEGMENTED

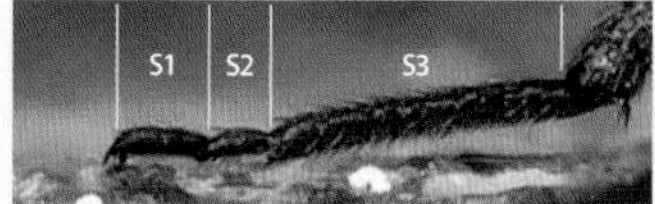

Mesopsocidae

Antennae with 13 segments in these families; the other 14 families have either 15 or 20+ segments.

FAMILY Mesopsocidae **1 GEN. | 3 spp. | 1 ILL.**

Larger species (>4 mm); tarsi 3-segmented.

Mesopsocus **3 spp. | 1 ILL.**

Woodland Barklouse
Mesopsocus immunis

BL 4–5 mm. **ID** ♂ black, with green and brown mottled areas. ♀ mottled pale yellowish and brown, with tiny wings. **Hab** Woodland, on various trees including Hawthorn; often found on tree trunks. **SS** Other *Mesopsocus* species, which are often darker but identification requires examination of genitalia.

FAMILY Ectopsocidae **1 GEN. | 7 spp. | 1 ILL.**

Small species (<3 mm); tarsi 2-segmented.

Ectopsocus **7 spp. | 1 ILL.**

Peters' Barklouse
Ectopsocus petersi

BL approx. 2 mm. **ID** Conspicuous resting position; spotted wings. ♀ has short-winged and long-winged forms. **Hab** Woodland, on various trees, including coniferous, also shrubs. Sometimes recorded on tree trunks. **SS** Other *Ectopsocus* species, but these have less prominent dark marks on the forewings.

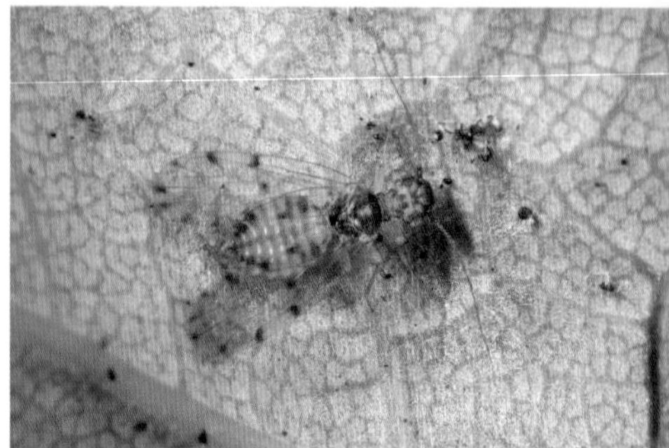

FAMILY Caeciliusidae — 4 GEN. | 8 spp. | 1 ILL.

Commoner species (such as *below*) on trees. Other species prefer low vegetation; tarsi 2-segmented; crossvein near pterostigma.

Valenzuela — 5 spp. | 1 ILL.

Yellow Barklouse
Valenzuela flavidus

BL approx. 3 mm. **ID** Yellow with brown markings and central stripe on head. Mainly parthenogenetic ♀s, although bisexual populations are known from continental Europe. **Hab** Mainly various deciduous trees, uncommon on conifers. In winter and spring, in leaf-litter. **SS** None (this is the only yellow-bodied species).

Common

J F M A M J J A S O N D

FAMILY Stenopsocidae — 2 GEN. | 4 spp. | 2 ILL.

Forewing crossveins differ from the Caeciliusidae (lacks crossvein near pterostigma); tarsi 2-segmented. Commoner (large and pale) species are widespread on trees and bushes.

Gnaphopsocus — 1 sp.

Blotched Barklouse
Graphopsocus cruciatus

BL approx. 3 mm. **ID** Head pale yellow with brown markings, brown also on thorax. Abdomen greenish-yellow, with brown tip. Conspicuous bold dark brown blotches, mainly on forewings. **Hab** Woodland and gardens on trees and bushes. **SS** None.

Common

J F M A M J J A S O N D

Stenopsocus — 3 spp. | 1 ILL.

Stigma Barklouse
Stenopsocus stigmaticus

Local

BL approx. 4 mm. **ID** Black bumps on thorax. Abdomen usually green, with white markings and band. Bold dark brown blotches. Forewing with distinctive black mark. **Hab** Trees and bushes in woodland, including coniferous, although Hawthorn preferred in some areas. **Dist** Probably under-recorded. **SS** Other *Stenopsocus* species, but Stigma Barklouse readily distinguished by abdomen colour and wing stigma.

J F M A M J J A S O N D

FAMILY Psocidae — 9 GEN. | 13 spp. | 3 ILL.

Larger species (>4 mm); tarsi 2-segmented.

Loensia — 3 spp. | 1 ILL.

Brown-banded Barklouse
Loensia fasciata

BL 4–5 mm. **ID** Spotted forewings strongly marked with dark brown bands. **Hab** Woodland, on various trees. **Dist** Apparently widespread but only occasionally recorded in GB and Ireland. **SS** Two other *Loensia* species, which both lack dark brown bands on forewings.

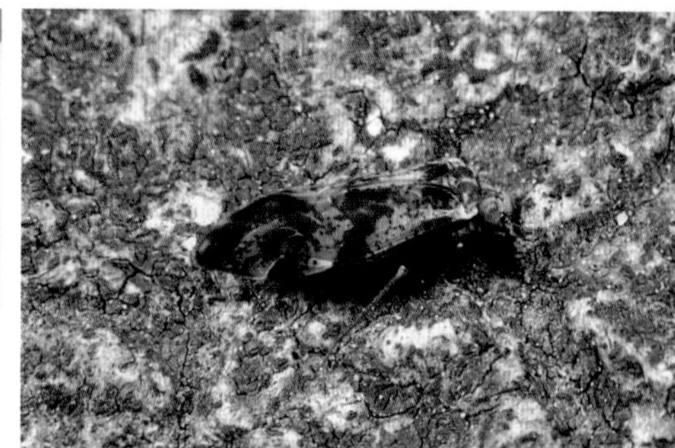

Metylophorus — 1 sp.

Cloudy Brown Barklouse
Metylophorus nebulosus

BL approx. 5 mm. **ID** Brown, cloudy wings (in ♀ more patterned). **Hab** Woodlands, branches of deciduous and coniferous trees. **SS** None.

Psococerastis — 1 sp.

Large Barklouse
Psococerastis gibbosa

BL 6–7 mm. **ID** Large and robust, yellowish and black with bold black veins with small dark brown markings on wings. **Hab** Woodland, on various trees, deciduous and coniferous; mainly on branches. **SS** None.

ORDER HEMIPTERA

Bugs

[Greek: *hemi* = half; *ptera* = wing]

BI | 63 families [approx. 688 genera, almost **2,000 species**] **W** | approx. 100 families [**approx. 104,200 spp.**]

Whilst non-entomologists call insects 'bugs', Hemiptera are the 'true' bugs. There are three suborders (see *pages 140–141*) which, although different in appearance, all have piercing, beak-like mouthparts (the rostrum) adapted for sucking juices from plants or other insects. The species range in size from less than 1 mm, to the elusive New Forest Cicada (up to 27 mm), or even longer Water Stick-insect (approx. 50 mm, including its tail). Many gardeners are familiar with, and dislike, aphids, which are serious agricultural pests. Some have personal experience of Bed Bugs, which feed on human blood! However, most bugs are harmless and often rather colourful. Shieldbugs and leatherbugs are reasonably well studied and popular with naturalists. Other families are yet to attract such interest.

Spiked Shieldbug with Buff-tip moth larva prey. The bug attracted a mate while being observed, started mating but refused to give up its meal.

FINDING BUGS Although some bugs can be seen on flowers or resting on vegetation, hemipterists (people who study bugs) often find many more species than would normally be seen, by using a sweep net on grasses or low vegetation and/or a beating tray beneath vegetation, after dislodging insects hidden in trees or large shrubs. Grubbing around on bare, dry ground is another technique, often productive in coastal areas. Water bugs can be examined in a tray before being returned to the water via a small net. Photography is straightforward except on hot, sunny days when bugs are often active and will fly; predatory bugs also provide plenty of opportunity for good behavioural images when piercing prey.

Structure of bugs

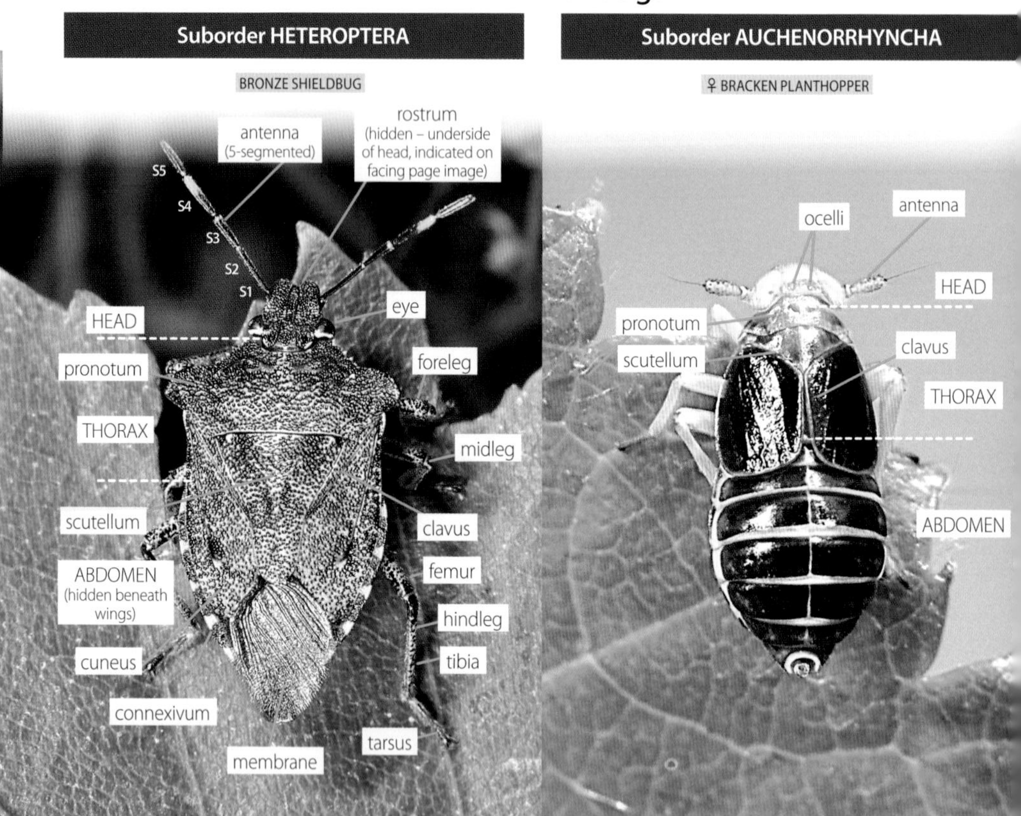

FURTHER READING AND USEFUL WEBSITES

www.britishbugs.org.uk British Bugs
Very comprehensive, well-illustrated resource, including website links for all UK bug recording schemes and useful publications and maps.

Bantock, T. 2016. ***A review of the Hemiptera of Great Britain: the shieldbugs and allied families Coreoidea, Pentatomoidea & Pyrrhocoroidea***. Species Status No. 26. Natural England commissioned Reports, Number 190.
Latest Red List (conservation) status of species, with up-to-date information on rarities.

Southwood, T.R.E & Leston, D. 1959. ***Land and water bugs of the British Isles***. F. Warne & Co.
Dated, but this rare book (only 400 printed) is still the standard reference work on British bugs. Also available on CD-ROM and as facsimile reprint, from Pisces Conservation, 2005 (www.pisces-conservation.com/softlwb2.html larger format, with some poorer quality images).

ALPHABETICAL LIST OF HEMIPTERAN FAMILIES COVERED

KEY: = aquatic; = terrestrial **A** = **AUCHENORRHYNCHA** **S** = **STERNORRHYNCHA** H = **HETEROPTERA**

HETEROPTERA TERRESTRIAL BUG INFRAORDERS: = Cimicomorpha; = Leptopodomorpha; = Pentatomomorpha

			Family Guide	Main accounts
Acanthosomatidae	**Keeled shieldbugs**	H	*145*	***149***
Aleyrodidae	**Whiteflies**	S	*222*	***226***
Alydidae	**Broad-headed bugs**	H	*147*	***173***
Anthocoridae	**Flower bugs**	H	*147*	***191***
Aphelocheiridae	**River bugs**	H	*143*	N/I
Aphididae	**Aphids**	S	*222*	***223***
Aphrophoridae	**Froghoppers**	A	*210*	***212***
Aradidae	**Flatbugs**	H	*146*	***148***
Berytidae	**Stiltbugs**	H	*147*	***184***
Cercopidae	**Froghoppers**	A	*210*	***212***
Cicadellidae	**Leafhoppers**	A	*210*	***214***
Cicadidae	**Cicadas**	A	*210*	***211***
Cimicidae	**Bed bugs**	H	*146*	***190***
Cixiidae	**Lacehoppers**	A	*210*	***218***
Coccidae	**Soft scales**	S	*222*	***227***
Coreidae	**Leatherbugs**	H	*147*	***165***
Corixidae	**Lesser water boatmen**	H	*143*	***208***
Cydnidae	**Burrowing shieldbugs**	H	*145*	***151***
Delphacidae	**Planthoppers**	A	*210*	***220***
Gerridae	**Pondskaters**	H	*143*	***206***
Hebridae	**Sphagnum bugs**	H	*143*	N/I
Hydrometridae	**Water measurers**	H	*143*	***205***
Issidae	**Issid planthoppers**	A	*210*	***221***
Lygaeidae	**Ground bugs**	H	*175*	***175***
Membracidae	**Treehoppers**	A	*210*	***211***
Mesoveliidae	**Pondweed bugs**	H	*143*	N/I
Miridae	**Plant bugs**	H	*146*	***194***
Nabidae	**Damsel bugs**	H	*147*	***191***
Naucoridae	**Saucer bugs**	H	*143*	***207***
Nepidae	**Water scorpions**	H	*143*	***207***
Notonectidae	**Backswimmers**	H	*143*	***208***
Pentatomidae	**Typical shieldbugs**	H	*145*	***156***
Piesmatidae	**Beetbugs**	H	*146*	***183***
Plataspidae	**Oval shieldbugs**	H	*145*	***145***
Pleidae	**Pygmy backswimmers**	H	*143*	N/I
Pseudococcidae	**Mealybugs**	S	*222*	***227***
Psyllidae	**Jumping plant lice**	S	*222*	***226***
Pyrrhocoridae	**Firebugs**	H	*146*	***173***
Reduviidae	**Assassin bugs**	H	*147*	***188***
Rhopalidae	**Rhopalid bugs**	H	*170*	***170***
Saldidae	**Shorebugs**	H	*147*	***204***
Scutelleridae	**Tortoise shieldbugs**	H	*145*	***154***
Stenocephalidae	**Spurgebugs**	H	*147*	***174***
Tettigometridae	**Tettigometrid planthoppers**	A	*210*	***222***
Tingidae	**Lacebugs**	H	*146*	***186***
Thyreocoridae	**Scarab shieldbugs**	H	*145*	***165***
Veliidae	**Water crickets**	H	*143*	***205***

ORDER Hemiptera

FOREWINGS divided into two sections: a tough and leathery base and membranous tip; HINDWINGS membranous.
All wings folded flat over the abdomen, except for flight; as a result; **SCUTELLUM** (triangular area) is conspicuous;
HEAD + ROSTRUM can point forward; rostrum beak-like.

[Where wingless, still distinguished from other insect orders by the mouthparts formed into a beak-like biting structure known as a rostrum, held beneath the body.]

HEAD typically points forwards

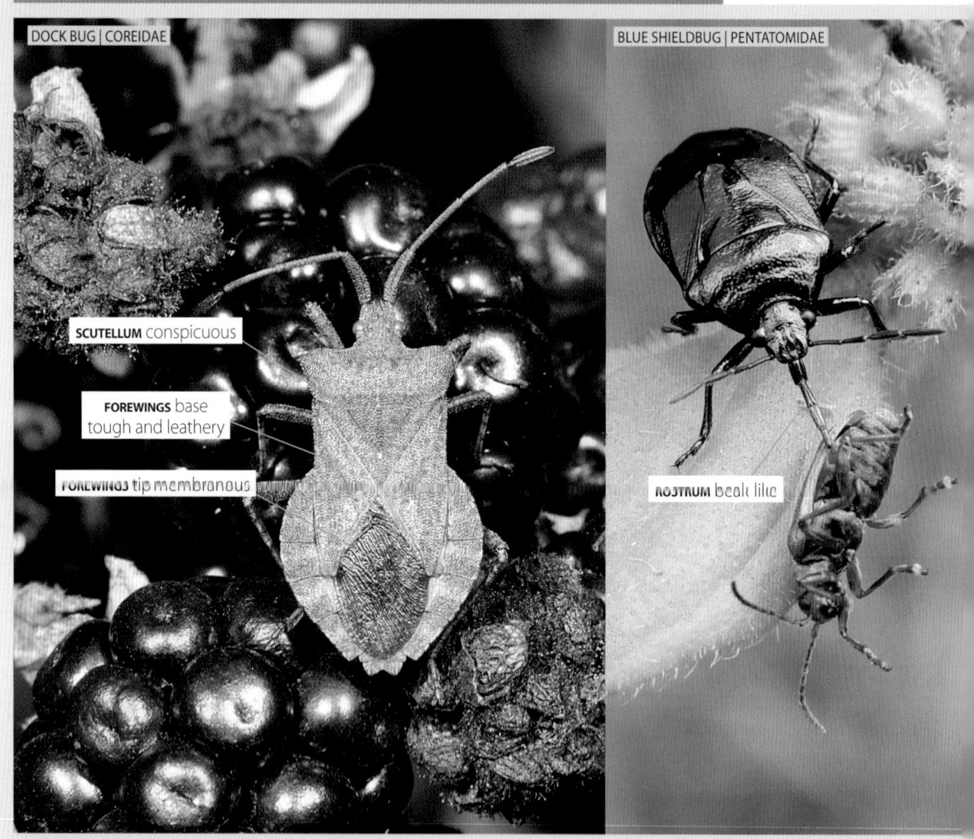

HETEROPTERA (Bugs)

35 FAMILIES | 33 ILL.

p. 142

GUIDE TO SUBORDERS | 3 BI

FOREWINGS (where present) **held like a roof over the body**; structure membranous or hardened (but not divided into two sections); **HINDWINGS** membranous; **SCUTELLUM** triangular area (where present) less conspicuous than most hemiptera; **HEAD + ROSTRUM** point down and back (rostrum may be lacking).

HEAD points down and back

NB Both **Auchenorryncha** and **Sternorrhyncha** were formerly combined and known as a single suborder HOMOPTERA. The **Cicadomorpha** and **Fulgoromorpha** combined are regarded by some authors (as here) as the single suborder AUCHENORRHYNCHA. The higher classification of bugs is still a contentious subject.

FORM winged, body not covered with wax or froth; **TARSAL SEGMENTS** 3; **ANTENNAE** short, comprising a few short basal segments, followed by a much longer slim bristle (arista)

COMMON ISSID BUG | ISSIDAE

AUCHENORRHYNCHA

9 FAMILIES | 9 ILL.

[= CICADOMORPHA[1] and FULGOROMORPHA[2]]

(Cicadas[1], leafhoppers[1], froghoppers[1], treehoppers[1] and **planthoppers[2])**

p. 210

FORM most wingless (but if winged, venation simple), soft-bodied (but a minority hard-bodied) insects (often covered with wax or froth) **ROSTRUM** held between fore coxae; **TARSAL SEGMENTS** 1 or 2; **ANTENNAE** longer and thread-like, with 4 or 5 similar segments (no bristle)

GIANT WILLOW APHID | APHIDAE

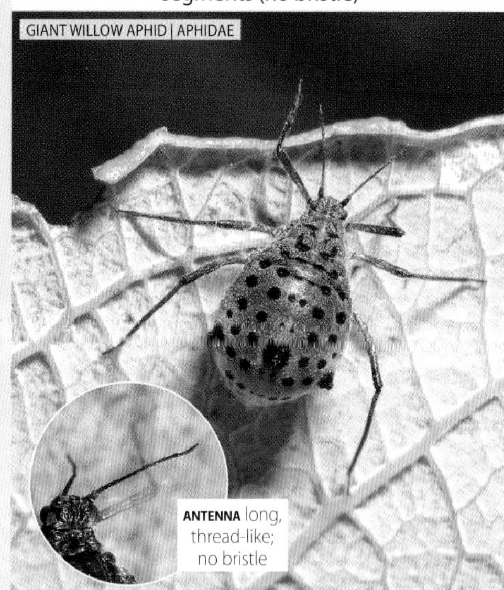

STERNORRHYNCHA

5 FAMILIES | 5 ILL.

(Aphids, jumping plant lice (psylloids)**, scale insects** (coccoids) and **whiteflies)**

p. 222

Guide to suborders *pp. 140–141*

SUBORDER Heteroptera [AQUATIC INFRAORDER NEPOMORPHA]

ANTENNAE SMALL, HIDDEN BENEATH HEAD WHEN VIEWED DORSALLY

AQUATIC SPECIES, LIVING UNDERWATER

TAIL long

FORM adults broad or elongate; large; **BL** 18–35 mm

Water Scorpion (BL 18–22 mm)

Water scorpions
Nepidae
[2 genera | 2 species]

p. 207

TAIL absent

SCUTELLUM only visible on some species (with **BL** < 2·5 mm); **ROSTRUM** short and broad; **LEGS** each pair of different size and shape, **FORE TARSI** 1-segmented; **BL** up to 14 mm

Common Water Boatman (BL 12–14 mm)

Lesser water boatmen
Corixidae
[9 genera | 39 species]

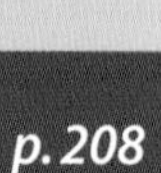

p. 208

FORM FLAT AND OVAL

SCUTELLUM visible; **BL** > 8 mm; **ROSTRUM** pointed; **LEGS** at least two pairs similar, **FORE TARSI** usually at least 2-segmented; **FOREWINGS** short

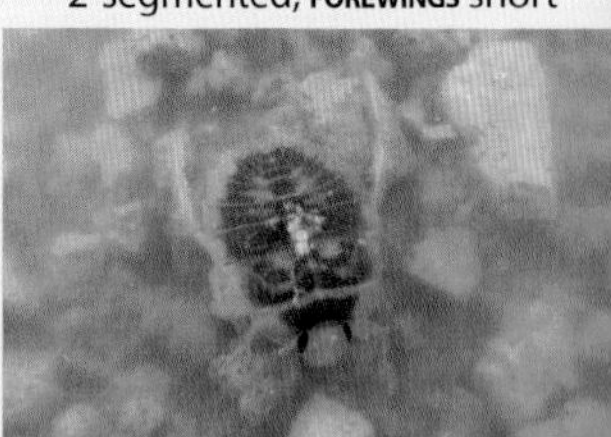

River Bug
Aphelocheirus aestivalis (BL 9–12 mm)

River bugs
Aphelocheiridae
[1 species]

ANTENNAE LONG, EASILY VISIBLE DORSALLY

AQUATIC SPECIES, MOSTLY ON WATER SURFACE (often with long antennae if on surface)

FORM stick-like; EYES well away from pronotum

HEAD elongated; **BL** 7–12 mm.

Water Measurer (BL 9–12 mm)

Water measurers
Hydrometridae
[1 genus | 2 species]

p. 205

FORM more robust; EYES near pronotum;

FORE TARSI CLAW TIPS GREENISH

BL 3–4 mm; **FORM** often wingless, but some individuals winged

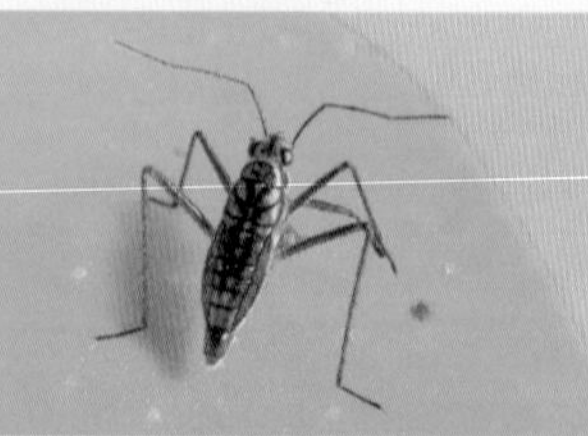

Pondweed Bug
Mesovelia furcata (BL approx. 3 mm)

Pondweed bugs
Mesoveliidae
[1 species]

FORE TARSI CLAW TIPS NOT GREENISH

BL 6–18 mm; **LEGS** midlegs and hindlegs elongated, **HIND FEMORA** reaching well beyond tip of abdomen

Common Pondskater (BL 8–10 mm)

Pondskaters
Gerridae
[3 genera | 10 species]

p. 206

GUIDE TO SELECTED FAMILIES (two small families omitted) | 33 OF 35 BI

1/1

FORM FLAT AND OVAL
SCUTELLUM visible; **WINGS** developed, large; **BL** 12–16 mm

FORM BOAT-SHAPED, BACKSWIMMERS
PRONOTUM and **FOREWINGS** smooth, lacking pits; **BL** large, ≥11 mm

FORM BOAT-SHAPED, BACKSWIMMERS
PRONOTUM and **FOREWINGS** pitted; **BL** small, <3 mm

Saucer Bug
(BL 12–16 mm)

Saucer bugs
Naucoridae
[2 genera | 2 species]

p. 207

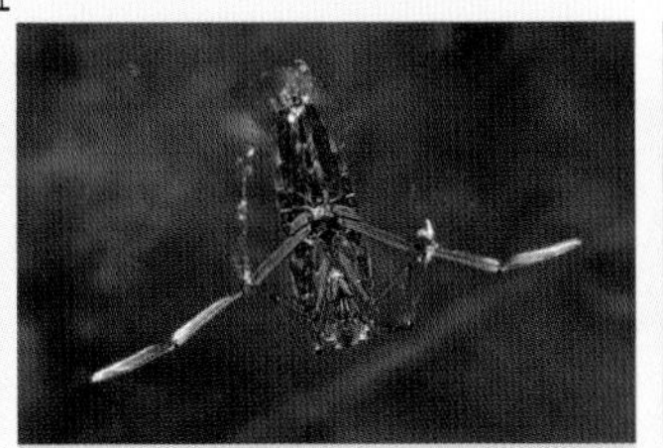

Common Backswimmer
(BL 13–16 mm)

Backswimmers
Notonectidae
[1 genus | 4 species]

p. 208

Pygmy Backswimmer
Plea minutissima (BL 2–3 mm)

Pygmy backswimmers
Pleidae
[1 species]

1/3

BL 2–7 mm; **LEGS** midlegs elongated, **HIND FEMORA** broad, not reaching tip of abdomen

BL <2mm; **LEGS** all elongated, but not greatly; **ANTENNAE** with 4 segments, but appearing to have 5 (the 4th is constricted in centre); **ABDOMEN** underside with dense silver hairs. **HABITAT** semiaquatic *Sphagnum* mosses

Water Cricket
(BL 5–7 mm)

Water crickets
Vellidae
[2 genera | 5 species]

p. 205

Sphagnum Bug
Hebrus ruficeps (BL approx. 1 mm)

Sphagnum bugs
Hebridae
[1 genus | 2 species]

TERRESTRIAL SPECIES

All with long, easily visible antennae. Families are grouped within colour-coded infraorders

GUIDE TO SELECTED FAMILIES OF TERRESTRIAL BUGS

p. 144

SUBORDER Heteroptera [TERRESTRIAL INFRAORDERS]

GUIDE TO SELECTED FAMILIES OF TERRESTRIAL BUGS

A visual guide to typical representatives of land bug infraorders

The colour code to the infraorder to which a family belongs is shown against each group.

INFRAORDER PENTATOMOMORPHA *p. 148*

Closely related to Cimicomorpha, comprises the majority of terrestrial bugs.

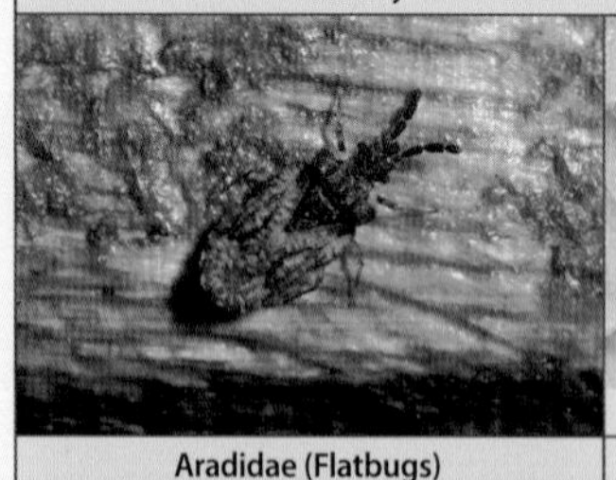

Aradidae (Flatbugs)

Acanthosomatidae (Keeled shieldbugs)

Rhopalidae (Rhopalid bugs)

Pyrrhocoridae (Firebugs)

Stenocephalidae (Spurgebugs)

Berytidae (Stiltbugs)

INFRAORDER CIMICOMORPHA *p. 186*

Rostrum often adapted for feeding on prey, for example in bed bugs and assassin bugs.

Reduviidae (Assassin bugs)

Cimicidae (Bed bugs)

Miridae (Plant bugs)

INFRAORDER LEPTOPODOMORPHA *p. 204*

The strange-looking, large-eyed shorebugs [*shown*] are predators, most species partly living beneath water. One marine bug [Aepophilidae, not illustrated] was formerly placed in the shorebug family Saldidae.

Saldidae (Shorebugs)

GUIDE TO SELECTED FAMILIES (two small families omitted) | 33 OF 35 BI

ANTENNAE LONG, EASILY VISIBLE DORSALLY 2/3

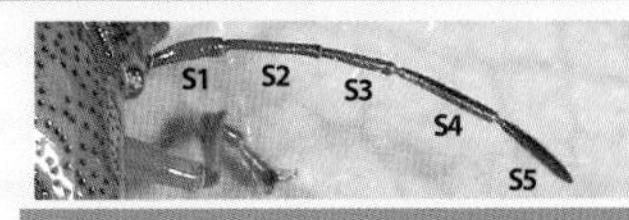

ANTENNAE 5-segmented [Shieldbugs]

TARSI 2-segmented

FORM typical shieldbug shape; **SCUTELLUM** does not reach tip of abdomen; **BL** 7–16 mm

Juniper Shieldbug (BL 8–11 mm)

Keeled shieldbugs
Acanthosomatidae
[4 genera | 5 species (1†)]
p. 149

FORM very round, usually shiny; **SCUTELLUM** covers almost the entire abdomen; **BL** approx. 4 mm

First recorded 2019 from Hastings, East Sussex

Trapezium Shieldbug *Coptosoma scutellatum* (BL 3–5 mm)

Oval shieldbugs
Plataspidae
[1 species]

TARSI 3-segmented; TIBIA with rows of large, dark spines

Black or brown, usually metallic bugs, with pale markings in some species; **BL** 3–9 mm

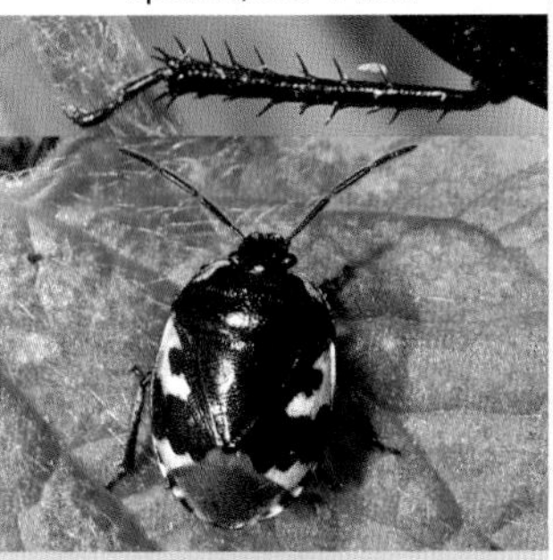

Pied Shieldbug (BL 6–8 mm)

Burrowing shieldbugs
Cydnidae
[6 genera | 9 species]
p. 151

TARSI 3-segmented; TIBIA lacking large spines (may have small spines, but not in rows)

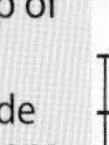

SCUTELLUM triangular or subtriangular, not reaching tip of abdomen; **BL** 5–16 mm
(NB in one species **PRONOTUM** side margins have two bent projections reaching near eyes)

Hairy Shieldbug (BL 10–13 mm)

Typical shieldbugs
Pentatomidae
[22 gen. (2†) | 27 spp. (2†)]
p. 156

SCUTELLUM reaching tip of abdomen; **PRONOTUM** side margins lacking projections

FORM brown, non-metallic, some species with black markings; **BL** 4–11 mm

Tortoise Shieldbug (BL 9–11 mm)

Tortoise shieldbugs
Scutelleridae
[2 genera | 5 species (1†)]
p. 154

FORM dark shining black or bronze; small; **BL** 3–4 mm

Scarab Shieldbug (BL 3–4 mm)

Scarab shieldbugs
Thyreocoridae*
[1 species]
p. 165

*regarded by some authors as Corimelaenidae

SUBORDER Heteroptera [TERRESTRIAL INFRAORDERS]

ANTENNAE LONG, EASILY VISIBLE DORSALLY 3/3

ANTENNAE 4-SEGMENTED

TARSI 2-segmented

FORM flat, leathery. Distinctively shaped head; **OCELLI** absent; **BL** 3·5–9·0 mm

Pale-shouldered Flatbug (BL 5–6 mm)

Flatbugs
Aradidae
[2 genera | 7 species]
p. 148

FOREWINGS with lace-like veins; **OCELLI** absent; **BL** 2–5 mm

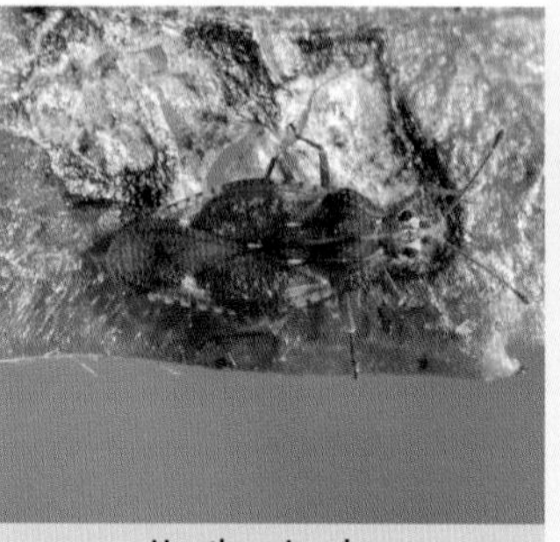

Hawthorn Lacebug (BL 3 mm)

Lacebugs
Tingidae
[14 genera | 25 species]
p. 186

FOREWINGS with punctures except veined tip, **OCELLI** present; **BL** 2–3 mm

Beet Leaf Bug (BL 3 mm)

Beetbugs
Piesmatidae
[2 genera | 2 species]
p. 183

TARSI 3-segmented; OCELLI absent

Reddish-brown. Flat, broadened; wingless or wings short and scale-like; **BL** 3–5 mm

Bed Bug (BL 4–5 mm)

Bed bugs
Cimicidae
[2 genera | 4 species]
p. 190

Various colours; wings fully formed, with cuneus; **BL** 2–11 mm

Oak Plant Bug *Psallus varians* (BL 4 mm)

Plant bugs
Miridae
[94 genera | 220 species]
p. 194

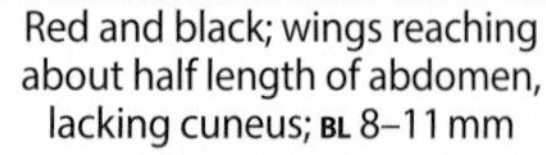

Red and black; wings reaching about half length of abdomen, lacking cuneus; **BL** 8–11 mm

Firebug (BL 8–11 mm)

Firebugs
Pyrrhocoridae
[Firebug]
p. 173

GUIDE TO SELECTED FAMILIES (two small families omitted) | 33 OF 35 BI

TARSI 3-segmented; OCELLI present

ROSTRUM 3-segmented

FORM stout or elongate, including legs; BL 5–18 mm

Heath Assassin Bug (BL 9–12 mm)

Assassin bugs
Reduviidae
[5 genera | 9 species]
p. 188

FOREWINGS with cuneus and costal break; BL < 5 mm

Common Flower Bug (BL 3–4 mm)

Flower bugs
Anthocoridae
[12 genera | 32 species]
p. 191

FORM often short-winged; FOREWINGS lack cuneus; BL 2–6 mm

Common Shorebug (BL 4–5 mm)

Shorebugs
Saldidae
[8 genera | 22 species]
p. 204

ROSTRUM 4-segmented

BL 6–16 mm; FORM broad and elongate; WING MEMBRANE with > 6 veins; FEMORA tips not expanded or darkened; SCENT GLAND OPENING (if present) visible from below, not side

Hypericum Rhopalid (BL 7 mm)

Rhopalid bugs
Rhopalidae
[7 gen. | 11 spp.]
p. 170

BL 5–15 mm; FORM often broad; HEAD front not rounded, projecting ahead; ANTENNAE + LEGS indistinct bands; SCENT GLAND OPENING between mid and hind coxae, visible from side

Dock Bug (BL 13–15 mm)

Leatherbugs
Coreidae
[10 gen. | 11 spp.]
p. 165

BL 10–12 mm; FORM elongate; HEAD front rounded; ANTENNAE + LEGS indistinct bands; ABDOMEN with orange patch

Ant Bug (BL 10–12 mm)

Broad-headed bugs
Alydidae
[1 species]
p. 173

BL 6–10 mm; ROSTRUM held away from body. FOREWINGS lacking cuneus and costal fracture

Heath Damsel Bug (BL 7mm)

Damsel bugs
Nabidae
[3 gen. | 14 spp.]
p. 191

BL 8–14 mm; FORM oval; HEAD front with two points; ANTENNAE + LEGS **strongly banded**

Wood Spurgebug (BL 8–11 mm)

Spurgebugs
Stenocephalidae
[1 gen. | 2 spp.]
p. 174

BL 2–8 mm; FORM broad; WING MEMBRANE with ≤5 longitudinal veins

Heather Groundbug (BL 4 mm)

Groundbugs
Lygaeidae
[40 gen. | 91 spp.]
p. 175

BL 2–8 mm; FORM elongate, (including legs); ANTENNAE with swollen tips. FEMORA tips expanded and often darkened

Common Stiltbug (BL 5–7 mm)

Stiltbugs
Berytidae
[4 gen. | 9 spp.]
p. 184

Guide to families pp. 142–147

FAMILY **Aradidae** (Flatbugs) — 2 GEN. | 7 spp. | 2 ILL.

Small, rather flattened, winged bugs found under bark, most likely otherwise to be seen on nearby posts and fences. **Antennae 4-segmented; tarsi 2-segmented; ocelli absent**. Only the common species are illustrated; others are rarer and difficult to find. *Aradus* and *Aneurus* species are easily distinguished – the wing membrane is membranous, except at the tip in *Aneurus*.

FOREWINGS developed, all except extreme base membranous	***Aneurus***
FOREWINGS membrane not reaching beyond tip of scutellum	***Aradus***

Aneurus — 2 spp. | 1 ILL.

Common Barkbug
Aneurus laevis

Common

J F M A M J J A S O N D

BL 4–5 mm. **ID** Broad, flattened. Wings transparent; hindwings greatly reduced (unable to fly), but forewings developed. Dark reddish-brown, with lighter patches. **Cyc** At least one generation a year, reaching adult stage in 2–4 months. **Hab** Woodland under bark of fallen trees **Fp** Fungi; on mycelia.**SS Confused Barkbug** *Aneurus avenius* [N/I] (**BL** 5 mm), in which the ♀ has flattened genital segments at tip of abdomen (bilobed in Common Barkbug).

Aradus — 5 spp. | 1 ILL.

Pale-shouldered Flatbug
Aradus depressus

Common

J F M A M J J A S O N D

BL 5–6 mm. **ID** Broad, flattened. Forewing membrane not reaching beyond tip of scutellum. Brown with whitish, reddish and black markings. Front corners of pronotum pale. **Cyc** Eggs are laid throughout the year, but mainly in summer. **Hab** Woodland on stumps or under bark. **Fp** Fungi, including *Polyporus*; on mycelia and fruiting bodies. **SS** Other rarer *Aradus* species, which have uniformly coloured pronotum.

SHIELDBUGS (known as stinkbugs in some countries, *e.g.* the USA) are the most popular British bugs, comprising five families. These are typically robust insects with a shield-like shape. They are able to produce a pungent liquid from glands in the thorax, located between the first two pairs of legs. The scutellum is often triangular. **Antennae 5-segmented; tarsi 2- or 3-segmented; ocelli present.** In most species, adults are more likely to be seen in spring or late summer to early autumn before they overwinter. There may be a gap when adults are scarce before nymphs mature.

FAMILY Acanthosomatidae (Keeled shieldbugs) 4 GEN. | 5 spp. (1†)

Includes some of the most attractive of the larger shieldbugs in Britain and Ireland. This family is easily distinguished from relatives by having **2-segmented tarsi**. Species can be separated as follows:

ANTENNAE S1 short

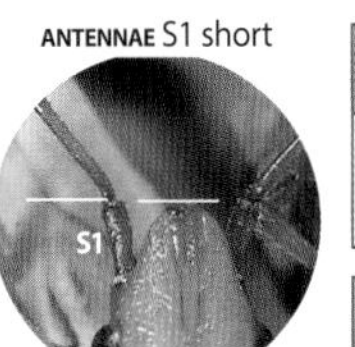

S1

ANTENNAE S1 long

ANTENNAE S1 short (not reaching front of head)		
HEAD + PRONOTUM with variously coloured punctuation	*Cyphostethus* [Juniper Shieldbug]	*p.149*
ANTENNAE S1 longer (exceeding front of head)		
HEAD + PRONOTUM **with black punctuation**		
CONNEXIVUM black-and-white; SCUTELLUM with black patch	*Elasmucha* [Parent Shieldbug]	*p.150*
CONNEXIVUM plain; SCUTELLUM lacking black		
FORM large; BL 12–16 mm; PRONOTUM with pointed 'shoulders'; SCUTELLUM green	*Acanthosoma* [Hawthorn Shieldbug]	*p.150*
FORM smaller; BL 8–12 mm; PRONOTUM lacking pointed 'shoulders'; SCUTELLUM green with some brown	*Elasmothethus* [Birch Shieldbug]	*p.150*

KEELED SHIELDBUGS – ANTENNAE S1 < TIP OF HEAD

Cyphostethus 1 sp.

Juniper Shieldbug
Cyphostethus tristriatus

BL 8–11 mm. **ID** Pronotum and forewings with characteristic curved pinkish-red markings on a green background. **Hab** Woodland, gardens and parks. **Fp** Juniper and Lawson's Cypress, on berries and cones. **Dist** Restricted to areas where foodplants occur; has spread widely since *ca.* 1970. **SS** Birch Shieldbug (*p. 150*), which lacks the colourful pinkish-red markings.

Guide to Acanthosomatidae p. 149

KEELED SHIELDBUGS – ANTENNAE S1 >TIP OF HEAD

Elasmucha — 2 spp. (1†) | 1 ILL.

NB **Bilberry Shieldbug** *Elasmucha ferrugata* [N/I] (extinct, last recorded 1950) has pointed pronotum edges.

Parent Shieldbug
Elasmucha grisea

BL 7–9 mm. **ID** Ranges from reddish, through grey to brown; scutellum and in some individuals forewings with black patch. Connexivum black-and-white, with dark punctures. ♂ noticeably smaller than ♀. **Hab** Woodland, gardens and parks. **Fp** Alder and birches. **Beh** ♀s tend the eggs and young nymphs. **SS** None; darker than other species in the family.

Common

CONNEXIVUM black-and-white; SCUTELLUM black patch

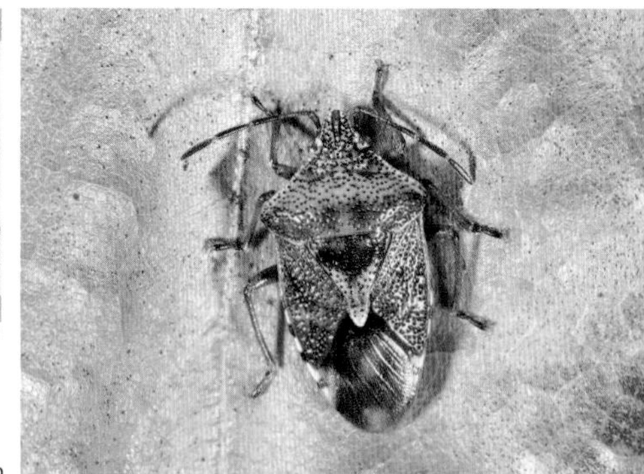

Elasmostethus — 1 sp.

Birch Shieldbug *Elasmostethus interstinctus*

BL 8–12 mm. **ID** Green; pronotum and forewings with reddish-brown markings. Scutellum green with some brown areas. Dark punctures. **Hab** Woodland, gardens and parks. Adults often overwinter in moss or under bark. **Fp** Birches, Aspen and Hazel. **SS Hawthorn Shieldbug**, **Juniper Shieldbug** (*p. 149*).

Common

CONNEXIVUM plain; PRONOTUM unpointed 'shoulders'; SCUTELLUM green + some brown; smaller than Hawthorn Shieldbug

Acanthosoma — 1 sp.

Hawthorn Shieldbug
Acanthosoma haemorrhoidale

BL 12–16 mm. **ID** Bright green with reddish-brown markings; pronotum with pointed 'shoulders'; scutellum all-green. Numerous dark punctures. **Hab** Woodland, gardens and parks. **Fp** Hawthorn berries, many other trees and shrubs. **SS Birch Shieldbug**.

Common

CONNEXIVUM plain; PRONOTUM pointed 'shoulders'; SCUTELLUM all-green; larger than Birch Shieldbug

FAMILY **Cydnidae** (Burrowing shieldbugs) — 6 GEN. | 9 spp.

Black with white markings or borders in some species; one with bluish body colour. **Tibiae with strong spines; tarsi 3-segmented**. As the name implies, well known for burrowing in sand or soil. Genera can be separated as follows:

Geotomus
HEAD + PRONOTUM margins with hairs

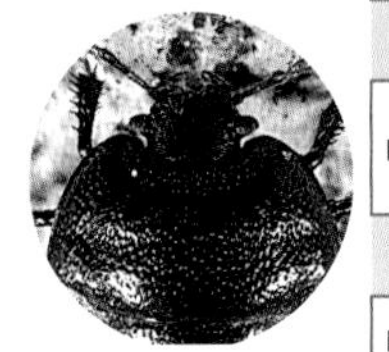

HEAD + PRONOTUM not hairy

HEAD + PRONOTUM margins with several long hairs	***Geotomus*** [2 species]	*p. 151*
HEAD + PRONOTUM margins not hairy		
FORM bluish-black		
BL 6–7 mm; FOREWINGS narrow white margin	***Canthophorus*** [Down Shieldbug]	*p. 152*
FORM black; BL >7 mm		
BL ≥ 7 mm		
BL 7–9 mm; FOREWINGS plain	***Sehirus*** [Forget-me-not Shieldbug]	*p. 152*
BL usually ≤ 7 mm		
FOREWINGS with white central spot and white margin; PRONOTUM white on margin	***Adomerus*** [Cow-wheat Shieldbug]	*p. 152*
FOREWINGS with white margin/border	***Legnotus*** [2 species]	*p. 153*
FOREWINGS with whitish patches; PRONOTUM white on margin	***Tritomegas*** [2 species]	*p. 153*

BURROWING SHIELDBUGS – PRONOTUM MARGIN WITH HAIRS

Geotomus — 2 spp.

Cornish Shieldbug — CR — Rare
Geotomus punctulatus

BL approx. 4 mm. **ID** Margins of head, pronotum and forewings with several long hairs. Black except hind margin of pronotum and forewings with hint of reddish. **Hab** Sparsely vegetated sand dunes. **Fp** Ladies bedstraw. **Dist** Restricted to Sennen Cove in Cornwall. **SS** Petite Shieldbug.

Petite Shieldbug *Geotomus petiti*

BL 3–4 mm. **ID** Inseparable from Cornish Shieldbug in the field; identification requires examination of dissected ♂ genitalia. **Dist** First recorded in 2019 from Dungeness, Kent.

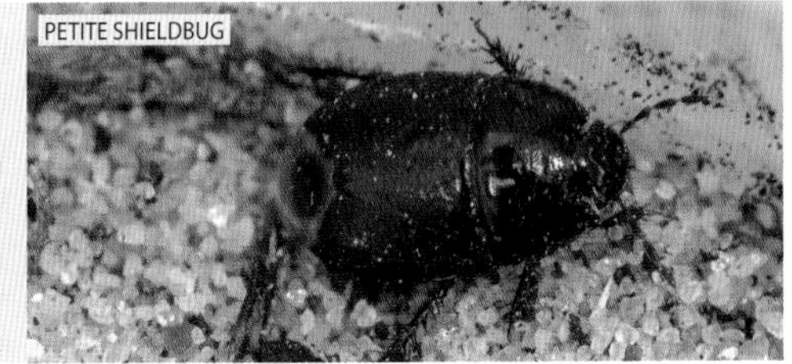
PETITE SHIELDBUG

Guide to Cydnidae p. 151

BURROWING SHIELDBUGS – PRONOTUM MARGIN WITHOUT HAIRS

Canthophorus 1 sp.

Down Shieldbug
Canthophorus impressus

BL 6–7 mm. **ID** Dark metallic bluish-black; pronotum and forewings with white margin. Connexivum with black and white bands. The nymphs are red. **Hab** Chalk grassland; best searched for around the foodplant, although possible on tracks or in flight. **Fp** Bastard-toadflax. **SS** None.

Nationally Scarce

PRONOTUM + FOREWING white margin

Sehirus 1 sp.

Forget-me-not Shieldbug
Sehirus luctuosus

BL 7–9 mm. **ID** Large, plain, black. **Hab** Grassland, particularly chalk, also gardens and woodland edges. Sometimes seen on the ground by gardeners when tidying up areas, although most likely to be encountered when the bugs are particularly active in May. **Fp** Forget-me-not seeds. **SS** None.

Common

PRONOTUM + FOREWING black margin

Adomerus 1 sp.

Cow-wheat Shieldbug
Adomerus biguttatus

BL 5–6 mm. **ID** Black; pronotum and forewings with white margin; forewings also with distinctive white central spots. **Hab** Woodland rides and clearings, favouring coppiced sites; overwinters in moss. **Fp** Common Cow-wheat. **SS** None.

Nationally Scarce

PRONOTUM + FOREWING white margin; white spots on forewing

Legnotus 2 spp.

Black, forewings without white patches

FOREWINGS white margin almost entire	Bordered
FOREWINGS whitish border on basal ⅓	Heath

Bordered Shieldbug
Legnotus limbosus

J F M A M J J A S O N D

BL 4–5 mm. **ID** Black; forewings with narrow white margin almost reaching to brown wing membrane. **Hab** Grassland, including gardens and waste ground. **Fp** Cleavers and bedstraws. **SS** Heath Shieldbug.

Heath Shieldbug
Legnotus picipes

J F M A M J J A S O N D

BL 3–4 mm. **ID** Black; forewings with narrow whitish margin on front ⅓ only; wing membrane brown. **Hab** Coastal sand dunes, heathland and Breckland. **Fp** Bedstraws. **SS** Bordered Shieldbug.

Tritomegas 2 spp.

Black, forewings with white patches; connexivum with black and white bands

PRONOTUM white margin front ½ only	Pied
PRONOTUM white margin extensive	Rambur's

Pied Shieldbug
Tritomegas bicolor

J F M A M J J A S O N D

BL 6–8 mm. **ID** Pronotum margin white on front ½ only; wing membrane pale. **Hab** Woodland rides, hedgerows; often noticed high up on vegetation when the bugs are active. **Fp** White Dead-nettle, occasionally Black Horehound. **SS** Rambur's Pied Shieldbug.

Rambur's Pied Shieldbug
Tritomegas sexmaculatus

J F M A M J J A S O N D

BL 6–8 mm. **ID** Pronotum margin white, almost reaching hind part of segment; wing membrane black. **Hab** Open areas. **Fp** Black Horehound. **Dist** First recorded at two sites in Kent in 2011 and has since spread in that county and been found at Purfleet, Essex. **SS** Pied Shieldbug.

Guide to families pp. 142–147

FAMILY Scutelleridae (Tortoise shieldbugs) 2 GEN. | 5 spp. (1†)

Scutellum reaches tip of abdomen. Tarsi 3-segmented. Genera/species can be separated as follows:

Non-hairy	FORM tortoise-like, larger; BL 9–11 mm; CONNEXIVUM banded	*Eurygaster* [3 species (1†)]
Hairy	FORM streaked appearance, smaller; BL 4–8 mm; CONNEXIVUM plain	*Odontoscelis* [2 species]

Odontoscelis 2 spp.

Greater-streaked Shieldbug
Odontoscelis fuliginosa VU

BL 6–8 mm. **ID** Rounded, brown; body with dark brown hairs; scutellum variably streaked. **Hab** Coastal sand dunes. **Fp** Stork's-bills. **Dist** Recently recorded only from Kent and Pembrokeshire. **SS** Lesser-streaked Shieldbug.

Rare

Lesser-streaked Shieldbug
Odontoscelis lineola

BL 4–6 mm. **ID** Small, rounded; pronotum and scutellum with bands of distinctive silver hairs; scutellum variably streaked. **Hab** Coastal dunes and some sandy inland heathland and Breckland areas; most likely on sandy tracks or on sand beneath the foodplant on hot days. **Fp** Stork's-bills. **SS** Greater-streaked Shieldbug.

Nationally Scarce

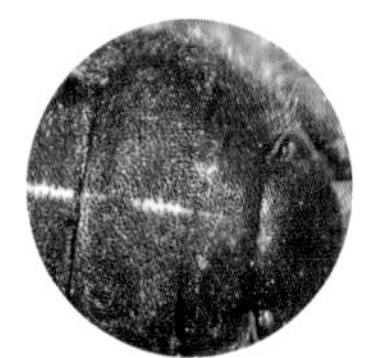

Greater-streaked Shieldbug
larger; BODY dense, dark brown hairs

Lesser-streaked Shieldbug
smaller; PRONOTUM + SCUTELLUM silver hairs

Eurygaster 3 spp. (1†) | 2 ILL.

NB **Austrian Tortoise Shieldbug** *Eurygaster austriaca* [N/I] (extinct) has not been recorded in GB since 1885.

Tortoise Shieldbug
Eurygaster testudinaria

Common

BL 9–11 mm. **ID** Front of head with slight central depression. Brown, often mottled darker brown, some with white markings. **Hab** Grassland, including damp areas. **Fp** Tall grasses, sedges and rushes; often visible on the tips. **SS** Scarce Tortoise Shieldbug, with which it is easily confused [see comparison *below*].

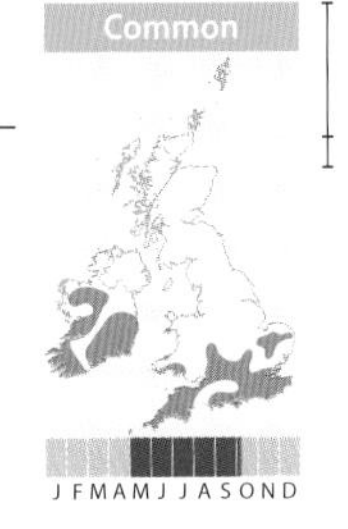

Scarce Tortoise Shieldbug
Eurygaster maura

Nationally Scarce

BL 9–11 mm. **ID** Front of head rounded. Brown, often mottled darker brown, some with white markings. **Hab** Grassland, often calcareous sites. **Fp** Grasses. **SS** Tortoise Shieldbug, with which it is easily confused [see comparison *below*].

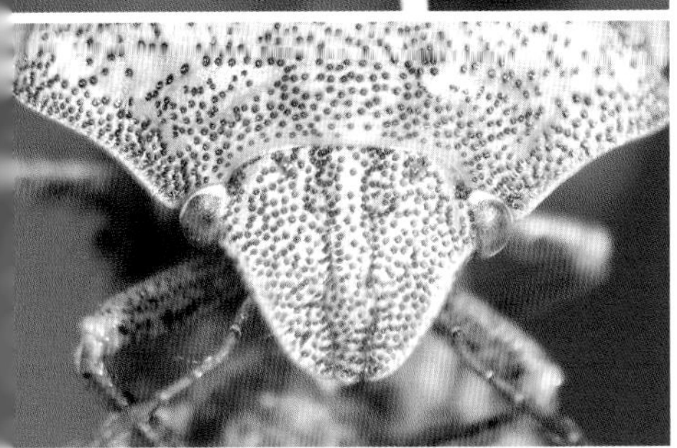

FRONT OF HEAD with central depression (hind part of pronotum slightly more protruding)

FRONT OF HEAD rounded

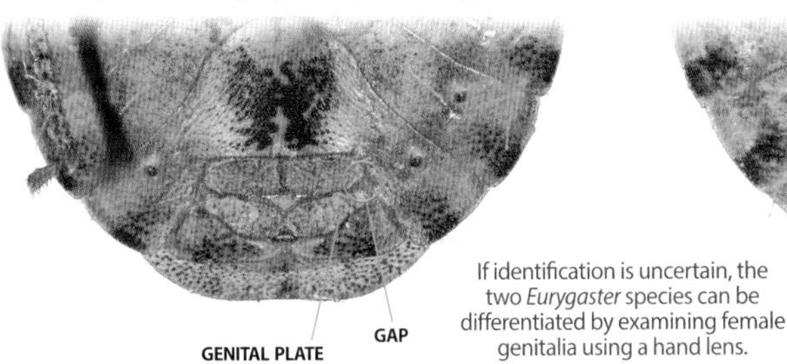

If identification is uncertain, the two *Eurygaster* species can be differentiated by examining female genitalia using a hand lens.

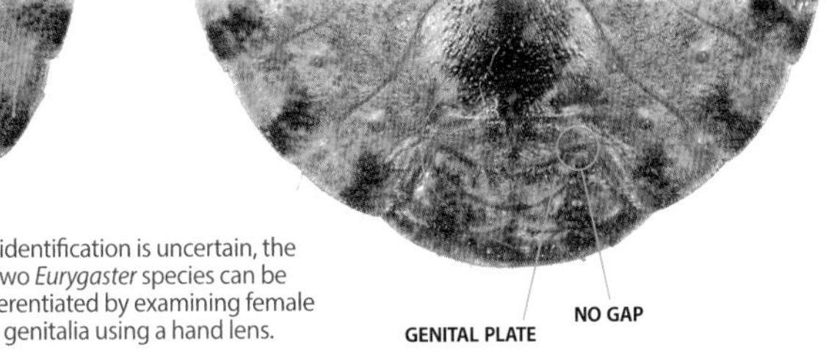

Guide to families *pp. 142–14*

FAMILY **Pentatomidae** (Typical shieldbugs)

23 GEN. (2†) | 27 spp. (2†) | 25 ILL.

Tarsi 2-segmented; lacking strong spines. As the name implies, features some of the more typical shieldbugs, including several species predatory on other insects. Genera can be separated as follows:

NB **Scarce Juniper Shieldbug** *Chlorochroa juniperina* [N/I] (extinct) has not been recorded in GB since 1925. **Orange-striped Shieldbug** *Jalla dumosa* [N/I] (extinct) was last recorded in the late 19th century.

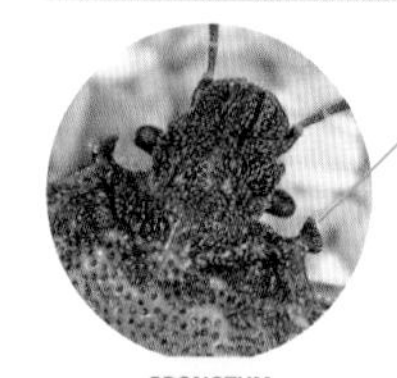

PRONOTUM hooked projection

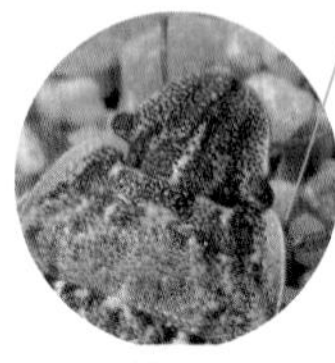

PRONOTUM sides expanded

PRONOTUM sides not expanded

SCUTELLUM covering most of the abdomen; PRONOTUM with a hooked projection on each side		***Podops*** [Knobbed Shieldbug]	*p. 164*
SCUTELLUM triangular, not covering abdomen; PRONOTUM lacking hooked projection on sides			
PRONOTUM sides expanded; FORM 'flattened' appearance			
BL 5–9 mm; FOREWINGS narrow white margin **Large Sandrunner** *Sciocoris homalonotus* (BL 6–9 mm) was first found in Chatham, Kent in 2016, then in Surrey 2020; EYES more stalk-like. **Purfleet Sandrunner** *Sciocoris sideritidis* (BL 5–6 mm) was first found at Purfleet, Essex in 2019; HEAD more elongated.		***Sciocoris*** [3 species]	*p. 163*
PRONOTUM sides not expanded; FORM not 'flattened' in appearance			1/2
HEAD pointed at the front and curved downwards (viewed laterally); EYES small			
HEAD short; BL 4–5 mm		***Neottiglossa*** [Small Grass Shieldbug]	*p. 164*
HEAD elongated, front narrowed	BODY not elongated; BL 8–10 mm	***Aelia*** [Bishop's Mitre Shieldbug]	*p. 164*
	BODY elongated; BL 10–12 mm First found in moth traps in Dorset, Devon and Hampshire in Dec. 2015	***Mecidea*** [Elongate Shieldbug]	*[inset below]*

Nymphs of Blue Shieldbug piercing prey (beetle larva).

Elongate Shieldbug
Mecidea lindbergi

FOREWINGS + PRONOTUM brightly spotted

PRONOTUM sides not expanded; FORM not 'flattened' in appearance				2/2
HEAD not pointed at the front (viewed laterally); EYES larger				
FOREWINGS + PRONOTUM brightly spotted black-and-red or black-and-white			*Eurydema* [3 species]	p. 158
FOREWINGS + PRONOTUM usually plain or mottled	BODY metallic blue		*Zicrona* [Blue Shieldbug]	p. 159
	smaller species; BL ≤ 7 mm; BODY brown or grey	SCUTELLUM metallic copper	*Stagonomus* [Woundwort Shieldbug]	p. 159
		SCUTELLUM white spot at basal angles	*Eysarcoris* [New Forest Shieldbug]	p. 159
	LONGER SPECIES; BL typically ≥ 10 mm ▶			

LONGER SPECIES ▶

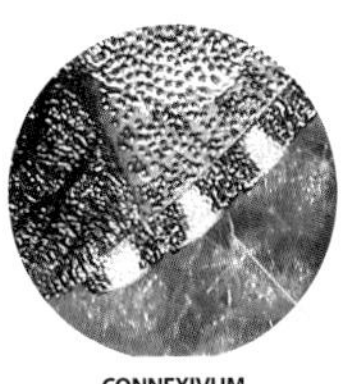

CONNEXIVUM with black spots

CONNEXIVUM with black spots			
HEAD + PRONOTUM margin hairy			
FORM brownish with purple patches; ANTENNAE banded black-and-white		*Dolycoris* [Hairy Shieldbug]	p. 160
HEAD + PRONOTUM margin not hairy; FORM brown or green			
CONNEXIVUM with a narrow black and yellowish, orange or white band; BODY numerous dark metallic punctures; TIBIA with orange bands; PRONOTUM often with pale line		*Rhacognathus* [Heather Shieldbug]	p. 161
CONNEXIVUM with a broad band, black and white or yellowish	BODY variable in colour; PRONOTUM striped	*Carpocoris* [Black-shouldered Shieldbug]	p. 161
	PRONOTUM rounded at sides; SCUTELLUM white 'shoulders' and tip	*Dyroderes* [White-shouldered Shieldbug]	p. 161
	PRONOTUM 'shoulders' hooked; SCUTELLUM orange to cream spot at tip; LEGS often reddish	*Pentatoma* [Red-legged Shieldbug]	p. 160
	PRONOTUM with narrow white border; SCUTELLUM pale tip; ANTENNAE mostly reddish	*Peribalus* [Vernal Shieldbug]	p. 160
	BODY heavily mottled; ANTENNAE banded black-and-white	*Rhaphigaster* [Mottled Shieldbug]	p. 160
	SCUTELLUM plain; ANTENNAE S4 partly orange	*Troilus* [Bronze Shieldbug]	p. 161
CONNEXIVUM without black spots			
MEMBRANE dark; BODY green, may change to brown in winter		*Palomena* [Green Shieldbug]	p. 162
PRONOTUM with yellow spots; MEMBRANE pale		*Nezara* [Southern Green Shieldbug]	p. 163
PRONOTUM + FOREWINGS late-season adults purplish-red		*Piezodorus* [Gorse Shieldbug]	p. 162
PRONOTUM 'shoulders' spiked		*Picromerus* [Spiked Shieldbug]	p. 163

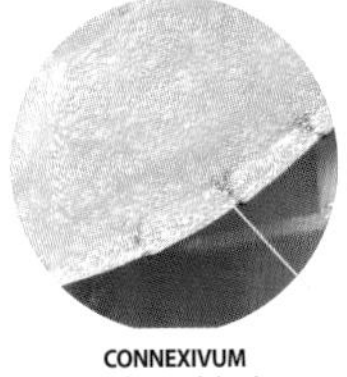

CONNEXIVUM without black spots

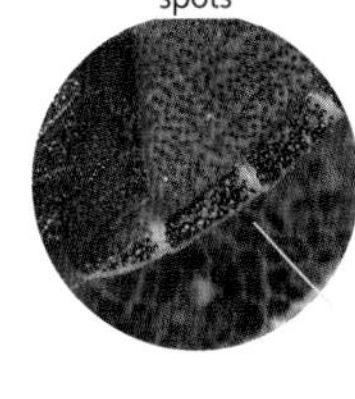

Guide to Pentatomidae pp. 156–15

TYPICAL SHIELDBUGS – Scutellum not covering abdomen | head not pointed

Eurydema	3 spp.
PRONOTUM with two large black blotches	Crucifer Shieldbug
PRONOTUM with several black spots; CORIUM (FOREWING) partly dark	Ornate Shieldbug
PRONOTUM with several black spots; CORIUM (FOREWING) plain	Scarlet Shieldbug

Crucifer Shieldbug
Eurydema oleracea

BL 6–7 mm. **ID** Bluish-black with red, white or yellow spots and markings that change to orange over winter. **Hab** Open areas. **Fp** Brassicaceae. **SS** Superficially similar to other *Eurydema* species [see table *above*].

Common

Ornate Shieldbug
Eurydema ornata

BL 7–9 mm. **ID** Red with black markings. Edge of forewings partly dark (black central spot at least). Orange and whitish colour forms also occur. **Hab** Open, sheltered coastal sites. **Fp** Brassicaceae. **Dist** First recorded in Dorset in 1997 and since found at a few sites from Dorset to Sussex; recently recorded in Surrey and the Isle of Wight, where most likely to be found. **SS** Scarlet Shieldbug.

Local (recent colonist)

Scarlet Shieldbug
Eurydema dominulus

EN

BL 5–8 mm. **ID** Red with black markings. Edge of forewings red. **Hab** Woodland rides. **Fp** Brassicaceae. **SS** Ornate Shieldbug.

Rare

| body length typically ≤ 10 mm

Zicrona 1 sp.

Blue Shieldbug
Zicrona caerulea

BL 6–7 mm. **ID** Metallic bluish-green. Nymphs black with vivid red abdomen with black markings (see *p. 156*). **Hab** Open areas including heathland, grassland and woodland rides. **Fp** Predatory mainly on larvae of leaf beetles *Altica* species [N/I] (**BL** 3–5 mm) (also metallic blue), also Heather Beetle (*p. 321*). **SS** None.

Stagonomus 1 sp.

Woundwort Shieldbug
Stagonomus venustissimus

BL 5–7 mm. **ID** Whitish-grey, with bronze punctures. Head, front of pronotum and part of scutellum metallic copper. **Hab** Open areas, including gardens. **Fp** Hedge Woundwort. **SS** New Forest Shieldbug.

Eysarcoris 1 sp.

New Forest Shieldbug EN
Eysarcoris aeneus

BL 4–6 mm. **ID** Brown or grey; scutellum with white spot at basal angles. **Hab** Bogs; also drier grassland. **Fp** possibly on heathers, Slender St. John's-wort. **Dist** New Forest, Hampshire and Isle of Wight. Rediscovered in Cornwall (Sennen Cove) in 2016. **SS** Woundwort Shieldbug, Small Grass Shieldbug (*p. 164*).

TYPICAL SHIELDBUGS – Scutellum not covering abdomen | head not pointed

Dolycoris 1 sp.

Hairy Shieldbug
Dolycoris baccarum

Common

BL 10–13 mm. **ID** Margin of head and pronotum noticeably hairy. Purplish-brown; antennae and connexivum black-and-white. **Hab** Grassland, gardens, parks and other open areas. **Fp** Fruits and flowers of various trees and shrubs. **SS** None.

Rhaphigaster 1 sp.

Mottled Shieldbug
Rhaphigaster nebulosa

Rare (not Red Listed) (recent colonist)

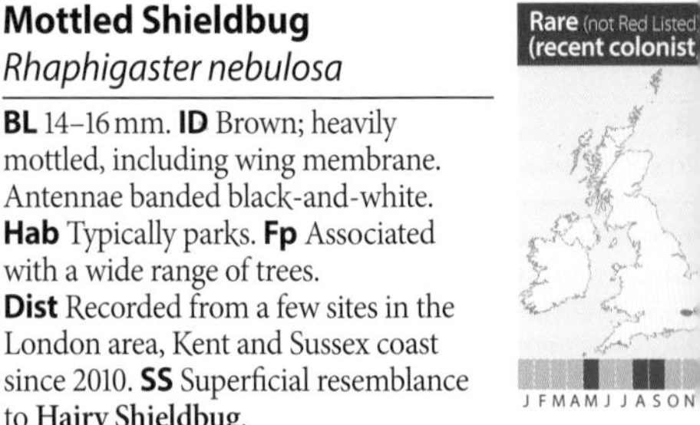

BL 14–16 mm. **ID** Brown; heavily mottled, including wing membrane. Antennae banded black-and-white. **Hab** Typically parks. **Fp** Associated with a wide range of trees. **Dist** Recorded from a few sites in the London area, Kent and Sussex coast since 2010. **SS** Superficial resemblance to **Hairy Shieldbug**.

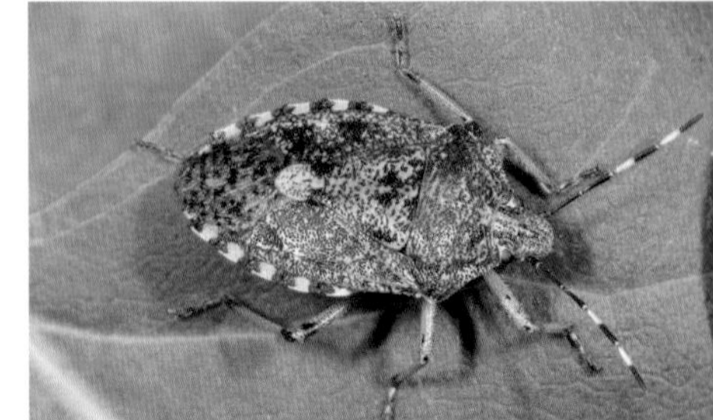

Peribalus 1 sp.

Vernal Shieldbug
Peribalus strictus

Rare

BL 9–11 mm. **ID** Yellowish-brown; pronotum rounded, with narrow white border; scutellum with pale tip. Antennae part-reddish, S4 and S5 with some black. Connexivum banded yellow or whitish and black. **Hab** Sheltered coastal sites near shrubs; possibly in woodland. **Fp** Many plants and trees. **Dist** Currently breeding around Bournemouth, Dorset and in Sussex. **SS Bronze Shieldbug**; superficial resemblance to **Hairy Shieldbug**.

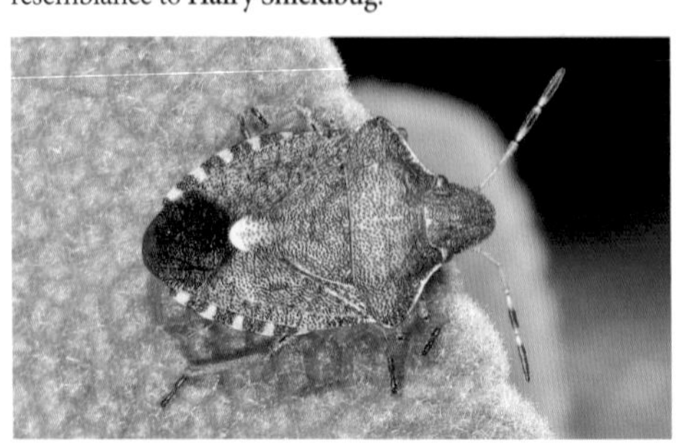

Pentatoma 1 sp.

Red-legged Shieldbug
Pentatoma rufipes

Common

BL 12–14 mm. **ID** Brown; pronotum hooked; scutellum with orange or cream spot at tip; legs often reddish. **Hab** Woodland, gardens, orchards and parks. **Fp** Oaks, Alder, fruit and other trees. Adults are also predatory on Lepidoptera larvae. **Beh** Nymphs (rarely adults) overwinter, maturing in about July. **SS Spiked Shieldbug** (*p. 163*), **Bronze Shieldbug**.

| body length typically ≥ 10 mm | connexivum with black spots

Carpocoris 1 sp.

Black-shouldered Shieldbug
Carpocoris purpureipennis

Rare vagrant

BL 11–13 mm. **ID** Ranges from purple to yellowish-brown. Pronotum striped; 'shoulders' black-tipped. **Hab** Open habitats. **Fp** Umbellifers and mulleins. **Dist** Scattered records from S England with no evidence of breeding, but may become established in future. **SS** None, although there are very similar species in continental Europe.

Dyroderes 1 sp.

White-shouldered Shieldbug
Dyroderes umbraculatus

Rare (not Red Listed) (recent colonist)

BL 7–9 mm. **ID** Pronotum rounded at sides; brown with white 'shoulders'. Scutellum with white spot at tip; connexivum bold black and white. **Hab** Open places. **Fp** Bedstraws. **Dist** Known from parts of London since 2013 and recorded from Hampshire in 2015. **SS** None.

Troilus 1 sp.

Bronze Shieldbug
Troilus luridus

Common

BL 10–12 mm. **ID** Brown with dark metallic punctures. Antenna S4 with orange or cream tip. **Hab** Woodland. **Fp** Predatory, mainly on butterfly, moth, sawfly and beetle larvae; sometimes feeds on plant juices. **SS** Heather Shieldbug, Red-legged Shieldbug and Vernal Shieldbug.

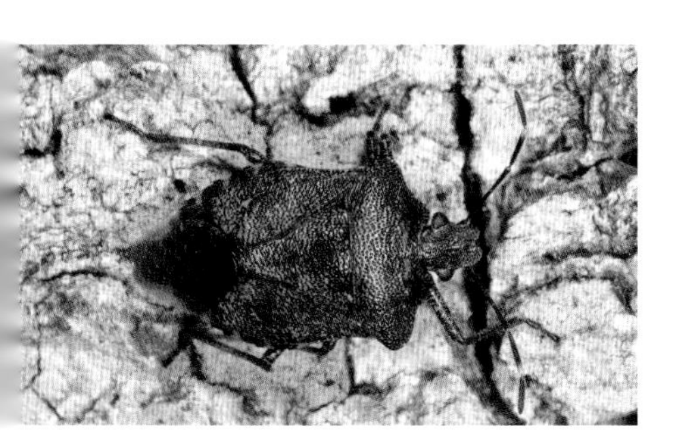

Rhacognathus 1 sp.

Heather Shieldbug
Rhacognathus punctatus

Local

BL 7–10 mm. **ID** Dark metallic bronze, often with pale central line on pronotum. All tibiae with orange or pale band. **Hab** Mainly wet heathland and moorlands. **Fp** Predatory on adults and larvae of beetles associated with heathers, mainly Heather Beetle (*p. 321*). **SS** Bronze Shieldbug.

Guide to Pentatomidae pp. 156–157

TYPICAL SHIELDBUGS – Scutellum not covering abdomen | head not pointed | body length typically ≥ 10 mm | connexivum without black spots

Piezodorus 1 sp.

Gorse Shieldbug
Piezodorus lituratus

Common

BL 10–13 mm. **ID** Yellowish-green; yellow margin to pronotum and adjoining forewing; rest of margin bluish. When adults mature later in the year (from July), the lower half of pronotum and forewings is reddish-brown. **Hab** Open places, including heathland and gardens, where well camouflaged. **Fp** Gorse or Broom seed pods. **SS** None.

Palomena 1 sp.

Green Shieldbug
Palomena prasina

Common

BL 12–14 mm. **ID** Green; wing membrane dark. Darkens and may become brown when overwintering. **Hab** Open places, often gardens and parks. **Fp** Various trees and shrubs, on berries; often seen on Bramble. **SS** Southern Green Shieldbug, which has a pale wing membrane.

Nezara 1 sp.

Southern Green Shieldbug
Nezara viridula

Local (recent colonist)

BL 11–15 mm. **ID** Front of scutellum with 3–5 whitish spots, otherwise green; wing membrane pale.
Hab Often gardens and allotments.
Fp Beans and tomato (where it can be a pest). **Dist** First reported breeding in the London area in 2003, although imported periodically into various parts of Britain. **SS** Green Shieldbug, which has a dark wing membrane.

Picromerus 1 sp.

Spiked Shieldbug
Picromerus bidens

Common

BL 10–14 mm. **ID** Brown with reddish legs. Distinctive spiked pronotum.
Hab Open places, including heathland, grassland and woodland. **Fp** Predatory on various larvae (butterflies, moths and beetles) but also suck sap.
SS Red-legged Shieldbug (*p. 160*).

TYPICAL SHIELDBUGS – Scutellum not covering abdomen | 'flattened' look

Sciocoris 3 spp. | 1 ILL.

Sandrunner
Sciocoris cursitans

Nationally Scarce

J F M A M J J A S O N D

BL 5–6 mm. **ID** Brown. Head rounded; connexivum broad, expanded.
Hab Mainly coastal sheltered sandy sites and chalk downlands; likely on tracks in hot weather. **Fp** Mouse-ear-hawkweed and cinquefoils, possibly stork's-bills. **SS** Other recently discovered *Sciocoris* species, which need to be examined under a microscope to separate with certainty: **Large Sandrunner** *Sciocoris homalonotus* [N/I] (**BL** 6–9 mm)usually much longer and with more stalk-like eyes; **Purfleet Sandrunner** *Sciocoris sideritidis* [N/I] (**BL** 5–6 mm) has a more elongated head.

TYPICAL SHIELDBUGS – Scutellum not covering abdomen | head pointed

Aelia 1 sp.

Bishop's Mitre Shieldbug
Aelia acuminata

BL 8–10 mm. **ID** Narrow, with pointed head and ridged pronotum; brown, with pale longitudinal stripes. **Hab** Grassland. **Fp** Seeds of various grasses. **SS** None.

Neottiglossa 1 sp.

Small Grass Shieldbug
Neottiglossa pusilla

BL 5–6 mm. **ID** Pale brown, small white spot at basal angles of scutellum. Whitish central line from head extending to scutellum. **Hab** Grassland, including damper places. **Fp** Various grasses. **SS** New Forest Shieldbug (*p. 159*).

TYPICAL SHIELDBUGS – Scutellum covering most of the abdomen

Podops 1 sp.

Knobbed Shieldbug
Podops inuncta

BL 5–6 mm. **ID** Large scutellum and projections at sides of pronotum; brown. **Hab** Grassland. **Fp** Grasses; sometimes common on sandy tracks. **SS** None.

FAMILY **Thyreocoridae** (Scarab shieldbugs) 1 sp.

The small British representative of this family has its scientific name derived from 'scarab' and the resemblance to a beetle can even confuse coleopterists! However, as with all bugs, it has a rostrum.

Thyreocoris 1 sp.

Scarab Shieldbug

Thyreocoris scarabaeoides

BL 3–4 mm. **ID** Rather rounded; black or bronze, slightly metallic. **Hab** Grassland, often chalk, on bare ground, but also coastal dunes; possible on tracks, on or under foodplants, or in moss or leaf-litter. **Fp** Hairy Violet and Field Pansy. **Dist** As well as the GB distribution shown, there are records from SE Ireland, where this species was last recorded in 1934. **SS** None.

FAMILY **Coreidae** (Leatherbugs) 10 GEN. | 11 spp.

Large, brown, fruit- or seed-feeders, also known as squash bugs (**the scent gland opening is in the thorax, viewed laterally between front and mid pair of legs**). Leatherbugs resemble shieldbugs in stature and size but most species are more elongated (although the abdomen is broadened in some); however, the **antennae are 4-segmented**, **tarsi 3-segmented** and **ocelli are present**. Adults are more likely to be seen in spring, or late summer to early autumn before they overwinter.

Guide to Coreidae (Leatherbugs)

LARGE – BL 15–20 mm		
FORM large and elongate; FOREWINGS centre with white dusting or zigzag mark; HIND TIBIAE with leaf-like expansion	*Leptoglossus* [Western Conifer Seed Bug]	p. 166
MEDIUM-SIZED – BL 9–15 mm		
ABDOMEN usually noticeably broadened laterally; ANTENNAE long		
ABDOMEN rounded; ANTENNAE S3 one colour; BL 13–15 mm	*Coreus* [Dock Bug]	p. 167
ABDOMEN tip squarish; ANTENNAE S3 end dark; BL 10–12 mm	*Enoplops* [Boat Bug]	p. 167
ABDOMEN diamond-shaped; ANTENNAE S3 one colour; BL 10–11 mm	*Syromastus* [Rhombic Leatherbug]	p. 167
ABDOMEN less broadened laterally; PRONOTUM 'shoulders' pointed	*Gonocerus* [Box Bug]	p. 168
HEAD dark eye stripe and long nose; THORAX, PRONOTUM (margin) + FOREWINGS with numerous brown tubercles; base of forewing margin pale	*Ceraleptus* [Slender-horned Leatherbug]	p. 168
SMALL – BL UP TO 9 mm		
ABDOMEN elongated or less broadened laterally; FORM smaller; ANTENNAE shorter		
PRONOTUM margin two-thirds whitish	*Bathysolen* [Cryptic Leatherbug]	p. 168
PRONOTUM margin with white tubercles; BODY hairy	*Coriomeris* [Denticulate Leatherbug]	p. 168
PRONOTUM with pale margin; SCUTELLUM heart-shaped, with two black marks	*Spathocera* [Dalman's Leatherbug]	p. 169
PRONOTUM with spines	*Arenocoris* [2 species]	p. 169

LEATHERBUGS – Large (BL 15–20 mm)

Leptoglossus — 1 sp.

Western Conifer Seed Bug
Leptoglossus occidentalis

BL 15–20 mm. **ID** Large, elongate, reddish-brown; forewings often with whitish dusting or zigzag pattern. Hind tibiae with leaf-like expansion. Abdomen yellow; connexivum banded black and yellow. **Hab** Coniferous woodland; often attracted to light. **Fp** Seeds and unripe cones of Scots Pine and other conifers. **Dist** First recorded in Britain in 2007. Numbers bolstered by migrants late in the year. **SS** None.

LEATHERBUGS – Medium-sized (BL 9–15 mm) 1/2

Coreus 1 sp.

Dock Bug *Coreus marginatus*

Common

BL 13–15 mm. **ID** Abdomen broad; **rounded**. Rather mottled; brown. **Hab** Open dry and damp sites, including gardens and hedgerows. **Fp** Leaves of docks and related plants. Larger nymphs and adults feed on the seeds and are possible in numbers on vegetation, such as Bramble. Adults readily feed on blackberries in September. **SS Boat Bug**, **Rhombic Leatherbug** [see table *opposite*]; superficially like the much narrower **Box Bug** (*p. 168*).

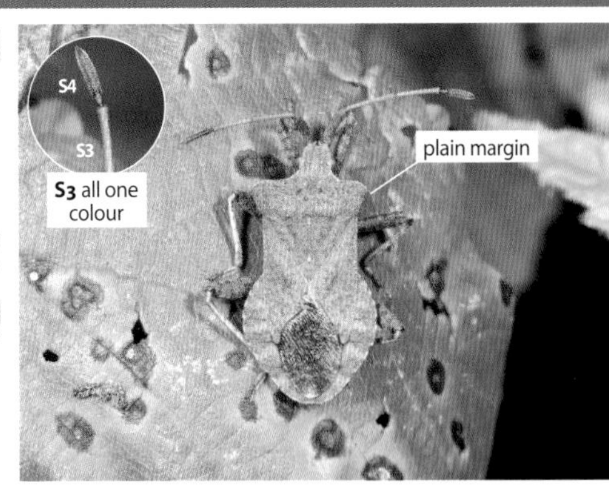

Syromastus 1 sp.

Rhombic Leatherbug *Syromastus rhombeus*

Local

BL 10–11 mm. **ID** Abdomen broadened, diamond-shaped. Pronotum and base of forewing with narrow whitish border. Generally brown, mottled. **Hab** Open places, including grassland and sand dunes; visits flowers and sometimes seen in flight. **Fp** Spurreys and sandworts. **SS Dock Bug**, **Boat Bug** [see table *opposite*].

Enoplops 1 sp.

Boat Bug *Enoplops scapha*

Nationally Scarce

BL 10–12 mm. **ID** Abdomen tip squarish, Pronotum and base of forewing with whitish margin, otherwise dark grey. Connexivum with whitish markings. Tip of antenna S3 dark. **Hab** Mainly sparsely vegetated coastal cliffs and sand dunes; likely on foodplants, nearby vegetation, walls or cliffs, as well as on the ground. **Fp** Composites including Scentless Mayweed. **SS Dock Bug**, **Rhombic Leatherbug** [see table *opposite*].

Guide to Coreidae *p. 166*

LEATHERBUGS – Medium-sized (BL 9–15 mm) 2/2

Gonocerus 1 sp.

Box Bug
Gonocerus acuteangulatus

Common

BL 11–14 mm. **ID** Brown. Pronotum rather slender with **pointed sides**. **Hab** Broad range, *e.g.* woodland edges, hedgerows, gardens, parks. **Fp** Hawthorn, Bramble, Buckthorn berries. **Dist** Has expanded its range of foodplants beyond Box and spread from Surrey since 1990. **SS** Superficially like the much broader **Dock Bug** (*p. 167*).

Ceraleptus 1 sp.

Slender-horned Leatherbug
Ceraleptus lividus

Nationally Scarce

BL 9–11 mm. **ID** Brown, with dark eye-stripe and long 'nose'. Thorax, forewings and margin of pronotum with many brown tubercles; base of forewing margin pale. **Hab** Grassland and sandy sites. **Beh** Likely on vegetation near foodplants or flowers, as well as on the ground. **Fp** Clovers and trefoils. **SS** None.

LEATHERBUGS – Small (BL up to 9 mm)

Bathysolen 1 sp.

Cryptic Leatherbug
Bathysolen nubilus

Nationally Scarce

BL 5–7 mm. **ID** Brown, stout. Pronotum lacks spines and ⅔ of margin whitish. **Hab** Grassland. **Beh** Found on the ground. **Fp** Black Medick. **SS** General resemblance to other small leatherbugs but told by details of pronotum.

Coriomeris 1 sp.

Denticulate Leatherbug
Coriomeris denticulatus

Common

BL 7–9 mm. **ID** Brown; whole body hairy. Pronotum margin with white tubercles. **Hab** Open grassland. **Beh** Likely to be seen on vegetation, possibly in flight. **Fp** Black Medick and other legumes. **SS** None.

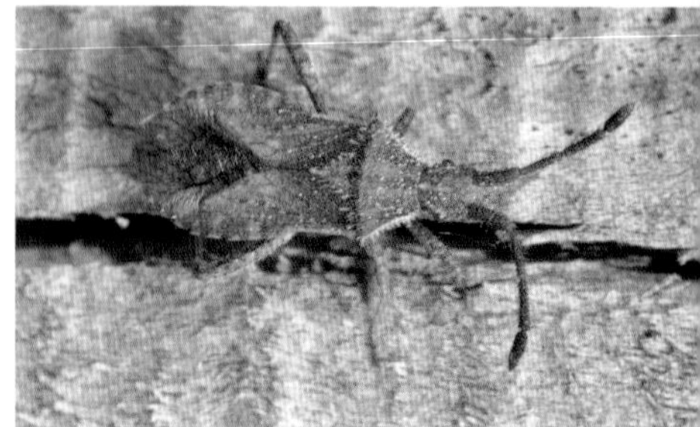

Spathocera 1 sp.

Dalman's Leatherbug
Spathocera dalmanii

Nationally Scarce

BL 5–7 mm. **ID** Brown. Pronotum with pale margin. Scutellum heart-shaped, with two black marks. Antennae held in a characteristic 'folded-back' pose. **Hab** Sandy heathland, acid grassland and Breckland; often selects foodplants with moss or grass nearby; well camouflaged on the ground. **Fp** Sheep's Sorrel seeds. **SS** General resemblance to other small leatherbugs but markings distinctive.

Arenocoris 2 spp.

PRONOTUM centre with pale spines forming inverted 'V'; **ANTENNAE** S3 not thickened towards tip	Fallén's Leatherbug
PRONOTUM lacks backward-pointing rows of pale spines; **ANTENNAE** S3 thickened towards tip	Breckland Leatherbug

Fallén's Leatherbug
Arenocoris fallenii

Nationally Scarce

BL 6–7 mm. **ID** Centre of pronotum with pale spines forming inverted 'V'; antenna S3 not thickened towards tip; brown. **Hab** Mainly coastal sandy sites, including dunes; likely on tracks in hot weather or under the foodplant. **Fp** Common Stork's-bill. **SS** **Breckland Leatherbug**, which may be present at the same site(s).

Breckland Leatherbug
Arenocoris waltlii

CR

Rare

BL 7–8 mm. **ID** Pronotum spiny; antenna S3 thickened towards tip; overall body dark brown. **Hab** Inland sandy sites with sparse vegetation; may be found under the foodplant. **Fp** Common Stork's-bill. **Dist** Only known from Breckland (Norfolk and Suffolk). Thought to be extinct until rediscovered in Suffolk in 2011; also recorded in Cambridgeshire in 2015. **SS** **Fallén's Leatherbug**, which may be present under same plant.

Guide to families pp. 142–147

FAMILY **Rhopalidae** (Rhopalid bugs) **7 GEN. | 11 spp. | 5 GEN., 9 SPP. ILL.**

Includes both drab and colourful species, and one that is remarkably elongated. Resemble the less robust-looking leatherbugs but have membranous and at least partly transparent forewings. Seed or fruit-feeders, often overwintering as adults. **Antennae 4-segmented, tarsi 3-segmented, ocelli present.**

Stictopleurus 2 spp.

PRONOTUM with two large black blotches	Knapweed Rhopalid
PRONOTUM with several black spots; CORIUM (FOREWING) plain	Banded Rhopalid

Knapweed Rhopalid *Stictopleurus abutilon*

BL 7–8 mm. **ID** Greenish-brown. Pronotum with line just behind the front margin that ends in a **full circle** near the margin; abdomen yellowish, but varies in colour between individuals; connexivum banded orange or greenish and black. **Hab** Open areas, brownfield sites. **Fp** Asteraceae, including Yarrow and knapweeds. **Dist** Has spread since 1992, after recolonizing Britain. **SS** Banded Rhopalid.

Banded Rhopalid *Stictopleurus punctatonervosus*

BL 7–8 mm. **ID** Brown. Pronotum with line just behind the front margin that ends in a **half circle** near the margin; connexivum banded orange and black. **Hab** Open areas, brownfield sites. **Fp** Creeping Thistle, Fleabane and others. **Dist** Has spread since 1997, after recolonizing GB. **SS** Knapweed Rhopalid.

PRONOTUM markings can be hard to discern

Rhopalus			4 spp.
FOREWING 3 visible cells			Marsh Rhopalid
FOREWING 6 visible cells	CONNEXIVUM banded black-and-white; SCUTELLUM with whitish tip		Hypericum Rhopalid
	CONNEXIVUM reddish or pale brown, with or without black spots	CONNEXIVUM edge with black spots; FOREWING VEINS spotted black; PRONOTUM often yellowish-brown	Transparent Rhopalid
		CONNEXIVUM edge plain; FOREWING VEINS not spotted black; PRONOTUM reddish-brown	Red Rhopalid

Hypericum Rhopalid
Rhopalus subrufus

BL 7 mm. **ID** Reddish-brown or brown. Connexivum banded black and pale brown. Scutellum with whitish, bifid tip (*i.e.* ends in two points). Forewings part reddish with black spots. **Hab** Woodland clearings and scrub. **Fp** St John's-worts. **SS** Other *Rhopalus* species [see table *above*].

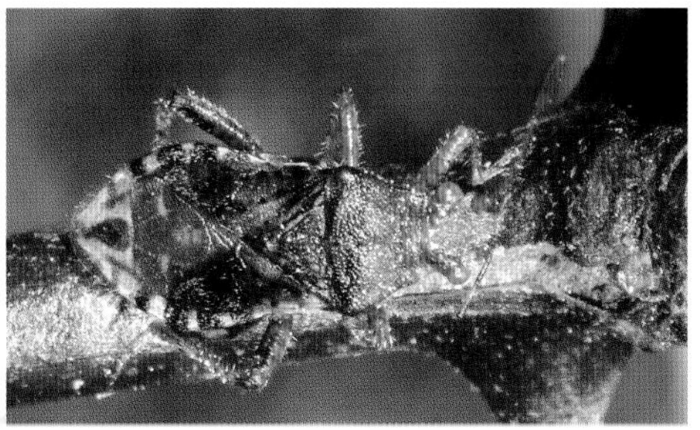
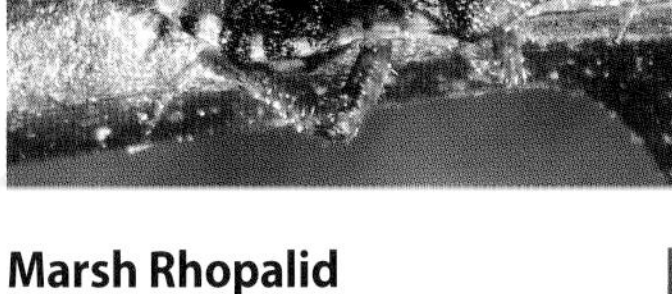

Transparent Rhopalid
Rhopalus parumpunctatus

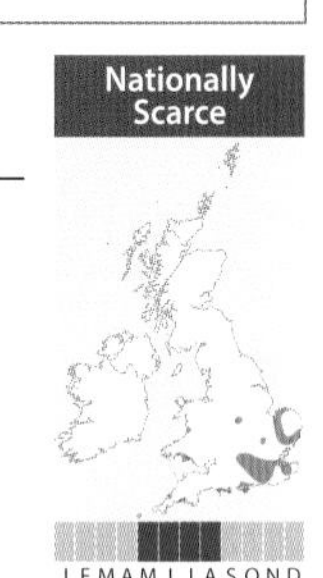

BL 7 mm. **ID** Brown. Connexivum banded or plain. Forewings membranous; veins brown or reddish with black spots. Abdomen black-and-orange. **Hab** Heathland and sandy sites. **Fp** Common Mouse-ear, Common Stork's-bill. **SS** Other *Rhopalus* species [see table *above*].

Marsh Rhopalid
Rhopalus maculatus

BL 8 mm. **ID** Brown. Connexivum usually with only small black spots. Forewings distinctively reddish. Abdomen orange with black marks. **Hab** Damp (but sometimes dry) heathland and fens. **Fp** Marsh Thistle, Marsh Cinquefoil. **SS** Other *Rhopalus* species [see table *above*].

Red Rhopalid
Rhopalus rufus

BL 6–7 mm. **ID** Pronotum reddish-brown. Connexivum reddish, edge plain. Forewing veins without black spots. Abdomen black-and-orange; margin entirely pale. **Hab** Heathlands, sand dunes. **Fp** Sand Spurrey. **SS** Other *Rhopalus* species [see table *above*].

Guide to families pp. 142–147

Myrmus 1 sp.

Short-winged Rhopalid
Myrmus miriformis

BL 6–10 mm. **ID** Wings short (both sexes); approx. 5% of individuals long-winged. ♂ green with red markings, or brown; ♀ green, longer, plumper. **Hab** Dry and damp grassland. **Fp** Grasses: leaves and seeds. **Cyc** Overwinters as egg. **SS** Superficially like some elongated plant bugs [Miridae, *p. 194*].

Corizus 1 sp.

Red & Black Rhopalid
Corizus hyoscyami

BL 9 mm. **ID** Distinctive red and black. **Cyc** Overwinters as egg. **Hab** Mainly open areas, including gardens. **Fp** Composites. **Dist** Spread from the S coast since the 1990s and also to SE Ireland. **SS** Superficially like **Firebug** [Pyrrhocoridae]; much brighter than the smaller (6–7 mm) **Plane Groundbug** *Arocatus longiceps* [Lygaeidae] (*bottom image*).

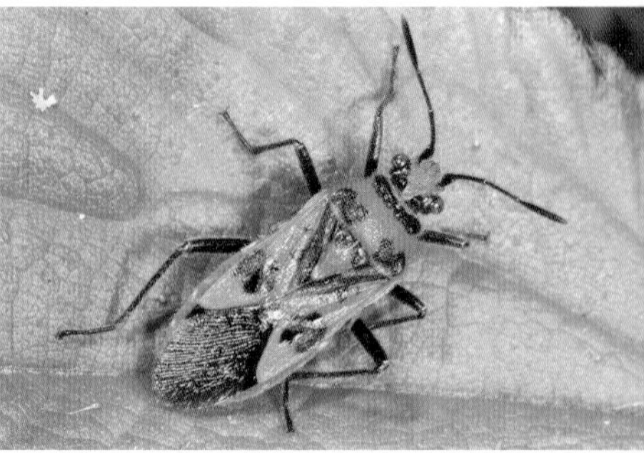

Plane Groundbug *Arocatus longiceps* [Lygaeidae] is similar to Red & Black Rhopalid, but is not as extensively red and has a black head.

Chorosoma 1 sp.

Schilling's Rhopalid
Chorosoma schillingi

BL 14–16 mm. **ID** Straw-coloured. Elongated body, legs and antennae; abdomen with black streaks. Wings reach just over halfway down abdomen. **Cyc** Overwinters as egg. **Hab** Sand dunes and grassland. **Fp** Marram and other tall grasses. **SS** Superficially similar to some plant bugs [Miridae, *p. 194*].

FAMILY **Pyrrhocoridae** (Firebugs) — 1 sp.

Easily recognized red-and-black bug with a slight resemblance to groundbugs (Lygaeidae). Adults overwinter. **Antennae 4-segmented; tarsi 3-segmented; ocelli absent.**

Pyrrhocoris

Firebug *Pyrrhocoris apterus*

BL 8–11 mm. **ID** Black with bright red markings. Forewings short, with a pair of large black spots. **Hab** Open, warm sites. **Fp** Tree Mallow and limes. **Dist** Established on Oarstone Rock, a rocky island off Torquay since *ca.* 1800, otherwise in scattered locations, including Folkestone, Kent, probably via imported plants. **SS** Superficially like **Red & Black Rhopalid**.

FAMILY **Alydidae** (Broad-headed bugs) — 1 sp.

Brown bug, resembling a particularly slender leatherbug (*p. 165*). **Antennae 4-segmented, with S4 curved; tarsi 3-segmented; ocelli present; hind femora spiny.**

Alydus

Ant Bug *Alydus calcaratus*

BL 10–12 mm. **ID** Slender, brown. Abdomen with large, bright orange patch. Hind femora spiny. The reddish-brown nymph resembles an ant. **Hab** Usually sandy heathland, coastal grassland and brownfield sites; sometimes seen in flight in hot weather. **Fp** Broom and Gorse seeds. **Beh** Although the nymph is reported to live in ants' nests, this has not been confirmed; nymphs have been observed being attacked and killed by similar-looking Southern Wood Ants (*p. 496*). **SS** None.

Ant Bug nymphs resemble an ant.

Guide to families pp. 142–147

FAMILY Stenocephalidae (Spurgebugs) — 2 spp.

Easily recognized brown-coloured spurge-feeders, resembling leatherbugs. **Antennae 4-segmented, banded black-and-white; tarsi 3-segmented; ocelli present. All femora and tibiae are distinctly banded**. Both British species overwinter as adults.

Dicranocephalus — 2 spp.

Two species with thorax, scutellum and forewings heavily punctured. Antennae, legs (femora and tibiae) and connexivum with yellowish bands; also small yellow spot on base of scutellum. Identified by habitat, shape and the extent of the pale area at the base of the antennae:

FORM elongate; BL 12–14 mm; ANTENNAE S3 pale ≤ basal ⅓ (always > ¼). HABITAT coastal	**Portland**
FORM smaller, broader; BL 8–11 mm; ANTENNAE S3 pale on basal ¼. HABITAT woodland clearings	**Wood**

Portland Spurgebug
Dicranocephalus agilis

Nationally Scarce

BL 12–14 mm. **ID** Brown, rather elongate. Antenna S3 pale on more than basal ¼. **Hab** Coastal sand dunes and cliffs. **Fp** Portland Spurge and Sea Spurge. **SS** Wood Spurgebug.

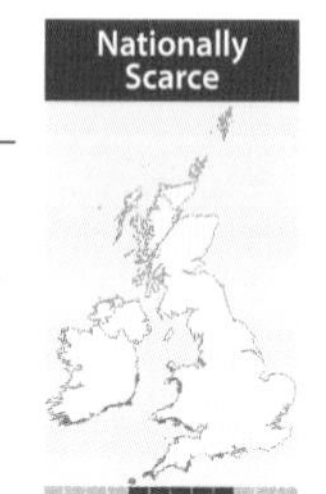

Wood Spurgebug
Dicranocephalus medius

Nationally Scarce

BL 8–11 mm. **ID** Brown, rather broad. Antenna S3 pale on basal ¼. **Hab** Woodland clearings and edges, **Fp** Wood Spurge and other spurges planted nearby. **SS** Portland Spurgebug.

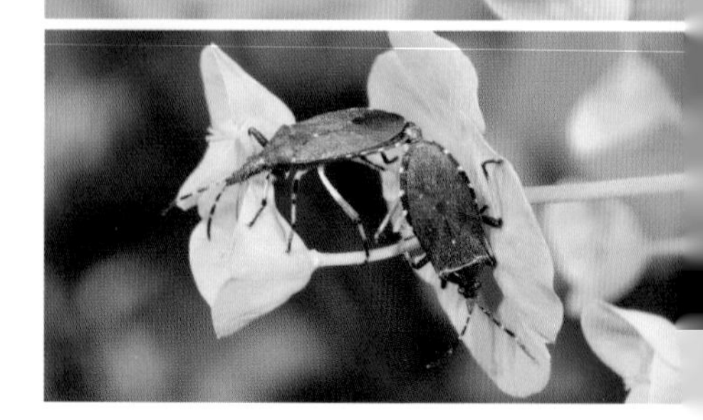

FAMILY Lygaeidae (Groundbugs) 40 GEN. | 91 spp. | 23 GEN., 33 SPP. ILL.

Although there are a few colourful species, most groundbugs are brown or black, some rather small and most ground-dwelling, feeding on seeds. Females have a saw-like ovipositor. Groundbugs often overwinter as adults. **Antennae 4-segmented; tarsi 3-segmented; ocelli present**. The following are examples of the more readily identified species, due to appearance and/or habitat.

Beosus 1 sp.

Coastal Groundbug

Beosus maritimus

BL 6–7 mm. **ID** Brownish with black head, pronotum, much of scutellum and part of forewings; the latter with large white marks; pronotum and forewings with whitish margin. Legs brownish-orange with broad black-banded femora. **Hab** Coastal sandy or rocky sites **Fp** Thrift. **Dist** S England and Wales. **SS** None.

Henestaris 2 spp. | 1 ILL.

Stalked-eyed Groundbug

Henestaris laticeps

BL 5–6 mm. **ID** Brown, with various darker marks, including on forewings and wing membrane. Eyes on long stalks. **Hab** Coastal areas on cliffs. **Fp** Buck's-horn Plantain. **SS Saltmarsh Groundbug** *Henestaris halophilus* [N/I] (**BL** approx. 5 mm) has shorter eye-stalks; uncommon wetland species, mainly Kent and Essex.

Raglius 1 sp.

White-spotted Groundbug

Raglius alboacuminatus

BL 5–6 mm. **ID** Pronotum and forewings with paired white spots; another on wing membrane. **Hab** Waste ground and verges by stands of foodplant **Fp** Black Horehound. **SS** None.

Heterogaster 2 spp. | 1 ILL.

Nettle Groundbug
Heterogaster urticae

Common

J F M A M J J A S O N D

BL 6–7 mm. **ID** Head and pronotum with long hairs. Brown with darker areas; connexivum and tibiae black and whitish banded. **Hab** Grassland, gardens, hedgerows. **Fp** Nettles, often in numbers on the vegetation. **Dist** Also one old record in Ireland. **SS** None.

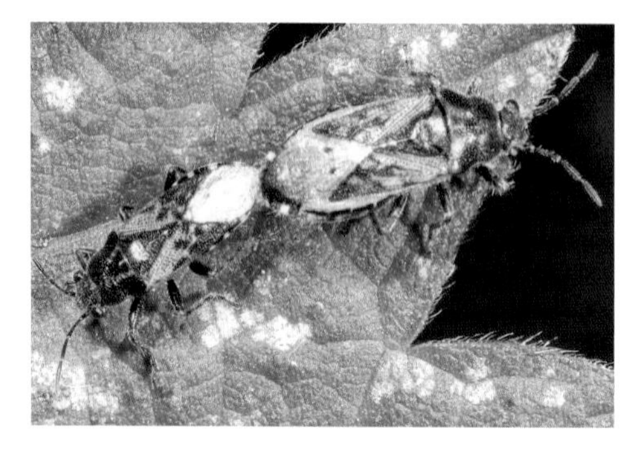

Ischnodemus 2 spp. | 1 ILL.

European Cinchbug
Ischnodemus sabuleti

Common

J F M A M J J A S O N D

BL 5–6 mm. **ID** Elongate, brown, with short, intermediate (least common) or long paler forewings. **Hab** Wetlands, also dry grassland; sometimes found in huge numbers. **Fp** Grasses and reeds. **SS** Folkestone Groundbug *Ischnodemus quadratus* [N/I] (**BL** 4–5 mm) is smaller; restricted to the Folkestone Warrens.

short-winged form

Chilacis 1 sp.

Reedmace Bug
Chilacis typhae

Common

J F M A M J J A S O N D

BL 4–5 mm. **ID** Brown, with darker markings; wing membrane pale. **Hab** Wetlands. **Fp** Greater Reedmace (Bulrush); overwinters in the seedheads. **SS** None.

Gastrodes 2 spp. | 1 ILL.

Pine-cone Bug
Gastrodes grossipes

Common

J F M A M J J A S O N D

BL 6–7 mm. **ID** Broad, rather flat. Reddish-brown; head, pronotum and scutellum black. Fore femora very swollen and often darker. **Hab** Coniferous woodland, gardens, parks. **Fp** Scots Pine. **SS** Localized Spruce-cone Bug *Gastrodes abietum* [N/I] (**BL** 6–7 mm) is paler and has a pale front margin to the pronotum; found on Norway Spruce.

Emblethis 2 spp.

Oval Groundbug
Emblethis griseus

Rare (RDB3)

BL 6 mm. **ID** Grey or brown with dark punctures; suboval, not elongate. 1st hind tarsal segment 1·5 × combined length of other two segments. **Hab** Sand dunes. **Fp** Common Stork's-bill. **SS** Straw Groundbug, which is distinguished by relatively longer 1st hind tarsal segment.

Straw Groundbug
Emblethis denticollis

Local

BL 5–6 mm. **ID** Straw-coloured, slightly elongate. 1st hind tarsal segment 2 × combined length of other two segments. **Hab** Sandy areas. **Fp** Common Stork's-bill. **SS** Oval Groundbug, which is distinguished by relatively shorter 1st hind tarsal segment.

Pachybrachius 2 spp.

Wetland Groundbug
Pachybrachius fracticollis

Local

BL 5 mm. **ID** Pronotum rear margin distinctively constricted, without keels, smooth. Brown; head, pronotum and at least upper part of scutellum darker. **Hab** Wetlands including bogs. **Fp** Sedges. **SS** Sphagnum Groundbug.

Sphagnum Groundbug
Pachybrachius luridus

Rare (RDB3)

BL 5–6 mm. **ID** Pronotum rear margin distinctively constricted, without keels; hairy. Brown; head, pronotum and scutellum darker. **Hab** Bogs. **Fp** Probably sedges. **Dist** Stronghold is the New Forest, Hampshire. **SS** Wetland Groundbug.

Megalonotus 6 spp. | 3 ILL.

Megalonotus species have an all-dark, heavily punctured pronotum.

Curved Groundbug
Megalonotus sabulicola

BL 5 mm. **ID** All-black with brown forewings, tibiae and part of antennae. Pronotum all-dark, heavily punctured, with erect hairs; hind part curved. **Hab** Sandy areas, often coastal. **Fp** Unknown. **SS** Other *Megalonotus* species, from which told by pronotum with long erect hairs and curved hind margin.

Storksbill Groundbug
Megalonotus praetextatus

BL 4–5 mm. **ID** Black; legs paler brown except fore femora; forewings upper part brown. Pronotum all-dark, heavily punctured, smooth. **Hab** Sheltered, well-drained areas. **Fp** Common Stork's-bill. **SS** Other *Megalonotus* species, from which told by hairless, shiny body.

Black Groundbug
Megalonotus dilatatus

BL 5–6 mm. **ID** Black. Pronotum with coarse punctures. **Hab** Heathlands, dunes and grasslands. **Fp** Unknown. **SS** Other *Megalonotus* species, from which told by black tibiae and antennae (partly pale in other species).

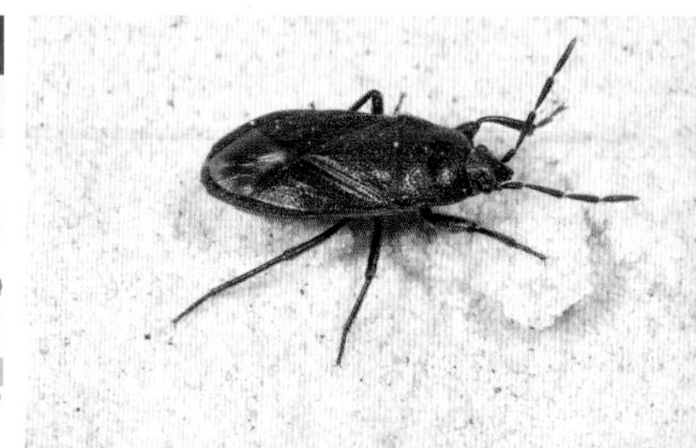

Drymus 7 spp. | 1 ILL.

Brown Groundbug
Drymus brunneus

BL 4–5 mm. **ID** Black with rear part of pronotum and wingcases brown. **Hab** Damp areas, often in leaf-litter. **Fp** Mosses and fungi. **SS** Other *Drymus* species, which do not have brown at rear of pronotum and have long, erect hairs on the tibiae (lacking in Brown Groundbug).

Trapezonotus 4 spp. | 3 ILL.

Trapezonotus species are somewhat oval; the pronotum has narrow keels.

Dune Groundbug
Trapezonotus arenarius

BL 4–5 mm. **ID** Brown with black markings; head, pronotum and scutellum black. Mid and hind tibiae black. **Hab** Sand dunes and Breckland. **Fp** Heather. **SS** Other *Trapezonotus* species, which can only be distinguished by ♂ genitalia.

Heath Groundbug
Trapezonotus desertus

BL 4–5 mm. **ID** Brown with black markings; head, pronotum and scutellum black. Mid tibiae often brown. **Hab** Heathland and chalk grassland. **Fp** Heathers. **SS** Other *Trapezonatus* species, which can only be distinguished by ♂ genitalia. However, mid tibiae of ♂ is often pale in Heath Groundbug.

Ullrich's Groundbug
Trapezonotus ullrichi

BL 6 mm. **ID** Pale brown with black markings; head, part of pronotum and scutellum black. Legs mostly brown, femora partly black; antenna S2 & S3 brown. **Hab** Coastal cliffs. **Fp** Ox-eye Daisy. **SS** Superficially like other *Trapezonatus* species, but larger and paler.

Macrodema 1 sp.

Short-winged Groundbug
Macrodema micropterum

BL 3 mm. **ID** Glossy black with rear part of pronotum and forewings brown. **Hab** Heathland. **Fp** Heathers. **SS** None.

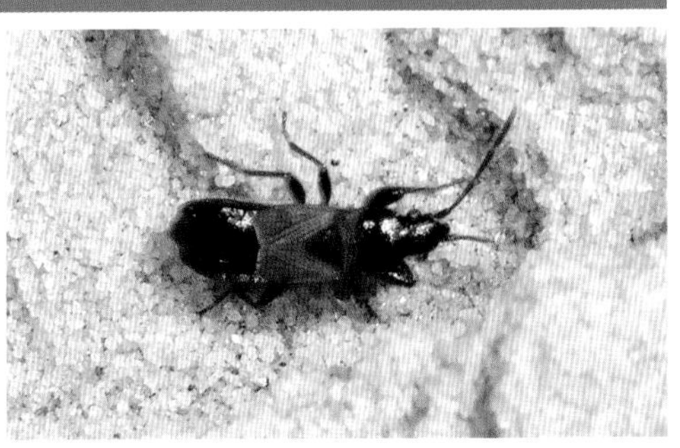

Kleidocerys 2 spp.

Heather Groundbug
Kleidocerys ericae

BL 4 mm. **ID** Reddish-brown; scutellum with broad whitish markings; corium with few dark marks. Forewings partly transparent (shorter than those of Birch Catkin Bug). **Hab** Heathland. **Fp** Heathers. **SS** Birch Catkin Bug.

Birch Catkin Bug
Kleidocerys resedae

BL 5 mm. **ID** Reddish-brown; scutellum with white markings; corium with several dark marks. Forewings very long and partly transparent (longer than those of Heather Groundbug). **Hab** Woodland, parks and gardens. **Fp** Birches and Alder, often on catkins. **SS** Heather Groundbug.

Scolopostethus 6 spp. | 2 ILL.

Decorated Groundbug
Scolopostethus decoratus

BL 4 mm. **ID** Brightly coloured, black, reddish and white; antennae black except base of 2nd segment, which is reddish. **Hab** Heathland. **Fp** Heather. **SS** Other *Scolopostethus* species, although these have shorter wings and are less colourful.

Thomson's Groundbug
Scolopostethus thomsoni

BL 4 mm. **ID** Brightly coloured, black, brown and white; usually short-winged. At least tip of antenna S2 dark, also S3 & S4. **Hab** Wide range, usually associated with nettles. **Fp** Presumably nettles. **SS** Other *Scolopostethus* species, which are longer winged and do not have dark-tipped antenna S2.

Rhyparochromus 2 spp.

3right-spotted Groundbug
Rhyparochromus vulgaris

Local

3L 7–8 mm. **ID** Elongate, with dark ving membrane; paler base colour han Black-spotted Groundbug; vingcases and pronotum with vhitish margins. **Hab** Grassland. 'p Unknown. **SS** Black-spotted Groundbug.

Black-spotted Groundbug
Rhyparochromus pini

Nationally Scarce

BL 7–8 mm. **ID** Elongate, with dark wing membrane; darker base colour than Bright-spotted Groundbug; pronotum all-dark. **Hab** Heathland and sand dunes. **Fp** Unknown. **SS** Bright-spotted Groundbug.

Nysius 7 spp. | 2 ILL.

'leabane Groundbug
Jysius ericae

Common

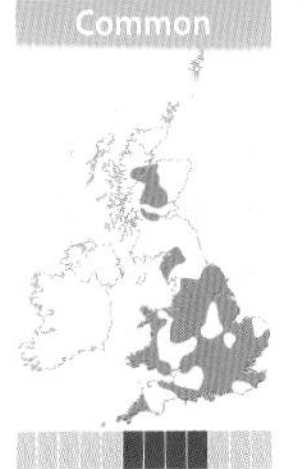

3L 4 mm. **ID** Brown, wing membrane vith dark markings. **Hab** Bare ground. 'p Composites including fleabanes nd Ploughman's-spikenard. **SS** Very imilar to **Wolff's Groundbug**, ♂ only eparable by genitalia differences; ♀ has longer hairs on wing veins.

Wolff's Groundbug
Nysius thymi

Common

BL 4 mm. **ID** Brown, wing membrane with dark markings. **Hab** Open habitats, especially coastal. **Fp** Composites including fleabanes and Ploughman's-spikenard. **SS** Very similar to **Fleabane Groundbug**, ♂ only separable by genitalia differences; ♀ has shorter hairs on wing veins.

Pionosomus 1 sp.

Bristly Groundbug
Pionosomus varius

Rare (RDB3)

BL 2–3 mm. **ID** Bristly dark hairs; white mark on wing membrane.
Hab Sand dunes. **Fp** Common Stork's-bill, Little Mouse-ear. **SS** None.

Stygnocoris 3 spp. | 1 ILL.

Hairy Groundbug
Stygnocoris sabulosus

Common

BL 3 mm. **ID** Brownish-black; pronotum and short forewings hairy.
Hab Sandy and chalky areas.
Fp Unknown. **SS** Other *Stygnocoris* species, from which distinguished by shiny body, hairiness and pale legs.

Eremocoris 4 spp. | 1 ILL.

Spiny Groundbug
Eremocoris plebejus

Nationally Scarce

BL 5–7 mm. **ID** Black; forewings reddish-brown, membrane with two semicircular whitish spots; fore femora broad, with two large and several small spines. **Hab** Coniferous woodlands.
Fp Unknown. **SS** Other *Eremocoris* species have different patterning and much shorter hairs on hind tibiae (in Spiny Groundbug, hairs are up to 2× tibial width).

Graptopeltus 1 sp.

Eyed Groundbug
Graptopeltus lynceus

Nationally Scarce

BL 6–7 mm. **ID** Broad, brown; pronotum broad, with pale margins; scutellum with raised 'V'-shaped mark, pale towards tip; forewings with large 'eye-spots'. **Hab** Dunes, sandy areas, Breckland. **Fp** Viper's-bugloss.
SS None.

Plinthisus 1 sp.

Glossy Groundbug
Plinthisus brevipennis

BL 2–3 mm. **ID** Small, black and glossy; forewings usually short, some individuals with brownish markings. **Hab** Grasslands, heathlands, dunes. **Fp** Unknown. **SS** None.

Pterotmetus 1 sp.

Red-winged Groundbug
Pterotmetus staphyliniformis

BL 5 mm. **ID** Black; forewings short, red. **Hab** Clifftops, on boulders with stonecrops. **Fp** Unknown. **SS** None.

FAMILY **Piesmatidae** (Beetbugs) 2 GEN. | 2 spp. | 1 ILL.

Small bugs with **net-like forewings. Antennae 4-segmented; tarsi 3-segmented; ocelli absent**. Feed on plants, sometimes with two generations a year; adults overwinter. Beetbugs need to be carefully searched for on vegetation. Although worldwide in distribution, few species have been recorded.

Parapiesma 1 sp.

Beet Leaf Bug
Parapiesma quadratum

BL 3 mm. **ID** Pale, brown or green-tinged, or black, with lace-like wings. **Hab** Coastal saltmarshes. **Fp** Oraches and Sea-purslane. **SS Spotted Beetbug** *Piesma maculatum* [N/I] (**BL** 2–3 mm), which is often darker, more heavily spotted and the pronotum has indented side margins.

FAMILY **Berytidae** (Stiltbugs) — **4 GEN. | 9 spp. | 6 SPP. ILL.**

Mainly brown, well-camouflaged bugs, characterized by **long, thin body and legs** with swollen femora tips. **Antennae long and thin, 4-segmented, S4 clubbed; tarsi 3-segmented; ocelli present**. Feed on seeds of host plants, often on or near the ground; adults overwinter. Genera and species can be identified as follows:

STILTBUGS – Clypeus not projecting; thorax humped [see *opposite*]

Metatropis — 1 sp.

Enchanter's Stiltbug *Metatropis rufescens*

BL 9–11 mm. **ID** Reddish-brown. Antenna S1 club brown; scutellum without spines; either short or long-winged. **Hab** Woodland. **Fp** Enchanter's-nightshade. **SS** None.

Gampsocoris — 1 sp.

Spined Stiltbug *Gampsocoris punctipes*

BL 4–5 mm. **ID** Black with green abdomen. Scutellum with long, curved spine. **Hab** Sandy areas, especially dunes. **Fp** Common Restharrow. **SS** None.

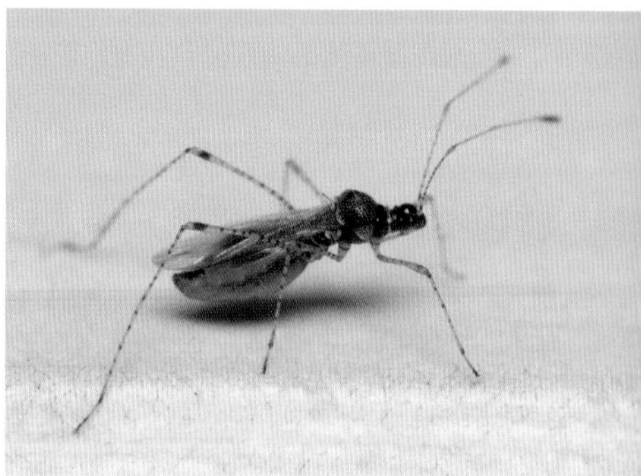

STILTBUGS – Clypeus projects forwards; thorax not humped [see *opposite*]

Neides — 1 sp.

Straw Stiltbug *Neides tipularius*

BL 10–12 mm. **ID** Straw-coloured; darker when overwintering. Antennae longer than body length; hind femora reaching beyond tip of abdomen. Either short or long-winged. **Hab** Weedy arable fields, heathland, sand dunes. **Fp** Various plants, including grasses and Tansy. **SS** None.

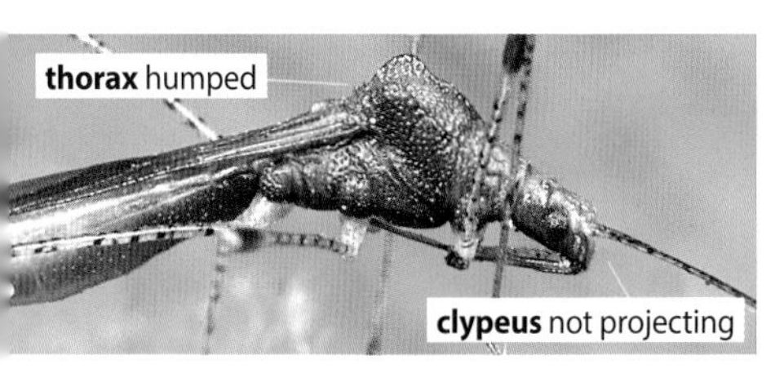

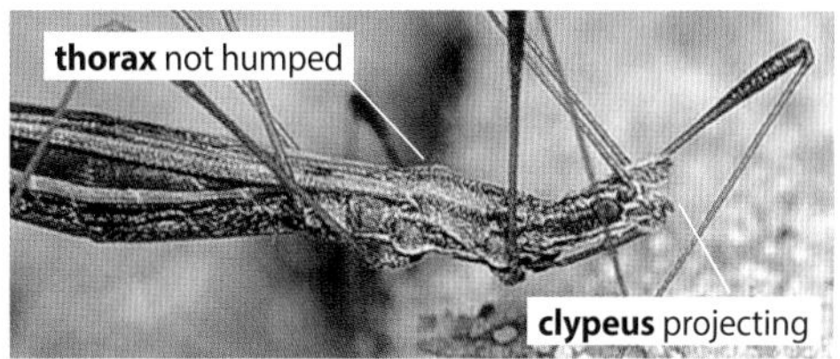

Berytinus 6 spp. | 3 ILL.

ANTENNAE Antennae shorter than body length; hind femora not reaching end of abdomen

ANTENNAE segment S1 with some long hairs; **FORM** particularly elongate	**Hairy Stiltbug**
ANTENNAE club of segment S1black	**Common Stiltbug**
ANTENNAE club of segment S1black; **WINGS** small black mark on basal veins	**Signoret's Stiltbug**

·lairy Stiltbug
ˈerytinus hirticornis

ˈL 7–11 mm. **ID** Brown, notably ·longate; antenna S1 hairy. **·lab** Grassland. **SS** Other *Berytinus* ·pecies [see table *above*]; told from the ·nree species not illustrated by having ·airy antenna S1.

·ommon Stiltbug
ˈerytinus minor

ˈL 5–7 mm. **ID** Brown; antenna S1 club ·lack. **Hab** Grassland. **Fp** Common ·estharrow, clovers and Lesser Trefoil. **·S** Other *Berytinus* species [see table *·bove*]; told from the three species ·ot illustrated by having shorter hind ·mora, which do not reach tip of ·orium.

·ignoret's Stiltbug
·erytinus signoreti

·L 5–6 mm. **ID** Brown; antenna S1 club ·rown. Either short or long-winged. **·ab** Sandy and chalk grassland. **·p** Trefoils. **SS** Other *Berytinus* species ·ee table *above*]; told from the three ·ecies not illustrated by having ·iangular frontal process twice as long · width at base; also antenna S3 is a ·ird longer than pronotum.

Guide to families pp. 142–14

FAMILY **Tingidae** (Lacebugs) **14 GEN. | 25 spp. | 6 GEN., 7 SPP. ILL.**

Distinctive but small plant-feeding bugs with lace-like forewings; some species are short-winged and have a limited range of foodplants; adults of some species overwinter. **Antennae 4-segmented; tarsi 2-segmented; ocelli absent.**

Physatocheila **3 spp. | 2 ILL.**

Physatocheila species have three keels on the pronotum which reach the top of the head.

NOTE: **Harwood's Lacebug** *Physatocheila harwoodi* EN [N/I] (**BL** 3 mm) has only been recorded from one tree in Witchampton, Dorset but was last recorded in 1956 and may now be extinct (it is identified by having a very short dark central part of marginal area compared to Apple Lacebug).

Hawthorn Lacebug
Physatocheila dumetorum

Local

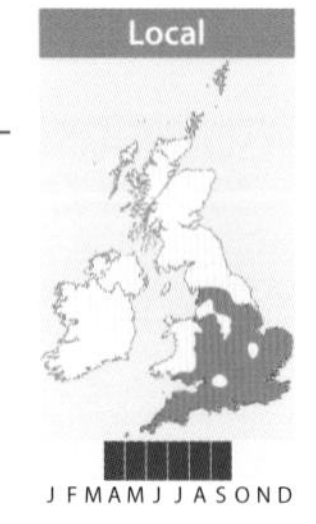

BL 3 mm. **ID** Margin of forewings in centre with two rows of meshes. As in other *Physatocheila* species, three keels on pronotum reach top of head. Brown, with darker and lighter areas. **Hab** Woodland. **Fp** Lichen-covered Hawthorn trees. **SS** Apple Lacebug [see *insets* for distinguishing feature].

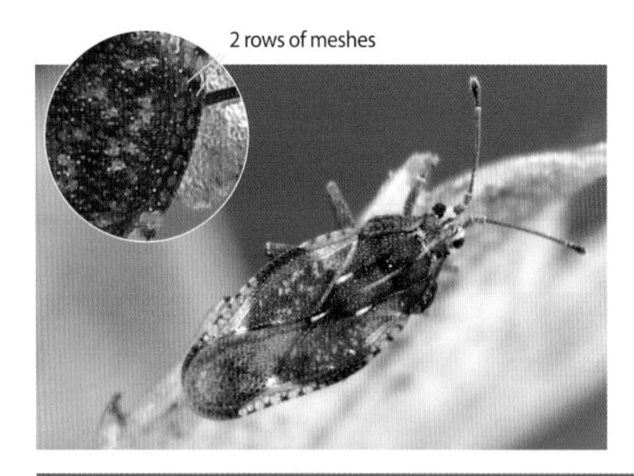
2 rows of meshes

Apple Lacebug
Physatocheila smreczynskii

Nationally Scarce

BL 4 mm. **ID** Margin of forewings in centre with three rows of meshes. Body brown, with darker and lighter areas that vary between individuals. **Hab** Woodland and orchards. **Fp** Lichen-covered apple trees. **SS** Hawthorn Lacebug [see *insets* for distinguishing feature].

3 rows of meshes

Stephanitis **2 spp. | 1 ILL.**

Andromeda Lacebug
Stephanitis takeyai

Local (recent colonist)

BL 4 mm. **ID** Black markings on head, thorax and otherwise transparent wings; antennae and legs yellowish. **Hab** Gardens and parks. **Beh** Overwinters as eggs. **Fp** Andromedas (*Pieris*), also rhododendrons. **Dist** Spreading N since first British record in 1998. **SS** None.
NOTE: **Rhododendron Lacebug** *Stephanitis rhododendri* [N/I] (**BL** 4 mm) is a delicate brown, and less boldly marked, and feeds on rhododendrons.

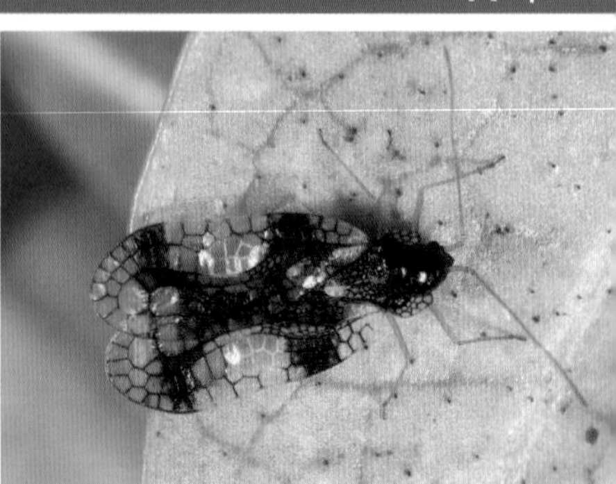

Acalypta 5 spp. | 1 ILL.

Moss Lacebug
Acalypta parvula

Common

BL 2 mm. **ID** Oval; brown with head and part of pronotum black; pronotum with three keels. **Hab** Sandy soils. **Fp** Associated with short moss. **SS** Other *Acalypta* species that are also found on moss, from which told by antenna S3 being thickened at base and basal antenna segment black.

Tingis 4 spp. | 1 ILL.

Spear Thistle Lacebug
Tingis cardui

Common

BL 3–4 mm. **ID** Greyish-brown with white powdery deposits of wax; antennae black-tipped. **Hab** Grassland and open areas. **Fp** Spear Thistle. **SS** Other *Tingis* species, from which told by black-tipped antennae and tibiae; also 2–3 rows of meshes at edge of pronotum and forewing margins.

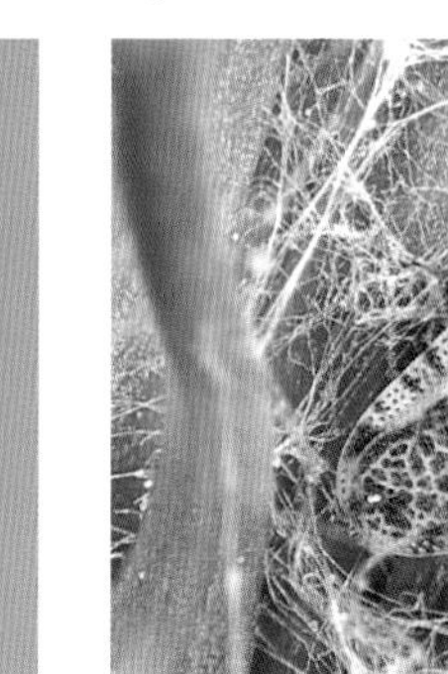

Dictyonota 2 spp. | 1 ILL.

Gorse Lacebug
Dictyonota strichnocera

Common

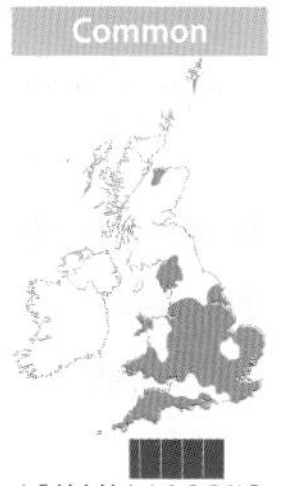

BL 4 mm. **ID** Pale, with black head and part of pronotum; margin of forewings with two rows of meshes, outer margin and some other veins yellowish; antennae black, without long hairs; legs dark. **Hab** Open places. **Beh** Overwinters as egg. **Fp** Gorse, sometimes also Broom. **SS** **Broom Lacebug** *Dictyonota fulginosa* [N/I] (**BL** 5 mm), which has brown antennae, and **Hairy Lacebug**.

Kalama 1 sp.

Hairy Lacebug
Kalama tricornis

Local

BL 3 mm. **ID** Brown, with head, part of pronotum and antennae black (with long hairs); margin of forewings with two rows of meshes. **Hab** Sandy sites or chalk downland. **Fp** Low-growing plants. **SS** **Broom Lacebug** *Dictyonota fulginosa* [N/I] (**BL** 5 mm), which has brown antennae, and **Gorse Lacebug**.

Guide to families pp. 142–14

FAMILY Reduviidae (Assassin bugs) 5 GEN. | 10 spp. | 3 GEN., 4 SPP. ILL.

Long-legged predators on other insects; some species are short-winged. Long curved rostrum. **Antennae long, 4-segmented, the first long; tarsi 3-segmented; ocelli present**. Eggs overwinter.

ASSASSIN BUGS – Elongate, antennae longer than body

Empicoris 3 spp. | 1 ILL.

Thread-legged Bug *Empicoris vagabundus*

BL 6–7 mm. **ID** Brown; side of connexivum cream with brown marking. Various darker narrow bands on legs and antennae. **Hab** Woodland, hunting insects in trees. **SS** Other *Empicoris* species, which are smaller (**BL** ≤5 mm) and have dark markings on the side of the connexivum.

ASSASSIN BUGS – Robust form, antennae shorter than body

Reduvius 1 sp.

Flybug *Reduvius personatus*

BL 16–18 mm. **ID** Large, elongate, black, except antennae and tibiae with paler bases. Body with dark hairs; wings long. **Hab** Woodland and gardens and often inside houses and outbuildings, hunting insects including bed bugs. **Beh** Nocturnal, attracted to lights; if handled, can cause a painful bite. **SS** None.

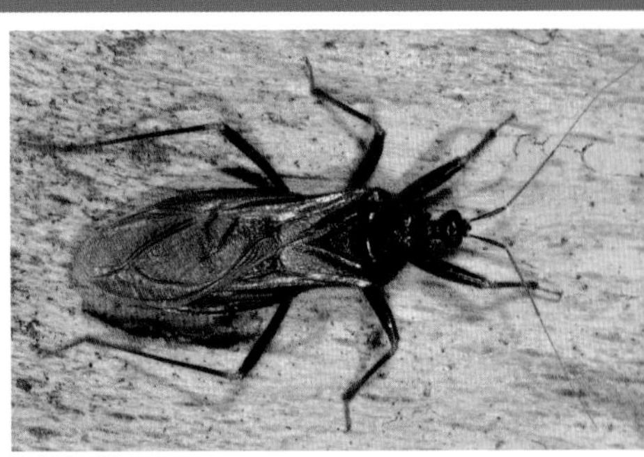

Ramparts Field, Suffolk is a well-known site for assassin bugs, including Woodroffe's Assassin Bug.

Coranus — 3 spp. | 2 ILL.

Brown with black and paler markings; body with pale hairs. Abdomen oval with connexivum banded black and whitish; wings usually short, but long-winged forms occur.

FOREWINGS short, but reaching back of abdomen S3 or S4; ♀ **ABDOMEN UNDERSIDE** pale	**Heath Assassin Bug**
FOREWINGS shorter, not reaching beyond front of abdomen S3; ♀ **ABDOMEN UNDERSIDE** pale with dark central line	**Woodroffe's Assassin Bug**
ABDOMEN UNDERSIDE all, or almost entirely black	**Thorne Assassin Bug** *Coranus aethiops* [N/I] (**BL** 9–14 mm)

Heath Assassin Bug
Coranus subapterus

Local

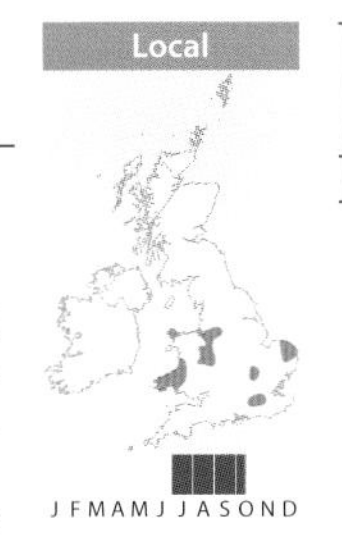

BL 9–12 mm. **ID** Greyish, bristly. Forewings short, reaching end of abdomen S3 or S4. **Hab** Sand dunes, Breckland, heathland. **Beh** Adults and nymphs stridulate if disturbed; handle with care, the rostrum can pierce skin. **SS** Other *Coranus* species [see table *above*]; also superficially similar to the brown *Himacerus* species (*p. 192*).

Woodroffe's Assassin Bug
Coranus woodroffei

Local

BL 9–12 mm. **ID** Greyish; bristly. Forewings short, not reaching beyond abdomen S3. ♀ abdomen with dark central line on underside. **Hab** Open heathland. **Dist** Much more extensive than shown on map; confusion in the past with Heath Assassin Bug. **Beh** Adults and nymphs stridulate if disturbed; handle with care, the rostrum can pierce skin. **SS** Other *Coranus* species [see table *above*]; also superficially similar to the brown *Himacerus* species (*p. 192*).

short-winged form

long-winged form

Heath Assassin Bug
WINGS reach end of abdomen S3 or S4

Woodroffe's Assassin Bug
WINGS only reach front of abdomen S3

Guide to families pp. 142–14

FAMILY **Cimicidae** (Bed bugs) — 2 GEN. | 4 spp. | 2 SPP. ILL.

Brown and black, round, blood-sucking bugs, primarily associated with birds and mammals. The main species featured is well known and has a worldwide distribution. **Antennae 4-segmented; tarsi 3-segmented; ocelli absent**. Can bite humans.

Cimex — 3 spp. | 2 ILL.

Bed Bug *Cimex lectularius*

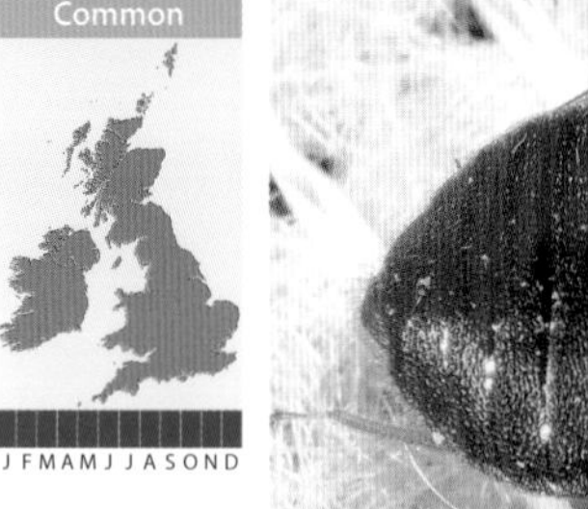

BL 4–5 mm. **ID** Brown, oval and flat, becoming bloated and more elongate after feeding. Reduced forewings present; hindwings absent. **Hab** Associated with man, likely near or inside beds or sleeping areas. **Beh** Bites result in skin rashes and other conditions in humans, but do not transmit diseases. Bed Bugs are hardy, surviving a few months without feeding. They feed on blood at night. **Dist** Largest numbers concentrated in areas of high human population, such as London, Dublin and popular holiday destinations. Pesticide resistance and international travel may be factors in an increasing number of records since the 1990s. **SS** Other *Cimex* species, *e.g.* Bat Bug *Cimex pipistrelli* (**BL** 5 mm) [BOTTOM RIGHT], which is found on bats, has longer hairs on pronotum margins. **NOTE**: the other species in this genus, Pigeon Bug *Cimex columbarius* (**BL** 5 mm), is only known in Britain from records in 1954.

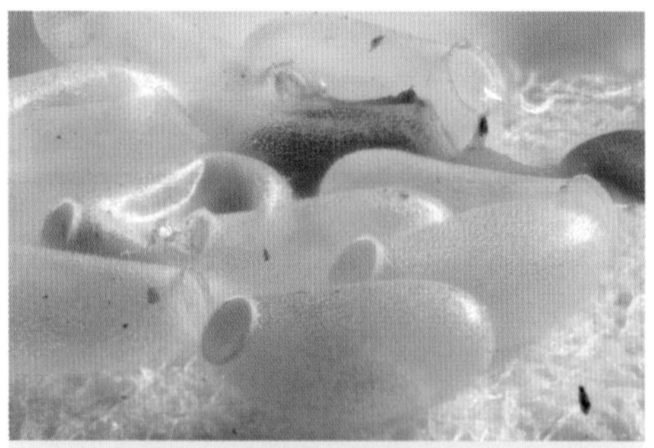
Bed Bug eggs

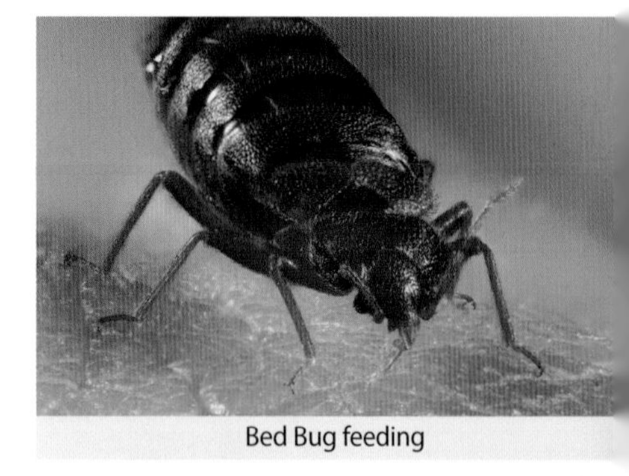
Bed Bug feeding

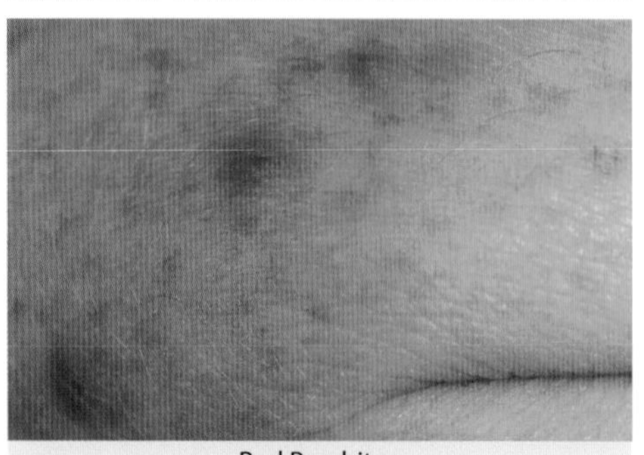
Bed Bug bites

Bed Bug (left) and Bat Bug (right)

FAMILY Anthocoridae (Flower bugs) — 12 GEN. | 32 spp. | 1 SP. ILL.

Small, fairly broad predators (body length ≤4 mm) of other soft-bodied insects, such as aphids. **Antennae long, 4-segmented, the first long; tarsi 3-segmented; ocelli present**. Can bite humans. Species are easily confused due to similar coloration and size; the commonest species only is illustrated. Species overwinter as adults.

Anthocoris — 12 spp. | 1 ILL.

Common Flower Bug
Anthocoris nemorum

Common

J F M A M J J A S O N D

BL 3–4 mm. **ID** Head and pronotum black; legs and most of Antennae segments S2 and S3 orange (except for black tips). Forewings reflective; boldly marked. Membrane with black hourglass-shaped mark. **Hab** Grassland, woodland, gardens. **Beh** Often on low vegetation, including flowers. **SS** Other *Anthocoris* species, from which told by combination of reflective forewings, black pronotum, orange-brown legs and pale antenna S2–3 (tips dark).

FAMILY Nabidae (Damsel bugs) — 3 GEN. | 14 spp. | 8 SPP. ILL.

Long-legged predators of other insects; often short-winged. **Rostrum long, 4-segmented; antennae 4-segmented, S2 the longest; tarsi 3-segmented; ocelli present**. Most species overwinter as eggs, except where stated. The genus *Prostemma* (1 sp.) is not included as there are few recent records. The other two genera can be distinguished as follows:

ROSTRUM long

CONNEXIVUM usually black with orange spots;
FEMORA + TIBIAE banded in darker species;
HIND TIBIAE broad black band at tip in some species

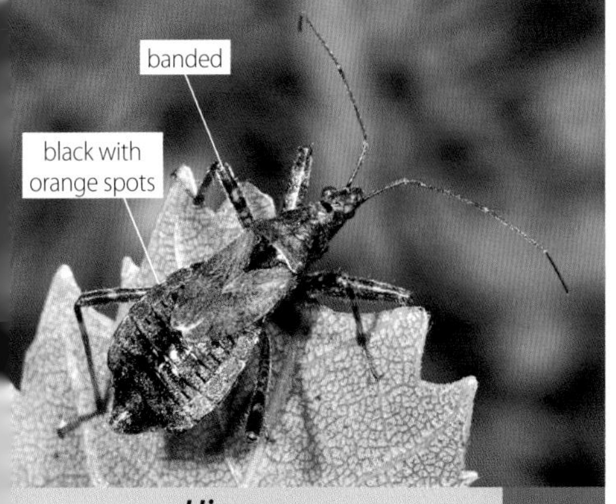

Himacerus
[4 species]
p. 192

CONNEXIVUM pale; often paler brown;
LEGS plain or spotted; not banded

Nabis
[9 species]
p. 193

Himacerus 4 spp.

Tree Damsel Bug
Himacerus apterus

Common

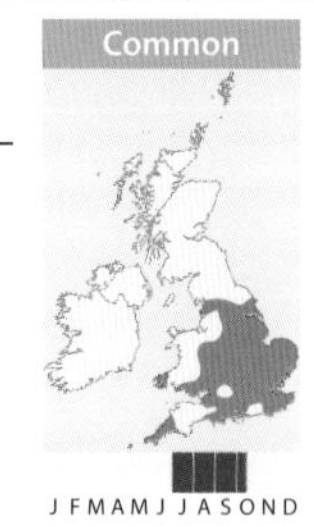

BL 8–11 mm. **ID** Reddish-brown. Forewings reach abdomen S3 or S4; membrane rarely reaches tip of abdomen. Antennae extend beyond end of wings if held back over body. **Hab** Woodland, gardens and parks, hunting insects in trees but rarely in conifers. **SS** Ant Damsel Bug.

Black-striped Damsel Bug
Himacerus boops

Local

BL 7 mm. **ID** Pronotum to abdomen with three distinct longitudinal black stripes; short-winged (long-winged form extremely rare). Hind femora with black band at tip. **Hab** Heathland and acid grassland; most likely on the ground. **SS** Grey Damsel Bug.

Ant Damsel Bug
Himacerus mirmicoides

Common

BL 7–8 mm. **ID** Brown. Forewings reach abdomen S4 or S5. Antennae shorter than most damsel bugs, not reaching end of wings if held back over body. Young nymphs are said to be ant-like, hence the common name. **Hab** Grassland, gardens. **Beh** Hunting insects in low vegetation; adults overwinter. **SS** Tree Damsel Bug.

Grey Damsel Bug
Himacerus major

Common

BL 8–9 mm. **ID** Brown or greyish with longitudinal dark stripes; always long-winged. Connexivum pale in some individuals. Hind femora with broad black band at tip and black spots below. **Hab** Grassland, including saltmarshes. **SS** Black-striped Damsel Bug. Superficially like *Nabis* species.

Nabis **9 spp. | 4 ILL.**

Heath Damsel Bug
Nabis ericetorum

BL 7 mm. **ID** Reddish-brown. Forewings almost reach tip of abdomen (extend beyond in long-winged examples (not uncommon). Antenna S2 length ± maximum width of pronotum. **Hab** Heathland. **Beh** Adults overwinter. **SS** Other *Nabis* species, in particular **Common Damsel Bug** (microscopic examination of male claspers may be needed to confirm identification).

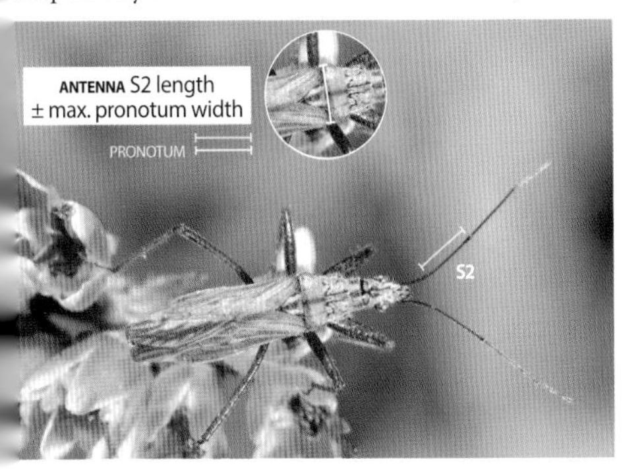

Broad Damsel Bug
Nabis flavomarginatus

BL 7–9 mm. **ID** Brown. Forewings short, reaching half-length of abdomen; the upper part of the abdomen is covered in golden hairs. **Hab** Grassland, mostly damper sites. **SS** Marsh Damsel Bug.

Common Damsel Bug
Nabis rugosus

BL 6–8 mm. **ID** Brown. Forewings pale yellowish brown, not quite or just reaching end of abdomen; longer-winged form rare; Antenna S2 length obviously greater than maximum width of pronotum. **Hab** Grassland. **Beh** Adults overwinter. **SS** Other *Nabis* species, in particular **Heath Damsel Bug** (microscopic examination of ♂ claspers may be needed).

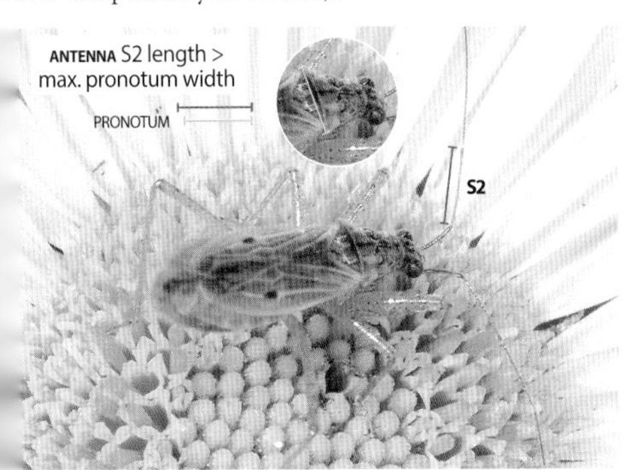

Marsh Damsel Bug
Nabis limbatus

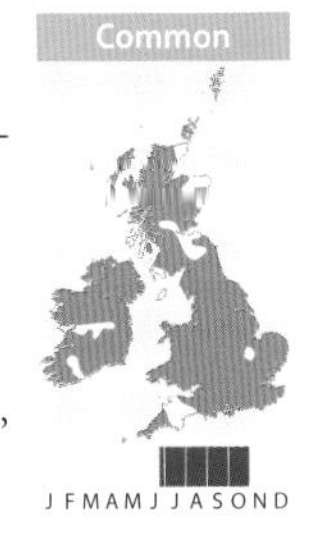

BL 7–9 mm. **ID** Brown with darker, in some individuals black, longitudinal lines down centre of body and either side. Forewings very short, only reaching part of abdomen S3. **Hab** Wetlands. **SS** Other *Nabis* species, in particular **Broad Damsel Bug** (the Marsh Damsel Bug's short wings are a useful indicator).

Guide to families pp. 142–14

FAMILY Miridae (Plant bugs)

94 GEN. | 220 spp. | 32 GEN., 38 SPP. ILL. (incl. 1 sp. on p. 146)

The largest family of land bugs, delicate-looking and mostly plant feeders. Some are predators of other insects, at least in part. **Antennae long, 4-segmented; tarsi 3-segmented; ocelli absent**. Species mainly overwinter as eggs. Representative species are illustrated, this family is sometimes known as mirid bugs. Care is needed with identification in some genera, which have various lookalike species and microscopic examination of genitalia or other characters may be required. **The following are examples of better-known colourful, and look-alike species**.

Leptopterna 2 spp.

Meadow Plant Bug
Leptopterna dolabrata

BL 8–10 mm. **ID** Pale, except forewings yellowish or reddish. The female is usually short-winged. In ♂, antenna segment S2 much longer than combined length of S3 + S4; in ♀, S2 segment thinner than base of fore tibiae. **Hab** Grassland including damp areas. **Fp** Various grasses. **SS** Red Meadow Plant Bug.

Red Meadow Plant Bug
Leptopterna ferrugata

BL 7–9 mm. **ID** Forewings reddish. In ♂, antenna segment S2 equal to combined length of S3 + S4; in ♀, S2 wider than base of fore tibiae. ♀ usually short-winged. **Hab** Grassland, preferring drier areas. **Fp** Various grasses. **SS** Meadow Plant Bug.

Heterotoma 1 sp.

Thick-antennae Plant Bug
Heterotoma planicornis

BL 5–6 mm. **ID** Dark brown with green legs; Antenna S2 broadened. **Hab** Hedgerows and grassland. **Fp** On nettles, where predatory on small insects, in addition to feeding on various buds and unripe fruits. **SS** None.

Deraeocoris 5 spp. | 1 ILL.

Red-spotted Plant Bug
Deraeocoris ruber

BL 6–8 mm. **ID** Orange-brown or black; forewings with pair of bold red spots. Antenna S1 and at least basal segment of S2 black. **Hab** Hedgerows and grassland. **Fp** Predatory on small insects, often on nettles or Bramble. **SS** None.

Capsus 2 spp. | 1 ILL.

Black Plant Bug *Capsus ater*

BL 5–6 mm. **ID** Black, rather oval; head and pronotum brown or black; antenna S2 with broadened tip. **Hab** Grassland **Fp** Lower parts of various grasses. **SS** None.

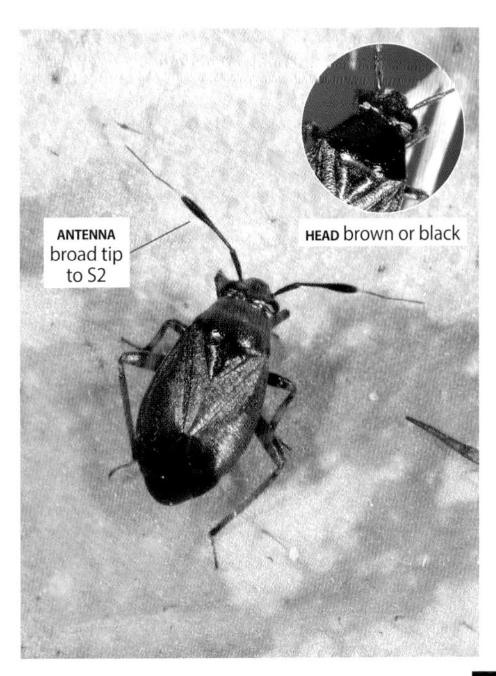

Adelphocoris 4 spp. | 1 ILL.

Lucerne Bug
Adelphocoris lineolatus

Common

BL 8–10 mm. **ID** Elongate, pale green; pronotum with pair of black spots; forewings and scutellum pale with brown markings. **Hab** Grassland. **Fp** Legumes. **SS** Potato Capsid and Timothy Grassbug. Others to a lesser extent, including Rosy Plant Bug.

J F M A M J J A S O N D

Calocoris 2 spp. | 1 ILL.

Rosy Plant Bug
Calocoris roseomaculatus

Common

BL 6–8 mm. **ID** Green, pronotum plain or suffused with short, dark marks; forewings with rose blotches; head, pronotum and legs often also rose-tinged; scutellum with dark central line. **Hab** Grassland. **Fp** Composites (flowers and fruits). **SS** Potato Capsid. Superficially other plant bugs.

J F M A M J J A S O N

Stenotus 1 sp.

Timothy Grassbug
Stenotus binotatus

Common

BL 6–7 mm. **ID** Black with yellowish pronotum and forewings with dark marks, much reduced in ♀. **Hab** Grassland. **Fp** Grasses, on flowering heads. **SS** Lucerne Bug, Potato Capsid, Orange-spotted Plant Bug.

J F M A M J J A S O N D

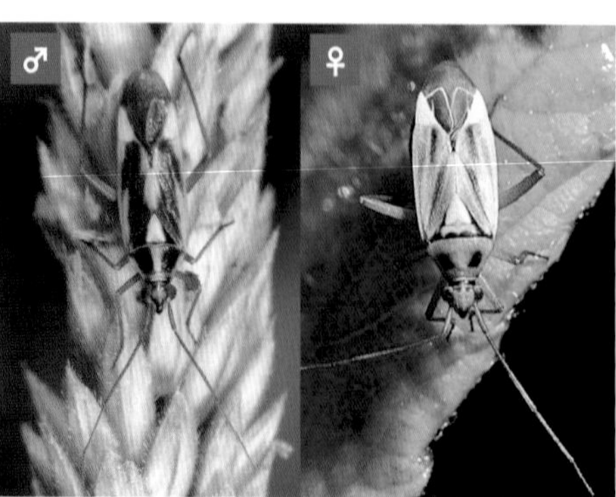

Grypocoris 1 sp.

Orange-spotted Plant Bug
Grypocoris stysi

Common

BL 6–8 mm. **ID** Black with yellow or whitish markings; cuneus orange. **Hab** Woodland. **Fp** Nettles (flower heads), as well as predating small insects. **SS** Timothy Grassbug.

J F M A M J J A S O N

Closterotomus 3 spp. | 2 ILL.

Potato Capsid
Closterotomus norwegicus

Common

BL 6–8 mm. **ID** Green, pale green with pair of black spots on pronotum; forewings green or suffused with brown; scutellum plain, or with two dark longitudinal lines. **Hab** Grassland, woodland, hedgerows. **Fp** Nettles, clovers and composites. **SS** Lucerne Bug, Rosy Plant Bug, Timothy Grassbug.

Trivial Plant Bug
Closterotomus trivialis

Common (recent colonist)

BL 6–8 mm. **ID** ♂ black with striking red patches; ♀ yellow and green, including head. Pronotum with two central and two lateral black spots. **Hab** Gardens, meadows, hedgerows. **Fp** St. John's-worts. **SS** Spotted Plant Bug *Closterotomus fulvomaculatus* [N/I] (**BL** 6–7 mm), which has smaller whitish spots with black tips.

Harpocera 1 sp.

Handsome Plant Bug
Harpocera thoracica

Common

BL 6–7 mm. **ID** ♂ rather elongated, black with whitish patterning; antenna S2 broadened. ♀ brown. Pronotum with pale central line. **Hab** Woodlands. **Fp** Oaks. **SS** None.

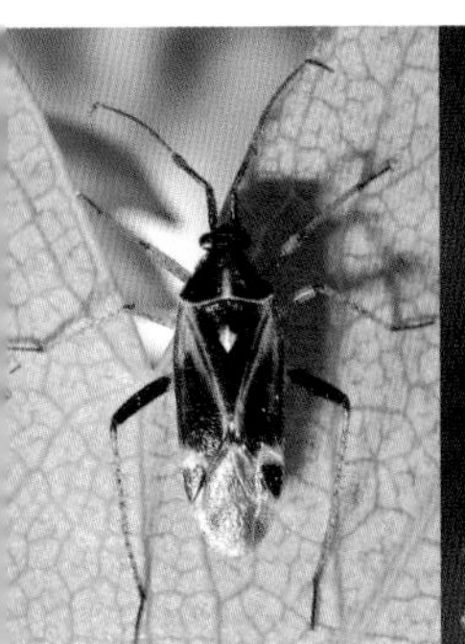

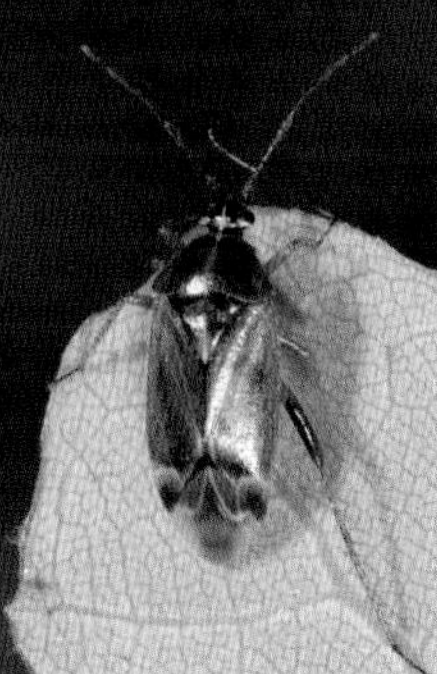

Liocoris 1 sp.

Nettle Plant Bug
Liocoris tripustulatus

Common

BL 4–5 mm. **ID** Variable, but distinctive, cream to deep yellow scutellum and cuneus. **Hab** Grassland. **Fp** Nettles. **SS** *Orthops* species (*p. 198*), which are, however, associated with umbellifers.

Phytocoris 9 spp. | 2 ILL.

Long-legged Plant Bug
Phytocoris varipes

BL 6–7 mm. **ID** Long hind femora and long antenna S1. **Hab** Grassland. **Fp** Flowers and fruits of various plants, including clovers and grasses. **SS** Other *Phytocoris* species: microscopic examination of ♂ genitalia and determination of length of antenna S4 required. (This is usually the commonest species.)

Lime Plant Bug
Phytocoris tiliae

BL 6–7 mm. **ID** Greyish-green, wings mottled black; pronotum usually with black sides and marks; hind femora long. **Hab** Woodlands. **Fp** Various trees, particularly oaks. **Beh** Feed mainly on small insects. **SS** Other *Phytocoris* species, although these lack black sides to pronotum.

Capsodes 3 spp. | 1 ILL.

Gothic Plant Bug
Capsodes gothicus

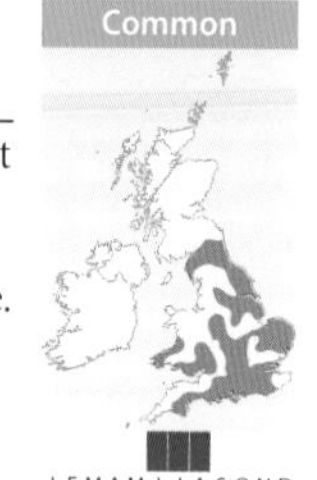

BL 6 mm. **ID** Black with distinctive, but variable orange markings; antennae and legs black. **Hab** Damp sites. **Fp** Greater Bird's-foot-trefoil. **SS** None.

Orthops 3 spp. | 1 ILL.

Kalm's Plant Bug
Orthops kalmi

BL 4 mm. **ID** Reddish brown or with paler areas (scutellum with variable yellow mark); antennae dark. **Hab** Grasslands. **Fp** Umbellifers. **SS** **Nettle Plant Bug** (*p. 197*); other similarly marked *Orthops* species, often on umbellifers, which may require microscopic examination of antennae (S2 length), rostrum length and subtle forewing marks.

Pantilius 1 sp.

Tunic Plant Bug
Pantilius tunicatus

Common

BL 8–10 mm. **ID** Rather large size and characteristic dull green to red colour, forewings mottled with dark spots; red patch at tip of abdomen. Can be short- or long-winged. **Hab** Woodland, gardens. **Fp** Hazel, Alder and birches. **SS** None.

Notostira 1 sp.

Elongated Grass Bug
Notostira elongata

Common

BL 8–9 mm. **ID** ♂ black with margins yellowish; ♀ paler. **Hab** Grasslands. **Fp** Grasses. **SS** Various other grass feeders, particularly *Stenodema* species (4) [N/I] (**BL** 6–9 mm), although these do not have long dark hairs on the basal segment of antennae and tibiae.

Lygocoris 2 spp. | 1 ILL.

Common Green Capsid
Lygocoris pabulinus

Common

BL 5–7 mm. **ID** Green, with smooth pronotum; under a microscope, small brown spines on tibiae. **Hab** Grassland, gardens and other sites with nettles. **Fp** Nettles. **SS** Green Willow Capsid *Lygocoris rugicollis* [N/I] (**BL** 6–7 mm), which has a wrinkled pronotum and is likely on willows; it lacks the brown tibial spines.

Oncotylus 1 sp.

Knapweed Plant Bug
Oncotylus viridiflavus

Common

BL 5–6 mm. **ID** Head and pronotum bluish-green with distinctive black patterning; forewings green, striped black. **Hab** Grasslands. **Fp** Knapweeds. **SS** None.

Miris 1 sp.

Streaked Plant Bug
Miris striatus

BL 9–11 mm. **ID** Pronotum mainly yellow. Forewings black with yellow streaks; cuneus yellow. **Hab** Woodland. **Fp** Oaks, Hawthorn. **SS Striped Plant Bug**, which is broader.

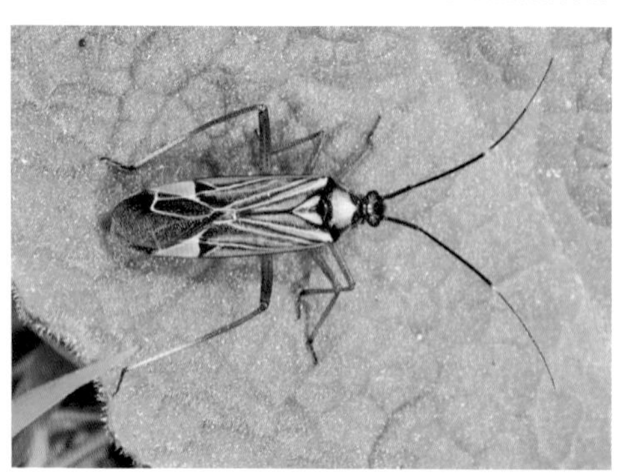

Dryophilocoris 1 sp.

Four-spotted Plant Bug
Dryophilocoris flavoquadrimaculatus

BL 6–7 mm. **ID** Black, with four extensive yellow markings on forewings. **Hab** Woodland. **Fp** Oaks. **SS** None.

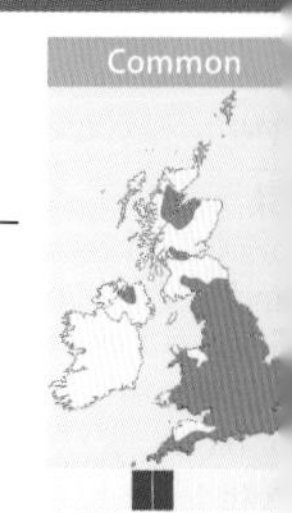

Rhabdomiris 1 sp.

Striped Plant Bug
Rhabdomiris striatellus

BL 7–8 mm. **ID** Pronotum yellow with black markings. Forewings black with yellow stripes; cuneus yellow with black tip. **Hab** Woodland. **Fp** Oaks. **SS Streaked Plant Bug**, which is more elongate.

Malacocoris 1 sp.

Delicate Apple Capsid
Malacocoris chlorizans

BL 3–4 mm. **ID** Forewings pale green, mottled darker green; underside of antenna S1 & S2 black. **Hab** Woodlands, hedgerows, parks; mainly on Hazel, also apples. **Beh** Mostly predatory. **SS** None.

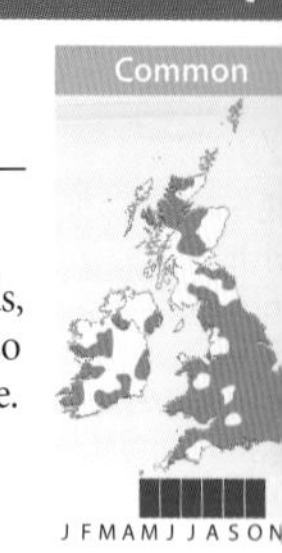

Monalocoris 1 sp.

3racken Bug
Vonalocoris filicis

3L 2–3 mm. **ID** Brown, head pale ›range. **Hab** Woodland. **Fp** Bracken ınd other, mainly on the sporangia. **;S** Fern Bug *Bryocoris pteridis* [N/I] **BL** 2–3 mm), which has a dark brown ›r black head and is usually short-›inged.

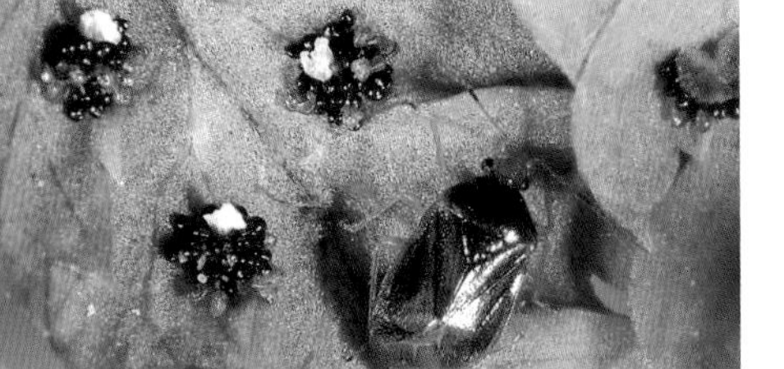

Pithanus 1 sp.

Short-winged Plant Bug
Pithanus maerkelii

BL 4–5 mm. **ID** Black with connexivum, forewing edges and antenna S1 pale; usually short-winged (occasionally long-winged); legs reddish. **Hab** Damp grasslands. **Beh** Predatory. **SS** None.

Lygus 5 spp. | 2 ILL.

:oastal Plant Bug
ygus maritimus

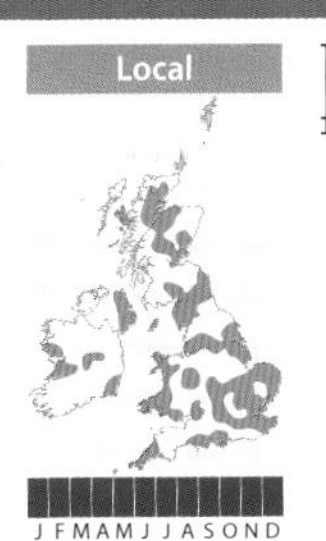

3L 5–6 mm. **ID** Pronotum broad, ›ith rosy markings. **Hab** Coastal ›nd sandy areas. **Fp** Common Sorrel, ;orse, clovers and others. **SS** Other *ygus* species, identification of which equires microscopic examination of etails of corium.

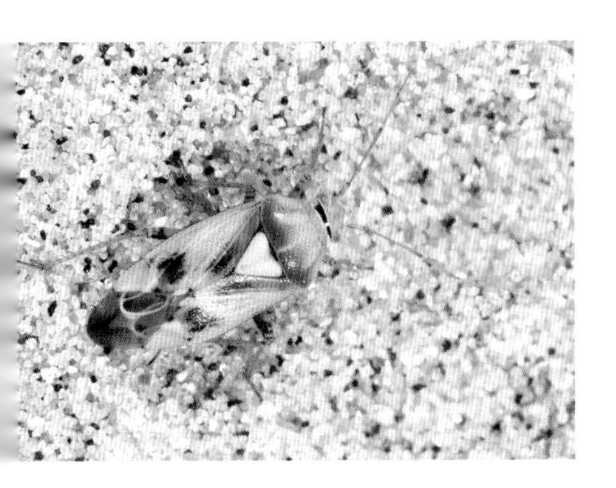

Tarnished Plant Bug
Lygus rugulipennis

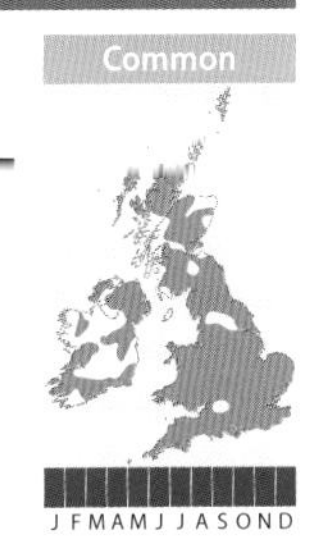

BL 5–6 mm. **ID** Mottled yellowish-brown to dark; . **Hab** Grasslands. **Fp** Common Nettle, docks, clovers and others. **SS** Other *Lygus* species, identification of which requires microscopic examination of details of corium.

Tuponia 3 spp. | 2 ILL.

Mediterraneran Tamarisk Plant Bug
Tuponia hippophaes

Local (recent colonist)

BL 3 mm. **ID** Green; tip of cuneus white. **Hab** Often coastal, on tamarisk. **Fp** Tamarisk. **Dist** First recorded 2010. **SS** Plain Tamarisk Plant Bug *Tuponia brevirostris* [N/I] (**BL** 3 mm), which has a shorter rostrum.

Colourful Tamarisk Plant Bug
Tuponia mixticolor

Local

BL 3 mm. **ID** Forewings part reddish-brown; front of scutellum orange, remainder pale yellowish. **Hab** Often coastal, on tamarisk. **Fp** Tamarisk. **Dist** First recorded 1979. **SS** None.

Systellonotus 1 sp.

Ant Plant Bug
Systellonotus triguttatus

Nationally Scarce

BL 4–5 mm. **ID** Boldly marked, ♂ with forewings brown with pale marks and black tip; ♀ short-winged, rather ant-like. **Hab** Sandy heaths and coastal dunes. **Beh** Partly predatory and believed to be associated with ants. **SS** *Hallodapus* species (2) [N/I] (**BL** 3–4 mm), in which eyes do not touch the pronotum.

Pilophorus 4 spp. | 1 ILL

Cinnamon Plant Bug
Pilophorus cinnamopterus

Common

BL 4·5 mm. **ID** Black; forewings partly rich medium-brown with narrow silvery bands, only the hind band continuous. **Hab** Coniferous woodlands, associated with Scots Pine. **Fp** Buds, needles and sap from pine. **Beh** Also predatory on aphids. **SS** None.

Orthocephalus 2 spp. | 1 ILL.

Hairy Plant Bug
Orthocephalus coriaceus

Local

BL 4–5 mm. **ID** Black and distinctly hairy with hint of yellowish on forewings (♂); usually long-winged (although some individuals short-winged); hindlegs long, adapted for jumping. **Hab** Grasslands. **Fp** Bedstraws, *etc.* **SS** Jumping Plant Bug *Orthocephalus saltator* [N/I] (**BL** 4–5 mm), which has red or yellowish-brown tibiae (black in Hairy Plant Bug).

J F M A M J J A S O N D

Megacoelum 2 spp. | 1 ILL.

Becker's Plant Bug
Megacoelum beckeri

Common

BL 7 mm. **ID** Black-and-brown; long legs (with tibial spines) and antennae. **Hab** Coniferous woodlands, associated with Scots Pine. **Beh** Predatory on aphids. **SS** The similar Oak Plant Bug *Megacoelum infusum* [N/I] (**BL** 7 mm) is found on oaks and tends to be paler, with shorter tibial spines.

J F M A M J J A S O N D

Platycranus 1 sp.

Gorse Plant Bug
Platycranus bicolor

Local

BL 4 mm. **ID** Green with distinct pale hairs; head, pronotum and clavus often brown (♂). **Hab** Scrubby areas, on Gorse. **Fp** Gorse. **SS** Some green *Orthotylus* species, but only Gorse Plant Bug is found on Gorse in August/September.

J F M A M J J A S O N D

Orthotylus 19 spp. | 1 ILL.

Glasswort Plant Bug
Orthotylus rubidus

Rare (RDB3)

BL 3 mm. **ID** Red. **Hab** Saline areas, on glassworts. **Fp** Glassworts. **SS** None.

J F M A M J J A S O N D

FAMILY Saldidae (Shorebugs) 8 GEN. | 22 spp. | 3 SPP. ILL.

Small, oval, predatory bugs with large eyes. They are quick-moving and can also fly well. Found in wetland margins, including coastal areas, where adults overwinter. **Antennae 4-segmented; tarsi 3-segmented; ocelli present**. Species easily confused and most, except those illustrated, require examination of pubescence, markings and antenna segment ratios under a microscope.

Saldula	14 spp. | 3 ILL.
MARKINGS white bands across corium; ANTENNAE S1 and tip of S2 white	White-banded Shorebug
MARKINGS no clear white bands; FOREWINGS pale; ANTENNAE dark	Pale Shorebug
MARKINGS very dark; FOREWINGS with pale marginal spots; ANTENNAE dark	Common Shorebug

White-banded Shorebug
Saldula arenicola

BL 4 mm. **ID** Dark, oval; forewings with whitish band across corium; clavus with two white spots, which also feature on forewing margin. Antenna S1 and tip of S2 whitish. Legs in part whitish. **Hab** Seepages on soft-rock cliffs; inland in sand and gravel pits. **SS** None.

Pale Shorebug
Saldula pallipes

BL 4 mm. **ID** Dark brownish; forewings pale, with brown markings. **Hab** Pools and flooded gravel pits. **Dist** Possibly more widespread than shown on map, but seldom recorded. **SS** Other *Saldula* species, particularly the pale form of the Nationally Scarce **Estuarine Shorebug** *Saldula palustris* [N/I] (**BL** 4–5 mm); specialist microscopic examination required.

Common Shorebug
Saldula saltatoria

BL 4–5 mm. **ID** Black or brownish, oval; forewings with pale marginal spots (individually variable). **Hab** Margins of ponds and ditches and slow-moving streams; well camouflaged on mud. **SS** Other *Saldula* species; specialist microscopic examination required.

FAMILY **Veliidae** (Water crickets) — 2 GEN. | 5 spp. | 1 SP. ILL.

Robust, predatory bugs inhabiting the water surface of swift-flowing streams; ready to pounce quickly when insects falling on the surface cause ripples. Antennae moderate length, 4-segmented. Adults overwinter. Other than the well-known species shown, some of the tiny (body length approx. 2 mm) water crickets require examination of minor abdomen details under a microscope.

Velia — 4 spp. | 1 ILL.

Water Cricket *Velia caprai*

Common

J F M A M J J A S O N D

BL 5–7 mm. **ID** Black; usually wingless but some individuals winged; abdomen with two broad orange longitudinal bands; pronotum and abdomen with paired white spots; pronotum with orange tinge. **Hab** Shaded parts of streams, rarely ponds; can occur in numbers. **Dist** Throughout GB; also most of Ireland. **SS** Saul's Water Cricket *Velia saulii* [N/I] (**BL** 6–7 mm), which occurs mainly in N Britain and N. Ireland (minor abdomen details need to be checked under a microscope to confirm identification).

FAMILY **Hydrometridae** (Water measurers) — 1 GEN. | 2 spp. | 1 ILL.

Elongate (body length 7–12 mm), slow, predatory bugs of water margins, which feed by piercing water fleas, mosquito larvae and other prey through the film of the water surface. The head is ≥5 × longer than wide, with eyes located low down. Antennae moderate length, 4-segmented – S3 the longest, ≥3 × length of S2. Adults overwinter.

Hydrometra — 2 spp. | 1 ILL.

Water Measurer *Hydrometra stagnorum*

BL 9–12 mm. **ID** Dark brown to black, head shape unmistakable, with the eyes located ⅓ of the way along from the base of the head; short-winged (rarely long-winged); abdomen with 5 pairs of pale spots. **Hab** Margins of slow-flowing streams, ponds, rivers, ditches, even seepages on cliffs. **SS** The Red Listed (Vulnerable) **Lesser Water Measurer** *Hydrometra gracilenta* [N/I] (**BL** 7–9 mm), restricted to few sites in the Norfolk Broads and Pevensey Levels, East Sussex, is smaller and the eyes are located almost halfway along the head.

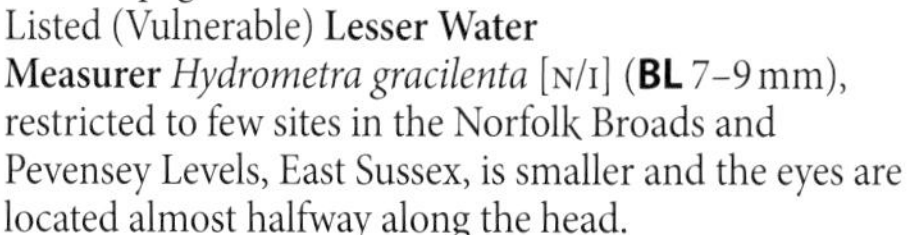

EYES located ⅓ of the way along the head from the base

Guide to families pp. 142–147

FAMILY Gerridae (Pondskaters) 3 GEN. | 10 spp. | 3 SPP. ILL.

Elongate (BL 6–18 mm), water surface predators. Mid and hindlegs long, adapted for 'rowing'; once invertebrate prey is detected on the water, the forelegs are used to grasp it. Antennae moderate length, 4-segmented. Adults overwinter on land. Some of the more easily identified species shown.

BL small ≤10 mm; WINGS short or long; PRONOTUM plain sides		Common Pondskater
BL large ±15 mm	WINGS usually wingless; PRONOTUM plain sides	River Pondskater
	WINGS short or long; PRONOTUM yellow line on sides	Lake Pondskater

Gerris 6 spp. | 1 ILL.

Common Pondskater
Gerris lacustris

BL 8–10 mm. **ID** Dark brown or black; short-winged, long-winged and intermediate forms frequent. Fore femora pale, with black lines. **Hab** Ponds, ditches, lakes, backwaters of rivers and streams. **SS** Other *Gerris* species, from which told by pale front to femora, with black band from tip to basal third.

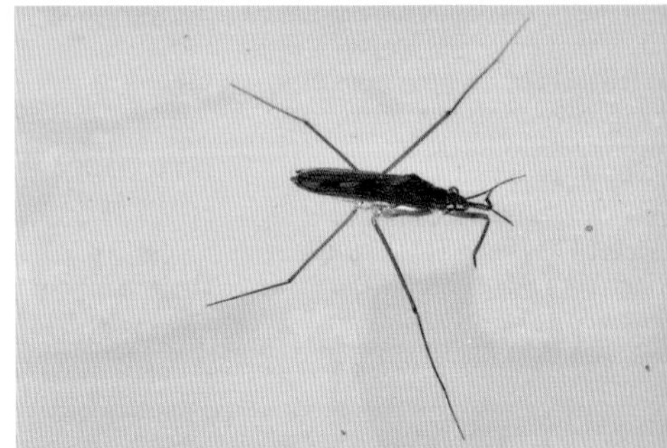

Aquarius 2 spp.

River Pondskater
Aquarius najas

BL 13–17 mm. **ID** Large, dark brown; nearly always wingless. **Hab** Margins of rivers, large streams and canals with flowing water; often seen in large numbers. **SS** Lake Pondskater.

Lake Pondskater
Aquarius paludum

BL 14–16 mm. **ID** Large, dark brown; short or long-winged; pronotum sides with yellow line. **Hab** Lakes, canals and large ponds. **SS** River Pondskater.

FAMILY **Nepidae** (Water scorpions) — 2 GEN. | 2 spp.

This family comprises two large species (BL 18–35 mm) of rather different appearance. They are easily recognized by their stiff, elongated tail, which is used for breathing. Prey includes invertebrates, tadpoles and even small fishes. Adults overwinter.

Nepa — 1 sp.

Water Scorpion *Nepa cinerea*

Common

BL 18–22 mm. **ID** Dark brown; oval; head small; eyes large. Tail ≥ half length of body. The common name originates from a scorpion-like appearance, the tail [similar to a stinger] and shape of the forelegs. Forewings cover the reddish abdomen. **Hab** Shallows of ponds, lakes, canals and streams. **Beh** Walks on land at night, sometimes by day, in search of prey. **SS** None.

Ranatra — 1 sp.

Water Stick-insect
Ranatra linearis

Local

BL 30–35 mm. **ID** Straw, but some individuals darker brown or black; elongate; head small, eyes large; fore coxae enlarged; tail well in excess of half length of body. Wings long. **Hab** Ponds and lakes with dense, emergent vegetation. **Beh** In defence, can play dead. When dispersing, may be seen on the ground, or in flight. **SS** None.

FAMILY **Naucoridae** (Saucer bugs) — 2 GEN. | 2 spp. | 1 SP. ILL.

Large, saucer-shaped; fore femora stout, curved. Prey includes invertebrates. Adults overwinter in water.

Ilyocoris — 1 sp.

Saucer Bug
Ilyocoris cimicoides

Common

BL 12–16 mm. **ID** Dark brown; oval; eyes large. Forewings cover abdomen; mid and hindlegs hairy. The rostrum can inflict a bite, similar in sensation to a wasp sting. **Hab** Muddy ponds, canals and ditches. **Beh** Usually flat on the bottom but sometimes seen walking rapidly on land by day, or in flight. **SS** None.

 Guide to families pp. 142–147

FAMILY **Corixidae** (Lesser water boatmen) 9 GEN. | 39 spp. | 1 SP. ILL.

Characteristic mainly yellow and black marked bugs (body length ≤3 to14 mm), with powerful legs. They mainly feed on algae, **swimming to the surface briefly right-side up**. Adults overwinter.

Corixa 5 spp. | 1 ILL.

Common Water Boatman
Corixa punctata

BL 12–14 mm. **ID** Light brown; head pale, eyes red. Pronotum pale at front, black at rear; scutellum black. Forewings dark with mottled areas. Usually winged but wingless forms known. **Hab** Ponds, lakes, canals, slow-moving rivers. **SS** Other *Corixa* species, from which told by slightly larger size (others <11 mm.) or, if same length, mid and hind femora simple, not notched at basal end of mid tibiae; distinguished from backswimmers by different resting position and pronotum shape.

FAMILY **Notonectidae** (Backswimmers) 1 GEN. | 4 spp. | 2 SPP. ILL.

Characteristic **backswimming** water surface bugs (body length ≥11 mm) with powerful hindlegs; hind tibiae flattened; tarsi 2-segmented. Prey includes invertebrates and tadpoles. Adults overwinter in some species. NB to observe the metanotum, the wings of a specimen need to be lifted gently – **beware, if handled, the short rostrum can inflict a painful wound.**

NOTE: traditionally known as Water Boatmen, with Corixidae as Lesser Water Boatmen, but the name backswimmers is gaining popularity and is preferred in order to avoid confusion.

Common Backswimmer as typically seen, swimming on its back.

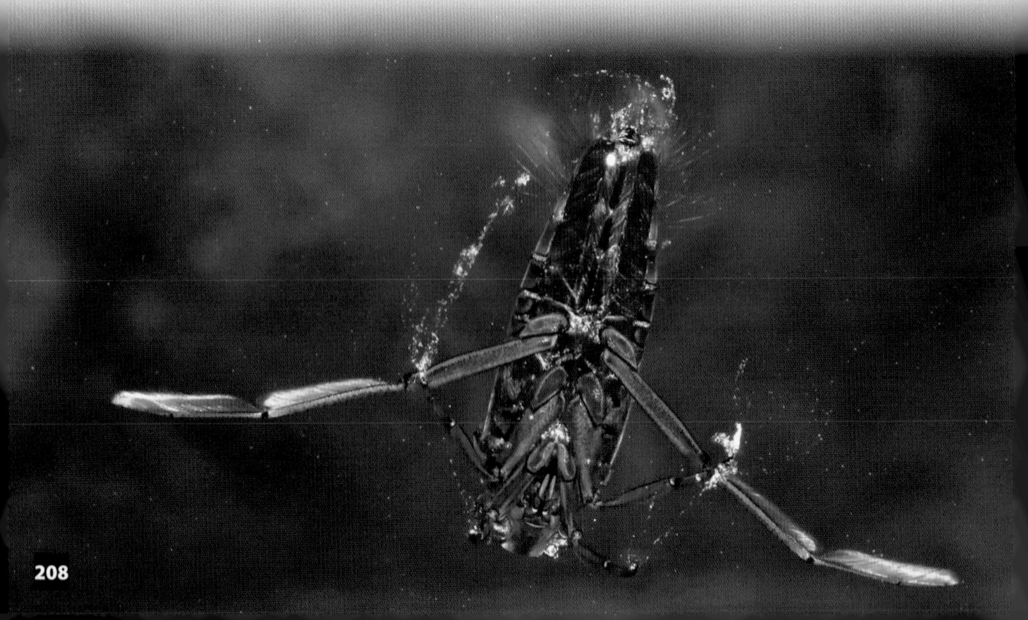

Notonecta 4 spp. | 2 ILL.

Common Backswimmer
Notonecta glauca

BL 13–16 mm. **ID** Light brown, head pale, eyes red. Pronotum has square front corners, and is pale at front and black at rear; scutellum black. Forewings light brown with a few dark marks. Metanotum black. Larger nymphs are pale green. **Hab** Ponds, canals and ditches. **SS** Other *Notonecta* species, particularly Green Backswimmer *Notonecta viridis* [N/I] (**BL** <14 mm), which also has pale wings, but pronotum has pointed front corners that extend partly around the eyes.

Peppered Backswimmer
Notonecta maculata

BL approx. 15 mm. **ID** Light brown, head pale, eyes red. Pronotum is pale at front and black at rear; scutellum black. Forewings dark with mottled areas. Metanotum orange with a pair of black spots. **Hab** Cattle troughs and small garden ponds. **Beh** Eggs overwinter, sometimes adults. **SS** Other *Notonecta* species, from which told by large size and mottled forewings.

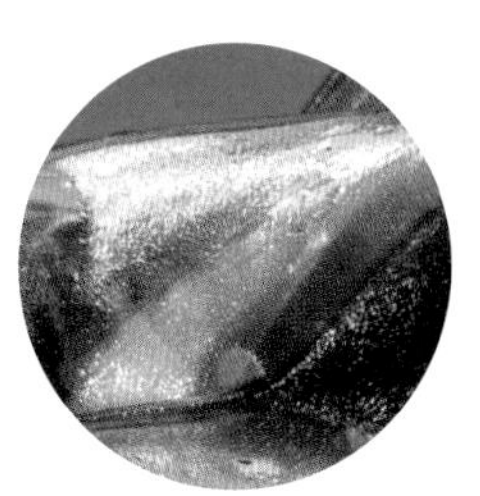

Common Backswimmer forewings light brown with a few dark marks

Peppered Backswimmer forewings dark with mottled areas

SUBORDER Auchenorrhyncha [= Cicadomorpha and Fulgoromorpha]

GUIDE TO SELECTED FAMILIES | 9 BI

Infraorder CICADOMORPHA
(Cicadas, leafhoppers, froghoppers, treehoppers)

Infraorder FULGOROMORPHA
(Lacehoppers, planthoppers, issid planthoppers, tettigometrid planthoppers)

BL large 16–27 mm; rarely<20 mm.

Cicadas
Cicadidae
[New Forest Cicada]
p. 211

BL mostly <12 mm.

PRONOTUM large, hood-like (unlike any others in this suborder)

Horned Treehopper

Treehoppers
Membracidae
[2 genera, 2 species]
p. 211

HIND TIBIAE with 1–3 rows of spines

Green Leafhopper

Leafhoppers
Cicadellidae
[105 genera, 285 species]
p. 214

HIND TIBIAE with a few well-spaced spines

FORM red-and-black

Froghoppers
Cercopidae
[Red-and-black Froghopper]
p. 212

FORM subdued colours

Common Froghopper

Froghoppers
Aphrophoridae
[3 genera, 9 species]
p. 212

ANTENNAE near lower margin of eyes

Bracken Planthopper

Planthoppers
Delphacidae
[40 genera, 75 species]
p. 220

ANTENNAE not near lower margin of eyes

FOREWINGS clear (veins distinctive) and membranous; **SCUTELLUM** with 3 or 5 keels

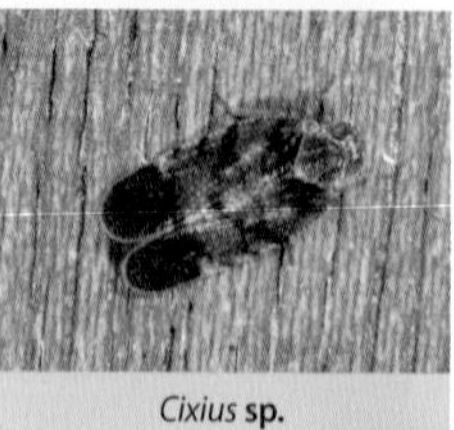

Cixius sp.

Lacehoppers
Cixiidae
[5 genera, 12 species]
p. 218

FOREWINGS bulging; many cross-veins; **SCUTELLUM** with 3 keels

Common Issid Bug

Issid Planthoppers
Issidae
[1 genus, 2 species]
p. 221

FOREWINGS hardened and pitted; **SCUTELLUM** smooth, without keels

Tettigometrid Planthoppers
Tettigometridae
[Chalk Planthopper]
p. 222

FAMILY **Cicadidae** (Cicadas) 1 sp.

Characteristic large (BL16–27 mm), robust, bugs with transparent wings; mainly associated with warm climates. Adults feed on twigs, while nymphs take 6–8 years to develop underground on roots. Adults sing loudly, although the New Forest Cicada has a high-pitched song that can be difficult to hear. Ash Cicada *Cicada orni* [N/I], a southern European species, was reported singing from trees in August 2018 (Essex, Sussex) and noted from Kent in 2005, but these may be short-lived, accidental imports.

Cicadetta 1 sp.

New Forest Cicada EN
Cicadetta montana

BL 16–27 mm. **ID** Black; abdomen orange-striped. Legs in part orange. Forewings well in excess of tip of abdomen; transparent. **Hab** Sunny woodland clearings. **Dist** Restricted to the New Forest, Hampshire, adult last recorded in 1992, but elusive and likely still to exist somewhere in the area. **SS** None.

FAMILY **Membracidae** (Treehoppers) 2 GEN. | 2 spp.

Pronotum enlarged and extended hood-like; backwards (body length 5–10 mm).

Centrotus 1 sp.

Horned Treehopper
Centrotus cornutus

BL Approx. 10 mm. **ID** Dark brown, pronotum with horn-like projections; large, wavy spine. **Hab** Woodland clearings and rides, on low vegetation. **SS** Broom Treehopper.

Gargara 1 sp.

Broom Treehopper
Gargara genistae

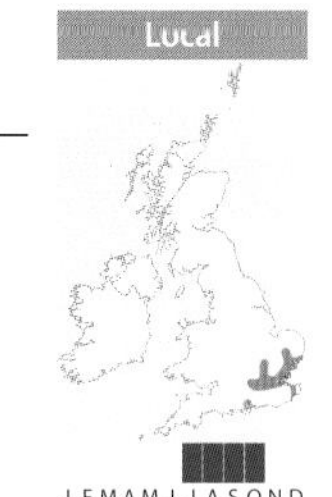

BL Approx. 6 mm. **ID** Dark brown, pronotum with large, straight spine. **Hab** Open areas with low Broom growth. **Fp** Broom. **SS** Horned Treehopper.

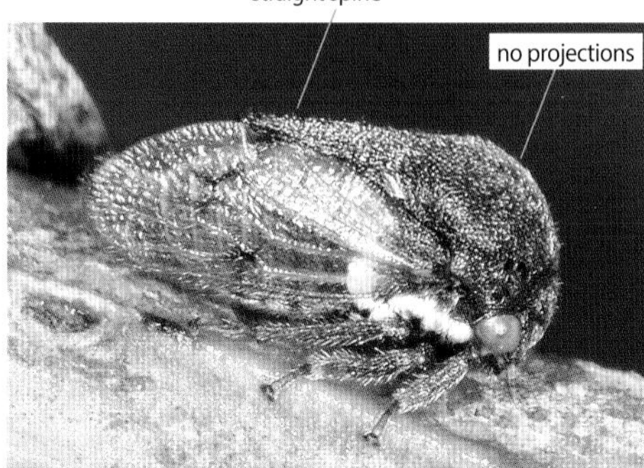

Guide to families *p. 21*

FAMILY **Cercopidae** (Froghoppers) — 1 sp.

Characteristic large, colourful red-and-black froghopper (BL approx. 10 mm), which can jump. Nymphs develop underground on roots.

Cercopis — 1 sp.

Red-and-black Froghopper
Cercopis vulnerata

BL 9–11 mm. **ID** Distinctive red-and-black species. **Hab** Woodland and more open habitats, often found on stems of low-growing vegetation and grasses. **SS** None.

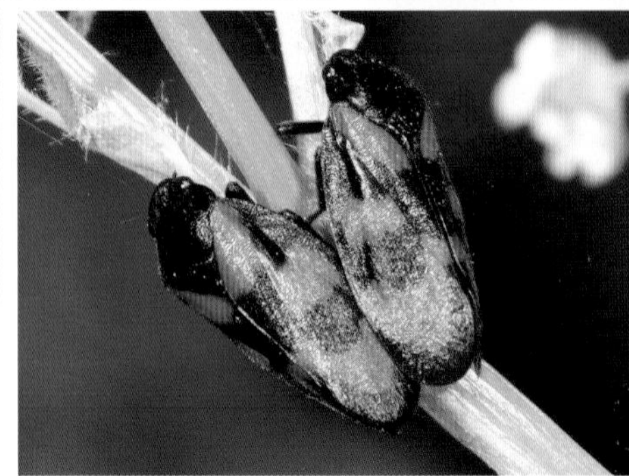

FAMILY **Aphrophoridae** (Froghoppers) — 3 GEN. | 9 spp. | 3 SPP. ILL.

Rather dull froghoppers (body length 3–10 mm), which can jump well. The plant-sucking nymphs are afforded protection by developing in froth on stems, commonly referred to as cuckoo-spit. Genera/species can be distinguished as follows:

BL 10 mm; **HEAD + PRONOTUM** with distinct keel on midline [see *inset opposite*]		***Aphrophora***
BL ± 6 mm	**HEAD + PRONOTUM** with central line [see *inset opposite*]; **FOREWINGS** with orange or brown veins, outer margin straight in final ⅔	***Neophilaenus***
	HEAD + PRONOTUM with no keel or central line [see *inset opposite*]; **FOREWINGS** with veins same colour as wings, outer margin fully rounded	***Philaenus***

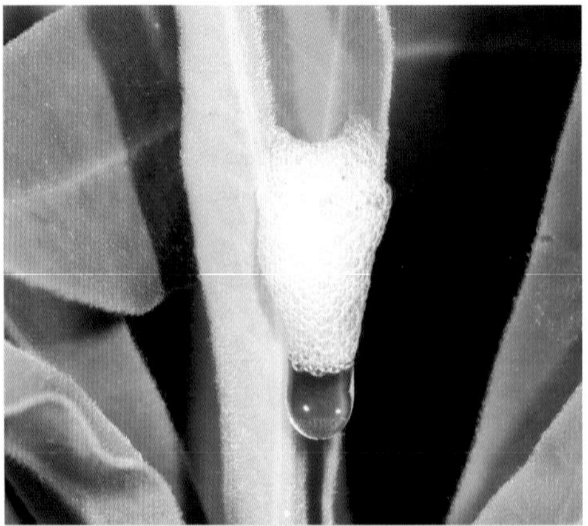

Common Froghopper nymph in cuckoo-spit

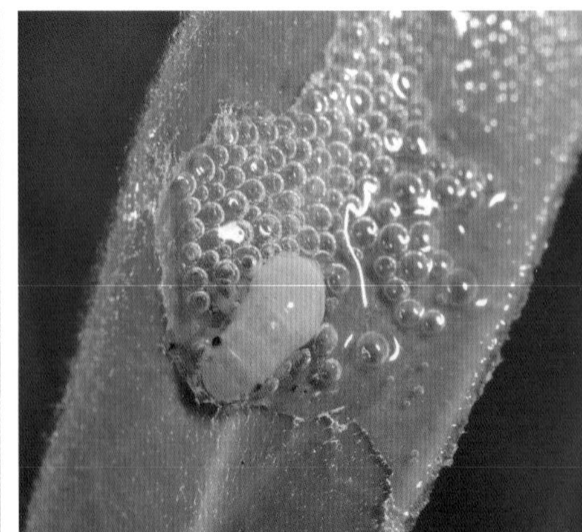

Common Froghopper nymph with cuckoo-spit removed

Aphrophora 4 spp. | 1 ILL.

Alder Spittlebug
Aphrophora alni

Common

BL 9–10 mm. **ID** Dark brown, often with white patches on forewing margins; body and smooth forewings have numerous dark tubercles. **Hab** Woodland, larger gardens and open habitats, on trees and bushes. **SS** Other *Aphrophora* species and colour forms of **Common Froghopper**, but larger and lacks fine hairs on body.

Aphrophora HEAD + PRONOTUM midline keel	keel
Neophilaenus HEAD + PRONOTUM central line	line
Philaenus HEAD + PRONOTUM no central line	no line

Neophilaenus 4 spp. | 1 ILL.

Lined Froghopper
Neophilaenus lineatus

Common

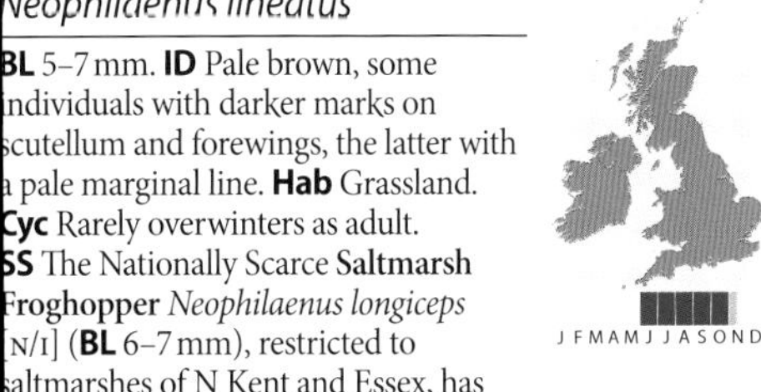

BL 5–7 mm. **ID** Pale brown, some individuals with darker marks on scutellum and forewings, the latter with a pale marginal line. **Hab** Grassland. **Cyc** Rarely overwinters as adult. **SS** The Nationally Scarce **Saltmarsh Froghopper** *Neophilaenus longiceps* [N/I] (**BL** 6–7 mm), restricted to saltmarshes of N Kent and Essex, has upper surface of head as long as the basal width (wider than long in Lined Froghopper).

Philaenus 1 sp.

Common Froghopper
Philaenus spumarius

Common

BL 5–7 mm. **ID** Brown, whitish, black or mixed colours; hairy. **Hab** Grassland, hedgerows and gardens. **SS** **Alder Spittlebug** [see images *above* for comparison with with other genera].

Guide to families p. 210

FAMILY Cicadellidae (Leafhoppers) 105 GEN. | 285 spp. | 16 SPP. ILL.

A varied range of elongate (BL 2–18 mm), jumping, plant-feeding bugs, some very colourful and distinctive. Hind tibiae bears 1–3 rows of spines; forewings very soft. Most species overwinter as eggs or nymphs, rarely as adults. A selection of common or particularly distinctive species is illustrated.

Acericerus 3 spp. | 1 ILL.

Maple Leafhopper
Acericerus vittifrons

Local

BL 5–6 mm. **ID** Brown; forewings with yellowish or pale marks on veins. **Hab** Parks and gardens. **Fp** Field Maple. **SS** Other *Acericerus* species, but only two feed on maple; ♂ Maple Leafhopper has a dark bar of the face, which lacks a pale midline.

Allygus 2 spp. | 1 ILL.

Mottled Leafhopper
Allygus mixtus

Common

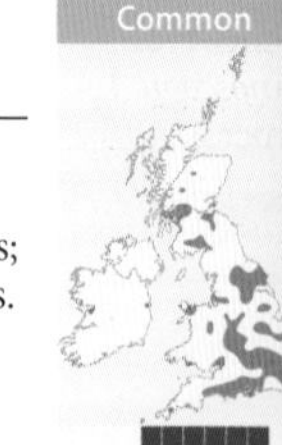

BL 3–4 mm. **ID** Brownish; forewings with large cells and whitish veins and cross-veins. **Hab** Woodland with oaks; grassland. **Fp** Nymphs feed on grasses. **SS** None.

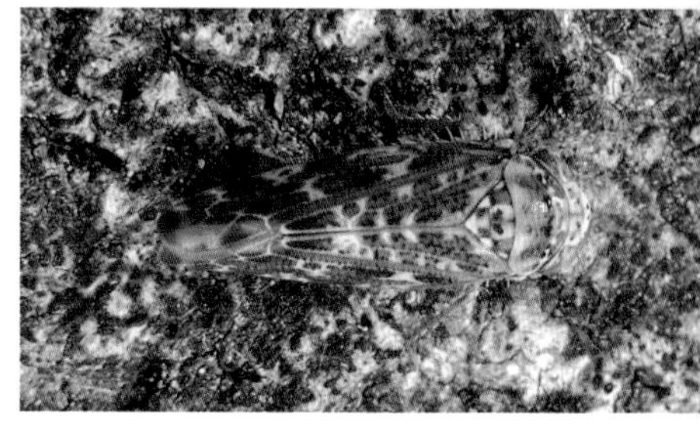

Adarrus 2 spp. | 1 ILL.

Eyed Leafhopper
Adarrus ocellaris

Common

BL 3–4 mm. **ID** Brownish; forewings with whitish cross-veins and dark-edged cells. Back of head with whitish cross. Forewings reduced or full length. **Hab** Grassland. **SS** None.

Alebra 4 spp. | 1 ILL.

Hazel Leafhopper
Alebra coryli

Local

BL 4 mm. **ID** Yellowish; head, pronotum and forewings with extensive broad, whitish stripes. **Hab** Woodland, hedgerows and others. **Fp** Hazel. **SS** Other *Alebra* species, from which told by dissection and microscopic examination of genitalia.

Anoscopus 7 spp. | 1 ILL.

White-banded Leafhopper
Anoscopus albifrons

BL 3–5 mm. **ID** Dark brownish. ♂ has black hind tibiae, and forewings with wavy whitish bands; larger ♀ has brown hind tibiae, and forewings mottled with whitish flecks on veins. **Hab** Grassland. **SS** Other *Anoscopus* species, from which told by microscopic examination of genitalia.

Cicadella 2 spp. | 1 ILL.

Green Leafhopper
Cicadella viridis

BL 6–8 mm. **ID** ♀ has a yellowish head with two black spots and a green pronotum and forewings; ♂ similar but forewings bluish-purple. **Hab** Wetlands, where can be abundant. **SS** None.

Eupelix 1 sp.

Large-headed Leafhopper
Eupelix cuspidata

BL 6–7 mm. **ID** Brown, head enlarged; head and pronotum with prominent central keel. **Hab** Grassland. **SS** None.

Evacanthus 2 spp. | 1 ILL.

Yellow-and-black Leafhopper
Evacanthus interruptus

BL 6–7 mm. **ID** Yellow-and-black, forewings rarely all-yellow. Forewings in ♀ much shorter than abdomen. **Hab** Grassland and scrub. **SS** None.

Graphocephala 1 sp.

Rhododendron Leafhopper
Graphocephala fennahi

BL 8–10 mm. **ID** Green, forewings with two long red streaks. Head, scutellum and part of pronotum yellowish. **Hab** Woodland, gardens and parks. **Fp** Rhododendron; can be abundant. **SS** None.

Iassus 2 spp. | 1 ILL.

Oak Leafhopper *Iassus lanio*

BL 7–8 mm. **ID** Head and pronotum brown, often rather mottled dark; eyes red; forewings green or pale brown. **Hab** Woodland. **Fp** Oaks. **SS** Elm Leafhopper *Iassus scutellaris* [N/I] (**BL** 7–8 mm), found on elms, mainly in SE England.

Ledra 1 sp.

Eared Leafhopper
Ledra aurita

BL 13–18 mm. **ID** Large, greyish; pronotum with ear-like projections. **Hab** Woodland. **Beh** If handled, can stridulate. **Fp** Lichen-covered trees, mainly oaks. **SS** None.

Linnavuoriana 2 spp. | 1 ILL.

10-spot Leafhopper
Linnavuoriana decempunctata

BL 3·5 mm. **ID** Whitish, pink or brown. Head with 2 black spots and pronotum with 6; scutellum with 2 larger spots. Forewings with dark markings, spots in some individuals, or suffused. **Hab** Woodland. **Fp** Birches, but hibernates in conifers and bushes. **SS** None (the other *Linnavuoriana* sp. is greenish).

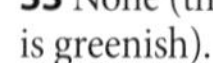

Opsius 1 sp.

Tamarisk Leafhopper

Opsius stactogalus

Local

BL 4–5 mm. **ID** Green, with paler head and pronotum; forewings either plain, with blue spots, or in some individuals tiny black dots; membrane dark. **Hab** Gardens and parks, also open areas (often coastal). **Fp** Tamarisk. **SS** None.

Pediopsis 1 sp.

Lime Leafhopper

Pediopsis tiliae

Nationally Scarce

BL 5–6 mm. **ID** Head and pronotum yellowish; scutellum and forewings light or dark brown. **Hab** Woodland. **Fp** Small-leaved Lime. **SS** None.

Typhlocyba 1 sp.

Orange-spotted Leafhopper

Typhlocyba quercus

Common

BL 3–4 mm. **ID** Easily recognized by the orange spots, which vary in number between individuals. **Hab** Woodland, gardens and parks. **Fp** Oaks. **SS** None.

Ulopa 1 sp.

Heather Leafhopper

Ulopa reticulata

Common

BL 3 mm. **ID** Stout, rounded; brown, forewings with at least two pale bands. **Hab** Heathland. **Fp** Heathers. **SS** The Nationally Scarce **Chalkhill Leafhopper** *Utecha trivia* [N/I] (**BL** 3 mm), in which ♀ is yellowish-brown with plain forewings; ♂ forewings with dark streaks.

FAMILY Cixiidae (Lacehoppers) 5 GEN. | 12 spp. | 5 SPP. ILL.

Forewings long, clear and lace-like, with conspicuous veins. Scutellum with 3 or 5 keels (body length 4–9 mm). A selection of species is illustrated, which can be distinguished as follows:

SCUTELLUM with 3 keels	SCUTELLUM with 5 keels	
	VERTEX broader than long; grassland	VERTEX longer than broad; saltmarshes
Cixius [8 species] ***Tachycixius*** [Spotted Lacehopper] ***Trigonocranus*** [Emma's Lacehopper *T. emmeae* [N/I] (BL 3–4 mm) Nationally Scarce]	***Reptalus*** [Five-ribbed Lacehopper]	***Pentastiridius*** [Saltmarsh Lacehopper]

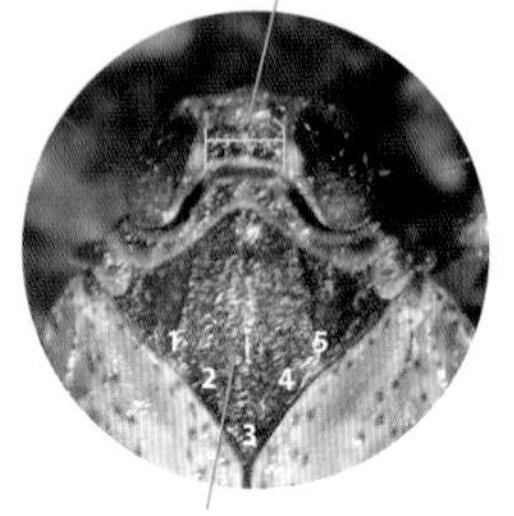

LACEHOPPERS – 3-keeled scutellum

Cixius 8 spp. | 2 ILL.

Common Lacehopper
Cixius nervosus

Common

BL 7–9 mm. **ID** Forewings with brown markings at base, a tranverse band on front half and spots and marks towards tip; margins brown-spotted. Some individuals are more yellowish. **Hab** Woodland. **Beh** On trees and shrubs. **SS** Other *Cixius* species, from which told by markings and larger size (but species in this genus are difficult to identify).

Banded Lacehopper
Cixius cunicularius

Local

BL 6–8 mm. **ID** Forewings with extensive brown markings; margins brown-spotted. **Hab** Near streams. **SS** Other *Cixius* species, from which told by the dark forewing bands and markings, where these are distinctive (but species in this genus are difficult to identify).

Tachycixius 1 sp.

Spotted Lacehopper
Tachycixius pilosus

Common

BL 5 mm. **ID** Plain to dark, with dark spots between veins near wingtips on forewings. **Hab** Woodland. **Fp** Larvae feed on grasses at their base. **Beh** Occurs on trees and shrubs. **SS** *Cixius* species, from which told by patterning on forewings.

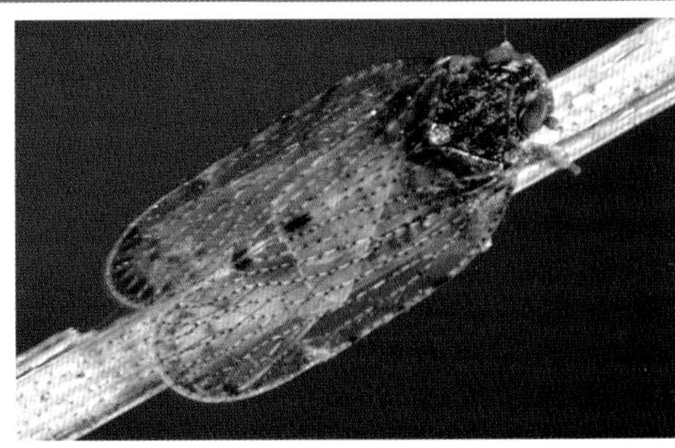

LACEHOPPERS – 5-keeled scutellum

Reptalus 1 sp.

Five-ribbed Lacehopper
Reptalus quinquecostatus

Nationally Scarce

BL 5–7 mm. **ID** Dark head and scutellum, the latter with 5 keels; vertex (area on head between eyes) broader than long. **Hab** Grassland; the nymphs feed on roots. **SS** Saltmarsh Lacehopper [see *opposite*].

Pentastiridius 1 sp.

Saltmarsh Lacehopper
Pentastiridius leporinus

Nationally Scarce

BL 6–8 mm. **ID** Pale with whitish abdomen tip often visible. Dark head and scutellum, the latter with 5 keels; vertex (area on head between eyes) longer than broad. **Hab** Coastal saltmarsh edge. **Fp** Grasses. **SS** Five-ribbed Lacehopper [see *opposite*].

Saltmarsh Lacehopper occurs in a range of colour forms.

FAMILY Delphacidae (Planthoppers) — 40 GEN. | 75 spp. | 4 SPP. ILL.

Small, seldom-studied bugs (body length 2–6 mm), with forewings often very shortened; antennae located beneath lower margin of eyes. **Hind tibiae with a moveable spur**. A small selection of distinctive, easily identified species is illustrated.

Ditropis — 1 sp.

Bracken Planthopper
Ditropis pteridis

Common

BL 2–4 mm. **ID** Black with yellow head, pronotum, legs and rings between abdomen segments; forewings short, ½ length of abdomen in ♂, rather shorter in ♀; Long-winged form rare. **Hab** Woodland. **Fp** Bracken. **SS** None.

Asiraca — 1 sp.

Broadened Planthopper
Asiraca clavicornis

Nationally Scarce

BL 3–4 mm. **ID** Dark brown, long-winged; antennae and fore legs remarkably broadened. **Hab** Grassland. **Fp** Unknown. **SS** None.

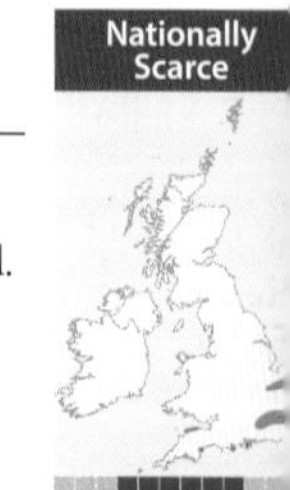

Conomelus — 1 sp.

Yellowish Planthopper
Conomelus anceps

Common

BL 4 mm. **ID** Yellowish-brown; short forewings with a broad, whitish central band; dark spots on veins; long-winged form with brownish markings. **Hab** Grassland. **Fp** Rushes. **SS Sedge Planthopper** *Euconomelus lepidus* [N/I] (**BL** 2–4 mm), which has forewing tips with white patches (S & C England).

Calligypona — 1 sp.

Rey's Planthopper
Calligypona reyi

Rare (RDBk)

BL 3–4 mm. **ID** Dark; short forewings brownish (orange abdominal band visible); pronotum white. **Hab** Saltmarshes. **Fp** Club-rushes. **SS** None.

FAMILY **Issidae** (Issid planthoppers) **1 GEN. | 2 spp.**

Robust bugs (body length 6–8 mm). Broadened forewings with numerous cross-veins. British and Irish species are flightless. Found on trees and shrubs.

Issus **2 spp.**

Common Issid Bug
'ssus coleoptratus

3L 6–8 mm. **ID** Brown, individually variable: forewings plain to patterned with numerous darker marks. Nymph abdomen tip has a tuft of bluish hairs. **Hab** Woodland, hedgerows, gardens. **Fp** Often Ivy or oaks, also many other deciduous trees. **SS** Scarce Issid Bug.

NOTE: the name '*coleoptratus*' refers to this species' beetle-like appearance.

Common Issid Bug occurs in a range of colour forms, from plain to with some banding, but never as strongly defined as in Scarce Issid Bug. In addition, the frons of Common Issid Bug is more extensively black than that of Scarce Issid Bug.

;carce Issid Bug
ssus muscaeformis

;L 6–8 mm. **ID** Brown, forewings trongly patterned with numerous vhitish and darker marks; black on rons relatively restricted. **Hab** Limestone pavement and oak voodland; on bushes or trees. **S** Common Issid Bug, which has nore extensive black on frons.

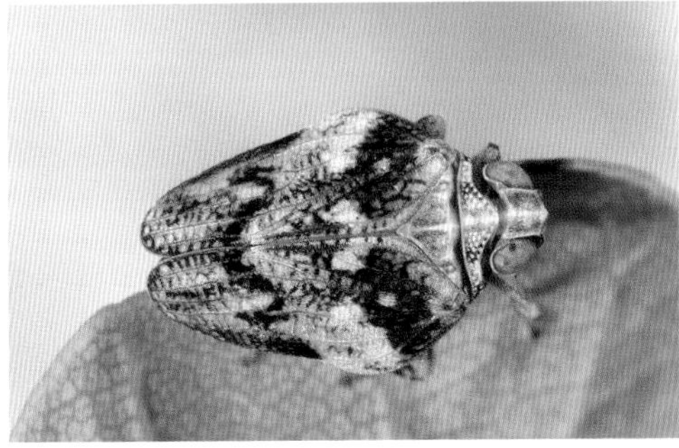

FAMILY Tettigometridae (Tettigometrid planthoppers) 1 sp.

Robust bug (body length 4–5 mm). Forewings hardened, with fine punctures. Found on the ground; adults overwinter.

Tettigometra 1 sp.

Chalk Planthopper
Tettigometra impressopunctata

BL 4–5 mm. **ID** Brown; abdomen reddish and black. Forewings hardened, with fine punctures.
Hab Chalk hills or calcareous dunes.
Beh Ground-dwelling and possibly associated with ants.
SS None.

Nationally Scarce

SUBORDER Sternorrhyncha (Aphids, jumping plant lice [= psylloids], scale insects [= coccoids] and whiteflies) GUIDE TO SELECTED FAMILIES | 5 OF 17 BI

Little-studied insects that include some of Britain's most important pests of plants. The suborder is distinguished by having the rostrum between the fore coxae. All, including the twelve omitted families, are small, rather specialist and difficult to identify.

FORM not covered in waxy powder

TARSI 2-segmented
Typical bug shape (could be confused for Cicadellidae *p. 214*); **ANTENNAE** long

Jumping Plant Lice
Psyllidae
[16 genera, 80 species]
p. 226

TARSI 3-segmented
Winged and wingless species/forms; **ANTENNAE** often long; a few species produce waxy coverings

Aphids
Aphididae
[160 genera, 630 species]
p. 223

FORM covered in waxy white powder; winged

Whiteflies
Aleyrodidae
[13 genera, 19 species]
p. 226

FORM often covered in wax; winged or wingless

Scales flat and often oval; legless and wingless; sometimes covered in wax

Soft Scales
Coccidae
[14 genera, 30 species]
p. 520

Covered in wax, secreted by the insect; ♂ two wings. ♀ often with waxy filaments protruding from the tip of the abdomen

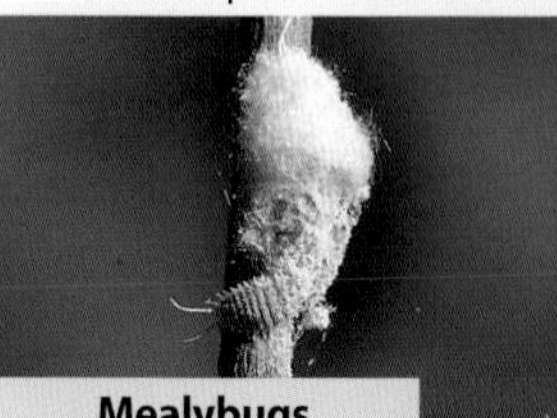

Mealybugs
Pseudococcidae
[16 genera, 50 species]
p. 227

FAMILY **Aphididae** (Aphids) **160 GEN. | 630 spp. | 9 SPP. ILL.**

Mainly small and soft-bodied bugs (body length 1–6 mm), including both winged and wingless species/forms; forewings (if present) have a vein near upper edge, from which longitudinal veins originate. Antennae long. A tail (cauda) is present; many have finger-shaped caudae, although there are numerous other shapes. A pair of tube-like cornicles is present in some species on abdomen segments S5 or S6 (these may be short or long, and on some species different in colour to the body); they are used in defence, secreting a waxy fluid to help protect against predation. Some aphids are parthenogenetic, or give birth to live young that grow quickly, enabling rapid reproduction. Aphids feed on the sap of plants and are often considered pests, controlled by ladybirds and many other predators. Sticky leaves often indicate the presence of aphids; some species are attended by protective ants, which feed on the honeydew the aphids excrete from their anus and may 'farm' them, moving the aphids to suitable locations.

Although some species are distinctive, aphids are generally not easy to identify. To complicate matters, some have different forms at different times of the year, including both winged and unwinged, which may feed on different plants. Photographing the adult, particularly at high resolution to show the main characters and noting the host plant may provide valuable identification clues. Whilst watching and photographing aphids, look out for predators such as ladybird larvae and adults, and also parasitic wasps; ladybirds are well-known gardeners' allies, eating numerous aphids. Specific identification of aphids is complex, requiring detailed examination of particular body parts, such as abdomen and antenna segments, tarsi and setae (although knowing the foodplant is often a help). Such coverage is beyond the scope of this book, but nine widespread and common species are shown to illustrate the range of variation within the family.

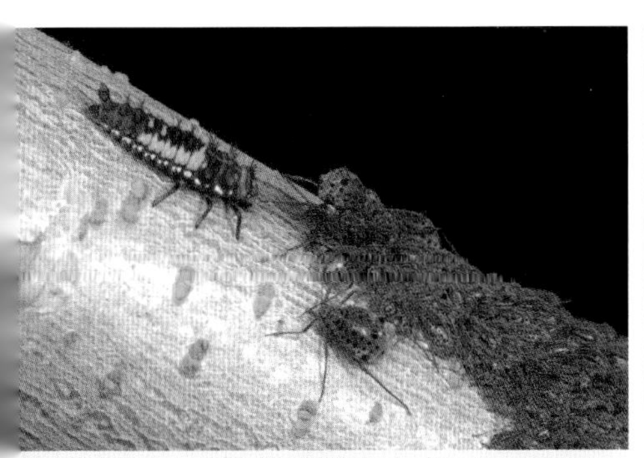

Harlequin Ladybird (*p. 311*) larva looking to prey on **Giant Willow Aphids**.

Southern Wood Ant (*p. 496*) attending to and protecting numerous **aphids**.

Common Sycamore Aphid adults, nymphs and shed skins on the underside of a Sycamore leaf.

Aphids secrete honeydew, which is 'milked' by the attendant ants.

Tuberolachnus 6 spp. | 1 ILL.

Giant Willow Aphid *Tuberolachnus salignus*

BL 5–6 mm. **ID** Brown with black patches; greyish sheen. **Hab** Willows.

Aphis 95 spp. | 1 ILL.

Dark Green Nettle Aphid *Aphis urticata*

BL 2 mm. **ID** Dark green; winged forms much darker. **Fp** Nettles.

Cavariella 8 spp. | 1 ILL.

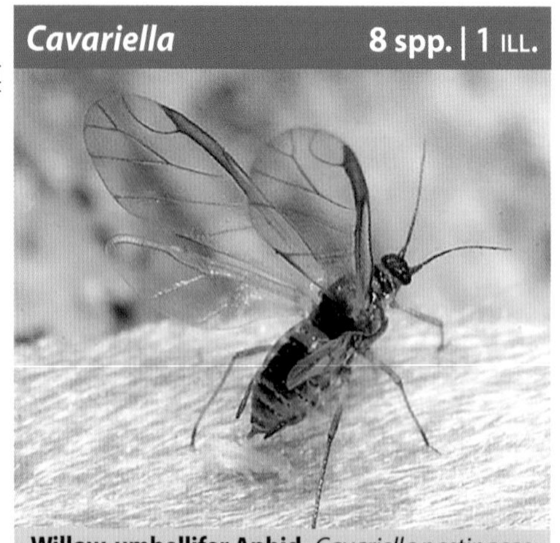

Willow-umbellifer Aphid *Cavariella pastinacae*

BL 2–3 mm. **ID** Green, legs pale except tarsi. Antennae, head, pronotum and abdomen with black markings. **Hab** Willows.

Macrosiphioniella 15 spp. | 1 ILL.

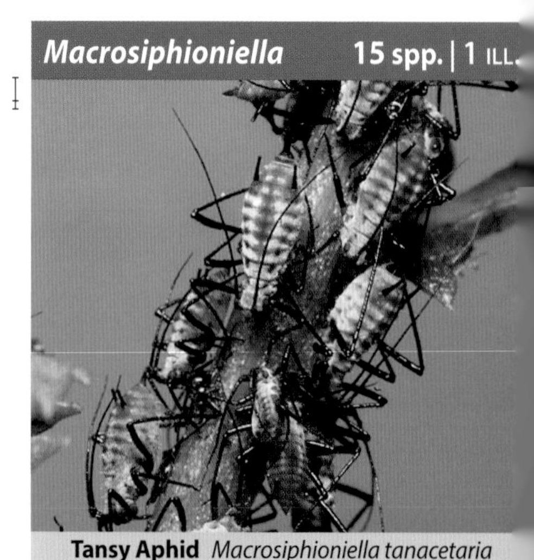

Tansy Aphid *Macrosiphioniella tanacetaria*

BL 3–4 mm. **ID** Green/pinkish-brown, wax-powdered antennae and most of legs black. **Fp** Tansy.

Eriosoma 9 spp. | 1 ILL.

Woolly Apple Aphid *Eriosoma lanigerum*

BL 1 mm. **ID** Brown or purplish-grey, covered in woolly wax; usually on bark. **Fp** Apples.

Eucallipterus 1 sp.

Common Lime Aphid *Eucallipterus tiliae*

BL 2–3 mm. **ID** Pale yellow; head and pronotum with black stripes; forewings black front margin and spots at tips of veins. **Fp** Limes.

Cinara 26 spp. | 1 ILL.

Large Pine Aphid *Cinara pinea*

BL 3–5 mm. **ID** Greyish-brown. **Fp** Pines.

Macrosiphum 16 spp. | 1 ILL.

Rose Aphid *Macrosiphum rosae*

BL 2–4 mm. **ID** Pink, green/reddish-brown. **Fp** Roses.

Drepanosiphum 4 spp. | 1 ILL.

Common Sycamore Aphid *Drepanosiphum platanoidis*

BL 3–4 mm. **ID** Head and thorax yellowish-brown with dark markings; abdomen green, black-banded; can be much paler. **Fp** Sycamore.

 Guide to families p. 222

FAMILY **Psyllidae** (Jumping plant lice) 16 GEN. | 80 spp. | 1 SP. ILL.

Small, soft-bodied bugs (body length 2–4 mm); forewings with central vein from which longitudinal veins originate. Antennae long. Psyllids have varied life-cycles, with nymphs of some species forming galls, often on specific host plants. Species are little studied but can often be identified by forewing characters, including shape, markings and venation, while others require examination of genitalia under a microscope.

Psyllopsis 4 spp. | 1 ILL.

Ash Psyllid *Psyllopsis fraxini*

BL 3–4 mm. **ID** Pale, except for black on head, black stripes on pronotum and on abdomen; forewings transparent, tip with dark clouds; eyes red. **Hab** Woodland, gardens and parks. **Fp** Ash; produces galls on the leaves. **SS** Other *Psyllopsis* species, some of which are almost identical and can only be told by microscopic examination of ♂ genitalia.

FAMILY **Aleyrodidae** (Whiteflies) 13 GEN. | 19 spp. | 1 SP. ILL.

Small, soft-bodied bugs (body length usually ≤2·5 mm); body covered in a white, waxy powder; wings also white. Antennae long. Whilst most of the similar species survive outdoors and are often regarded as 'pests', other tropical species exist only in greenhouses. The best-known species is shown below; in growing areas and allotments, when cabbage plants are disturbed, Cabbage Whiteflies can fly upwards in clouds, before settling back down on plants. Identification is challenging, with specialists often using pupal case differences to help place to species level.

Aleyrodes 2 spp. | 1 ILL.

Cabbage Whitefly *Aleyrodes proletella*

BL 1·5 mm. **ID** White; forewings each with four greyish spots; eyes red. **Hab** Gardens, allotments. **Fp** A pest of cabbage crops. **SS** Other whiteflies (but these are associated with other plants). Superficially similar to waxflies (*p. 236*), but these have longer antennae and hold their wings 'tent-like'.

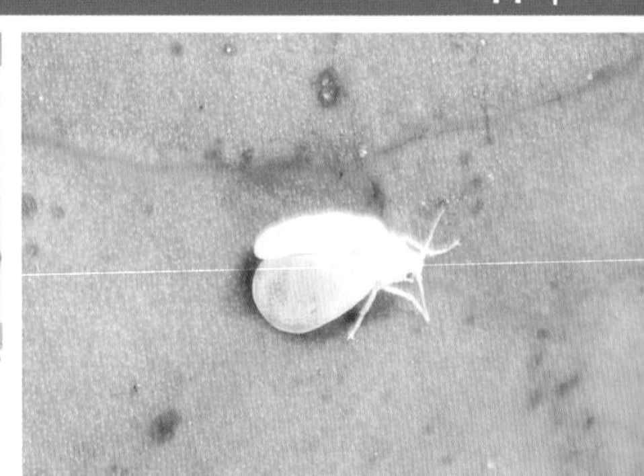

FAMILY **Coccidae** (Soft scales)

14 GEN. | 30 spp. | 2 SPP. ILL.

Small (scale length ≤ 5 mm), sap-sucking bugs, the scale often flat, oval and may be covered in wax; often legless beyond the crawler stage and do not really look like insects, except for a rostrum which is attached to plants. Antennae short or absent; males (if present) may be winged. Often attended by ants. Identification of species is difficult and beyond the scope of this book, but the host plant may help. Often horticultural pests, with some species occurring on fruit trees, others in greenhouses.

Pine Ladybird larva surrounded by, and covered in the wax from, its prey, Horse-chestnut Scale Insect *Pulvinaria regalis*.

Pulvinaria

5 spp. | 2 ILL.

Woolly Currant Scale Insect
Pulvinaria vitis

Introduced; common

BL ♀ 5–7 mm, ♂ 1·5 mm.
ID ♀ round or heart-shaped; wingless, dark brown, with whitish waxy surrounds. ♂ pink or red; winged.
Hab Woodland, gardens, parks.
Fp Various trees (including fruit trees) and bushes, including grape vines.
SS Other soft scales, which are, however, smaller.

Pine Ladybird eating Woolly Currant Scale Insect.

FAMILY **Pseudococcidae** (Mealybugs)

16 GEN. | 50 spp. | 1 SP. ILL.

Small bugs (<5 mm), covered in wax, which they secrete. Males winged; females wingless. Antennae long. Legs present and mobile to a certain extent. Approx. 20 of the 30 native species live on grasses. Introduced species in greenhouses or indoors feed on plant juices and can cause serious damage to plants (including cacti, succulents and bulbs), acting as a vector for plant diseases. Some gardeners and those in the horticultural trade use biological control techniques on these and other insects regarded as 'pests'. Identification of species is difficult and beyond the scope of this book. Additional tropical species are likely to occur in future (these are periodically intercepted during quarantine).

Example of a mealybug species on a houseplant (BL approx. 3 mm); found all-year.

ORDER THYSANOPTERA Thrips

[Greek: *thysanos* = fringe; *pteron* = wing]

BI | 3 families [67 genera, **176 species.** (incl. 19 spp. with short-lived, non-established populations under glass)]
W | 9 families [**approx. 6,200 spp**

Mostly small species less than 2 mm long, but some reaching up to 15 mm (7 mm in Great Britain and Ireland). The British and Irish total excludes 52 introduced species found during quarantine checks on imported plants. Typically elongate insects, eyes ranging in size from small to large, with 3 ocelli in winged species; antennae 4–9 segmented. The abdomen is 11-segmented (with 10 segments visible) with an ovipositor present in females of some species. The legs are short, particularly hindlegs; tarsi have one to two segments. Cerci are absent in females. Forewings and hindwings are similar in size, mostly with a fringe, known as a 'setal' fringe. Single and several individuals are known as 'thrips', which are capable of weak flight. These insects feed on plants, including flowers, but occasionally on fungi or small insects. Species associated with flowers are better known and may be regarded as pests. Thrips sometimes reproduce parthenogenetically. The life-cycle features an inactive pre-pupal and pupal stage and is often completed within a year.

FINDING THRIPS During peak season, thrips are occasionally seen in abundance on flowers, such as on hot, thundery days, when their dark bodies are obvious against bright backgrounds.

Structure of a thrips

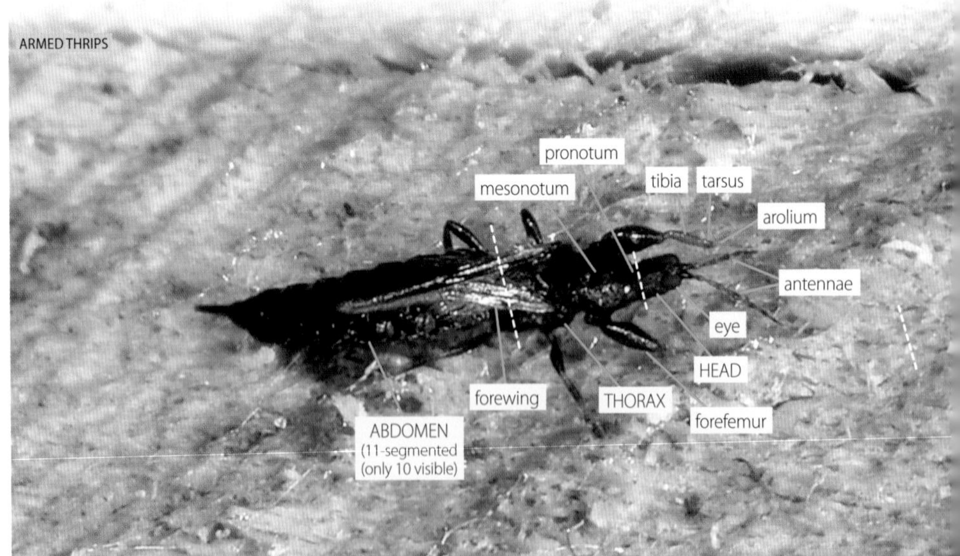

FURTHER READING AND USEFUL WEBSITES

Collins, D.W (2010) **Thysanoptera of Great Britain: a revised and updated checklist**. *Zootaxa* 2412: 21–41.

Kirk, W.D.J. (1996) ***Thrips***. Naturalists' Handbook 25. Richmond Publishing.

Thrips *iD* www.thrips-id.com/en/

Thrips are difficult to identify, particularly due to the considerable variation in body length and shape within some species, although knowledge of the foodplant can be a helpful pointer. Comprehensive coverage is beyond the scope of this book and two representative species are shown here. For those wishing to learn more, there are specialist keys available (see *Further reading*, *opposite*). These focus on a wide range of body parts, including antennae, eyes, thorax and abdomen, but may still only enable identification to genus. In order to use these keys, a specimen needs to be mounted on a microscope slide and viewed with a powerful (at least 400×) microscope.

FAMILY Thripidae — 37 GEN. | 105 spp. | 1 SP. ILL.

The largest of the families in the Thysanoptera. Most are flower feeders, many being of economic importance, causing damage to crops as well as commercially grown flowers.

Odontothrips — 8 spp. | 1 ILL.

Gorse Thrips
Odontothrips ulicis

BL approx. 2 mm. **ID** Dark but with paler-banded abdomen, antennae and, in some, legs. **Hab** Open areas. **Fp** Gorse. **SS** Various, including **Broom Thrips** *Odontothrips cytisi* [N/I] on Broom; the two are virtually identical, although some individuals are separable by slight differences in hairs on forewing vein 2 (16–25 in Gorse Thrips, only 13–19 in Broom Thrips). Otherwise, ♂s distinguishable by genitalia.

FAMILY Phlaeothripidae — 16 GEN. | 39 spp. | 1 SP. ILL.

Includes some leaf and flower feeders, also species associated with fungi.

Hoplothrips — 8 spp. | 1 ILL.

Armed Thrips
Hoplothrips pedicularius

BL approx. 2 mm. **ID** Dark, with paler markings. ♂ has pre-tarsal 'tooth' and some have greatly enlarged fore femora. Wings long or short. Larvae yellowish with red internal pigment visible. **Hab** Woodland; associated with dead wood, under bark. **Fp** *Stereum* fungi, on hyphae. **Beh** During the mating season, ♂s use the pre-tarsal 'tooth' in fights with rival ♂s. Larger, dominant ♂s mate with approx. [illegible]0% of ♀s. **SS** Other *Hoplothrips* species, from which told by dark colour and blunt bristles on upperside of abdomen S8 (hard to see even at high magnification).

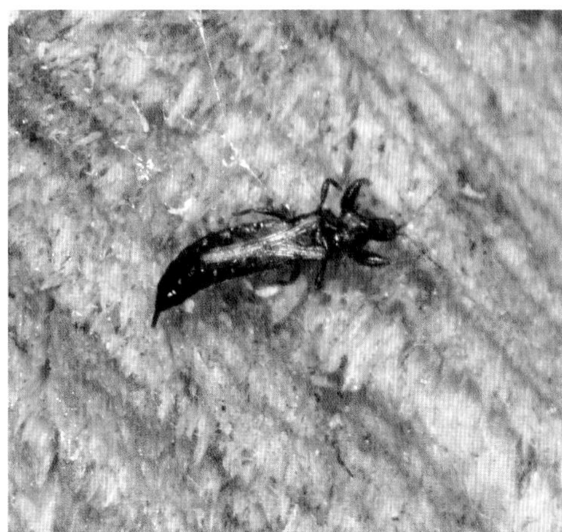

Further reading p. 236

ORDER NEUROPTERA

Lacewings & antlions

[Greek: *neuro* = nerve; *pteron* = wing]

BI | 6 families [25 genera, **70 species**]

W | 17 families [**approx. 5,940 species**]

Lacewings (from 3 mm) and the rare, larger (≥30 mm) antlions are sometimes seen in flight, which is distinctly fluttering. The two pairs of wings are lace-like, often cross-veined. Eyes are colourful, but ocelli are absent except in the Osmylidae. The wings are held roof-like over the soft body when at rest. Typically have long antennae and large eyes. Adult lacewings feed on small insects, including aphids; antlions prey on ants. A few species overwinter as adults. Unusually, eggs of lacewings in the Chrysopidae hang by threads from vegetation. Some larvae cover their body with debris, which possibly provides camouflage when they are hunting aphids. Antlions are fascinating due to their larval pits in sand; the ferocious larva uses its large jaws to grab an insect (often an ant) that falls into its pit. Some Neuroptera species are easy to identify, others require microscopic examination.

FINDING LACEWINGS & ANTLIONS **Lacewings** are attracted to lights and may appear on house windows. By day, they can be seen fluttering around in gardens, hedgerows and woodland. **Antlions** inhabit sandy areas, such as at the RSPB Minsmere reserve in Suffolk, where larval pits may be found in sand. However, the adults are nocturnal and are difficult to observe by day.

Structure of a lacewing

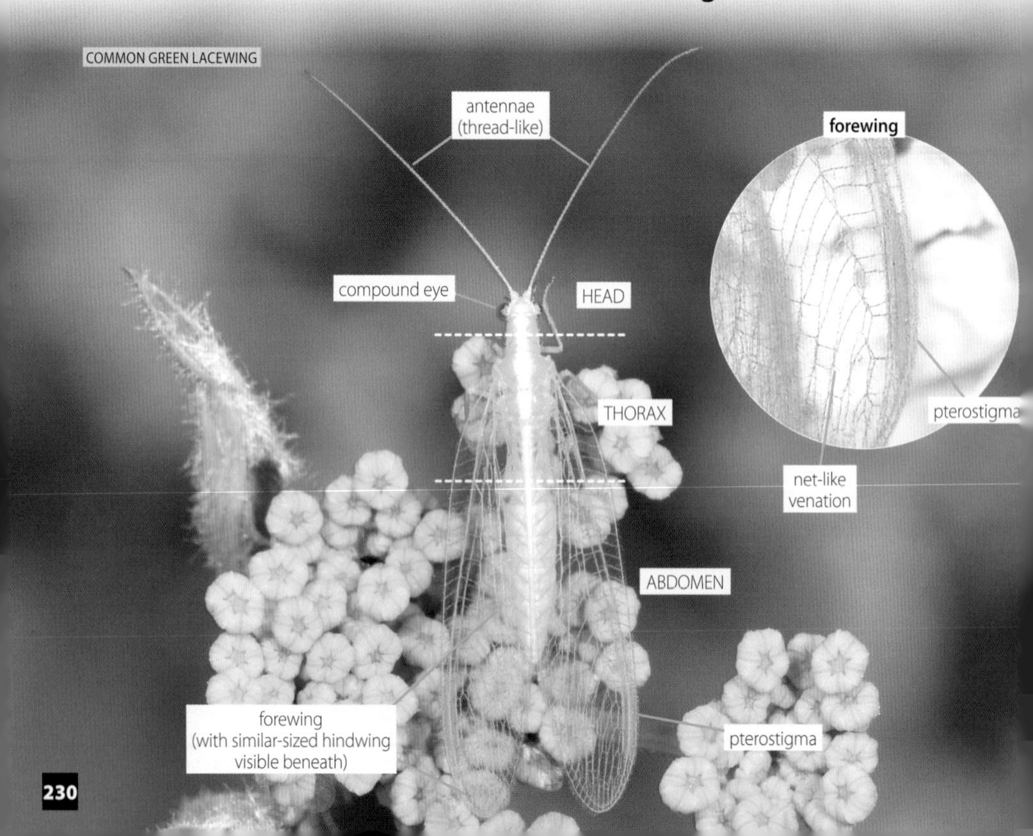

ORDER Neuroptera — GUIDE TO FAMILIES | 6 BI

Lacewings and Antlions are best identified to family group by the presence of waxy powder on the body and then by the presence of ocelli, body colour and wing details, as outlined *below*. **Length bars in the species accounts indicate forewing length (≈ ½ wingspan).**

BODY + WINGS white/grey, covered in a waxy powder

FORM small (wingspan approx. 5 mm)

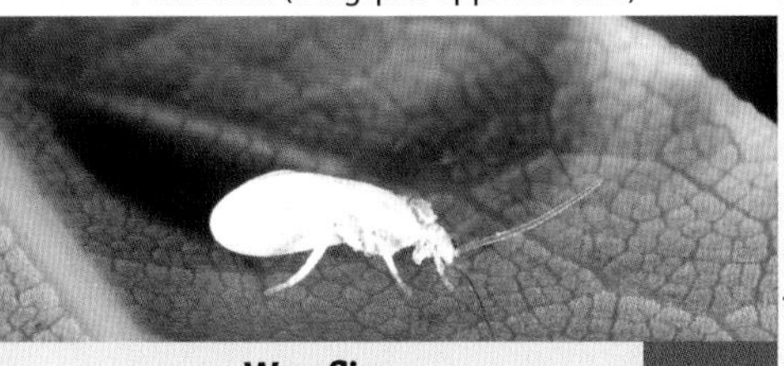

Wax-flies
Coniopterygidae
[6 genera | 12 species]
p. 236

BODY + WINGS not waxy — 1/2

FORM **large (wingspan ≥50 mm);** FOREWINGS **elongated;** ANTENNAE **tip thickened and curved**

Antlions
Myrmeleontidae
[2 genera | 2 species]
p. 232

BODY + WINGS not waxy — 2/2

FORM **smaller (wingspan ≤ 45 mm);** FOREWINGS **less elongated;** ANTENNAE **not thickened at tip**

HEAD **3 ocelli**; WINGSPAN 40–45 mm; WINGS **distinctive spotted pattern**

3 ocelli

Giant lacewings
Osmylidae
[Giant Lacewing]
p. 236

HEAD **no ocelli**; WINGSPAN ≤35 mm; FORM **mainly green (2 species brown with orange head)**

Green lacewings
Chrysopidae
[8 genera | 21 species]
p. 233

HEAD **no ocelli**; WINGSPAN ≤35 mm; FORM brown; FOREWINGS brown, often with dark shading or spots; **cross-veins forking near base**

Brown lacewings
Hemerobiidae
[7 genera | 31 species]
p. 234

HEAD **no ocelli**; WINGSPAN ≤12 mm; FORM brown; FOREWINGS uniform brown (some species with darker veins); **cross-veins not forking near base**

Sponge flies
Sisyridae
[1 genus | 3 species]
p. 235

Guide to families *p. 231*

FAMILY **Myrmeleontidae** (Antlions) — 2 GEN. | 2 spp.

Brown, with elongated wings, which allow for a fluttering flight and might, at first, appear like damselflies. The antennae have a clubbed tip. Antlions are best known for their ferocious larvae, which utilise cone-shaped pits in sand to feed on ants and other insects that fall in; although they usually live underground, they sometimes chase prey. Adults live for approx. three weeks.

Euroleon — 1 sp.

Suffolk Antlion
Euroleon nostras

Rare (not Red Listed) **(recent colonist)**

WS 52–68 mm, **BL** approx. 30 mm. **ID** Greyish-brown; wings transparent, with black marks; pterostigma black-and-white; forewing tips almost rounded. **Hab** Sandy soils, often coastal sites. **Dist** Good population at RSPB Minsmere, Suffolk, confirmed as breeding in 1996. **SS** European Antlion.

Myrmeleon — 1 sp.

European Antlion
Myrmeleon formicarius

Immigrant; rare

WS 70–80 mm, **BL** approx. 35 mm. **ID** Greyish-brown; wings transparent, plain; pterostigma white; forewing tips, in part, subtruncate. **Hab** Sandy coastal sites. **Beh** Mainly nocturnal. **Dist** Recorded at Freshwater, Isle of Wight in August 2013. **SS** Suffolk Antlion.

Suffolk Antlion
PTEROSTIGMA
black-and-white

European Antlion
PTEROSTIGMA
white

LEFT TO RIGHT: Typical antlion pits; Suffolk Antlion larva lying in wait for prey in a pit; Suffolk Antlion larva on the surface.

FAMILY **Chrysopidae** (Green lacewings) 8 GEN. | 21 spp. | 3 SPP. ILL.

The best known of our Neuroptera; green, bluish-green, or yellowish-green insects, some with black marks. Two species are large, brown with orange heads and also differ in wing venation. Identification to species level for most green species requires close examination under a microscope to confirm the species. Possibly only the Common Green Lacewing overwinters as adults.

Chrysoperla 3 spp. | 1 ILL.

ommon Green Lacewing
'hrysoperla carnea

Common

J F M A M J J A S O N D

/S 25–30 mm, **BL** 10–12 mm.
) Pale green; body with a yellow orsal stripe (body changes to straw-)loured during winter). Regarded as species aggregate, or a complex of)ecies, and may not be distinguishable ıorphologically. **Hab** Grassland, ardens, parks and woodland.
S Confused Green Lacewing *'hrysoperla lucasina* [N/I], which has brown marks n abdomen S1 + S2 when viewed laterally (check ıembrane at base, near thorax) and slightly pointed)rewing wingtip; green individuals found in winter are kely to be Common Green Lacewing.

Chrysopa 6 spp. | 1 ILL.

Pearl Lacewing
Chrysopa perla

Common

J F M A M J J A S O N D

WS 24–30 mm, **BL** 9–12 mm.
ID Bluish-green; head, pronotum and abdomen with black markings (underside of abdomen black); antenna S2 black. **Hab** Hedgerows, scrub and woodland edges. **SS** Other *Chrysopa* species, particularly **Pine Lacewing** *Chrysopa dorsalis* [N/I], a rare species in S England, associated with pines; the main distinction is that the forewing vein near margin is black, rather than green in Pearl Lacewing.

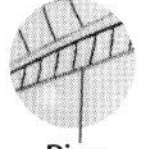

Pine Lacewing **VEIN** black

Pearl Lacewing **VEIN** green

Nothochrysa 2 spp. | 1 ILL.

)range-headed Lacewing
lothochrysa capitata

Local

J F M A M J J A S O N D

/S 28–35 mm, **BL** approx. 10 mm.
) Head orange-brown; remainder f body and pterostigma brown r greenish; abdomen with darker ıarkings on segments. **Hab** Woodland, ,pically pine although sometimes n oaks. **SS Scarce Orange-headed .acewing** *Nothochrysa fulviceps* [N/I] **WS** 40 mm), with very few British ecords, has a broad, pale stripe on pronotum.

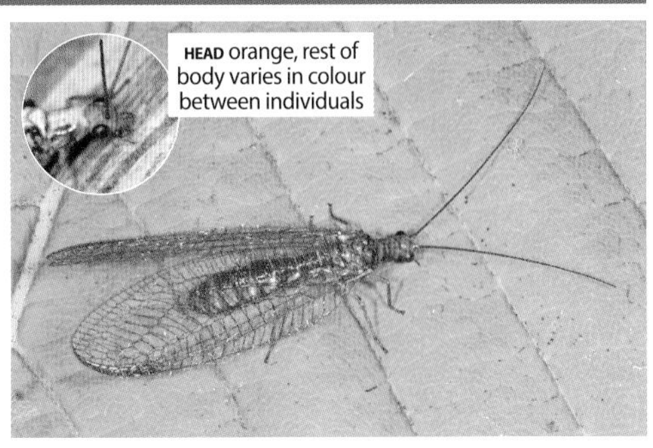

HEAD orange, rest of body varies in colour between individuals

Guide to families p. 2

FAMILY Hemerobiidae (Brown lacewings) 7 GEN. | 31 spp. | 5 SPP. ILL.

Brown insects, forewings brown with black shading or spots in some species; **cross-veins fork near base**, one species has hooked forewings. **Wing venation is important in identification to genus**, and then species level. **Most brown species require microscopic examination of slight differences in forewing venation to confirm the species.**

Hemerobius 13 spp. | 3 ILL.

Common Brown Lacewing
Hemerobius humulinus

WS 15–18 mm, **BL** approx. 5 mm.
ID Brown; forewings brownish with varied spotting and markings.
Hab Grassland, gardens, parks, woodland. **SS** Other *Hemerobius* species and other brown lacewings, from which can only be told by microscopic examination of genitalia.

Brown Pine Lacewing
Hemerobius stigma

WS 15–18 mm, **BL** approx. 5 mm.
ID Brown, top of the thorax same shade as sides. Forewings with dark spots; pterostigma orange. **Hab** Coniferous woodland with Scots Pine or Larch.
SS Other *Hemerobius* species, although these do not have have orange pterostigmata.

Dash Lacewing
Hemerobius micans

WS 11–18 mm, **BL** approx. 6 mm.
ID Body and wings yellowish. Forewings veins with pale-haired dashes. **Hab** Woodland. **SS** Other *Hemerobius* species, from which told by forewing dashes between veins, rather than spots.

Drepanepteryx — 1 sp.

Hook-winged Lacewing
Drepanepteryx phalaenoides

WS 30–32 mm, **BL** approx. 9 mm. **ID** Large, chestnut-brown. Forewings brown, hooked at tip. **Hab** Woodland. **SS** None, shape distinctive.

Micromus — 3 spp. | 1 ILL.

Spotted Brown Lacewing
Micromus variegatus

WS 10–13 mm, **BL** approx. 3 mm. **ID** Brown; forewings transparent with black or dark brown marks. **Hab** Grassland, gardens, parks, woodland, probably associated with bushes. **SS** None.

NOTE: the other *Micromus* species have different coloured wings (reddish-brown or yellowish-brown) and less distinct markings.

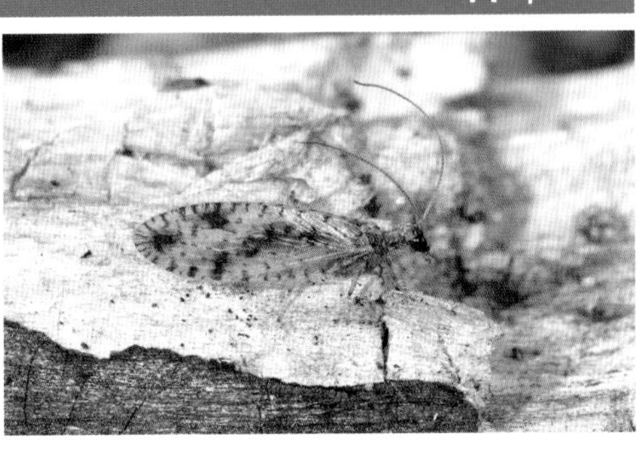

FAMILY Sisyridae (Sponge flies) — 1 GEN. | 3 spp. | 1 SP. ILL.

Uniform brown, although in some species forewing veins are darker; **forewings near base with cross-veins not forking**. This is the only aquatic Neuropteran that occurs in Great Britain and Ireland, the larvae developing in freshwater sponges. Adults frequent treetops, but are sometimes found on lower vegetation, or attracted to lights. Variable in number, they can be common in some years but occur in low numbers in others.

Sisyra — 3 spp. | 1 ILL.

Common Sponge Fly
Sisyra nigra

WS 10–12 mm, **BL** approx. 4 mm. **ID** Uniform brown, including antennae. **Hab** Still or slow-flowing water bodies, including ponds. **SS** Other *Sisyra* species, although Common Sponge Fly is the only one with all-dark antennae.

 Guide to families p. 231

FAMILY Osmylidae (Giant lacewings) 1 sp.

Represented in Great Britain and Ireland by a large (wingspan approx. 40 mm), distinctive species with three ocelli. Adults may be found by day by searching under bridges, but doing so can be a challenge; they mainly fly at dusk over the water's surface. The larvae develop in damp mosses at stream and river edges, overwintering as larvae and pupating in spring.

Giant Lacewing
Osmylus fulvicephalus

WS 40–45 mm, **BL** approx. 12 mm. **ID** Large size; black, except for red head; wings transparent with bold black markings. **Hab** Edges of streams and rivers, often by or near woodland. **SS** None.

FAMILY Coniopterygidae (Wax-flies) 6 GEN. | 12 spp. | 1 SP. ILL.

Little-known small, white or grey (wingspan approx. 5 mm), covered in a white, waxy powder. **Identification to species level is difficult, although wing venation narrows the options down to genera; examination of male internal genitalia may be required to confirm the species.**

Conwentzia 7 spp. | 1 ILL.

Common Wax-fly
Conwentzia psociformis

WS 5 mm, **BL** 2 mm. **ID** White, waxy body and wings. Hindwings narrow; ≤⅓ width of forewings. **Hab** Woodland and gardens on various trees and bushes. **SS** Other wax-flies. Superficially similar to some aphids (*p. 223*) and whiteflies (*p. 226*) (both Hemiptera) but these have shorter antennae and do not hold their wings in a 'tent-like' fashion.

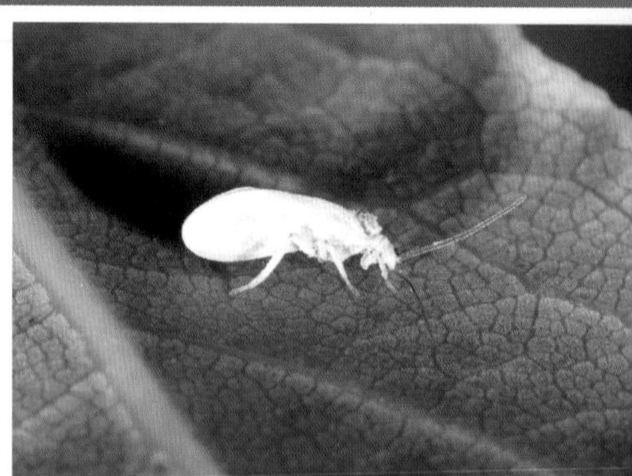

FURTHER READING AND USEFUL WEBSITES

Plant, C.W. 1997. ***A key to the adults of British lacewings and their allies (Neuroptera, Megaloptera, Raphidioptera and Mecoptera)***. Field Studies Council (AIDGAP).

Wachmann, E. & Saure, C. 1997. ***Netzflügler, Schlamm- und Kamelhalsfliegen: Beobachtung, Lebensweise***. Naturbuch-Verlag.
[in German, but this well-illustrated guide includes related orders in 'Neuropterida']

http://lacewings.myspecies.info Lacewings and Allies Recording Scheme

ORDER **MEGALOPTERA** Alderflies

[Greek: *megalo* = large; *pteron* = wing]

BI | 1 family [1 genus, **3 species**] **W** | 2 families [**380 species**]

This small order comprises two families, the Sialidae (alderflies) and Corydalidae (dobsonflies and fishflies). Although not closely related to Neuroptera (lacewings and antlions), they are often studied with them, along with Mecoptera (scorpionflies) and Raphidioptera (snakeflies) in a group termed 'Neuropterida'. Alderflies are robust-looking, dark, soft-winged and broad-headed, and are associated with water bodies. British and Irish species have a wingspan of 22–34 mm. The two pairs of wings are dark brown or black-veined and the forewings have several costal cross-veins. Flight action is rather clumsy. The typical life-cycle takes two years, almost all as aquatic larvae that feed on insect larvae, worms and crustaceans. The non-feeding adults live 2–3 days and lay egg masses on emergent vegetation. Although known as alderflies, they are not exclusively associated with Alder.

FINDING ALDERFLIES Most likely to be found flying in sunny weather by water bodies (lakes, ponds, large rivers and streams) or resting on waterside vegetation.

Structure of an alderfly

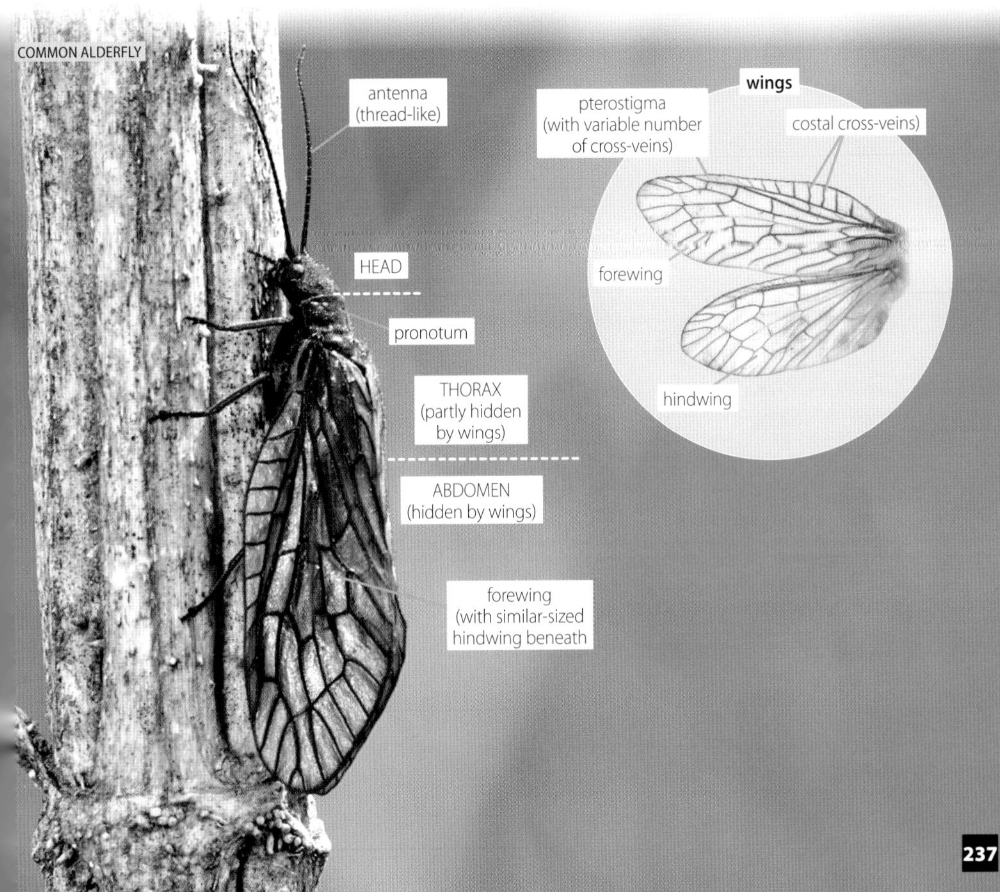

FAMILY **Sialidae** (Alderflies) — 1 GEN. | 3 spp. | 1 SP. ILL.

The Alderflies are large (wingspan 22–34 mm), dark brown to black insects, usually with darker wing veins. Antennae are shorter than forewing length. Those wishing to study alderflies to species level must either examine the genitalia in males, particularly the subgenital plate size, or check the anal plate shape in females [see table *opposite*], which is possible with a powerful hand lens in the field, but may need examination under a microscope. It is not possible to identify the species from photographs of adults at rest. **Length bars in the species accounts indicate forewing length (≈ ½ wingspan).**

Common Alderfly
Sialis lutaria

WS 22–30 mm, **BL** approx. 10 mm. **ID** Dark brown. **Hab** Still and flowing waters from small ponds to large lakes and rivers. **SS** Flowing Water Alderfly, Scarce Alderfly.

Flowing Water Alderfly
Sialis fuliginosa

WS 27–33 mm, **BL** approx. 10 mm. **ID** Dark brown. **Hab** Margins of running water. **SS** Common Alderfly, Scarce Alderfly.

COMMON ALDERFLY (both images)

Scarce Alderfly
Sialis nigripes

WS 27–34 mm, **BL** approx. 10 mm. **ID** Dark brown. **Hab** Lakes and rivers. **SS** Common Alderfly, Flowing Water Alderfly.

FURTHER READING AND USEFUL WEBSITES

Plant, C.W. 1997. ***A key to the adults of British lacewings and their allies (Neuroptera, Megaloptera, Raphidioptera and Mecoptera)***. Field Studies Council (AIDGAP).

http://lacewings.myspecies.info Lacewings and Allies Recording Scheme

View from side (*left*) and from rear (*right*) of ♂ Common Alderfly abdomen tip showing the features illustrated *below*.

	♂ **TIP OF ABDOMEN** **hinged, revealing genital structure; lateral view shows shape of subgenital plates**	♀ **TIP OF ABDOMEN** **rounded; underside reveals anal plates in live individuals**
Common Alderfly	**SUBGENITAL PLATE** **long**, reaching/exceeding tip of abdomen	**ANAL PLATE** **square-edged**
Flowing Water Alderfly	**SUBGENITAL PLATE** **shorter** than tip of abdomen	**ANAL PLATE** **large oval areas**
Scarce Alderfly	**SUBGENITAL PLATE** **much shorter** than tip of abdomen	**ANAL PLATE** **small eye-like tips**

ORDER COLEOPTERA Beetles

[Greek: *koleos* = sheath; *pteron* = wing]

BI | 102 families [1,295 genera, **4,131 species** (including extinct species)]
W | 175 families [**approx. 392,400 species**

The third largest order of insects in Britain and Ireland (but the largest in the world, by far, with around 40% of all described insects), represented by tiny (≤1 mm) to large species (up to 80 mm, Stag Beetle male). The name for the order relates to the hardened forewings (elytra) that cover the membranous hindwings, where present (a few species lack hindwings), and usually the whole abdomen. Structure of antennae helps in identification to family level. Biting jaws are used to take in a wide range of food, including live prey, plants (live or dead), live and rotting wood, and dung. Many species are of economic importance, particularly the longhorn beetles and others whose larvae damage valuable timbers; some are pests of stored products. Conversely, ladybirds are beneficial, as they devour numerous aphids. The text provides a brief overview of a wide range of families likely to be seen, and various species, but excludes some families of mainly tiny beetles.

FINDING BEETLES Found everywhere, but particularly in ancient woodland, which is why the New Forest National Park, Hampshire, and Windsor Forest & Great Park, Berkshire, are known as the best areas for saproxylic (dead wood) species. Localities such as these are uncommon, and many species inhabiting these areas are threatened. Grassland areas are also productive for beetles.

Structure of a beetle

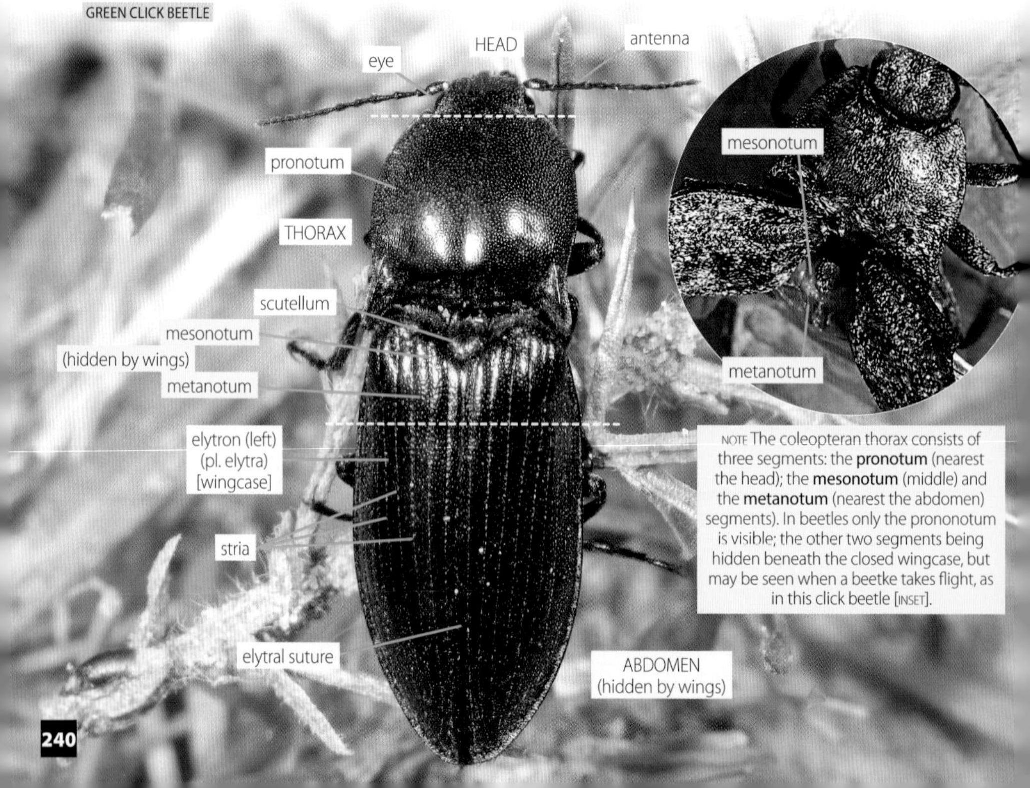

NOTE The coleopteran thorax consists of three segments: the **pronotum** (nearest the head); the **mesonotum** (middle) and the **metanotum** (nearest the abdomen) segments). In beetles only the prononotum is visible; the other two segments being hidden beneath the closed wingcase, but may be seen when a beetke takes flight, as in this click beetle [INSET].

ORDER **Coleoptera** VISUAL GUIDE TO SELECTED FAMILIES | 74 of 102 BI

This guide, which is presented in order of classification, omits minor families with little-known, often tiny, British and/or Irish species that are unlikely to be seen unless specifically searched for, or need microscopic examination of particular features for identification. Shape is important as well as a range of characters. **Families depicted by more than one species are listed alphabetically here**:

Family		Genera	Species	Covered	Page No.
Carabidae	**Ground beetles**	86	364	48	*243*
Cerambycidae	**Longhorn beetles**	51	73	47	*253*
Chrysomelidae	**Leaf beetles**	63	279	41	*253*
Coccinellidae	**Ladybirds**	31	55	34	*250*
Curculionidae	**Weevils**	34	494	15	*254*
Dytiscidae	**Diving beetles**	29	118	19	*242*
Elateridae	**Click beetles**	38	73	32	*246*
Lucanidae	**Stag beetles**	4	4 (1†)	3	*244*
Meloidae	**Blister beetles**	3	10 (3†)	7	*252*
Scarabaeidae	**Scarab beetles**	55	84	22	*244*
Staphylinidae	**Rove beetles**	275	1,129	19	*244*

Beetles sometimes appear in large groups, such as these Red Poplar Leaf Beetles feeding on Aspen leaves.

FURTHER READING AND USEFUL WEBSITES

Cooter, J. & Barclay, M.V.L. 2006. ***A Coleopterist's Handbook***. 4th Edition. The Amateur Entomologist Vol. 11. The Amateur Entomologists' Society. *Useful techniques.*

Duff, A.G. 2018. ***Checklist of Beetles of the British Isles***. 3rd Edition. Pemberley Books (Publishing).

Duff, A.G. ***Beetles of Britain and Ireland***. Published so far: *Vol. 1 Sphaeriusidae to Silphidae* 2012, *Vol. 3 Geotrupidae to Scraptiidae* 2020, *Vol. 4 Cerambycidae to Curculionidae* 2016. A.G. Duff (Publishing).
Comprehensive specialist works, with keys and colour illustrations featuring one species in each genus. Without these resources there is only old literature and books to refer to, such as N.H. Joy's **A Practical Handbook of British Beetles** *(1932), and some recent RES Handbooks.*

www.coleopterist.org.uk The Coleopterist
Issues a leading journal to subscribers. The website includes biographical information on British coleopterists and useful notes on where to find beetles, county by county.

www.coleoptera.org.uk/recording-schemes UK Beetle Recording
Details of who to contact for relevant schemes.

www.ukbeetles.co.uk UK Beetles
Includes photos of many species.

The two suborders are separated by minor morphological differences; the most notable being the position of the hind coxa when viewed from beneath: in **Adephaga** the hind coxae divide the first abdominal segment (sternite); in **Polyphaga** the hind coxae do not completely divide this same segment.

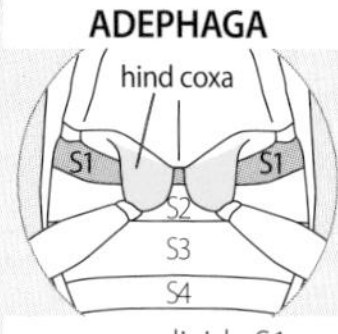

SUBORDER Adephaga (Water beetles, ground beetles) | 6 of 6 BI

IN WATER – legs adapted for swimming

BL 4–8 mm. **FORM** oval, shining, mainly black body; **LEGS** midlegs and hindlegs short, forelegs adapted for clasping prey; **ANTENNAE** short. **BEHAVIOUR** may be seen in numbers on the surface of streams, ponds and lakes, rarely stationary.

Common Whirligig
Gyrinus substriatus
(BL 5–7 mm)

Whirligig beetles
Gyrinidae
[2 genera | 12 species]

BL ≤5 mm. **FORM** oval, often black with yellowish markings; **SCUTELLUM** not visible; **FOREWINGS** elytra with large punctures; **LEGS** hind coxae (seen from underside), greatly enlarged. **BEHAVIOUR** swim using a clumsy-looking 'crawl' technique.

Common Crawling Water Beetle
Haliplus confinis
(BL 3 mm)

Crawling water beetles
Haliplidae
[3 genera | 19 species]

BL 1–38 mm. **FORM** small to very large, oval, head almost appearing to be part of thorax; **SCUTELLUM** present on many species (but not visible in one group with patterned elytra).

Black-bellied Diving Beetle
(BL 22–30 mm)

Diving beetles
Dytiscidae
[29 genera | 118 species]

p. 255

BL ≤10 mm. **FORM** stout, with conspicuous head, rounded above and below; **LEGS** hind tarsi elongated. **BEHAVIOUR** squeaks if handled.

Screech Beetle *Hygrobia hermanni*
(BL 9–10 mm)

Screech beetles
Hygrobiidae
[Screech Beetle]

BL ≤5 mm. **FORM** oval; **SCUTELLUM** not visible; **FOREWINGS** elytra plain, with large pits.

Larger Noterus *Noterus clavicornis*
(BL 4–5 mm)

Burrowing diving beetles
Noteridae
[1 genus | 2 species]

ON LAND (mostly) – legs not adapted for swimming

Ground Beetles
(including Tiger Beetles)
Carabidae ***p. 255***
[86 genera | 364 species]

see opposite page

SUBORDER Adephaga (Water beetles, ground beetles) | 6 of 6 BI

II➤ ON LAND (mostly) – legs not adapted for swimming

BL 2–35 mm. **FORM** small to very large; thorax broader than head; **FOREWINGS** elytra long, often long-oval and with longitudinal lines or sculptured; **ANTENNAE** located between eyes and mandibles.

ALL EXCEPT TIGER BEETLES

Moorland Ground Beetle (BL 16–20 mm)
INSET Caterpillar-hunter

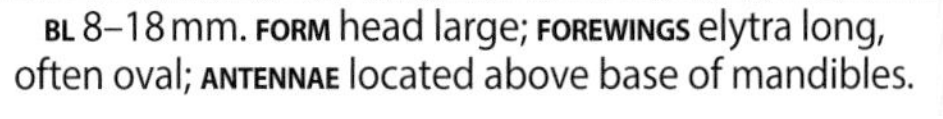

BL 8–18 mm. **FORM** head large; **FOREWINGS** elytra long, often oval; **ANTENNAE** located above base of mandibles.

TIGER BEETLES ONLY

Green Tiger Beetle (BL 12–17 mm)
INSET Dune Tiger Beetle

Ground beetles Carabidae
[86 genera | 364 species]

p. 262

SUBORDER Polyphaga (Huge variety of mainly non-water beetles and weevils) | 68 of 95 BI

WATER BEETLES [with some terrestrial species in rotting vegetation and/or dung]; legs adapted for swimming; palps long, at least ¾ of length of antennae

BL 2–7 mm. **FORM** oblong, several longitudinal grooves; **PALPS** long; **PRONOTUM** broadest in middle. **LIFESTYLE** on decaying vegetable matter.

Common Pool Helophorus
Helophorus griseus
(BL 3–4 mm)

Grooved water scavenger beetles
Helophoridae
[1 genera | 20 species]

BL 1–48 mm. **FORM** oval; **PALPS** long (in some genera longer than antennae); **PRONOTUM** broadest at base; **ANTENNAE** with clubbed tip. **LIFESTYLE** well equipped for swimming, using fringes of hairs on legs (aquatic and terrestrial species).

NT Great Silver Water Beetle
Hydrophilus piceus
(BL 34–48 mm)

Yellow-patched Dung Beetle
Sphaeridium scarabaeoides
(BL 5–7 mm)

Hydrophilid beetles
Hydrophilidae
[18 genera | 73 species]

SUBORDER Polyphaga

TERRESTRIAL BEETLES; legs not adapted for swimming; palps much shorter than length of antennae

1/5

BL 10–80 mm. **FORM** large, black; largest species has ♂ with huge reddish-brown mandibles and elytra; **ANTENNAE** elbowed, with clubbed tip. **LIFESTYLE** larvae feed for several years on rotting wood.

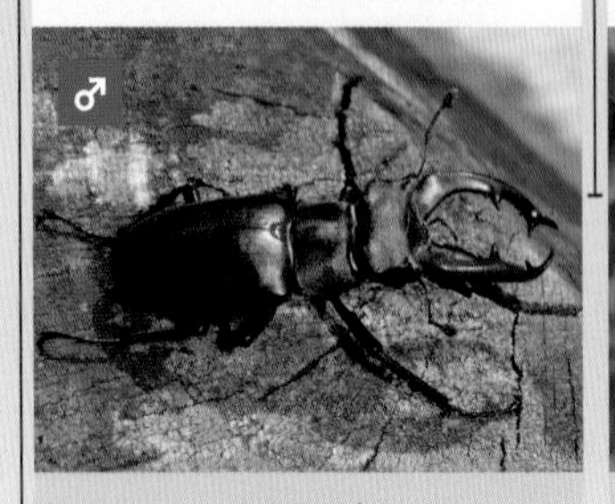

Stag Beetle
(BL 25–80 mm)

Stag beetles
Lucanidae
[4 genera | 4 species (1†)]

p. 280

BL 2–38 mm. **FORM** typically large and rounded, at least at front and back, coloration very variable; **ANTENNAE** short, with conspicuous clubbed tip. **LIFESTYLE** includes pests, larvae causing damage to cereals, grasses and gardens. Others mainly breed in dung.

Summer Chafer
(BL 15–20 mm)

Scarab beetles
Scarabaeidae
[55 genera | 84 species]

p. 282

BL 2–5 mm. **FORM** round or oval, soft-bodied, yellowish-brown; **ANTENNAE** long, thread-like. **LIFESTYLE** found near aquatic habitats; larvae develop underwater.

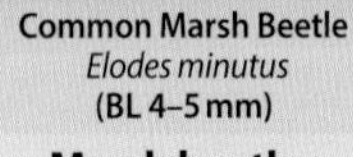

Common Marsh Beetle
Elodes minutus
(BL 4–5 mm)

Marsh beetles
Scirtidae
[7 genera | 20 species

BL 1–32 mm. **FORM** elongate or cylindrical; **WINGS** elytra short, leaving final 3–6 abdomen segments visible – in those with very short elytra, the folded hindwings are developed; **ANTENNAE** usually slender. **LIFESTYLE** mainly prey on invertebrates; some feed on fungi/dung/dead plant material.

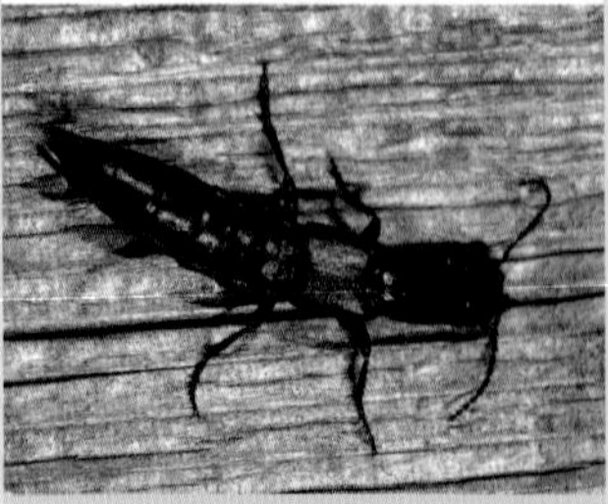

Black Coach-horse
Ocypus nitens
(BL 17–19 mm) NATIONALLY SCARCE

Rove beetles
Staphylinidae
[275 genera | 1,129 species]

p. 274

BL 2–13 mm. **FORM** elongate, sub-cylindrical, often metallic; **EYES** longer than wide; **ANTENNAE** thread-like, toothed or slightly club-tipped; inserted low. **LIFESTYLE** larvae of most species are woodborers and can damage trees, including oaks, hence are considered as pests by the timber trade.

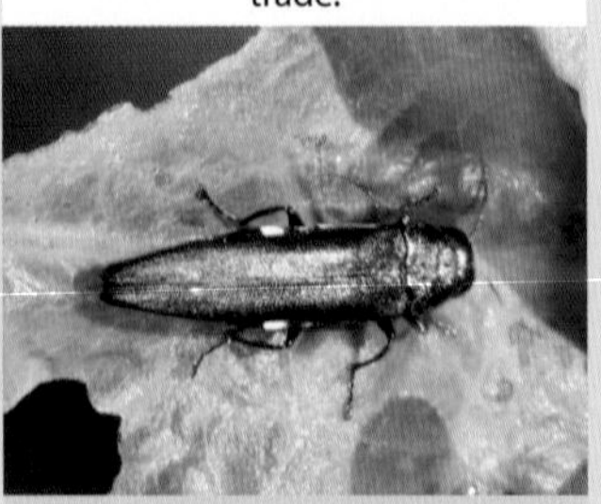

Oak Jewel Beetle *Agrilus biguttatus*
(BL 8–13 mm)

Jewel beetles
Buprestidae
[6 genera | 20 species (1†)]

BL 9–11 mm. **FORM** broad, oval, brown or greyish and hairy; **ANTENNAE** long and thin. **LIFESTYLE** larvae develop in roots.

Orchid Beetle *Dascillus cervinus*
(BL 9–11 mm)

Soft-bodied plant beetles
Dascillidae
[Orchid Beetle]

BL 1–11 mm. **FORM** compact, mainly glossy and very rounded; **ANTENNAE** with round, clubbed tip; **ELYTRA** short in most species, leaving last two abdomen segments visible. **BEHAVIOUR** predators that play dead when disturbed. **LIFESTYLE** associated with dung, fungi, seaweed, dead wood and birds' nests.

Garden Clown
Margarinotus merdarius
(BL 5–8 mm)

Hister beetles
Histeridae

[22 genera | 53 species]

BL 1–7 mm. **FORM** oval or round; **ANTENNAE** club-tipped or thread-like. **BEHAVIOUR** most likely under or on bark of trees. A superficially ladybird-like species is illustrated. **LIFESTYLE** feed on subterranean fungi, mainly on sandy soils.

Orange-patched Leiodid
Anisotoma humeralis
(BL 3–4 mm)

Leiodid beetles
Leiodidae

[22 genera | 94 species]

BL 8–30 mm. **FORM** variable: often broad and oval, most species black or with orange-and-black abdomen; **HEAD** narrow in some species; **WINGS** elytra ridged, short in some, with final 1–3 abdomen segments exposed; **ANTENNAE** tip often clubbed. **LIFESTYLE** many on dead birds, mammals, reptiles, smelling corpse and flying in to feed, some burying carcass; other species found in fungi, feed on vegetation or predate snails.

Common Snail-hunter
Phosphuga atrata (BL 10–15 mm)
Brown form not uncommon

Carrion beetles
Silphidae

[9 genera | 21 species]

BL 5–10 mm. **FORM** broad, rounded above and below, black, brown or grey; **ANTENNAE** with clubbed tip. **LIFESTYLE** larvae and adults feed on dry animal remains, with only one commoner species (shown) that can be found in birds' nests.

Common Hide Beetle
Trox scaber
(BL 6–8 mm) LOCAL

Hide beetles
Trogidae

[1 genus | 3 species]

BL 8–26 mm. **FORM** heavily built, oval, blackish with metallic coloration, underside often metallic blue or mauve; **ANTENNAE** short, with 3-segmented clubbed tip. **BEHAVIOUR** clumsy, often seen on the ground in pastures and woodland. **LIFESTYLE** larvae develop in large herbivore dung.

Common Dumble Dor
Geotrupes spiniger (BL 16–26 mm)
NB duller than related species

Dor beetles
Geotrupidae

[5 genera | 8 species]

BL 1–10 mm. **FORM** almost rounded, hard; **ANTENNAE** short, often with clubbed tip; **ELYTRA** have longitudinal lines. **BEHAVIOUR** when disturbed plays dead by tucking its short legs under the body; slow-moving. **LIFESTYLE**; feed on mosses, some on young plants.

Pill Beetle
Byrrhus pilula
(BL 7–9 mm)

Pill beetles
Byrrhidae

[7 genera | 13 species]

SUBORDER Polyphaga

TERRESTRIAL BEETLES; legs not adapted for swimming; palps much shorter than length of antennae 2/5

BL 3–5 mm. **FORM** small, elongate and hairy; **ANTENNAE** short; **PRONOTUM** sides arched. **LIFESTYLE** found near water; larvae semiaquatic.

Yellowish Long-toed Water Beetle *Dryops luridus* (BL 4 mm)

Long-toed water beetles Dryopidae

[2 genera | 9 species]

BL 3–11 mm. **FORM** elongate, and parallel-sided; brown or black and hairy; **ANTENNAE** long and often comb-like; **PRONOTUM** with pointed margins. **LIFESTYLE** larvae are only partly aquatic.

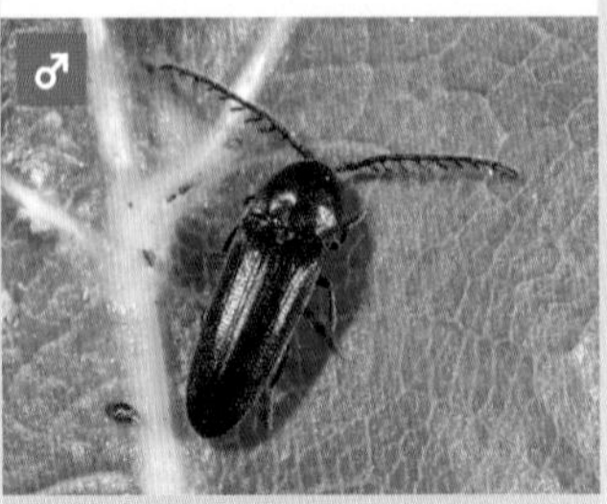

♂

Pygmy False Click Beetle *Microrhagus pygmaeus* (BL 4–6 mm) RED LIST (RDB3): RARE

False click beetles Eucnemidae

[6 genera | 7 species]

BL 2–3 mm. **FORM** small, brown; less elongate than other click beetle relatives; **ANTENNAE** short, with clubbed tip. **BEHAVIOUR** can jump (like click beetles), as well as playing dead when disturbed. **LIFESTYLE** in rotten logs, grass tussocks and fungoid soil; adults visit flowers, probably for pollen.

Common Throscid *Trixagus dermestoides* (BL 2–3 mm)

Throscid beetles Throscidae

[2 genera | 5 species]

BL 2–24 mm. **FORM** elongate and parallel-sided; coloration very variable; **ANTENNAE** long, thread-like, toothed in a few cases or feathery; **PRONOTUM** hind margins backward-pointing. **LIFESTYLE** larvae develop in dead wood, vegetation or in the soil; some prey on invertebrates.

♀

Copper Click Beetle (BL 11–16 mm)

Click beetles Elateridae

[38 genera | 73 species]

p. 290

BL 5–18 mm. **FORM** elongate; ♂ **ELYTRA** brownish + **ANTENNAE** comb-like; ♀ (rarely seen) wingless, much larger + **ANTENNAE** moderately long and thread-like. **LIFESTYLE** larvae feed on snails.

♂

Brown False Firefly *Drilus flavescens* (BL ♂ 5–8 mm; ♀ 16–18 mm) LOCAL

False firefly beetles Drilidae

[Brown False Firefly]

BL 5–13 mm. **FORM** elongate; red with black patches; **ELYTRA** soft with net-like pattern of lines across and down; **ANTENNAE** long, robust and serrate (toothed). **LIFESTYLE** larvae develop in well-decayed wood.

Small Net-winged Beetle *Platycis minutus* mating pair (BL 5–9 mm)

Net-winged beetles Lycidae

[4 genera | 4 species]

BL 5–20 mm. **FORM** ♂ long-oval, soft-bodied; ♀ wingless; larva-like; **HEAD** hidden beneath rounded pronotum; **ELYTRA** covering abdomen or reduced; **ANTENNAE** short. **BEHAVIOUR** ♀s 'glow' at night, to attract ♂s. **LIFESTYLE** larvae prey on invertebrates on the ground.

Glow-worm *Lampyris noctiluca*
(BL 10–20 mm)

Glow-worms
Lampyridae
[3 genera | 3 species (1†)]

BL 2–15 mm. **FORM** elongate, soft-bodied; often colourful black-and-red; **ANTENNAE** long and slender. **LIFESTYLE** larvae feed on arthropods; adults are mostly seen on flowers.

Common Red Soldier Beetle
Rhagonycha fulva
(BL 7–10 mm)

Soldier beetles
Cantharidae
[7 genera | 42 species]

BL 2–11 mm. **FORM** oval or round, but robust-looking and covered with fine hairs; **ANTENNAE** often with clubbed tip. **LIFESTYLE** larvae and adults feed on dried animal or sometimes plant material.

Two-banded Megatoma
Megatoma undata
(BL 4–6 mm) NATIONALLY SCARCE

Skin beetles
Dermestidae
[13 genera | 41 species]

BL 2–10 mm. **FORM** cylindrical, with head in some species hidden beneath pronotum; **ANTENNAE** often with clubbed tip. **LIFESTYLE** larvae bore tunnels in wood or plant pith; some are associated with stored grain.

Powder-post Beetle *Lyctus brunneus*
(BL 3–8 mm)
RED LIST (RDB3): RARE

Auger beetles
Bostrichidae
[4 genera | 5 species]

BL 2–9 mm. **FORM** compact to long-oval, with spider-like legs in some; **PRONOTUM** narrower than elytra; **ANTENNAE** long, comb-like in some. **LIFESTYLE** larvae develop in dead wood, fungi or dried animal or plant material.

Fan-bearing Wood-borer
Ptilinus pectinicornis
(BL 3–6 mm)

Anobiid and spider beetles
Ptinidae
[29 genera | 57 species]

BL 6–18 mm. **FORM** elongate, soft-bodied; **ANTENNAE** short; **ELYTRA** not reaching tip of abdomen. **LIFESTYLE** larvae bore tunnels in heartwood of trees or dead wood, where they have a symbiotic relationship with fungi.

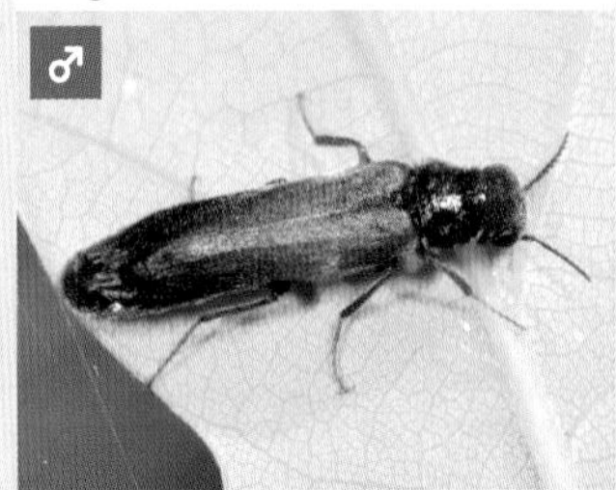

Red Ship-timber Beetle
Elateroides dermestoides
(BL 6–18 mm)

Ship-timber beetles
Lymexylidae
[2 genera | 2 species]

SUBORDER Polyphaga

TERRESTRIAL BEETLES; legs not adapted for swimming; palps much shorter than length of antennae 3/5

BL 3–11 mm. **FORM** round to elongate and rather flattened; **ANTENNAE** short, with clubbed tip. **LIFESTYLE** larvae of many species are predators of bark beetles (*p. 326*) living under bark or in galleries. The species shown *below* feeds on bracket fungi.

Tortoise Trogossitid
Thymalus limbatus
(BL 5–7 mm) NATIONALLY SCARCE

Trogossitid beetles
Trogossitidae

[5 genera | 5 species]

BL 3–17 mm. **FORM** sub-cylindrical and hairy, some species brightly coloured; **ANTENNAE** often with clubbed tip. **LIFESTYLE** larvae and adults of some species are predators of wood-boring beetles.

Black Chequered Beetle
Tillus elongatus (often all-black)
(BL 6–9 mm) NATIONALLY SCARCE

Chequered beetles
Cleridae

[8 genera | 13 species]

BL 2–7 mm. **FORM** often elongate and colourful; **ELYTRA** soft; **ANTENNAE** thread-like or toothed. **LIFESTYLE** larvae are thought to be scavengers or predators, while adults feed on flower pollen and nectar.

Common Malachite
Malachius bipustulatus
(BL 5–6 mm)

Soft-winged flower beetles
Melyridae

[15 genera | 26 species]

BL 3–5 mm. **FORM** small, cylindrical, brown or grey, hairy; **ANTENNAE** with clubbed tip. **LIFESTYLE** larvae feed in flowerheads.

Ochre Beetle
Byturus ochraceus
(BL 3–5 mm)

Fruitworm beetles
Byturidae

[1 genus | 2 species]

BL approx. 3 mm. **FORM** long-oval, dark, hairy; **PRONOTUM** with ridge on side; **ANTENNAE** with clubbed tip. **LIFESTYLE** larvae are associated with fungoid fruiting bodies on dead wood.

King Alfred's Beetle
Biphyllus lunatus
(BL 3 mm)

False skin beetles
Biphyllidae

[2 genera | 2 species]

BL 2–7 mm. **FORM** oval, often colourful; **ANTENNAE** with clubbed tip. **LIFESTYLE** larvae and adults feed on fungoid fruiting bodies on dead wood or decaying vegetation.

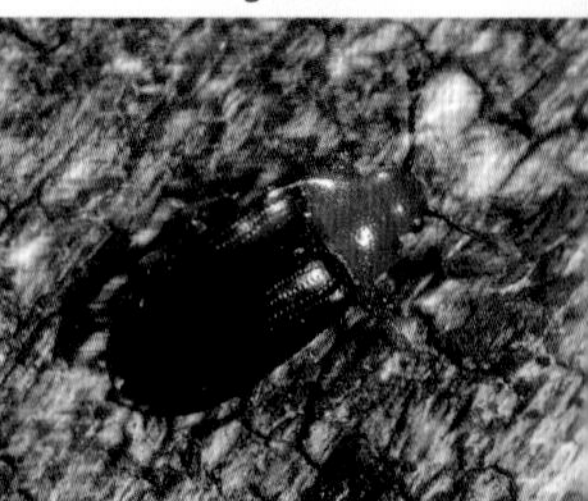

Red-and-blue Mushroom Beetle
Triplax aenea
(BL 3–4 mm)

Pleasing fungus beetles
Erotylidae

[4 genera | 8 species]

BL 2–6 mm. **FORM** small, elongate and flattened; **ANTENNAE** short, with clubbed tip. **LIFESTYLE** larvae are predators or associated with decaying animal or plant matter.

Red-and-blue Rhizophagus
Rhizophagus dispar
(BL 2–5 mm)

Root-eating beetles
Monotomidae
[2 genera | 23 species]

BL 1–5 mm. **FORM** small, oval or long-oval; **ANTENNAE** with clubbed tip. **LIFESTYLE** larvae and adults feed on fungal hyphae and spores in decaying vegetation. A few species are pollen feeders in flowers.

Lycoperd's Fungus Beetle
Cryptophagus lycoperdi
(BL 2–4 mm)

Silken fungus beetles
Cryptophagidae
[11 genera | 23 species]

BL 2–7 mm. **FORM** small, elongate, head narrower than thorax; **ANTENNAE** long, with weakly clubbed tip. **LIFESTYLE** larvae and adults occur under bark, in vegetation or in the soil, probably feeding on fungi.

Dark Silvanid
Uleiota planatus
(BL *c.* 5 mm) NATIONALLY SCARCE

Silvanid beetles
Silvanidae
[10 genera | 12 species]

BL approx. 4 mm. **FORM** small, elongate and flattened; **LEGS** short; **ANTENNAE** short, with clubbed tip. **LIFESTYLE** larvae and adults are found under bark.

Common Flat Bark Beetle
Pediacus dermestoides
(BL 4 mm)

Flat bark beetles
Cucujidae
[1 genus | 2 species]

BL 1–3 mm. **FORM** small, oval; **ANTENNAE** short, with clubbed tip. **LIFESTYLE** larvae and adults feed on fungoid vegetation, or just vegetation; some on pollen in flowers.

Common Shining Flower Beetle
Olibrus aeneus
(BL 2 mm)

Shining flower beetles
Phalacridae
[3 genera | 16 species]

BL 1–3 mm. **FORM** small, long-oval; **LEGS** short; **ANTENNAE** short, with clubbed tip; **ELYTRA** short, final 2–3 abdomen segments exposed. **LIFESTYLE** larvae and adults feed on flowers.

Orange-patched Flower Beetle
Kateretes pusillus
(BL 2–3 mm)

Short-winged flower beetles
Kateretidae
[3 genera | 9 species]

SUBORDER Polyphaga

TERRESTRIAL BEETLES; legs not adapted for swimming; palps much shorter than length of antennae 4/5

BL 1–8 mm. **FORM** small, long-oval; **ANTENNAE** short, with strongly clubbed tip; **ELYTRA** usually cover abdomen but fall short in some species, with final abdomen segment(s) showing; **LEGS** short, stout. **LIFESTYLE** larvae and adults mainly feed on sap runs on tree trunks, decaying fruit or fungi.

Four-spotted Sap Beetle
Glischrochilus hortensis
(BL 4–8 mm)

Sap beetles
Nitidulidae

[16 genera | 92 species]

BL 1–3 mm. **FORM** small, oval or elongate; **ANTENNAE** short, with 1–2 segmented, round, clubbed tip; **TARSUS** final segment typically longer than remaining tarsal segments combined. **LIFESTYLE** larvae and adults feed under fungoid bark on recently dead trees.

Common Cerylonid
Cerylon ferrugineum
(BL 2 mm)

Cerylonid beetles
Cerylonidae

[2 genera | 5 species]

BL 1–6 mm. **FORM** small, oval or long-oval (includes a ladybird lookalike); **ANTENNAE** with segments of equal length, or with clubbed tip. **LIFESTYLE** larvae and adults feed in rotten wood and under fungoid bark.

False Ladybird
Endomychus coccineus
(BL 4–6 mm)

Handsome fungus beetles
Endomychidae

[5 genera | 8 species]

BL 1–9 mm. **FORM** round or oval and often brightly coloured (red-and-black or yellow-and-black) with spots on elytra; **ANTENNAE** short, with clubbed tip. **LIFESTYLE** larvae and adults mostly prey on aphids and scale insects.

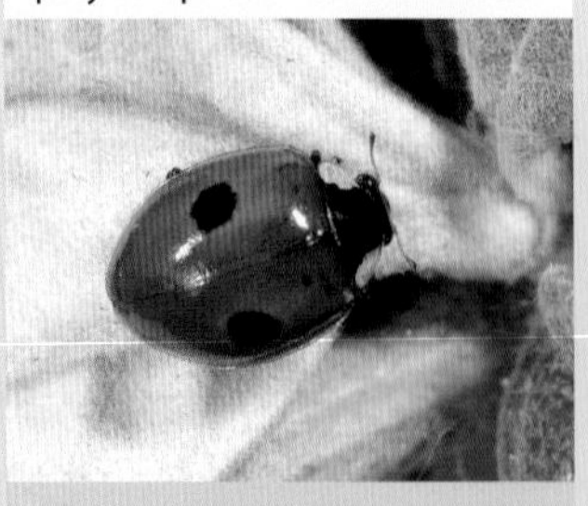

2-spot Ladybird
(BL 4–5 mm)

Ladybirds
Coccinellidae

[31 genera | 55 species]

p. 309

BL 1–3 mm. **FORM** small, elongate or long-oval and covered in hairs; **ANTENNAE** with clubbed tip. **LIFESTYLE** larvae and adults feed in mouldy vegetation.

Brown Mould Beetle
Stephostethus lardarius
(BL 3 mm)

Mould beetles
Latridiidae

[13 genera | 56 species]

BL 1–6 mm. **FORM** small, oval or slightly elongate; **ANTENNAE** typically with clubbed tip. **LIFESTYLE** larvae and adults feed in bracket fungi and under fungoid bark.

Orange-spotted Fungus Beetle
Mycetophagus quadripustulatus
(BL 5–6 mm)

Hairy fungus beetles
Mycetophagidae

[7 genera | 15 species]

(Huge variety of mainly non-water beetles and weevils) | 68 of 95 BI

BL 1–4 mm. **FORM** small, long-oval; pronotum barrel-shaped; **ANTENNAE** with clubbed tip. **LIFESTYLE** larvae and adults feed under fungoid bark and in fungi.

Common Ciid
Cis boleti
(BL 3–4 mm)

Ciid beetles
Ciidae

[7 genera | 22 species]

BL 3–6 mm. **FORM** small, glossy and cylindrical; **ANTENNAE** with clubbed tip. **LIFESTYLE** larvae and adults feed mainly in bracket fungi.

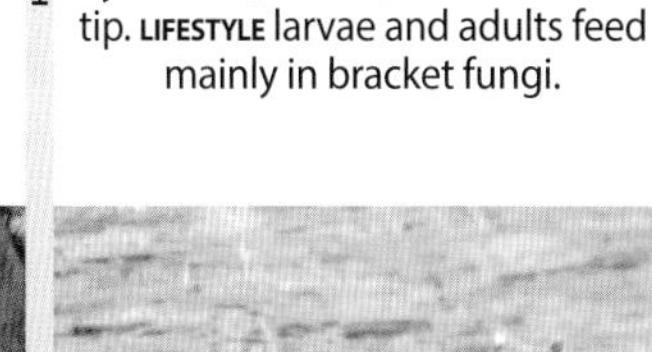

Anchor Beetle
Tetratoma ancora
(BL 3 mm) NATIONALLY SCARCE

Polypore fungus beetles
Tetratomidae

[2 genera | 4 species]

BL 2–16 mm. **FORM** small to medium-sized, elongate and hardened; **ANTENNAE** fairly long, positioned at side of head, visible from above. **LIFESTYLE** larvae feed in dead wood and bracket fungi.

Blue False Darkling Beetle
Melandrya caraboides
(BL 10–16 mm) LOCAL

False darkling beetles
Melandryidae

[10 genera | 17 species]

BL 2–9 mm. **FORM** long-oval and rather hunched; **HEAD + PRONOTUM** large; **ELYTRA** not reaching tip of pointed abdomen (pygidium visible); **ANTENNAE** short, toothed. **LIFESTYLE** larvae feed in stems of herbaceous plants, rotting wood and fungi. Adults feed on pollen.

Shaggy Tumbling Flower Beetle
Variimorda villosa
(BL 6–8 mm) NATIONALLY SCARCE

Tumbling flower beetles
Mordellidae

[5 genera | 17 species]

BL 8–12 mm. **FORM** elongate, hunched and wedge-shaped; **ELYTRA** pointed, but not reaching tip of abdomen, or meeting in centre; **ANTENNAE** feathery; **LEGS** long. **LIFESTYLE** larvae are parasitoids of social wasps, including Common Wasp (*p. 513*).

Wasp Nest Beetle
Metoecus paradoxus
(BL 8–12 mm)

Ripiphorid beetles
Ripiphoridae

[Wasp Nest Beetle]

BL 2–7 mm. **FORM** long-oval to elongate; **ANTENNAE** generally short with clubbed tip. **LIFESTYLE** larvae are associated with woodland fungi, but some are predators.

Saddle-backed Bitoma
Bitoma crenata
(BL 3 mm)

Zopherid beetles
Zopheridae

[9 genera | 13 species]

SUBORDER Polyphaga

TERRESTRIAL BEETLES; legs not adapted for swimming; palps much shorter than length of antennae 5/5

BL 2–26 mm. FORM small to large, shape variable, but dark; ANTENNAE situated beneath ridge at side of head (base not visible from above), segments equal length, toothed or club-tipped. LIFESTYLE **larvae and adults feed in dead wood, fungoid wood and fungi; some adults feed on nectar and pollen.**

Common Darkling Beetle
Nalassus laevioctostriatus
(BL 7–12 mm)

Darkling beetles
Tenebrionidae
[33 genera | 47 species]

BL 5–20 mm. FORM elongate; ELYTRA soft with 3 longitudinal lines; ANTENNAE long; LEGS long, in one species hind femora hugely swollen in ♂. BEHAVIOUR **some species cause blistering on skin.** LIFESTYLE **larvae develop in stems and roots of herbaceous plants, or rotting wood. Adults feed on flowers.**

Swollen-thighed Beetle
Oedemera nobilis
(BL 6–11 mm)

False blister beetles
Oedemeridae
[4 genera | 10 species]

BL 6–35 mm. FORM medium-sized to large, robust, mostly black with small pronotum; ELYTRA usually much shorter than abdomen, which is bloated in ♀; ANTENNAE fairly long, in some species kinked in middle. LIFESTYLE larvae are parasitoids on various bees. BEHAVIOUR Cantharidin is released by reflex bleeding in defence and can cause blistering on skin; it is very poisonous in high doses.

Black Oil Beetle
(BL 11–35 mm)

Blister beetles
Meloidae
[3 genera | 10 species (3†)]

p. 288

BL 8–18 mm. FORM large, mostly red, some black, flat; ELYTRA broadened towards hind part; ANTENNAE long, toothed or feathery. LIFESTYLE **larvae develop in decaying wood.**

Black-headed Cardinal
Pyrochroa coccinea
(BL 14–18 mm)

Cardinal beetles
Pyrochroidae
[2 genera | 3 species]

BL 2–5 mm. FORM small, glossy, some with rostrum; ANTENNAE long. LIFESTYLE **larvae develop under bark or in decaying vegetation.** **SS** weevils (*p. 323*) have a rostrum, and typically long and elbowed antennae.

Red-snouted Bark Beetle
Salpingus ruficollis
(BL 3–5 mm) Weevil-like species

Narrow-waisted bark beetles
Salpingidae
[6 genera | 11 species]

BL 2–5 mm. FORM superficially ant-like; PRONOTUM narrow; one species with pronotal horn; ANTENNAE medium-length, slightly clubbed tip. LIFESTYLE **larvae and adults are scavengers, often in litter on the ground.**

Monoceros Beetle
Notoxus monoceros
(BL 4–5 mm)

Ant-like flower beetles
Anthicidae
[7 genera | 14 species]

BL 2–4 mm. **FORM** small, slender; **ANTENNAE** fairly long, most species with yellow basal segment; **LEGS** long. **LIFESTYLE** **larvae develop under bark, in dead wood and bracket fungi.**

Hawthorn Anaspis
Anaspis frontalis
(BL 4 mm)

False flower beetles
Scraptiidae
[2 genera | 14 species]

BL 3–45 mm. **FORM** small to very large, robust or slender, but cylindrical; often colourful; **ANTENNAE** long to very long, in some cases exceeding body length; **ELYTRA** usually full length of abdomen, but shorter in a few species; **LEGS** long. **LIFESTYLE** larvae develop in wood, some species in plants.

Black-and-yellow Longhorn Beetle
(BL 13–20 mm)

Longhorn beetles
Cerambycidae
[51 genera | 73 species]

p. 296

BL 4–8 mm. **FORM** small, elongate, variable in colour within species; **PRONOTUM** narrow; **ANTENNAE** fairly long; **LEGS** short. **LIFESTYLE** **larvae feed on leaves, adults on anthers and pollen.**

Variable Orsodacne
Orsodacne cerasi
(BL 5–6 mm) NATIONALLY SCARCE

Orsodacnid leaf beetles
Orsodacnidae
[1 genus | 2 species]

BL 2–3 mm. **FORM** small, elongate, red-and-black; **PRONOTUM** small; **ANTENNAE** fairly long; **LEGS** short. **LIFESTYLE** **larvae and adults feed on poplar and willow leaves.**

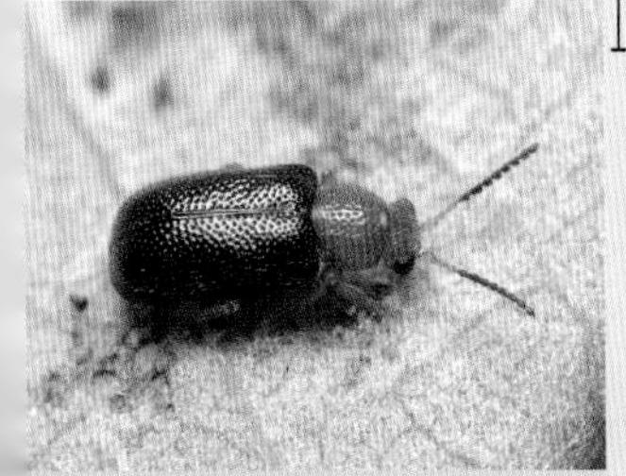

Aspen leaf-miner Beetle
Zeugophora subspinosa
(BL 3 mm)

Megalopodid leaf beetles
Megalopodidae
[1 genus | 3 species]

BL 1–20 mm. **FORM** small to large, broad and oval, often very colourful, which may be variable within species; **ANTENNAE** long; **LEGS** short. **LIFESTYLE** larvae and adults feed on plants. Larvae may feed on leaves or roots, some species regarded as pests of economic importance in agriculture, forestry and horticulture.

Dead-nettle Leaf Beetle
(BL 5–7 mm)

Lily Beetle
(BL 6–8 mm)

Leaf beetles
Chrysomelidae
[63 genera | 279 species]

p. 315

SUBORDER Polyphaga
(Huge variety of mainly non-water beetles and weevils) | 68 of 95 BI

WEEVILS and RELATIVES; with a rostrum (snout) 1/1

BL 2–15 mm. **FORM** small to medium-sized, broad and robust appearance; **ROSTRUM** short and broad; **ANTENNAE** short to long, often with clubbed tip. **LIFESTYLE** larvae develop in dead and dying branches or in fungi.

White-spotted Fungus Weevil
Platystomos albinus
(BL 8–15 mm) NATIONALLY SCARCE

Fungus weevils
Anthribidae
[9 genera | 10 species]

BL 2–8 mm. **FORM** broad; **ELYTRA** often metallic; **ROSTRUM** long and narrow; **ANTENNAE** with clubbed tip. **LIFESTYLE** larvae develop in trees and shrubs, often living communally in a tent.

Leaf-rolling Weevil
Byctiscus populi
(BL 4–5 mm) RED LIST (RDB3): RARE

Tooth-nosed snout weevils
Rhynchitidae
[9 genera | 18 species]

BL 4–8 mm. **FORM** broad, hunched, bright red-and-black; **HEAD** narrow; **EYES** positioned high up, **ROSTRUM** short; **ANTENNAE** with clubbed tip. **LIFESTYLE** larvae develop in rolled-up leaves of trees.

Hazel Leaf-roller Weevil
Apoderus coryli
(BL 6–8 mm) LOCAL

Leaf-rolling weevils
Attelabidae
[2 genera | 2 species]

BL 1–4 mm. **FORM** pear-shaped; **ROSTRUM** long; **ANTENNAE** with S1 short and a compact, clubbed tip; **LEGS** long. **LIFESTYLE** larvae develop inside host plants.

Violet Weevil
Perapion violaceum
(BL 3 mm)

Apionid weevils
Apionidae
[34 genera | 87 species]

BL 1–2 mm. **FORM** broad, rounded; **ROSTRUM** long; **ANTENNAE** elbowed with clubbed tip. **LIFESTYLE** larvae and adults feed on waterside plants. Larval activity can cause galls on stems.

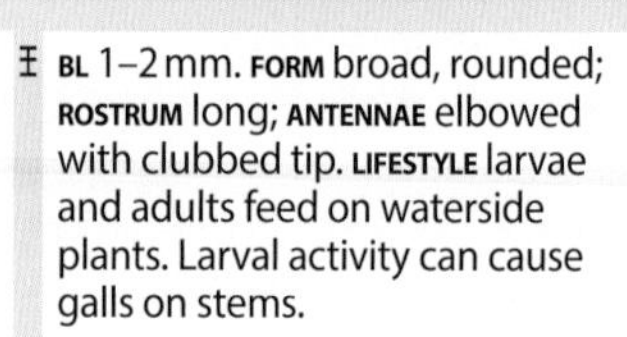

Loosestrife Weevil
Nanophyes marmoratus
(BL 2 mm)

Nanophyid weevils
Nanophyidae
[2 genera | 2 species]

BL 1–14 mm. **FORM** broad; **ROSTRUM** long; **ANTENNAE** often elbowed (S1 much longer than S2+S3 combined), also clubbed. **LIFESTYLE** feed on plants, including roots. **SS** some Narrow-waisted bark beetles (see *p. 252*) have a rostrum, but antennae are short and not elbowed.

Pine Weevil

Weevils
Curculionidae
[164 genera | 494 species]

p. 323

SUBORDER Adephaga (Water beetles, ground beetles) | 2 of 6 BI

FAMILY Dytiscidae (Diving beetles) 29 GEN. | 118 spp. | 19 ILL.

Small to very large, oval, water beetles (1–38 mm), with the head almost appearing to be part of the body. Fast-moving, streamlined beetles adapted for life in the water; many species hunt prey, helped by thrusting hindlegs; others feed on vegetation. They breathe by trapping air under the wing-cases. A selected range of species is covered; some require careful examination as there are similar species.

SCUTELLUM visible			
FORM large to very large (BL 22–38 mm); FORE TARSI expanded; PRONOTUM with yellow on all or just side margins		***Dytiscus*** [6 species]	*p. 256*
FORM medium-sized to large (BL 12–18 mm); FORE TARSI expanded			
ELYTRA with pale margins	ELYTRA base with transverse pale bars or small, often faint, yellowish marks; PRONOTUM yellow sides with large black crescent-shaped mark	***Hydaticus*** [2 species]	*p. 259*
ELYTRA with pale margins	ELYTRA blackish, mottled; margin yellow with black specks; PRONOTUM yellow with two black transverse bars	***Acilius*** [2 species]	*p. 259*
ELYTRA heavily mottled with partial narrow margin	PRONOTUM yellowish with two black transverse stripes	***Graphoderus*** [3 species (1†)]	*p. 258*
ELYTRA heavily mottled with partial narrow margin	PRONOTUM dark brown with yellowish sides; ELYTRA with pattern of transverse net-like lines	***Colymbetes*** [Large-grooved Diving Beetle]	*p. 258*
ELYTRA heavily mottled with partial narrow margin	PRONOTUM with smudged central mark; ELYTRA mottled yellow and black; pattern of longitudinal net-like lines	***Rhantus*** [5 species]	*p. 258*
FORM medium-sized (BL 6–15 mm); FORE TARSI not expanded			
A single key for the genera *Agabus* and *Ilybius* is traditionally used to identify species (usually requiring dissection of male genitalia), except where there are obvious colour or size differences, as in most species shown here.			
Body length 6–12 mm (only one species >10 mm) **NB – species shown 8–9 mm**		***Agabus*** [18 species]	*p. 260*
Body length 6–15 mm; usually larger than most *Agabus* species. **NB – species shown 10–15 mm**		***Ilybius*** [10 species]	*p. 261*
BL 6–8 mm; FORM narrow; reddish-brown		***Liopterus*** [Piles Beetle]	*p. 261*
SCUTELLUM absent			
FORM small (BL 4–6 mm)			
ELYTRA heavily spotted		***Stictotarsus*** [12-spotted Diving Beetle]	*p. 256*
ELYTRA plain		***Hyphydrus*** [1 species]	*p. 256*

Guide to Dytiscidae p. 255

DIVING BEETLES – SCUTELLUM ABSENT

Stictotarsus 1 sp.

12-spotted Diving Beetle
Stictotarsus duodecimpustulatus

BL 5–6 mm. **ID** Small to medium-sized, pronotum widest in centre. Yellowish ground colour, black at back of head, upper half of antennae, back of pronotum and on elytra, which also has 12 large yellowish spots.
Hab Ditches, lakes, ponds, rivers.
SS None.

Hyphydrus 1 sp.

Cherrystone Beetle
Hyphydrus ovatus

BL 4–5 mm. **ID** Small to medium-sized, rounded, reddish-brown.
Hab Ditches, canals, ponds, rivers, streams. **SS** None.

DIVING BEETLES – SCUTELLUM VISIBLE; BL 22–38 mm

Dytiscus 6 spp. | 3 ILL.

King Diving Beetle NT
Dytiscus dimidiatus

BL 30–38 mm. **ID** Largest *Dytiscus* species. Pronotum with yellow side margins only; body black with underside yellow; prothorax with yellow margin. ♀ elytra heavily ridged.
Hab Richly vegetated fenland drains.
SS Black-bellied Diving Beetle, which has a black underside; other *Dytiscus* species, which have pronotum with continuous yellowish margin.

Black-bellied Diving Beetle

Dytiscus semisulcatus

BL 22–30 mm. **ID** Very large. Pronotum with yellow side margins only; body black with underside black, except for prothorax with yellow margin. ♀ elytra heavily ridged. **Hab** Shallow stagnant waters. **SS** King Diving Beetle, which has a yellow underside; other *Dytiscus* species, which have a continuous yellowish margin to pronotum.

Great Diving Beetle

Dytiscus marginalis

BL 26–32 mm. **ID** Very large. Pronotum with narrow yellow margins around whole segment; body black but with underside yellow. ♀ elytra heavily ridged. **Hab** Ponds. **SS** Other *Dytiscus* species that have pronotum with continuous yellow margin, from which told by yellow (not yellow and black-banded) underside and lack of a yellow eye margin.

Black-bellied Diving Beetle underside black

Great Diving Beetle underside yellow

Guide to Dytiscidae p. 255

DIVING BEETLES – SCUTELLUM VISIBLE; BL 12–18 mm; FORE TARSI EXPANDED

Colymbetes 1 sp.

Large-grooved Diving Beetle
Colymbetes fuscus

BL 15–17 mm. **ID** Large; body greyish and brown; pronotum black with brownish side margins; elytra with distinctive transverse stripes. Legs reddish-brown. **Hab** Streams, ditches, ponds. **SS** None.

Rhantus 5 spp. | 1 ILL.

Supertramp
Rhantus suturalis

BL 10–13 mm. **ID** Large; body dark or yellowish-brown; pronotum with smudged central mark and at least broad yellowish margin; elytra with longitudinal lines; antennae all-yellow. **Hab** Ditches, ponds, pools. **SS** Other *Rhantus* species, which lack black central mark on pronotum and have dark-tipped antennae.

Graphoderus 3 spp. (1†) | 2 ILL.

Spangled Diving Beetle CR ●
Graphoderus zonatus

BL 12–16 mm. **ID** Large; pronotum yellowish with two black transverse stripes. **Hab** Lakes and pools. **Dist** Restricted to one site in North Hampshire. **SS** Other *Graphoderus* species, but distinguished by narrow yellow bands on pronotum at front (before first black stripe) and back.

Lesser Spangled Diving Beetle
Graphoderus cinereus VU

BL 14–15 mm. **ID** Large; pronotum yellowish with two black transverse stripes. **Hab** Ponds and fenland drainage ditches. **SS** Other *Graphoderus* species, but unlike **Spangled Diving Beetle** lacks yellow band at back of pronotum.

Acilius 2 spp. | 1 ILL.

Lesser Diving Beetle
Acilius sulcatus

BL 16–18 mm. **ID** Large; pronotum yellow with two black transverse bars; elytra black, margin yellow with black dots; hind femora dark at least on inner half. Underside of abdomen black with yellow spots. **Hab** Ditches, lakes, large ponds. **SS** Scarce Lesser Diving Beetle, from which distinguished by colour of underside and hind femora.

Scarce Lesser Diving Beetle
Acilius canaliculatus

BL 14–16 mm. **ID** Large; pronotum yellow with two black transverse bars; elytra black, margin yellow with black dots; hind femora pale. Underside of abdomen mainly pale. **Hab** Pools and fens. **SS** Lesser Diving Beetle, from which distinguished by colour of underside and hind femora.

Hydaticus 2 spp.

Semi-black Diving Beetle
Hydaticus seminiger

BL 13–15 mm. **ID** Elytra base black, except for small, often faint yellowish marks; margins of pronotum and elytra yellow. **Hab** Peat cuttings, woodland bogs. **SS** Transverse Diving Beetle, in which pronotum has transverse yellow bars on shoulders.

Transverse Diving Beetle
Hydaticus transversalis

BL 12–13 mm. **ID** Elytra black (apart from yellow margins) and transverse, pale bars at base. **Hab** Ponds, fenland ditches. **SS** Semi-black Diving Beetle, in which pronotum has transverse yellow bars on shoulders.

DIVING BEETLES – SCUTELLUM PRESENT; BL 6–15 mm

Agabus 18 spp. | 4 ILL.

Two-spotted Diving Beetle
Agabus biguttatus

Common

BL 8–9 mm. **ID** Elytra with two pale spots; pronotum black, with brownish side margins; antenna S11 darkened at tip. **Hab** Lime-rich streams and ponds. **SS** Other *Agabus* species, including **Spotted Diving Beetle**; the number and shape of spots on elytra is a helpful feature.

Spotted Diving Beetle
Agabus guttatus

Common

BL 8–9 mm. **ID** Elytra with four pale spots (also two spots at top of head); pronotum black, with brownish side margins; antenna S11 brown. **Hab** Shallow streams, often in woodland. **SS** Other *Agabus* species, including **Two-spotted Diving Beetle**; the number and shape of spots on elytra is a helpful feature.

Brown Diving Beetle
Agabus brunneus

VU

Rare

BL 8–9 mm. **ID** Completely brown or reddish-brown. **Hab** Heathland streams. **Beh** Can be buried in gravel, or under pebbles. **SS** None (other species are black, although a little caution is needed as teneral forms of some other *Agabus* spp. can appear brown).

Sturm's Diving Beetle
Agabus sturmii

Common

BL 8–9 mm. **ID** Orange edge to pronotum and elytral shoulders. Some have two orange dots on head readily visible. **Hab** Usually in lakes and ponds, also slower sections of ditches and rivers. **SS** Other *Agabus* species, from which told by pronotal margin colour; the elytra has double reticulation (net-like pattern) throughout, but microscope needed.

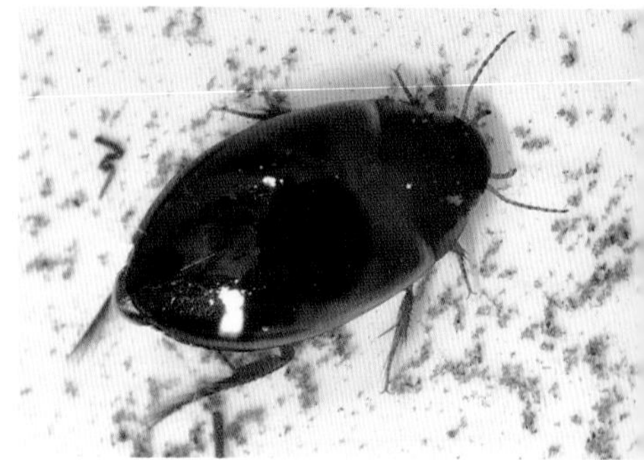

Ilybius — 10 spp. | 2 ILL.

Mud Dweller
Ilybius ater

Common

BL 13–15 mm. **ID** Largest *Ilybius* species, body matt black; antennae, mandibles, clypeus and legs reddish-brown, and also a small mark near tip of each wingcase. **Hab** Stagnant water, edges of ponds and ditches. **SS** Other *Ilybius* species (large size and colour distinguishes this species in the field).

Sooty Mud Dweller
Ilybius fuliginosus

Common

BL 10–12 mm. **ID** Large, body weakly shining black; pronotum and elytra with reddish or brownish margins. **Hab** Rivers, ditches, lakes, ponds. **SS** Other *Ilybius* species, although the pale margin is distinctive.

Liopterus — 1 sp.

Piles Beetle
Liopterus haemorrhoidalis

Common

BL 6–8 mm. **ID** Medium-sized, with narrow reddish-brown body. Antennae and legs reddish. **Hab** Ditches, ponds. **SS** None.

FAMILY **Carabidae** (Ground beetles) **86 GEN. | 364 spp. | 48 ILL.**

Small to very large (body length 2–35 mm) and mostly ground-dwelling; thorax broader than head; elytra long, often oval and with longitudinal lines or sculptured. Antennae located between eyes and mandibles. The medium-sized, metallic tiger beetles (8–18 mm) have a large head, eyes and jaws (for hunting prey), long elytra, and are often oval. Some species active by day, such as tiger beetles, are brightly coloured. Many species can be found by looking under stones and logs. **Examples of genera and species that are typically encountered are shown in this section, with a general description that avoids highly technical details and should at least help to narrow down the identification of some ground beetles. See *p. 241* for references to specialist identification resources.**

GROUND BEETLES – SCUTELLUM NOT VISIBLE, BODY ROUNDED

Omophron 1 sp.

Spangled Button Beetle
Omophron limbatum

BL 5·0–6·5 mm. **ID** Body rounded, tawny with metallic green marks, heavily punctured. **Hab** Subaquatic, flooded sand/gravel pit margins. **SS** None.

GROUND BEETLES – SCUTELLUM VISIBLE, BODY NOT ROUNDED

FORM **oval;** HEAD + JAWS **large;** ANTENNAE **located above base of mandibles;** ELYTRA **with metallic markings**

TIGER BEETLES
Cylindera
[Cliff Tiger Beetle]
Cicindela
[4 species]

Cylindera 1 sp.

Cliff Tiger Beetle VU
Cylindera germanica

BL 8–12 mm. **ID** Green, some bluish or bronze (rarely black) or two colours (as shown), but not dark brown; pronotum longer than wide. Elytra with three paired pale spots on margin. **Hab** Soft rock cliffs, near seepages. **SS Green Tiger Beetle**, which has pale central spots on elytra.

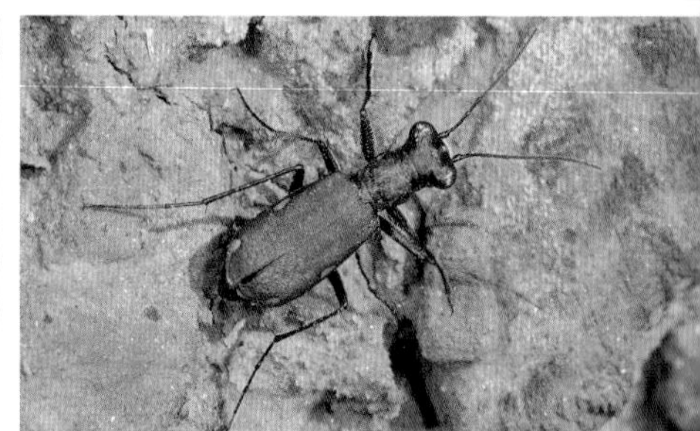

Cicindela 4 spp.

Northern Dune Tiger Beetle
Cicindela hybrida VU

Rare

BL 12–17 mm. **ID** Dark brown; elytra with whitish markings, those in centre being characteristic shape resembling the number 3. **Hab** Coastal sand dunes and heathland. **SS** Other brown *Cicindela* spp. but readily distinguished by shape of central markings on elytra.

Dune Tiger Beetle
Cicindela maritima NT

Rare

BL 11–15 mm. **ID** Dark brown; elytra with whitish markings, those in being characteristic shape resembling a musical note. **Hab** Coastal sand dunes. **SS** Other brown *Cicindela* spp. but readily distinguished by shape of central markings on elytra.

Green Tiger Beetle
Cicindela campestris

Common

BL 12–17 mm. **ID** Green; legs and ntennae metallic coppery or bronze; pronotum wider than long. Elytra with –6 pale yellow spots, some central. **Hab** Sandy heathland, hillsides. **Beh** On bare ground, often tracks. **SS** **Cliff Tiger Beetle**, which lacks pale central spots on elytra.

Heath Tiger Beetle
Cicindela sylvatica EN

Rare

BL 14–18 mm. **ID** Dark brown, elytra with narrow cream markings; labrum black. **Hab** Heathland. **SS** Other brown *Cicindela* spp. but readily distinguished by narrow central markings on elytra and black (not pale) labrum.

FORM oval; HEAD + JAWS mainly medium-sized; ANTENNAE located laterally, between eyes and mandibles; ELYTRA typically black, a few metallic.

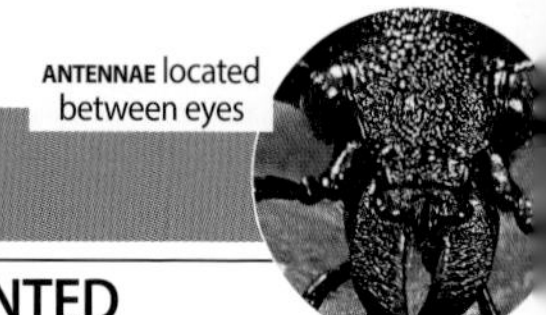

GROUND BEETLES – ABDOMEN UNDERSIDE 7/8-SEGMENTED

Brachinus — 2 spp. | 1 ILL.

Medium-sized (BL 5–10 mm); ABDOMEN underside **7-segmented in ♀, 8 in ♂**.

Bombardier Beetle
Brachinus crepitans

BL 6–10 mm. **ID** Elytra green or blue with truncate tip; head and thorax red or orange. **Hab** Chalk grassland and quarries, often under rocks. **Beh** In defence emits a noxious chemical spray from tip of abdomen. **SS** None.

Nationally Scarce

GROUND BEETLES – ABDOMEN UNDERSIDE 6-SEGMENTED — 1/

Small to very large (BL 2–35 mm); ABDOMEN underside **6-segmented**.

Eurynebria — 1 sp.

Beachcomber — EN
Eurynebria complanata

BL 16–23 mm. **ID** Large; cream to brown, elytra streaked black. **Hab** Sandy beaches above high tide line, under driftwood. **SS** None.

Rare

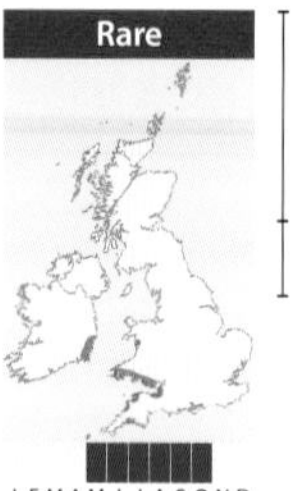

Cychrus — 1 sp

Snail Hunter
Cychrus caraboides

BL 14–19 mm. **ID** Large; body bulbous; pronotum narrow but rounded, black; hind tarsi with a few pale hairs. **Hab** Damp woodland and moorlands. **Beh** Stridulates with buzzing sound in defence. **SS** None.

Common

Elaphrus 4 spp. | 2 ILL.

FORM medium-sized (**BL** 7–11 mm); **EYES** large; **ELYTRA** heavily punctured with large round pits.

Copper Peacock
Elaphrus cupreus

BL 8–10 mm. **ID** Ground colour coppery-black; pronotum narrower than head; elytra glossy with large purplish punctures; tibiae brown. **Hab** Wetlands such as bogs, edge of streams, damp woodland. **SS** Other *Elaphrus* species, from which told by the extent of punctuation on elytra.

Green-socks Peacock
Elaphrus riparius

BL 7–8 mm. **ID** Ground colour olive green; pronotum slightly wider than head; elytra glossy with large shining punctures with dot in centre; tibiae brown. **Hab** Wetlands including marshes, bogs and ponds. **SS** Other *Elaphrus* species, from which told by the extent of punctuation on elytra.

Nebria 5 spp. | 1 ILL.

FORM medium-sized to large (**BL** 9–16 mm).

Common Heart-shield
Nebria brevicollis

BL 11–14 mm. **ID** Body and femora black/dark brown; rest of legs and antennae reddish-brown; pronotum rounded; hind tarsi have a few pale hairs. **Hab** Woodland, agricultural land, gardens, hedgerows. **SS** Other *Nebria* species, particularly **Bare-footed Heart-shield** *Nebria salina* [N/I], which has all-dark hind tarsi hairs and beaded edge to pronotum.

Notiophilus 8 spp. | 1 ILL.

FORM medium-sized (**BL** 5–7 mm).

Common Springtail-stalker
Notiophilus biguttatus

BL 5–6 mm. **ID** Dark brown; eyes very large; elytra with large gap between 2nd and 3rd longitudinal line. **Hab** Woodland, gardens, heathland, parks. **SS** Other *Notiophilus* species, from which told by markings on the elytra; leg colour and punctuation on head and pronotum may also be useful identification features.

GROUND BEETLES – ABDOMEN UNDERSIDE 6-SEGMENTED 2/6

Broscus 1 sp.

Strand-line Burrower
Broscus cephalotes

Common

BL 16–23 mm. **ID** Large to very large, including mandibles; black, but tips of antennae and palps brown; elytra with indistinct rows of fine punctures. **Hab** Sandy areas, mainly coastal. **Beh** Under stones and driftwood. **SS** None.

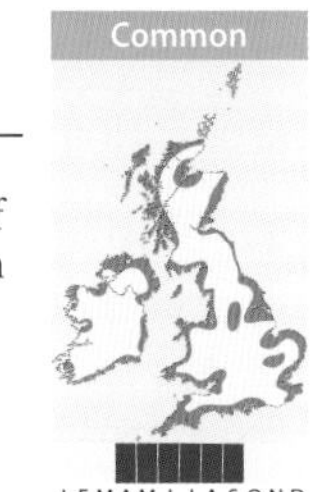

Calosoma 2 spp. | 1 ILL.

Caterpillar-hunter
Calosoma inquisitor

Nationally Scarce

BL 16–22 mm. **ID** Large to very large; dark metallic bronze; pronotum green or purple edges; elytra edge same colour as pronotum, with 15 longitudinal stripes each side and various cross-stripes. **Hab** Oak woodland, often on trees. **Beh** Feed on moth larvae; mainly nocturnal. **SS** None.

Pterostichus 18 spp. | 2 ILL.

FORM medium-sized to large (**BL** 5–21 mm); **BODY** black, shiny; **PRONOTUM** with depressions, often less broad than abdomen; **ABDOMEN** rounded, tapered or broad; **ELYTRA** heavily ridged.

Common Blackclock
Pterostichus madidus

Common

BL 14–18 mm. **ID** Pronotum rounded, with hind-angles cornered. Elytra with longitudinal lines, with fine punctures; black. Legs black or red. **Hab** Grasslands, heathlands, gardens, woodlands. **SS** Other *Pterostichus* species, from which told by rounded pronotum and often reddish femora.

Mitten Blackclock
Pterostichus nigrita

Common

BL 9–12 mm. **ID** Pronotum rounded, with 'tooth' at hind angles; deep punctured depressions. **Hab** Wetlands. **SS** Other *Pterostichus* species with sharp hind angles to pronotum, narrowed to a small 'tooth'; identification can only be confirmed by microscopic examination of the genitalia.

Poecilus 4 spp. | 3 ILL.

FORM large (**BL** 11–14 mm); **BODY** metallic, colour variable within species; **PRONOTUM** 2 large lateral depressions.

Copper Greenclock
Poecilus cupreus

BL 11–13 mm. **ID** Metallic coppery, green, blue to pinkish-brown, less commonly black, or two-coloured; antenna S1 & S2 red; head pitted; hind tibia inside with 8–11 pale bristles. **Hab** Grassland, arable land, gardens. **SS** Rainbow Greenclock; the similar Heath Greenclock *Poecilus lepidus* [N/I] (**BL** 11–14 mm) has all-black antennae.

Rainbow Greenclock
Poecilus versicolor

BL 11–13 mm. **ID** Metallic coppery, green, blue to pinkish-brown, less commonly black and some with 'rainbow'-like coloration; antenna S1 & S2 red; head nearly smooth, finely pitted; hind tibiae inside with 5–8 stout black bristles. **Hab** Heathland, moorlands, damp grassland. **SS** Copper Greenclock, Heath Greenclock [N/I] (see *above*).

Kugelann's Ground Beetle NT
Poecilus kugelanni

BL 12–14 mm. **ID** Head and pronotum netallic copper, elytra green (rarely, all-black); antenna basal segments partly brown. **Hab** Sandy heathland or coastal areas. **SS** Other *Poecilus* spp., in which basal antenna segments are all one colour (pale or dark). Usually, two-tone coloration is distincive.

Abax 1 sp.

Common Shoulderblade
Abax parallelepipedus

BL 17–22 mm. **ID** Large to very large, black, shiny (elytra, dull in ♀); pronotum as broad as abdomen, with two distinctive large depressions; elytra heavily ridged, with sharp, raised ridge at shoulder. **Hab** Woodland and damp moorlands. **SS** Superficially like many, mostly smaller, black ground beetles, particularly *Pterostichus* spp, from which told by large, wide body.

GROUND BEETLES – ABDOMEN UNDERSIDE 6-SEGMENTED 3/6

Carabus 12 spp. (1†) | 11 ILL.

FORM large to very large (**BL** 13–35 mm), often black or dark, with some species variable in colour.

Moorland Ground Beetle
Carabus arvensis

BL 16–20 mm. **ID** Body coppery, some individuals green, bluish or black. Elytra with raised, elongated granulations. **Hab** Heathland and moorlands. **SS** **Sausage Ground Beetle**, which has more extensive granulations on elytra; **Necklace Ground Beetle.**

Sausage Ground Beetle
Carabus granulatus

BL 16–23 mm. **ID** Black with coppery (occasionally greenish) reflection; elytra with raised, extensive 'sausage'-shaped granulations. **Hab** Marshes, bogs, damp woodland. **SS** **Moorland Ground Beetle**, which has less extensive granulations on elytra; **Necklace Ground Beetle.**

Bronze Ground Beetle
Carabus nemoralis

BL 20–26 mm. **ID** Head and pronotum black (margins of pronotum purple or copper); elytra black with greenish reflection, rough sculpturing, also three rows of 8–10 small punctures. **Hab** Grasslands, woodlands, gardens. **SS** **Violet Ground Beetle, Ridged Violet Ground Beetle.**

Necklace Ground Beetle EN
Carabus monilis

BL 22–26 mm. **ID** Metallic coppery or greenish. Elytra 3-ridged, with rows of small granules. **Hab** Grassland and scrub. **Dist** Also pre-1900 records from Ireland. **SS** **Moorland Ground Beetle, Sausage Ground Beetle**

Ridged Violet Ground Beetle
Carabus problematicus

BL 20–28 mm. **ID** Black, metallic blue or mauve reflection, extensive on pronotum. Elytra with extensive violet or blue margins and with distinctive ridges and granules. **Hab** Grassland, heathland, woodland. **SS** **Violet Ground Beetle**, which is less extensively blue on pronotum; **Bronze Ground Beetle**.

Violet Ground Beetle
Carabus violaceus

BL 20–30 mm. **ID** Black, with metallic violet or blue on sides of pronotum. Elytra granules forming indistinct lines; elytra also with narrow violet or blue margins. **Hab** Woodland, grassland, gardens. **SS** **Ridged Ground Beetle**, which is more extensively blue on pronotum; **Bronze Ground Beetle**.

Blue Ground Beetle NT
Carabus intricatus

BL 25–35 mm. **ID** Metallic blue; elytra with elongated sculpturing. **Hab** Oak and Beech woodland, on the ground or on tree trunks. **SS** None.

Smooth Ground Beetle
Carabus glabratus

BL 22–30 mm. **ID** Large; black with metallic bluish reflection; elytra fairly smooth with fine, small punctures. **Hab** Moorlands, bogs, woodland. **SS** None.

GROUND BEETLES – ABDOMEN UNDERSIDE 6-SEGMENTED 4/6

Carabus (cont'd)

Golden-dimpled Ground Beetle
Carabus clatratus

Nationally Scarce

BL 22–30 mm. **ID** Black with a greenish-gold reflection; elytra with 3 keels, with 8 to 10 regular, coppery punctures in between. **Hab** Bogs, wet moorlands. **SS** None.

Goldsmith
Carabus auratus

Introduced; Rare

BL 20–27 mm. **ID** Metallic green or gold. **Hab** Gardens. **Dist** Deliberate introduction, with occasional reports from GB and Ireland; most recently established in Berkshire and north Essex. **SS** Heath Goldsmith (note colour distinction).

Heath Goldsmith
Carabus nitens

Nationally Scarce

BL 13–18 mm. **ID** Smaller than other *Carabus* species. Metallic; pronotum coppery-red, elytra green with copper-red margins and black keels. **Hab** Wet heathland, bogs and dune slacks. **SS** Goldsmith (note colour distinction).

Drypta 1 sp.

Chine Beetle EN
Drypta dentata

Rare

BL 7–9 mm. **ID** Metallic green or bluish with most parts of legs and antennae reddish; pronotum cylindrical. **Hab** Soft-rock cliffs, at seepages. **SS** None.

Bembidion 52 spp. | 1 ILL.

FORM most small (**BL** 2–7 mm); **BODY** black to yellowish, often shining; **ELYTRA** spotted in many species.

Rock Bembidion
Bembidion saxatile

Nationally Scarce

BL 4–5 mm. **ID** Black but elytra with two large spots (paler bluish-brown form with very pale spots occurs commonly), legs yellowish; also occurs in a darker form, mainly black with brown legs. **Hab** Coastal soft rock cliffs around seepages. **SS** Other *Bembidion* species, from which told by extent of pale spots on elytra.

Badister 7 spp. | 1 ILL.

FORM small to medium-sized (**BL** 4–9 mm); **ELYTRA** iridescent, some with coloured spots.

Black-headed Bullatus
Badister bullatus

Common

BL 5–6 mm. **ID** Head black; pronotum and legs orange; elytra black-and-orange; antennae black with red base. **Hab** Heathland, grassland, dunes. **SS** Other *Badister* species, from which told by pale markings on elytra, combined with pale antenna S1 and dark scutellum.

Panagaeus 2 spp. | 1 ILL.

FORM medium-sized (**BL** 7–9 mm); **BODY** with punctures; **ELYTRA** hairy, with colourful spots.

2-spotted Ground Beetle
Panagaeus bipustulatus

Nationally Scarce

BL 7–8 mm. **ID** Black, elytra with two large red spots on each side; pronotum and elytra punctured. **Hab** Coastal grassland and dunes. **SS** The **Crucifix Ground Beetle** *Panagaeus cruxmajor* VU [N/I] (**BL** 7–9 mm), which has larger red spots on elytra, reaching side margins.

Chlaenius 4 spp. | 1 ILL.

FORM most large (**BL** 9–13 mm); **BODY** metallic; **ELYTRA** hairy, uniform in colour except in species illustrated.

Yellow-bordered Nightrunner
Chlaenius vestitus

Local

BL 9–11 mm. **ID** Head and pronotum metallic green; elytra greenish or dark brownish with pale yellowish border; legs and antennae also yellowish. **Hab** Near water, in damp clay or sand. **SS** Other *Chlaenius* species, although these lack yellow sides to the elytra

GROUND BEETLES – ABDOMEN UNDERSIDE 6-SEGMENTED 5/6

Amara 29 spp. | 1 ILL.

FORM medium-sized to large (**BL** 5–13 mm); **BODY** oval. Includes variable but similar species, ID difficult.

Common Sun Beetle
Amara aenea

Common

J F M A M J J A S O N D

BL 7–9 mm. **ID** Shiny, brassy or coppery. Antennae black, but three basal segments yellow; tibiae brown. **Hab** Grassland, agricultural land, gardens. **SS** Other *Amara* species, from which told by the combination of fine elytral lines, two short, deep streaks in centre and towards base of pronotum, and three pale basal antenna segments.

Calathus 8 spp. | 1 ILL.

FORM medium-sized to large (**BL** 7–14 mm); **BODY** black, brown/reddish-brown; **TARSUS** claws toothed.

Red-thorax Ground Beetle
Calathus melanocephalus

Common

J F M A M J J A S O N D

BL 6–9 mm. **ID** Head and elytra black; pronotum and legs reddish (in some individuals body and legs all-black). **Hab** Grassland, heathland, gardens, arable fields. **SS** None. Other *Calathus* species are less brightly coloured; if in doubt (only likely with dark forms), legs dark (not pale).

Anisodactylus 3 spp. | 1 ILL.

Saltmarsh Shortspur
Anisodactylus poeciloides

Nationally Scarce

J F M A M J J A S O N D

BL 9–13 mm. **ID** Black, with metallic copper or greenish sheen. **Hab** Coastal saltmarshes. **SS** *Harpalus* species, but hind tarsal spur shorter than tarsal segment S1. (Other *Anisodactylus* species are black.)

Platynus 1 sp.

Common Woodland Ground Beetle *Platynus assimilis*

Common

J F M A M J J A S O N D

BL 9–13 mm. **ID** Body shining black; pronotum distinctively shaped with raised, sharp hind angles; elytra broad. Base of antennae and femora black, remainder of legs brown. **Hab** Woodland. **SS** None, although superficially like other black ground beetles, pronotum shape distinctive, antenna S3 smooth and tarsi not grooved.

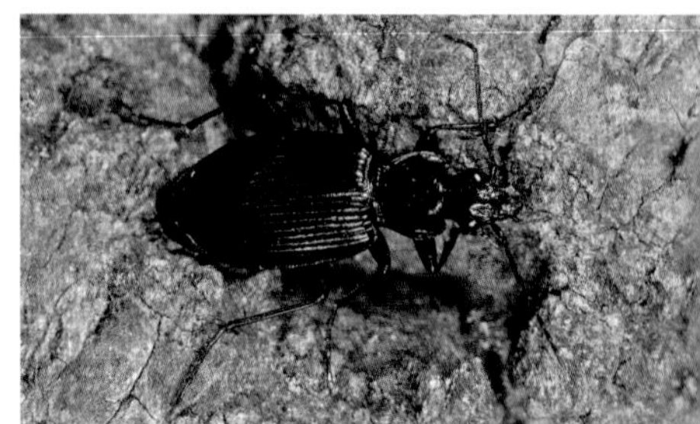

Harpalus 22 spp. | 2 ILL.

FORM medium-sized to large (**BL** 5–16 mm) robust, broad; **LEGS** hind tarsal spur longer than tarsal segment S1.

Metallic Harpalus
Harpalus affinis

BL 9–12 mm. **ID** Metallic golden green, copper, purple or black; elytra with hairy outer margin. Legs pale or dark. **Hab** Grassland, arable fields, dunes. **SS** Other metallic green *Harpalus* species, from which told by punctured elytra with hairy outer margin.

Strawberry Seed Beetle
Harpalus rufipes

BL 11–16 mm. **ID** Body black, dull; pronotum rounded. Legs and antennae red. **Hab** Grassland, arable fields. **SS** Other *Harpalus* species, from which told by large size and lack of hairs on head and pronotum; pronotum of other species have sharp hind angles.

Agonum 17 spp. | 2 ILL.

FORM medium-sized (**BL** 6–10 mm); **PRONOTUM** hind-angles rounded; **ELYTRA** one-coloured or variegated, except one species shown below, which has pale margins.

Shining Green Ground Beetle
Agonum marginatum

BL 9–10 mm. **ID** Coppery green; pronotum margins pale yellow. Legs dark brown, tibiae and base of antennae paler. **Hab** Riverbanks, margins of lakes and pondsides; often on clay soils. **SS** None.

Müller's Ground Beetle
Agonum muelleri

BL 7–9 mm. **ID** Head and pronotum coppery green; elytra darker. Legs dark brown to black, tibiae and antenna S1 paler. **Hab** Damp grassland. **SS** Other *Agonum* species, from which told by combination of fine lines on elytra and pale antenna S1.

GROUND BEETLES – ABDOMEN UNDERSIDE 6-SEGMENTED 6/6

Dromius 4 spp. | 1 ILL.

FORM medium-sized (**BL** 5–7 mm); **HEAD + PRONOTUM** narrow; **ELYTRA** broad and truncate at tip.

Great Four-spot Treerunner
Dromius quadrimaculatus

Common

BL 5–6 mm. **ID** Head dark brown; pronotum reddish with darker central mark; elytra black with four pale yellowish spots. **Hab** Woodland. **Beh** Often on tree trunks, on or under bark. **SS** Related four-spotted ground beetles in other genera [N/I], however these are rather smaller. Other *Dromius* species lack well-defined pale spots.

Lebia 5 spp. | 1 ILL.

FORM small to medium-sized (**BL** 4–8 mm); **BODY** metallic; **HEAD** neck constricted; **ELYTRA** broadened.

Green-headed Lebia
Lebia chlorocephala

Common

BL 6–8 mm. **ID** Red with metallic green head and elytra; antennae and tarsi dark. **Hab** Grassland. **SS** The **Blue-headed Lebia** *Lebia cyanocephala* VU [N/I] (**BL** 5–8 mm) (Surrey only, although recorded from Ireland, pre-1900) is usually bright metallic blue, femora darkened towards tip. Two of the other *Lebia* species have not been recorded since the 19th century.

SUBORDER Polyphaga
(Huge variety of mainly non-water beetles and weevils) | 68 of 95 BI

FAMILY Staphylinidae (Rove beetles) 275 GEN. | 1,129 spp. | 19 ILL.

Elongate beetles, representing approx. 25% of the British and Irish beetle fauna; **their key feature is short elytra, with the final 3–6 abdomen segments visible**. Often black, some species have reddish and/or yellow coloration. They mainly prey on various invertebrates, although some feed on fungi, dung or dead plant material. Examples of typically encountered genera are given; initial separation is by position of the antennae.

ORANGE-TIPPED ROVE BEETLE

ANTENNAE attached to **top** of head; insertion point clearly seen from above

ANTENNAE attached to **side** of head; insertion point obscured from above

ANTENNAE **attached to top of head**		
FORM narrow-headed; broad, colourfully marked body; **ELYTRA** colourfully marked	***Lordithon*** [5 species]	*p. 276*
FORM broadened, stout body; **ELYTRA** black with four red marks	***Scaphidium*** [Four-spotted Rove Beetle]	*p. 280*
MANDIBLES large and triangular; **HEAD** black; **PRONOTUM + SCUTELLUM** dark red; **ABDOMEN** orange to dark brown	***Atrecus*** [Bark Rove Beetle]	*p. 276*
HEAD small rounded head; **PRONOTUM** broad, rounded; **EYES** often large	***Quedius*** [44 species]	*p. 276*
FORM parallel-sided, typically black; **HEAD** quadrangular; **PRONOTUM** elongated	***Ocypus*** [7 species]	*p. 279*
HEAD + PRONOTUM typically glossy; **ELYTRA** with punctures	***Philonthus*** [46 species]	*p. 278*
Rather like *Ocypus*, slight differences in mandibles (narrow, lacking teeth – stout and with teeth in *Ocypus*)	***Tasgius*** [6 species]	*p. 278*
ELYTRA red; **HEAD** short yellow hairs on front; **ABDOMEN** short yellow hairs laterally	***Staphylinus*** [3 species]	*p. 277*
HEAD expanded towards hind angles; **ABDOMEN** basal half of at least segments S7 and S8 with bands of small yellow hairs	***Platydracus*** [3 species]	*p. 277*
BODY black-and-white	***Creophilus*** [Hairy Rove Beetle]	*p. 278*
BODY with golden-brown variegated pubescence (short hairs)	***Ontholestes*** [2 species]	*p. 278*
FORM large; **BODY** black; **ELYTRA** whitish basal half; **HEAD, PRONOTUM** and end of **ABDOMEN** with extensive short yellow hairs	***Emus*** [Pride of Kent Rove Beetle]	*p. 280*
ANTENNAE **attached to side of head**		
BODY orange-red and bluish-black	***Paederus*** [4 species]	*p. 279*

Guide to Staphylinidae pp. 274–27

ROVE BEETLES – ANTENNAE ATTACHED TO TOP OF HEAD

Quedius 44 spp. | 2 ILL.

HEAD small, rounded; **PRONOTUM** broad, rounded; **EYES** usually large.

Orange-tipped Rove Beetle
Quedius cruentus

Common

BL 8–10 mm. **ID** Head, pronotum and scutellum black; elytra red; tip of abdomen orange; tips of abdomen S3–7 with narrow orange band. **Hab** Woodland. **Beh** Often under bark or in old bracket fungi. **SS** None.

Hornet Rove Beetle
Quedius dilatatus

EN Rare (RDB1)

BL 15–24 mm. **ID** Black; head and pronotum glossy; elytra large. **Hab** larvae develop in the base of Hornets' (*p. 511*) nests; adults attracted to sap runs on trees damaged by Goat Moth (*p. 422*). **SS** Devil's Coach-horse (*p. 279*), which has a matt pronotum and head.

Atrecus 1 sp.

Bark Rove Beetle
Atrecus affinis

Common

BL 6–8 mm. **ID** Mandibles large and triangular. Head black, pronotum and scutellum dark red; abdomen segments orange to dark brown. **Hab** Woodland. **SS** None.

Lordithon 5 spp. | 1 ILL.

Lunar Rove Beetle
Lordithon lunulatus

Common

BL 5–7 mm. **ID** Head narrow, black; pronotum and much of abdomen red; elytra glossy black with pale marks, including crescent shapes (lunules). **Hab** Woodland. **Beh** Inhabit fungi, where they prey on other insects. **SS** None. Each *Lordithon* species has distinctive marks on the elytra and different colour combinations, Lunar Rove Beetle being the most colourful.

Platydracus — 3 spp. | 2 ILL.

FRONT OF HEAD lacking yellow pubescence (hairs).

Dung Platydracus
Platydracus stercorarius

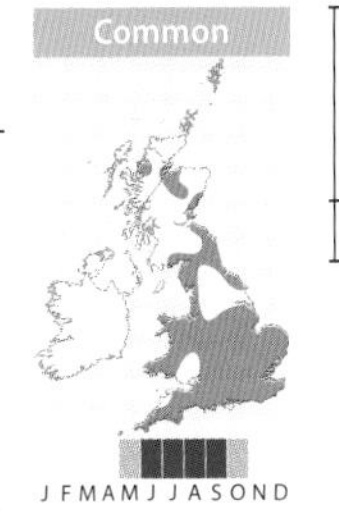

BL 13–16 mm. **ID** Head expanded towards hind angles; head, pronotum and scutellum black; elytra, legs, antennae and palps red. Abdomen black, S4–6 and basal half of S7 and S8 with small yellow hairs. **Hab** Grassland, heathland, woodland. **SS** Other *Platydracus* species, from which told by its larger size (if elytra red) and broader head; *Staphylinus* species.

Blue-green Platydracus
Platydracus fulvipes

BL 13–17 mm. **ID** Head expanded towards hind angles; head and pronotum metallic green; elytra metallic bluish-green. Basal half of abdomen S7 and S8 with bands of small yellow hairs. **Hab** Woodland. **SS** None (elytra are bluish-green, not red).

Staphylinus — 3 spp. | 2 ILL.

FRONT OF HEAD with yellow pubescence (hairs)

Yellow-marked Staphylinus
Staphylinus erythropterus

BL 14–18 mm. **ID** Head and pronotum black with short bright yellow hairs on front of head and on scutellum. Antennae dark except for 2–3 red basal segments. Elytra red; abdomen black, S3–8 laterally with thick, small yellow hairs. **Hab** Damp sites. **SS** Other *Staphylinus* species and *Platydracus* species.

Common Staphylinus
Staphylinus dimidiaticornis

BL 16–23 mm. **ID** Head and pronotum black with short yellowish hairs on front of head, neck and pronotum; scutellum covered in short black hairs; elytra red, antennae also red, except final five or six segments darkened. Abdomen black, S3–7 laterally with small yellow hairs. **Hab** Damp sites. **SS** Other *Staphylinus* species and *Platydracus* species.

ROVE BEETLES – ANTENNAE ATTACHED TO TOP OF HEAD 2/3

Creophilus 1 sp.

Hairy Rove Beetle
Creophilus maxillosus

BL 15–21 mm. **ID** Body black; elytra and abdomen S4–6 with white bands of hairs. **Hab** Grassland, in carrion. **SS** None.

Ontholestes 2 spp. | 1 ILL.

Greyish Chequered Rove Beetle *Ontholestes murinus*

BL 8–14 mm. **ID** Overall greyish appearance, body with variegated short, golden-brown hairs (looks rather chequered), including abdomen S3–5. Antennae yellow; legs dark except fore tarsi reddish.

Hab Grassland, in carrion and dung. **SS** Chequered Rove Beetle *O. tessellatus* [N/I] (**BL** 15–22 mm), which has largely reddish legs.

Philonthus 46 spp. | 1 ILL.

Beautiful Philonthus
Philonthus decorus

BL 11–13 mm. **ID** Head and rounded pronotum black, with slight greenish metallic sheen; elytra with metallic coppery sheen. Antennae dark. **Hab** Damp sites. **SS** Other *Philonthus* species, from which told by the presence of five punctures on the pronotum.

Tasgius 6 spp. | 1 ILL.

Red-legged Tasgius
Tasgius morsitans

BL 12–16 mm. **ID** Body matt black; legs and palps red, along with much of antennae; tip of abdomen orange; tips of abdomen S3–7 with narrow orange band. **Hab** Woodland and grassland. **Dist** England, Wales and Ireland. **SS** Other *Tasgius* species, but matt black body and red legs are distinctive.

Ocypus 7 spp. | 2 ILL. (see *p. 244*)

Devil's Coach-horse
Ocypus olens

BL 23–32 mm. **ID** Body black (head and pronotum matt black); antenna tip and tip of maxillary palps paler. **Hab** Woodland, grassland and parks. **Beh** Abdomen curled scorpion-like in defence; ♂ can emit a fluid from mouth and release chemicals from tip of abdomen. **SS** **Hornet Rove Beetle** (*p. 276*), which has a glossy pronotum and head.

ROVE BEETLES – ANTENNAE ATTACHED TO SIDE OF HEAD 1/1

Paederus 4 spp. | 3 ILL.

Coastal Paederus
Paederus littoralis

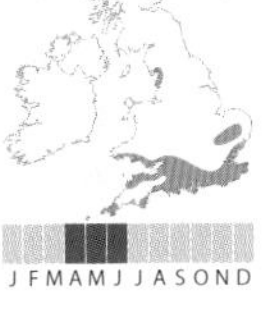

BL 8–9 mm. **ID** Body elongated, black with orange-red pronotum and broad band on abdomen. Pronotum and elytra almost quadrate. Mandibles black; legs red except femora with black knees. **Hab** Grassland, most plentiful on coast. **SS** Other *Paederus* species, from which told by black mandibles.

Wetland Paederus
Paederus riparius

BL 7–8 mm. **ID** Body elongated, black with pronotum orange-red, also broad band on abdomen. Pronotum and elytra elongated. Mandibles yellow (labrum black); legs orange-red except femora with black knees. **Hab** Fens and marshes. **SS** Other *Paederus* species.

Dusky Paederus
Paederus fuscipes

BL 6–7 mm. **ID** Body, including elytra, broad; black with orange-red pronotum and broad band on abdomen. Mandibles yellow. Elytra much wider than pronotum. Hind tibiae orange-red with dark area reaching to tip or approx. ⅓ of length. **Hab** Wet mires, saltmarshes. **SS** Other *Paederus* species; the **Rare Paederus** *P. caligatus* [N/I] (RDB3) has darker legs.

Guide to Staphylinidae pp. 274–27

ROVE BEETLES – ANTENNAE ATTACHED TO TOP OF HEAD 3/3

Scaphidium 1 sp.

Four-spotted Rove Beetle
Scaphidium quadrimaculatum

Local

BL 5–6 mm. **ID** Body broad and rounded, very different in shape compared to other rove beetles; elytra black with four red marks.
Hab Woodland. **Beh** Associated with fungi growing on dead branches.
SS None.

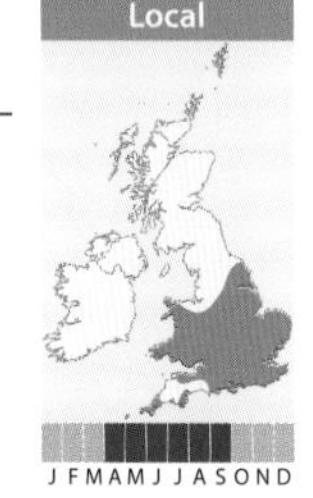

Emus 1 sp.

Pride of Kent Rove Beetle EN
Emus hirtus

Rare (RDB1)

BL 16–30 mm. **ID** Large; body black, elytra with whitish basal half; head, pronotum and end of abdomen with extensive short yellow hairs.
Hab Grassland, on dung near livestock. **SS** None.

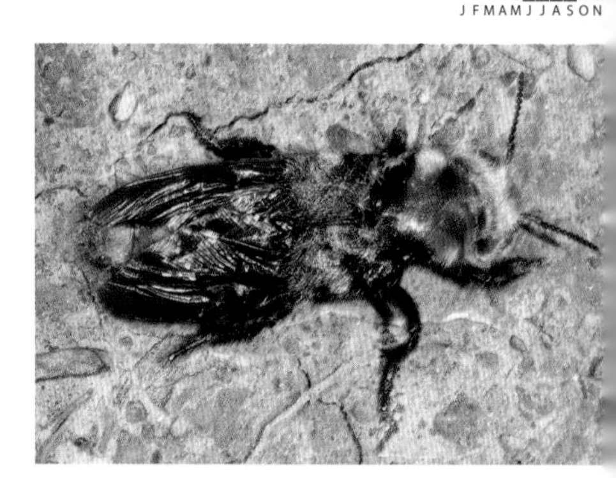

FAMILY Lucanidae (Stag beetles) 4 GEN. | 4 spp. (1†) | 3 ILL.

Body large to very large (body length 10–80 mm), black or dark brown; male with huge reddish-brown mandibles in the largest species (wide range in sizes; female usually smallest, with small mandibles); antennae 10-segmented, elbowed, with clubbed tip. Larvae feed for several years on rotting wood.

Stag Beetle larva – sometimes found in old tree stumps by gardeners.

Sinodendron 1 sp.

Rhinoceros Beetle
Sinodendron cylindricum

Common

BL 10–18 mm. **ID** Robust, black; horn at front of head (♂ only, ♀ with small bump); pronotum and elytra heavily punctured. **Hab** Woodland. **SS** None.

Lucanus 1 sp.

Stag Beetle
Lucanus cervus

BL 25–80 mm. **ID** ♂ has huge reddish-brown antler-like mandibles. Huge size range, some ♂s half the size of others. Black with shiny elytra that have at least a hint of dark reddish-brown. **Hab** Gardens, parks, woodland. **Beh** ♂s use 'antlers' in fights over ♀s. Larvae are found in dead tree stumps, taking 3–4 years to mature. **SS** Lesser Stag Beetle.

Dorcus 1 sp.

Lesser Stag Beetle
Dorcus parallelipipedus

BL 18–32 mm. **ID** Mandibles small, slightly larger in ♂; elytra matt greyish-black. **Hab** Woodland. **SS** ♀ Stag Beetle, which has shiny brown or maroon elytra.

Guide to families pp. 242–25

FAMILY **Scarabaeidae** (Scarab beetles) 55 GEN. | 84 spp. | 22 ILL.

Small to very long (BL 2–38 mm), often large and rounded, at least at front and back; antennae short, with conspicuous clubbed or fan-like tip. Some are pests, such as Cockchafer, which causes damage to cereals, grasses and gardens. Smaller dung beetles mainly breed in dung, but also rotting fungi.

CHAFERS

FORM medium to large (**BL** 5–38 mm), stout, long-legged and often brightly coloured; larvae feed on plants, often roots, while some adults feed on leaves and may be associated with flowers.

Trichius 2 spp.

Bee Chafer *Trichius fasciatus*

BL 12–16 mm. **ID** Thorax with brownish or whitish hairs; elytra strongly marked with large black spots on yellowish-orange background, the basal spots often form an uneven band. **Hab** Woodland. **SS** **French Bee Chafer** *Trichius gallicus* (**BL** 11–14 mm) [RIGHT] (recent colonist, spreading in London area), has isolated black spots at base of elytra (not a band).

Bee Chafer

French Bee Chafer

Omaloplia 1 sp.

Downland Chafer *Omaloplia ruricola*

BL 5–8 mm. **ID** Hairy; head, pronotum, scutellum and borders of elytra black; rest of elytra brown. **Hab** Chalk downland. **SS** **Garden Chafer** (*p. 285*), in which head and pronotum are green, not black.

Protaetia 1 sp.

Scottish Chafer *Protaetia cuprea*

BL 16–22 mm. **ID** Body dull coppery-green; elytra notched laterally, with variable number of white flecks; scutellum large. **Hab** grassland; on dung near livestock. **SS** **Rose Chafer** is brighter, more metallic and has more rounded apices to elytra. Somewhat like **Noble Chafer** and **Dune Chafer**.

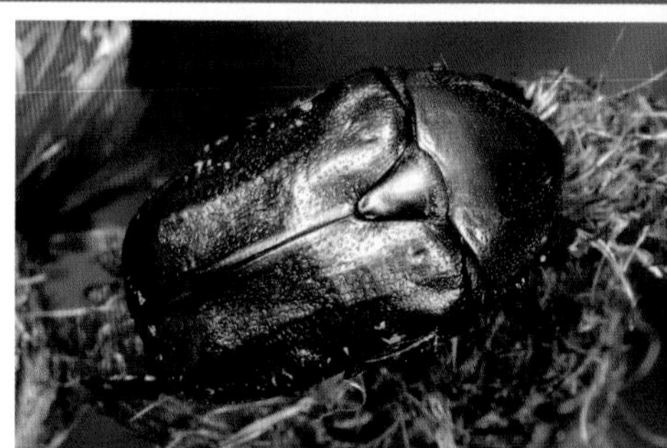

Gnorimus 2 spp.

Variable Chafer
Gnorimus variabilis

EN

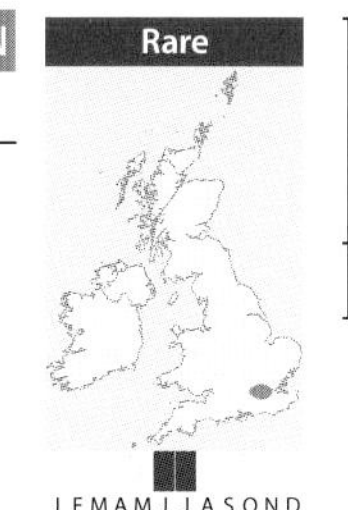

BL 17–22 mm. **ID** Body black; elytra with white spots; scutellum small. **Hab** Ancient woodland. **Dist** Stronghold Windsor Forest, Berkshire. **SS** None.

Noble Chafer
Gnorimus nobilis

VU

BL 15–20 mm. **ID** Body metallic coppery-green; elytra wrinkled with variable number of white spots; scutellum small. **Hab** Old orchards, developing in rot holes (other trees, including old oaks in some woodlands). **SS** Rose Chafer and Scottish Chafer, which have a much larger scutellum and a different body shape.

Anomala 1 sp.

Dune Chafer *Anomala dubia*

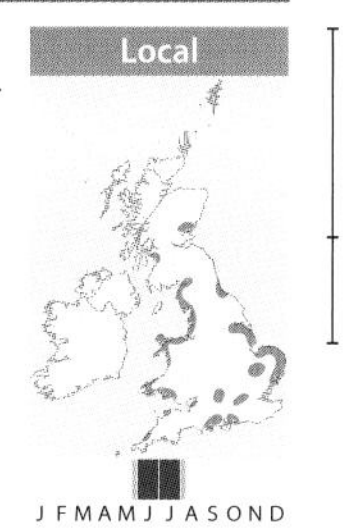

BL 11–15 mm. **ID** Body all-green (but elytra often brown with a green hue), some individuals bluish. **Hab** Sandy sites, particularly coastal dunes. **Dist** Recorded in fewer than 100 tetrads since 1989 and would normally be regarded as Nationally Scarce, but believed to be under-recorded. **SS** None, except green individuals have a superficial resemblance to **Noble Chafer**.

Cetonia 1 sp.

Rose Chafer *Cetonia aurata*

BL 14–21 mm. **ID** Body metallic golden-green; elytra with white flecks (number and extent of which individually variable); scutellum large. **Hab** Woodland, gardens. **SS** **Scottish Chafer**, which is less metallic; **Noble Chafer**, which has a small scutellum; superficially like **Dune Chafer**, which lacks white marks on elytra.

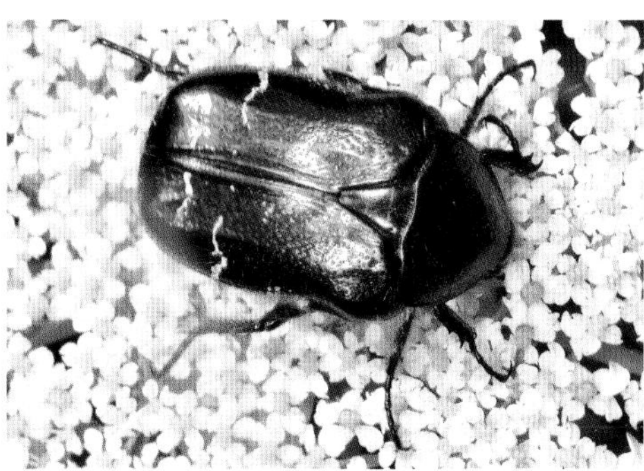

Melolontha 2 spp.

Cockchafer
Melolontha melolontha

BL 20–30 mm. **ID** Large, feathery antennae; abdomen with pointed tip; brown. **Hab** Woodland, gardens, farmlands. **Beh** Can swarm in huge numbers; regarded as a forestry and agricultural pest as larvae develop in roots of trees and shrubs. **SS** The **Northern Cockchafer** *Melolontha hippocastani* NT (**BL** 22–27 mm) (woodland in Scotland and Northern Ireland during May) has a shorter pygidium.

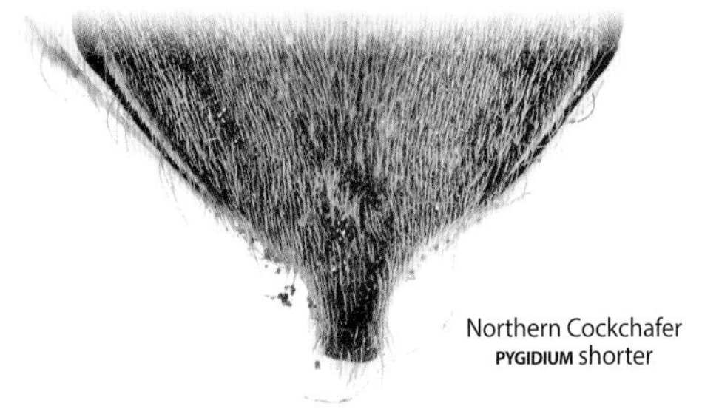
Northern Cockchafer
PYGIDIUM shorter

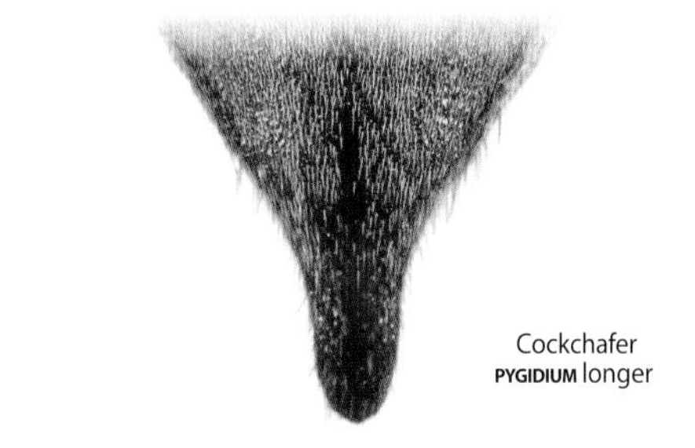
Cockchafer
PYGIDIUM longer

Amphimallon 2 spp. | 1 ILL.

Summer Chafer
Amphimallon solstitiale

BL 15–20 mm. **ID** Hairy; head black; thorax and elytra brown, the latter with weak longitudinal dark stripes. **Hab** Grassland, gardens. **SS** Scarce **Summer Chafer** *Amphimallon fallenii* NT [N/I] (**BL** 14–17 mm), which has much longer hairs on pronotum than around tips of wingcases (about same length in Summer Chafer).

Serica 1 sp.

Brown Chafer
Serica brunnea

BL 8–10 mm. **ID** Body brown. **Hab** Woodland. **Beh** Sometimes swarms; this species is readily attracted to lights. **SS** None.

Phyllopertha 1 sp.

Garden Chafer
Phyllopertha horticola

Common

BL 7–12 mm. **ID** Head and thorax metallic green, black in some individuals; elytra brown. Rather hairy. **Hab** Grassland, gardens. **SS** None.

Hoplia 1 sp.

Welsh Chafer
Hoplia philanthus

Common

BL 8–9 mm. **ID** Thorax non-metallic, ashy black; elytra chestnut, with hint of bluish which also features on abdomen. **Hab** Open areas, including grassland. **SS** None.

DUNG BEETLES

FORM small to medium-sized (**BL** 3–13 mm), stout; **LEGS** broad, adapted for digging; **ELYTRA** full length; **HEAD + PRONOTUM** invariably black or dark, although elytra may be brightly coloured in some species; associated with dung, sometimes fungi. **NOTE:** the genus *Aphodius*, which until recently comprised 46 species, has now been split into 29 genera (5 illustrated).

Nimbus 2 spp. | 1 ILL.

Black-streaked Dung Beetle
Nimbus contaminatus

Common

BL 5–7 mm. **ID** Pronotum margins laterally fringed with long yellowish hairs; elytra brown with black streaks, also hairy. **Hab** Pastures, heathland. **Beh** Associated with dung, particularly of horses, and fungi; note this is a late season species. **SS** Lesser-streaked Dung Beetle *Nimbus obliteratus* [N/I] (**BL** 5–7 mm) lacks yellow hairs on side of pronotum and some individuals have obliterated streaks.

Acrossus 3 spp. | 1 ILL.

Reddish-brown Dung Beetle
Acrossus rufipes

BL 9–13 mm. **ID** Body dark reddish-brown to black, slightly bluish-looking from some angles; pronotum with mainly fine punctures. **Hab** Pastures around horses and cattle, heathland. **SS** Other *Acrossus*, *Aphodius* species and relatives, from which told by body colour, large size and semi-circular front of head.

Teuchestes 1 sp.

Gravedigger Dung Beetle
Teuchestes fossor

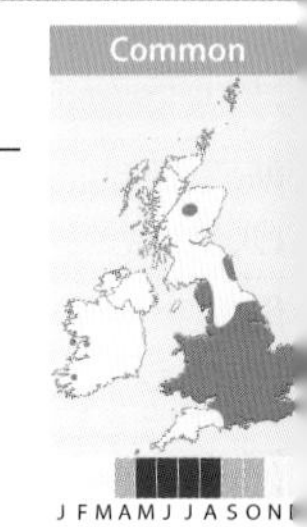

BL 9–13 mm. **ID** Body shining black, slightly bluish-looking from some angles; pronotum with scattered, distinct coarse punctures. **Hab** Grassland around horses and cattle. **SS** Other closely related small dung beetles, from which told by broad, shining black body and large size.

Bodilopsis 2 spp. | 1 ILL.

Pale-margined Dung Beetle
Bodilopsis rufa

BL 5–7 mm. **ID** Elytra shining yellowish-brown or black; pronotum with paler margins. **Hab** Pastures, heathland, gardens. **SS** Other closely related small dung beetles, from which told by colour of body (including pronotum margins) and tip of elytra shining, with fine punctures.

Aphodius 3 spp. | 1 ILL.

Red Dung Beetle
Aphodius fimetarius

BL 5–8 mm. **ID** Head and thorax black. Elytra and antennae red; front angles of pronotum yellow. **Hab** Pastures around herbivores, heathland. **SS** Other *Aphodius* species, from which told by colour of body and appendages; also tip of elytra smooth (not rough), with fine (not coarse) micro punctures.

Onthophagus 8 spp. (2 †) | 3 ILL.

FORM small to medium, bulkier, dung beetles (body length 4–11 mm); back of head often with horn (in ♂); **ELYTRA** shorter, **PYGIDIUM** at tip of abdomen visible when viewed from above. **HABITAT** associated with dung.

Metallic Horned Dung Beetle
Onthophagus coenobita

Common

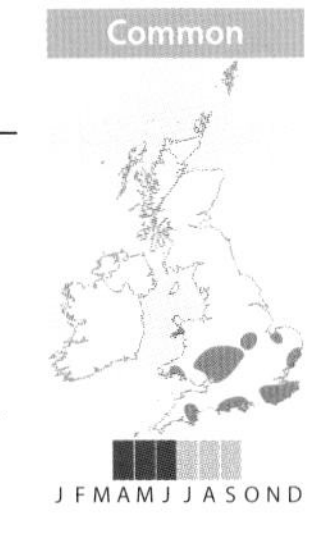

BL 6–9 mm. **ID** Head (horned in ♂) and pronotum metallic green or copper (front angle wavy); elytra yellowish-brown with few marks. **Hab** Pastures, heathland. **SS** Other *Onthophagus* species, from which told by combination of elytra having few marks and brighter metallic head and pronotum.

Common Horned Dung Beetle
Onthophagus similis

Common

BL 5–8 mm. **ID** Head and pronotum black. Elytra yellowish-brown and heavily mottled with dark marks. **Hab** Grassland, heathland, woodland edges. **SS** Other *Onthophagus* species, from which told by mottled elytra, combined with ♂ head horn smoothly contoured at base; punctuation on head irregular.

Common Horned Dung Beetle rolling dung.

Joanna's Dung Beetle
Onthophagus joannae

Common

BL 4–6 mm. **ID** ♂ not horned; wholly black but covered in short yellow hairs. **Hab** Chalk grassland, or sandy soils. **SS** Other *Onthophagus* species, from which told by dark body colour and lack of horns on head of ♂.

Guide to families pp. 242–254

FAMILY **Meloidae** (Blister and oil beetles) **3 GEN. | 10 spp. (3†) | 7 ILL.**

Medium to large (6–35 mm), robust, flightless oil beetles are mostly black with a small pronotum; head wider than thorax; elytra usually well short of end of abdomen (female very bloated); antennae fairly long, in some species kinked in middle. Blister beetles are equally unusual in appearance, bright green or black and orange. In defence adults can release an oily liquid. They are parasitic on solitary bees and have unusual life-histories. The larvae are known as 'triungulins' and have hook-like forelegs to attach themselves to bees, after climbing vegetation. If carried to its nest by a female solitary bee, the larva becomes more maggot-like and feeds on the bee's egg and pollen store. Adults can be identified as follows:

Flame-shouldered Blister Beetle triungulins hibernating. In spring they hope to attach themselves to the **Hairy-footed Flower Bee** (*p. 550*) in order to enter the nest of this solitary bee; in some cases they can probably just crawl directly into the nest

<table>
<tr><th colspan="3">BLISTER BEETLES – BODY elongated; ELYTRA long</th><th></th></tr>
<tr><td colspan="3">metallic emerald green</td><td>Spanish Fly</td></tr>
<tr><td colspan="3">black with orange 'shoulders'</td><td>Flame-shouldered Blister Beetle</td></tr>
<tr><th colspan="3">OIL BEETLES – BODY broadened, often very plump; wholly bluish or black; ELYTRA short</th><th></th></tr>
<tr><td rowspan="3">THORAX 'rectangular'; ANTENNAE straight</td><td rowspan="2">BODY dull black, rough-texture;</td><td>THORAX heavily punctured with conspicuous central groove</td><td>Rugged Oil Beetle</td></tr>
<tr><td>THORAX not heavily punctured and lacking central groove</td><td>Mediterranean Oil Beetle</td></tr>
<tr><td colspan="2">BODY shiny bluish-black; ANTENNAE tips thickened.</td><td>Short-necked Oil Beetle</td></tr>
<tr><td rowspan="2">THORAX 'square'; ANTENNAE kinked</td><td colspan="2">BODY shiny bluish-black; THORAX base indented</td><td>Violet Oil Beetle</td></tr>
<tr><td colspan="2">BODY black or with bluish reflection; THORAX base straight</td><td>Black Oil Beetle</td></tr>
</table>

BLISTER BEETLES

Sitaris 1 sp.

Flame-shouldered Blister Beetle VU
Sitaris muralis

BL 7–15 mm. **ID** Body black with orange 'shoulders'. **Hab** Old brick walls, including cob. **Dist** Rediscovered in Brockenhurst, Hampshire in 2010 and since then recorded elsewhere. **Beh** A parasitoid of Hairy-footed Flower Bee (*p. 550*). **SS** None.

Lytta 1 sp.

Spanish Fly *Lytta vesicatoria*

BL 9–21 mm. **ID** Body metallic emerald green. **Hab** Grassland where ground-nesting bees breed in profusion, with some trees nearby; adults feed on Ash and privet leaves. **SS** None.

OIL BEETLES

Meloe 8 spp. (3†) | 5 ILL.

THORAX 'square'; ANTENNAE kinked

Black Oil Beetle
Meloe proscarabaeus

Local

BL 11–35 mm. **ID** Body black, some have a bluish sheen; elytra very shortened. Antennae kinked, particularly in ♂s. Thorax square-shaped, base straight [*inset*] (close examination of base shows a slight rounded tooth). **Hab** Grassland. **SS** Violet Oil Beetle.

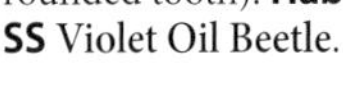

Violet Oil Beetle
Meloe violaceus

Local

BL 10–32 mm. **ID** Body with violet-bluish sheen, but some individuals black; elytra very shortened. Antennae kinked, particularly in ♂. Thorax square-shaped, base indented [*inset*] (close examination of base shows a sharp tooth). **Hab** Grassland. **SS** Black Oil Beetle.

THORAX 'rectangular'; ANTENNAE straight

Short-necked Oil Beetle *Meloe brevicollis* VU

BL 7–24 mm. **ID** Body shiny bluish-black. Antennae tips thickened. **Hab** Heathland, dunes. **Dist** Only known from a few sites: one in SW England, four in W Scotland (Coll), and SE Ireland.

Rare

Mediterranean Oil Beetle VU
Meloe mediterraneus

BL 6–20 mm. **ID** Dull black, rough-texture. Thorax not heavily punctured and lacking central groove. **Hab** Heathland, dunes. **Dist** Only known from one small site in SW England, and from E Sussex.

Rare

Rugged Oil Beetle *Meloe rugosus*

Nationally Scarce

BL 6–18 mm. **ID** Dull black, rough texture; thorax heavily punctured, with conspicuous central groove. **Hab** Chalk grassland, sandy soils. **Beh** Nocturnal.

central groove

FAMILY **Elateridae** (Click beetles) — 38 GEN. | 73 spp. | 32 ILL.

Small to large (3–24 mm), elongate, drab (brown, black, or mottled) or metallic beetles with bright orange, red or greenish areas. The pronotum has pointed hind margins. When disturbed, click beetles have an amazing ability to jump backwards with a clicking sound and right themselves, taking more than one flip to land the correct way up. The larvae, known as 'wireworms', are liquid feeders, associated with dead wood (where they are predators), or roots. The species shown here include common representatives and several rarer, often brightly coloured click beetles.

A coleopterist in Windsor Great Park looking for click beetles in an area good for Rusty Red Click Beetle, Cardinal Click Beetle and many others

A 'wireworm', the larva of Common Melanotus or Confused Melanotus; these species have very similar larvae as well as adults.

Ischnodes — 1 sp.

Red-thorax Click Beetle
Ischnodes sanguinicollis

BL 8–11 mm. **ID** Black; pronotum bright red. **Hab** Woodland. **SS** None.

Nationally Scarce

J F M A M J J A S O N D

Elater — 1 sp.

Rusty Red Click Beetle EN
Elater ferrugineus

BL 17–24 mm. **ID** Reddish; head black. **Hab** Woodland. **Beh** Elusive, but can fly in numbers to a commercially produced pheromone. **SS** None.

Rare (RDB1)

J F M A M J J A S O N

Ctenicera 2 spp.

FORM metallic green or, purple; **ELYTRA** as body, or yellow and black; **ANTENNAE** feathery in ♂.

Copper Click Beetle *Ctenicera cuprea*

BL 11–16 mm. **ID** Body metallic green or purple, elytra may be yellow and partly black; ♂ with large feathery antennae, those of ♀ shorter and plainer. **Hab** Grassland. **SS** Pectinate Click Beetle.

Pectinate Click Beetle *Ctenicera pectinicornis*

BL 13–18 mm. **ID** Body metallic green; ♂ antennae very feathery. **Hab** Upland hay meadows, woodland valleys. **SS** Copper Click Beetle, but ♂ easily told by more extreme feathery antennae; ♀ antennae more toothed.

Actenicerus 1 sp.

Marsh Click Beetle *Actenicerus sjaelandicus*

BL 12–15 mm. **ID** Body black with bronze or copper sheen; elytra with slight mottled appearance. **Hab** Wetlands. **SS** None.

Calambus 1 sp.

Orange-shouldered Click Beetle *Calambus bipustulatus*

BL 6–8 mm. **ID** Black, with orange-spotted 'shoulders'. **Hab** Woodland. **SS** None.

Adrastus 2 spp. | 1 ILL.

Small Click Beetle *Adrastus pallens*

BL 4–5 mm. **ID** Head and pronotum black; elytra, legs, antennae and pronotal spines orange-brown, elytra with darker central stripe. **Hab** Wetlands, grassland. **SS** Pale Agriotes (*p. 293*) but antenna segments all equal length.

Limoniscus 1 sp.

Violet Click Beetle *Limoniscus violaceus*

EN

BL 10–12 mm. **ID FORM** Body with violet sheen. **Hab** Ancient woodland. **SS** None.

CLICK BEETLES 2/3

Ampedus 13 spp. | 8 ILL.

HEAD + PRONOTUM black; **ELYTRA** red, red-and-black or black. Species identified by eltyra colour or critical examination of pronotum punctures/hair colour and shape of antennae.

Belted Click Beetle *Ampedus balteatus*

BL 7–10 mm. **ID** Body black, except approx. first ⅔ of elytra, which is red; tarsi reddish. **Hab** Woodland, heathland. **SS** Other *Ampedus* species; broad black band across tip of elytra distinctive.

Black-tailed Click Beetle *Ampedus elongatulus*

BL 7–10 mm. **ID** Elytra red with black tip. **Hab** Old woodland, on Beech, oaks and pines. **SS** Other *Ampedus* species; black-tipped elytra distinctive.

Black-centered Click Beetle *Ampedus sanguinolentus*

BL 9–12 mm. **ID** Elytra with black central patch. **Hab** Woodland and heathland, on birches. **SS** Other *Ampedus* species; black central patch on elytra (variable in extent) is diagnostic.

Dull Red Click Beetle *Ampedus pomorum*

BL 8–12 mm. **ID** Elytra dull brownish-red, often with two orange spots at front; tarsi brown. **Hab** Old woodland, on oaks and birches. **SS** Other 'red' *Ampedus* species; dull elytra characteristic.

Cardinal Click Beetle *Ampedus cardinalis* VU

BL 12–16 mm. **ID** Elytra red, slightly darker around suture; pronotum only with dull shine. **Hab** Ancient woodland, on oaks. **SS** Other 'red' *Ampedus* species but pronotum duller.

Red-bodied Click Beetle *Ampedus quercicol*

BL 9–11 mm. **ID** Elytra bright red, patterned with slightly darker red; pronotum hairs normally black. **Hab** Old woodland, on Beech and birches. **SS** Other 'red' *Ampedus* species but pronotum hairs usually black; elytra shiny.

Yellow-haired Click Beetle *Ampedus cinnabarinus*

BL 11–15 mm. **ID** Pronotum and margins of elytra with yellowish hairs, including underside. **Hab** Ancient woodland, on oaks and Beech. **SS** Other 'red' *Ampedus* species but pronotum hairs yellowish; elytra shiny.

Black Click Beetle *Ampedus nigerrimus* EN

BL 8–10 mm. **ID** Head and pronotum and elytra black, pronotum strongly punctured; tarsi brownish. **Hab** Ancient woodland, on oaks. **SS** Pronotal angle point helps distinguish from most black click beetles.

Dalopius 1 sp.

Margined Click Beetle *Dalopius marginatus*

BL 6–9 mm. **ID** Head and pronotum dark brown or reddish; elytra orange-brown, with broad, dark brown central stripe that is broader at base, narrowing immediately. **Hab** Woodland. **SS** Other click beetles.

Agriotes 6 spp. | 3 ILL.

FORM broad-bodied; brown; **ANTENNAE** longer than head + pronotum. Species identified by body colour/shape.

Obscure Agriotes *Agriotes obscurus*

BL 7–10 mm. **ID** Body mid-brown with dense paler brown hairs; elytra with ridged lines; pronotum narrower at front than in middle. **Hab** Grassland. **SS** Other *Agriotes* species.

Common Brown Agriotes *Agriotes sputator*

BL 6–9 mm. **ID** Body dark brown, with lighter brown hairs; pronotum more parallel-sided; elytra with ridged lines. **Hab** Grassland. **SS** Other *Agriotes* species.

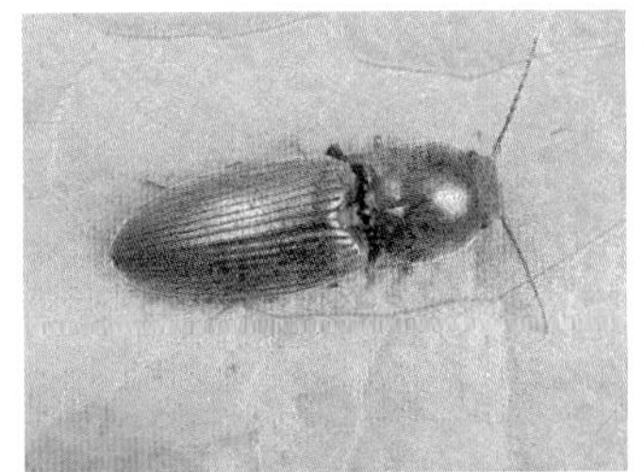

Pale Agriotes *Agriotes pallidulus*

BL 4–6 mm. **ID** Head and pronotum black or dark brown (pronotal angles brown); elytra hairy with ridged lines, lighter brown with darker 'mid-line'; legs and antennae brown, 2nd & 3rd segments being more elongate than others. **Hab** Grassland. **SS** Small Click Beetle (*p. 291*).

CLICK BEETLES **3/3**

Athous **5 spp. | 2 ILL.**

FORM elongated; **PRONOTUM** angles with short spines; **ANTENNAE** S4 and above weakly triangular, not serrated.

Common Brown Click Beetle
Athous haemorrhoidalis

BL 10–15 mm. **ID** Body finely hairy, dark brown with pronotum darker than elytra; antennae all-black; legs brown. **Hab** Grassland, hedgerows, woodland. **SS** Other *Athous* species.

Two-coloured Brown Click Beetle
Athous bicolor

BL 9–12 mm. **ID** Pronotum blackish, sides reddish-brown; elytra brownish-yellow, with dark margin/narro black central stripe on some (body occasionally all-brown in ♀). **Hab** Grassland. **SS** Other *Athous* species.

Denticollis **1 sp.**

Bordered Click Beetle *Denticollis linearis*

BL 9–13 mm. **ID** Eyes convex; pronotum orange/brown, heavily pitted; elytra yellowish-orange to black, heavily punctured, margin pale. **Hab** Woodland. **SS** None.

Stenagostus **1 sp.**

V Click Beetle *Stenagostus rhombeus*

BL 15–21 mm. **ID** Orange-brown; elytra with darker cross-lines, including a 'V' shape towards tip. **Hab** Woodland. **SS** None.

Hemicrepidius **1 sp.**

Hairy Click Beetle *Hemicrepidius hirtus*

BL 11–14 mm. **ID** Metallic black, elytra hairy. **Hab** Woodland. **SS** Other black click beetles duller, not hairy.

Limonius **1 sp**

Ponel's Click Beetle *Limonius poneli*

BL 6–8 mm. **ID** Metallic black, with fine hairs. **Hab** Woodland. **SS** Other black click beetles.

Melanotus 3 spp. | 2 ILL.

FORM elongate, black; **ELYTRA** hairy. The two species shown are best distinguished by measuring the pronotum and elytra along the central line.

Common Melanotus *Melanotus castanipes*

BL 12–19 mm. **ID** Elytra typically ±4× length of pronotum. **Hab** Woodland. **SS** Confused Melanotus.

Confused Melanotus *Melanotus villosus*

BL 13–20 mm. **ID** Elytra ≤3·5× length of pronotum; tarsi brown. **Hab** Woodland. **SS** Common Melanotus.

Agrypnus 1 sp.

Orange-bodied Click Beetle *Agrypnus murinus*

BL 10–18 mm. **ID** Broad; brown or greyish, heavily mottled with light patches; abdomen orange; antennae fairly short, ≤ length of pronotum. **Hab** Grassland, parks, woodland. **SS** Chequered Click Beetle.

Prosternon 1 sp.

Chequered Click Beetle *Prosternon tessellatum*

BL 9–13 mm. **ID** Broad-bodied (pronotum broader than long), dark brown with golden-brown hairs in chequered-like pattern; abdomen black. **Hab** Grassland. **SS** Orange-bodied Click Beetle.

Megapenthes 1 sp.

Windsor Click Beetle *Megapenthes lugens* EN

BL 8–10 mm. **ID** Black, rather dull; corners of rear margin of pronotum sharply angled. **Hab** Woodland, on Beech and elms. **Dist** Strongholds Windsor Forest, Berks; New Forest, Hants. **SS** Other black click beetles.

Selatosomus 3 spp. | 1 ILL.

Green Click Beetle *Selatosomus aeneus*

BL 11–17 mm. **ID** Metallic greenish-gold to greenish-purple. **Hab** Grassland. **SS** Irish Click Beetle *S. melancholicus* [N/I] has a metallic green body.

Guide to families *pp. 242–2*

FAMILY Cerambycidae (Longhorn beetles) 51 GEN. | 73 spp. (5†) | 47 ILL.

Small to very large (body length 3–45 mm), robust or slender, but cylindrical; often very colourful; antennae long to very long, in some cases exceeding body length; elytra usually full length, but shortened in a few species; legs long. In general, easily recognized; the total includes five long-extinct species and accidental imports in timber. Adults of some species are plentiful on flowers in woodland, including Hawthorn, umbellifers and Bramble. The larvae develop mainly in dead wood and may take several years to mature.

LONGHORN BEETLES – HEAD CONSTRICTED, FORMING DISTINCT NECK 1/

FORM rather elongated species, often tapered towards tip; **ANTENNAE** often almost length of body, or > half length. Recognizable from body shape, size, colour and markings.

Grammoptera 3 spp.

FORM black or golden-haired, with darker black abdomen tip.

Black Grammoptera
Grammoptera abdominalis

BL 5–10 mm. **ID** black; antennae, S2 ± long as wide; femoral base may be reddish. **Hab** Woodland. **SS** Common Grammoptera.

Common Grammoptera
Grammoptera ruficornis

BL 3–7 mm. **ID** Black; antenna segments basal half brownish. Antennae S2 longer than wide. Legs black or in part reddish. **SS** Black Grammoptera.

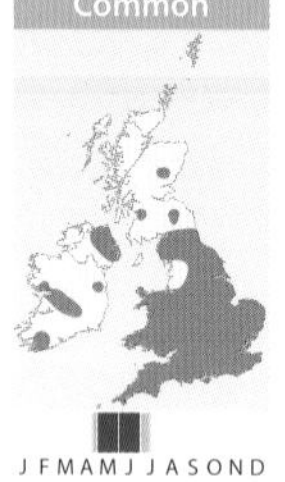

Burnt-tip Grammoptera
Grammoptera ustulata

BL 5–9 mm. **ID** Pronotum and elytra golden-haired; tip black. Legs pale, except tarsi. **Hab** Woodland. **SS** None.

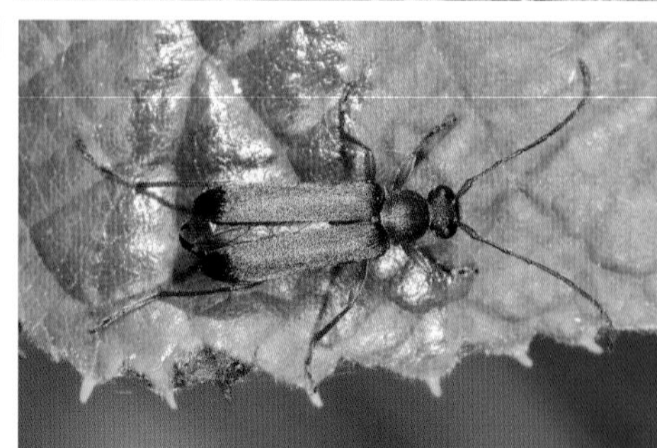

Pachytodes 1 sp.

Speckled Longhorn Beetle
Local

Pachytodes cerambyciformis

BL 7–12 mm. **ID** Unmistakable: elytra narrow; yellow with black spots. **Hab** Woodland. **SS** None.

J F M A M J J A S O N D

Anoplodera 1 sp.

Six-spotted Longhorn Beetle
Rare

Anoplodera sexguttata NT

BL 7–12 mm. **ID** Black, with six reddish spots on elytra (rarely these are obscure). **Hab** Woodland. **SS** None.

J F M A M J J A S O N D

Paracorymbia 1 sp.

Tawny Longhorn Beetle
Nationally Scarce

Paracorymbia fulva

BL 9–14 mm. **ID** Black, except for tawny elytra with black tip. **Hab** Woodland. **SS** Blood-red Longhorn Beetle (♂ only), but that species has a northern distribution.

J F M A M J J A S O N D

Anastrangalia 1 sp.

Blood-red Longhorn Beetle CR
Rare

Anastrangalia sanguinolenta

BL 8–13 mm. **ID** Body black, except for elytra, which are yellow with a black tip in ♂, red in ♀. **Hab** Woodland in Scottish Highlands. **SS** Red Longhorn Beetle (*p.* 299), but pronotum black not red; Tawny Longhorn Beetle (♂ only), but that species has a southern distribution.

J F M A M J J A S O N D

LONGHORN BEETLES – HEAD CONSTRICTED, FORMING DISTINCT NECK 2/3

Stenurella 2 spp.

Black-striped Longhorn Beetle
Stenurella melanura

BL 6–10 mm. **ID** Black, except for reddish-yellow elytra; ♀ with black tip and broad central band; ♂ with much narrower band; pronotum bell-shaped. **Hab** Woodland. **Dist** Doubtful Ireland. **SS** None.

Common

J F M A M J J A S O N D

Small Black Longhorn Beetle
Stenurella nigra

BL 6–9 mm. **ID** Body and appendages black (abdomen orange); body strongly tapered towards tip; pronotum bell-shaped. **Hab** Woodland. **SS** None.

Rare

J F M A M J J A S O N D

Alosterna 1 sp.

Tobacco-coloured Longhorn Beetle *Alosterna tabacicolor*

BL 6–10 mm. **ID** Head, pronotum and scutellum black; elytra tawny (with blackish tip), mouthparts and legs tawny. Pronotum hind angles blunt or sharp. Antennae black, basal segment reddish. **Hab** Woodland. **SS** None.

Common

J F M A M J J A S O N D

Pseudovadonia 1 sp.

Fairy-ring Longhorn Beetle
Pseudovadonia livida

BL 5–9 mm. **ID** Head, pronotum, scutellum, antennae and most of legs black; elytra tawny or brownish (well punctured), also fore and mid tibiae. Pronotum hind angles short, rounded. **Hab** Woodland, heathland, grassland. **Dist** Also 1 old record Ireland. **SS** None.

Common

J F M A M J J A S O

Stictoleptura 3 spp.

FORM large; **ELYTRA** tip truncate; **PRONOTUM** hind angles present, rounded.

Red Longhorn Beetle
Stictoleptura rubra

BL 10–20 mm. **ID** ♂ pronotum black, elytra tawny or yellowish; ♀ pronotum and elytra red; tibiae yellowish to reddish. **Hab** Coniferous woodland. **SS** Blood-red Longhorn Beetle (*p. 297*), but pronotum red not black.

Woodland habitat for Red Longhorn Beetle at Brandon, Suffolk.

Heart Longhorn Beetle
Stictoleptura cordigera

BL 14–20 mm. **ID** Body black, except elytra, which are red with black mark narrowing sharply to black tip. **Hab** Woodland and nearby marshes; visits flowers. **Dist** Naturalized introduction, first recorded 2007. **SS** None.

Large Black Longhorn Beetle
Stictoleptura scutellata

BL 12–20 mm. **ID** Body black, heavily punctured; scutellum and hind part of pronotum with short yellow hairs. **Hab** Woodland. **SS** None.

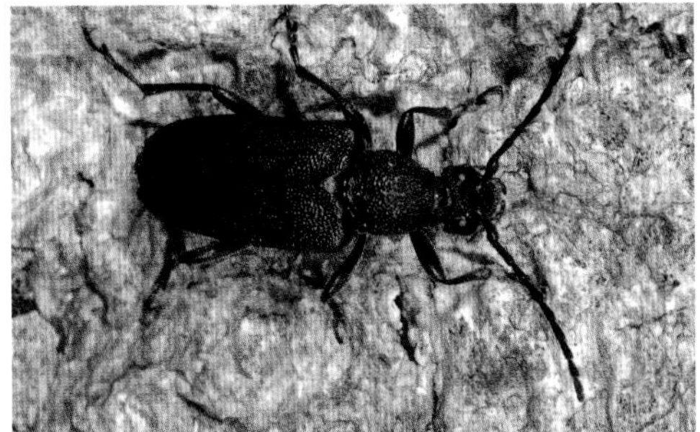

LONGHORN BEETLES – HEAD CONSTRICTED, FORMING DISTINCT NECK 3/

Rhagium 3 spp.

Black-spotted Longhorn Beetle *Rhagium mordax*

BL 13–23 mm. **ID** Elytra black and light brown, mottled with large black, buff-edged marks; antennae often < ½ length of body. **Hab** Woodland. **SS** Other *Rhagium* species, distinguished by elytra marks.

Common

J F M A M J J A S O N D

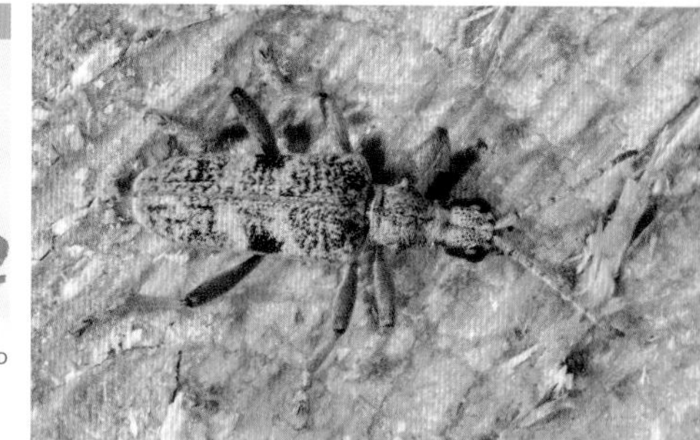

Ribbed Pine Borer *Rhagium inquisitor*

BL 12–22 mm. **ID** Elytra with several black marks and whitish hairs; antennae often < ½ length of body. **Hab** Coniferous woodland. **SS** Other *Rhagium* species, distinguished by elytra marks.

Nationally Scarce

J F M A M J J A S O N D

Two-banded Longhorn Beetle *Rhagium bifasciatum*

BL 12–22 mm. **ID** Elytra dark brown with pair of slanting buff-coloured bands; antennae often < ½ length of body. **Hab** Woodland. **SS** Other *Rhagium* species, distinguished by elytra marks.

Common

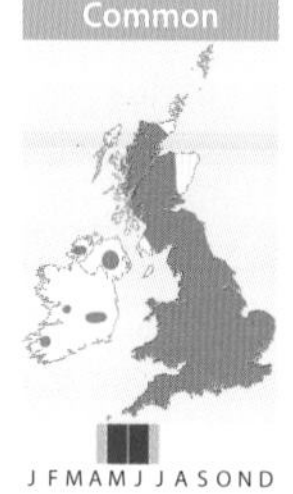

J F M A M J J A S O N D

Stenocorus 1 sp.

Variable Longhorn Beetle *Stenocorus meridianus*

BL 15–27 mm. **ID** Elytra black, or with tawny or yellowish patches; femora often tawny, but some all-black. **Hab** Woodland. **SS** None.

Common

J F M A M J J A S O N D

Leptura 2 spp.

Four-banded Longhorn Beetle *Leptura quadrifasciata*

BL 15–27 mm. **ID** Elytra black with four oloured bands; pronotum front and ack lacking golden-yellow hairs; legs lack. **Hab** Woodland. **SS** Golden-aired Longhorn Beetle.

Golden-haired Longhorn Beetle *Leptura aurulenta*

BL 12–25 mm. **ID** Elytra with four wavy, ellow or red bands; pronotum with olden-yellow hair fringe on front and ack; legs tawny. **Hab** Woodland. **SS** Four-banded Longhorn Beetle.

Rutpela 1 sp.

Black-and-yellow Longhorn Beetle *Rutpela maculata*

BL 13–20 mm. **ID** Elytra; yellow with lack spots; unmistakable. **Hab** Woodland. **SS** None.

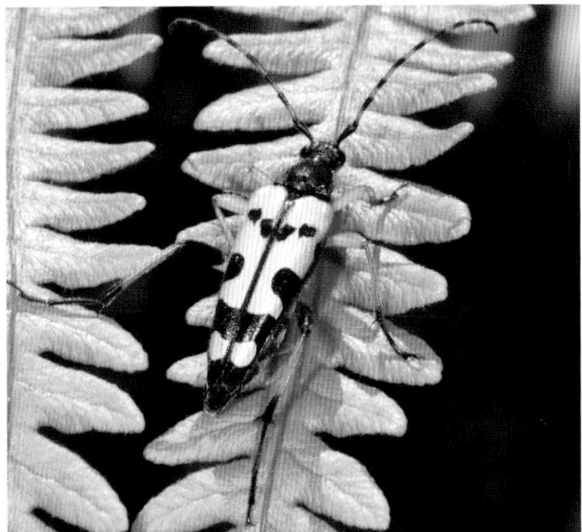

LONGHORN BEETLES – NO DISTINCT NECK, DECIDUOUS WOODLAND 1/

Saperda 3 spp. | 2 ILL.

FORM medium-sized to large, black or grey; **ELYTRA** either ladder-marked, spotted or with numerous tiny dots yellow lines and marks; **TARSUS** claws simple; **ANTENNAE** long.

Ladder-marked Longhorn Beetle *Saperda scalaris*

Nationally Scarce

J F M A M J J A S O N D

BL 13–35 mm. **ID** Body black, except pronotum and elytra which have extensive yellow hairs that form ladder-like marks and spots; antennae grey, banded. **Hab** Woodland. **SS** None.

Small Poplar Borer *Saperda populnea*

Nationally Scarce

J F M A M J J A S O N

BL 9–15 mm. **ID** Body black, except pronotum; elytra with several yellow spots and heavy punctures; antennae banded black and grey. **Hab** Woodland. **SS** None.

Clytus 1 sp.

Wasp Beetle *Clytus arietis*

Common

J F M A M J J A S O N D

BL 6–15 mm. **ID** Body black, except pronotum and elytra which have distinctive yellow lines and marks; antenna basal segments reddish; legs reddish, femora mostly black. **Hab** Woodland. **SS** None.

Phymatodes 1 sp.

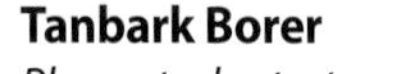

Tanbark Borer *Phymatodes testaceus*

Local

J F M A M J J A S O N

BL 6–18 mm. **ID** Bluish-black or reddish to dark brown; pronotum reddish; eyes smooth. **Beh** Nocturnal. **Hab** Woodland. **SS** None.

Pyrrhidium 1 sp.

Welsh Oak Longhorn Beetle *Pyrrhidium sanguineum*

Local

BL 6–15 mm. **ID** Head, antennae and legs black; pronotum and elytra reddish-orange. **Hab** Woodland. **SS** None.

Leiopus 2 spp. | 1 ILL.

Black-clouded Longhorn Beetle *Leiopus nebulosus*

Common

BL 6–10 mm. **ID** Brown, with greyish-brown hairs and black and whitish marks, spots and bands; scutellum black (hand lens needed). **Hab** Woodland. **SS** Linnaeus's Longhorn Beetle *Leiopus linnei* DD [N/I] (split in 2009), often confused, usually darker, with more black markings and spots (spp. best distinguished by genitalia examination or molecular work).

Anaglyptus 1 sp.

Rufous-shouldered Longhorn Beetle *Anaglyptus mysticus*

Local

BL 8–15 mm. **ID** Body black, except elytra, which are deep reddish-brown at base and have three whitish transverse bands and a broad, grey-haired tip. **Hab** Woodland. **SS** None.

Poecilium 1 sp.

White-banded Longhorn Beetle *Poecilium alni*

Nationally Scarce

BL 3–7 mm. **ID** Body black, except elytra, which are reddish at base and have two white bands towards posterior; legs reddish-brown, apices darker. **Hab** Woodland. **SS** None.

LONGHORN BEETLES – NO DISTINCT NECK, DECIDUOUS WOODLAND 2/2

Pogonocherus 4 spp. | 2 ILL.

FORM small to medium-sized, body dark brown; **ELYTRA** with large whitish or lighter patch, tip sharp-spined or truncate; **PRONOTUM** with sharp lateral teeth.

Greater Thorn-tipped Longhorn Beetle *Pogonocherus hispidulus*

Common

BL 5–8 mm. **ID** Scutellum mostly white; elytra tip with outer angles forming sharp teeth. **Hab** Woodland. **SS** Other *Pogonocherus* species, which have black scutellum and shorter teeth on elytra tip.

Lesser Thorn-tipped Longhorn Beetle *Pogonocherus hispidus*

Common

BL 4–6 mm. **ID** Scutellum black (hand lens needed); elytra tip with outer angles forming sharp teeth. **Hab** Woodland. **SS** Other *Pogonocherus* species, which have longer or lack teeth on elytra tip.

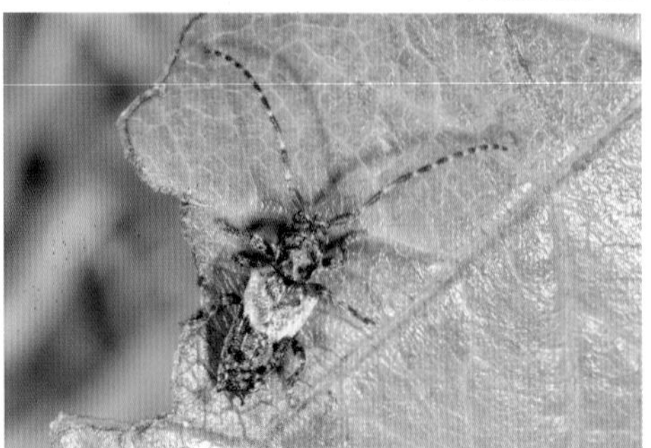

Mesosa 1 sp.

White-clouded Longhorn Beetle *Mesosa nebulosa*

Nationally Scarce

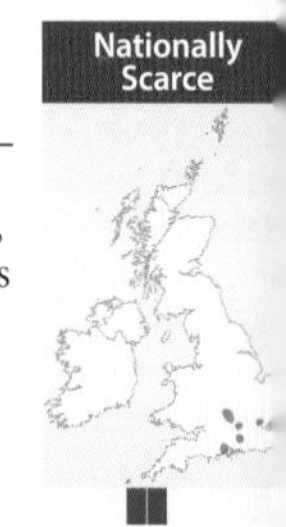

BL 9–16 mm. **ID** Brownish, heavily punctured; pronotum and elytra black, brown and whitish, with various marks and stripes; antennae grey and brown. **Hab** Woodland. **SS** None.

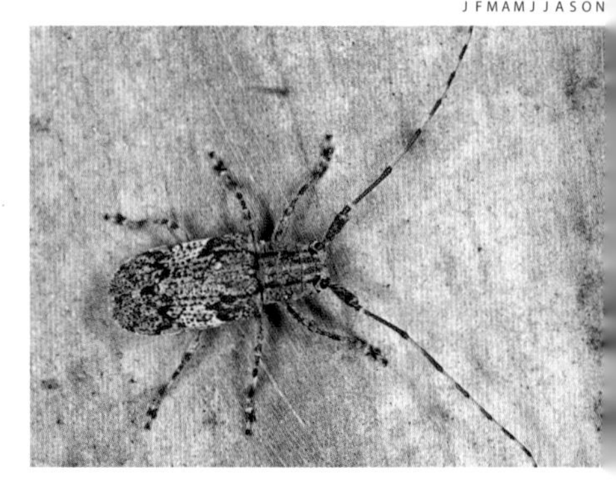

Stenostola 1 sp.

Lime Longhorn Beetle *Stenostola dubia*

Nationally Scarce

BL 8–14 mm. **ID** Black, with short greyish hairs; tarsus claws divided by a deep notch into two parts. **Hab** Woodland. **SS** None.

Tetrops 2 spp. | 1 ILL.

ʼlum Longhorn Beetle
ʼetrops praeustus

Common

ʼL 3–6 mm. **ID** Head, pronotum and antennae black; elytra hairy, ʼrown, tip black; legs brown. **ʼab** Woodland. **SS** The rare and ʼttle-known **Ash Longhorn Beetle** *ʼetrops starkii* [N/I] (**BL** 3–6 mm), ʼhich has more extensive black tip ʼo elytra.

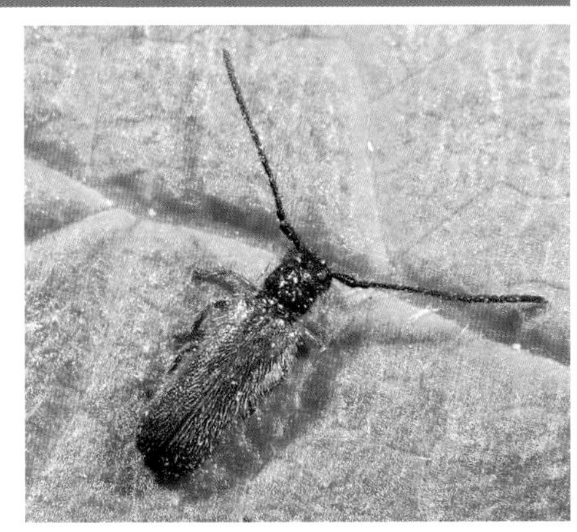

Aromia 1 sp.

ʼlusk Beetle
ʼromia moschata

Local

ʼL 13–35 mm. **ID** Metallic green, ʼme ♂s mauve. **Hab** Wetlands; ʼrvae develop in willows. **SS** None.

Prionus 1 sp.

Tanner Beetle
Prionus coriarius

Nationally Scarce

BL 18–45 mm. **ID** Very large and robust; dark reddish-brown to black; antnennal segments particularly thickened and serrated (toothed). **Hab** Woodland. **SS** None.

LONGHORN BEETLES – NO DISTINCT NECK, CONIFEROUS WOODLAND

Coniferous woodland specialists: FORM often broad, usually plain brown or black; ANTENNAE > length of body

Acanthocinus 1 sp.

Timberman Beetle NT
Acanthocinus aedilis

Rare

BL 12–20 mm. **ID** Grey, with brownish marks; antennae very much longer than body length; hind femora swollen. **Hab** Coniferous woodland, sawmills. **SS** None.

J F M A M J J A S O N D

♂

Molorchus 1 sp.

Spruce Shortwing Beetle
Molorchus minor

Local

BL 6–16 mm. **ID** Body black, elytra reddish-brown with two yellow marks; antennae longer than body length; hind femora swollen. **Hab** Coniferous woodland. **SS** The Nationally Scarce **Pear Shortwing Beetle** *Glaphyra umbellatarum* [N/I] (**BL** 5–9 mm), which lacks yellow marks on elytra and has much shorter antennae.

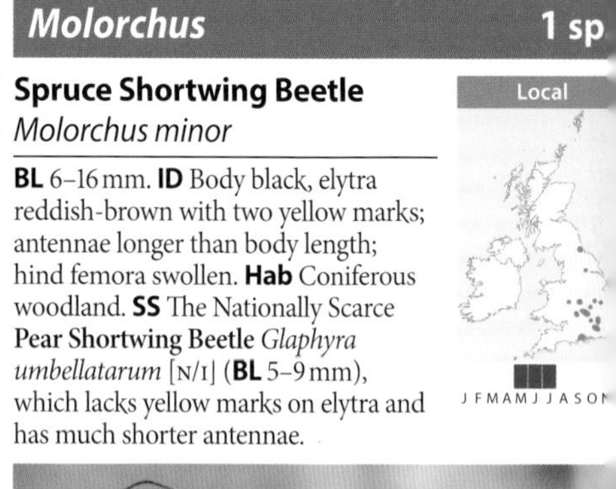

J F M A M J J A S O N D

Obrium 2 spp. (1†) | 1 ILL.

Brown Longhorn Beetle
Obrium brunneum

Local

BL 4–7 mm. **ID** Body brown with short, sparse, yellowish hairs. **Hab** Coniferous wodland. **SS** Rare **Brown Longhorn Beetle** *Obrium cantharinum* [N/I] (**BL** 5–10 mm) was last recorded in 1929 and hence is considered extinct; eyes larger than those of Brown Longhorn Beetle.

J F M A M J J A S O N D

Asemum 1 sp.

Pine-stump Borer
Asemum striatum

Local

BL 4–7 mm. **ID** Body broad, black; antennae short ± ½ body length; elytra dark brown or greyish. **Hab** Coniferous woodland. **SS** None.

J F M A M J J A S O N D

♀

Tetropium 3 spp. | 2 ILL.

FORM robust, rather flattened; **EYES** almost divided into two parts.

Black Spruce Longhorn Beetle *Tetropium castaneum*

Introduced; local

BL 8–23 mm. **ID** Body black or dark brown; elytra, tibiae and tarsi reddish brown; pronotum with sparse punctures. **Hab** Coniferous woodland. **SS** Other *Tetropium* species, from which distinguished by paler elytra and broader body.

Larch Longhorn Beetle *Tetropium gabrieli*

Local

BL 8–18 mm. **ID** Body matt black but elytra can be dark or reddish-brown; pronotum with dense punctures. **Hab** Coniferous woodland. **SS** Other *Tetropium* species, from which distinguished by body colour combined with pronotum punctuation.

Arhopalus 2 spp.

Burnt Pine Longhorn Beetle *Arhopalus ferus*

Local

BL 10–27 mm. **ID** Body blackish-brown or dark brown; eyes smooth. **Hab** Coniferous woodland. **SS** Dusky Longhorn Beetle.

Dusky Longhorn Beetle *Arhopalus rusticus*

Local

BL 10–30 mm. **ID** Body brown or reddish-brown; eyes with fine hairs (confirm using a hand lens with at least ×20 magnification). **Hab** Coniferous woodland. **SS** Burnt Pine Longhorn Beetle.

LONGHORN BEETLES – NO DISTINCT NECK, GRASSLAND

Agapanthia 2 spp.

Grassland specialists: ANTENNAE banded black-and-grey.

Golden-bloomed Grey Longhorn Beetle *Agapanthia villosoviridescens*

BL 11–19 mm. **ID** Body black; elytra grey mottled with short yellowish hairs. **Hab** Grassland. **SS** None.

Striped Thistle Longhorn Beetle *Agapanthia cardui*

BL 6–14 mm. **ID** Body black; elytra margin grey, sutural line pale. **Hab** Grassland. **Dist** First recorded in Folkestone, Kent in 2017; now spreading. **SS** None.

Phytoecia 1 sp.

Umbellifer Longhorn Beetle *Phytoecia cylindrica*

BL 6–14 mm. **ID** Body black; fore femora and tibiae reddish. **Hab** Grassland. **SS** None.

Stockbridge Down, Hampshire. Chalk downland with plentif flowers where several longhorn species occur, includir Golden-bloomed Grey Longhorn Beetle and Umbellif Longhorn Beet

FAMILY **Coccinellidae** (Ladybirds) 31 GEN. | 55 spp. | 34 ILL.

More than half of the British and Irish species in this family look like typical brightly coloured, spotted beetles known to practically everyone and popular enough to feature in several standalone field guides and publications, as well as featuring strongly in children's literature. Other species are drab, but mainly typically 'ladybird'-shaped. Ladybirds overwinter as adults, sometimes appearing in houses or outbuildings in numbers. Aphid-eating ladybirds are gardeners' friends, although the invasive Harlequin Ladybird is unpopular and blamed for declines in some native species. The main ladybird season is April to September; note that the number of spots can vary within a species, some, such as the Harlequin Ladybird, being particularly variable. Always check the thorax markings (see *below*) when dealing with non-typical colour forms, as these usually help; size is also important. In this section, species are presented by colour combination, regardless of their taxonomic relationship.

Ladybird larvae (*left*) and pupae (*right*) may be confused for other insects but are told by their shape (often spiny) and some have bright colours (photos show **Harlequin Ladybirds** (*p. 311*)).

Freshly emerged ladybirds of any species may lack their mature colours, which appear as their elytra harden (photo shows **Kidney-spot Ladybird** (*p. 313*)).

LADYBIRDS – RED OR BROWN WITH BLACK SPOTS 1/2

Adalia 2 spp.

2-spot Ladybird *Adalia bipunctata*

BL 4–5 mm. **ID** 2 spots; legs black. **Hab** Generalist, includes grassland, woodland, gardens. **SS** 10-spot Ladybird.

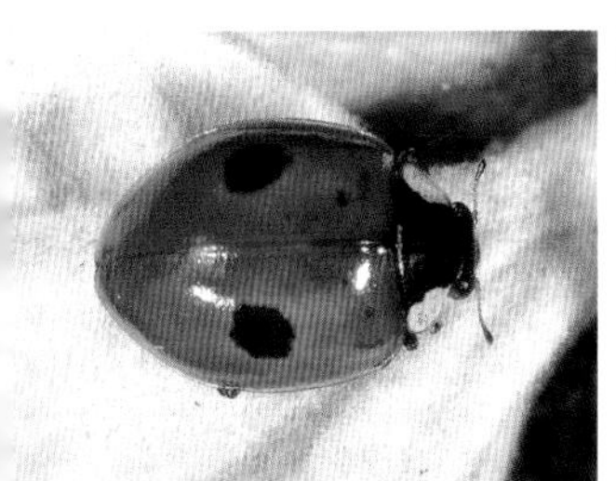

10-spot Ladybird *Adalia decempunctata*

BL 4 mm. **ID** Often small black spots (variable, can be 0–15 spots, or even elytra black with red spots); legs brown. **Hab** Woodland, gardens. **SS** 2-spot Ladybird.

LADYBIRDS – RED OR BROWN WITH BLACK SPOTS 2/2

Coccinella 5 spp.

7-spot Ladybird *Coccinella septempunctata*

BL 5–8 mm. **ID** 7 small to modest-sized black spots; white marks below forelegs only. **Hab** Generalist, including grassland, woodland, dunes, gardens. **SS** Scarce 7-spot Ladybird.

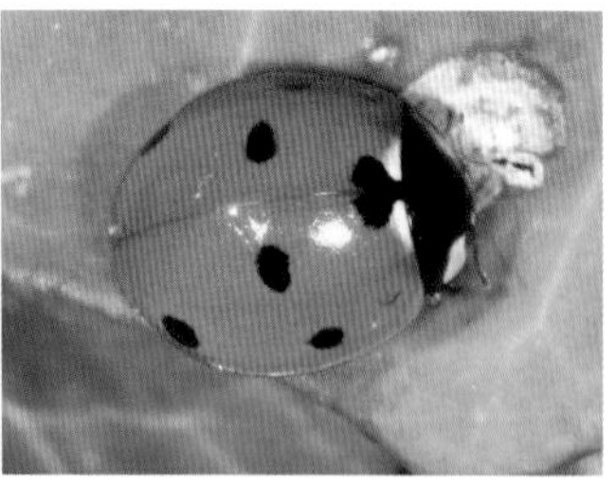

Scarce 7-spot Ladybird *Coccinella magnifica*

BL 6–8 mm. **ID** 7 large black spots; underside with white marks below both fore and midlegs [*inset*]. **Hab** Woodland, heathland, on or around Southern Wood Ant (*p. 496*) nests. **SS** 7-spot Ladybird.

11-spot Ladybird *Coccinella undecimpunctata*

BL 4–5 mm. **ID** 11 spots. **Hab** Often coastal grassland and dunes; on sandy soils inland. **SS** Adonis Ladybird.

5-spot Ladybird *Coccinella quinquepunctata*

BL 4–5 mm. **ID** 5 large black spots. **Hab** River shingle, on low-growing vegetation. **SS** None.

Hieroglyphic Ladybird *Coccinella hieroglyphica*

BL 4–5 mm. **ID** Brown with 5 black spots; black form, plain or with reddish spots, not uncommon. **Hab** Heathland. **SS** None.

Anatis 1 sp.

Eyed Ladybird *Anatis ocellata*

BL 7–8 mm. **ID** 15 modest-sized black, often pale-ringed spots. **Hab** Coniferous woodland. **SS** None.

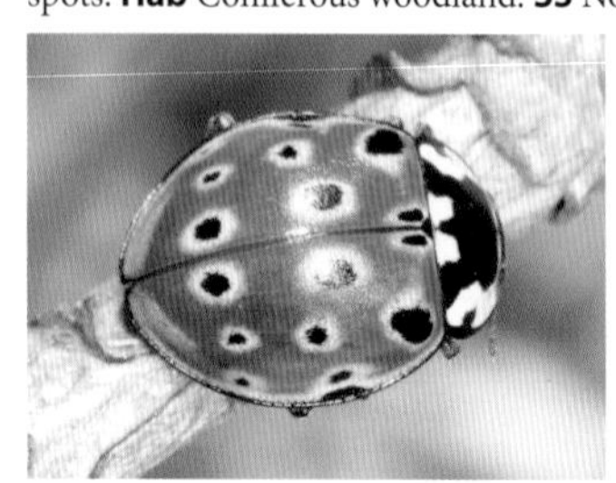

Hippodamia 2 spp.

Adonis Ladybird *Hippodamia variegata*

BL 4–5 mm. **ID** Body elongated; 7 black spots. **Hab** Often coastal dunes, but also on grassland inland. **SS** 11-spot Ladybird.

13-spot Ladybird *Hippodamia tredecimpunctata*

BL 5–7 mm. **ID** 13 spots. **Hab** Wetlands. **SS** None.

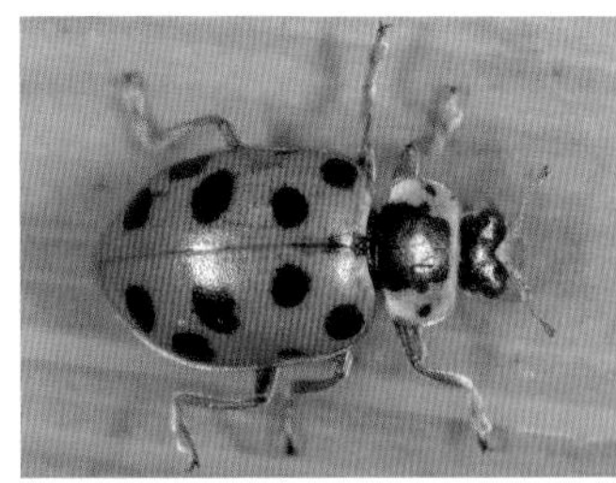

Harmonia 2 spp.

Harlequin Ladybird *Harmonia axyridis*

BL 5–8 mm. **ID** 16 modest-sized black spots (but remarkably variable, with 0–21 spots and including black forms with large red marks). **Hab** Woodland, gardens, parks, scrub. **SS** None.

Cream-streaked Ladybird *Harmonia quadripunctata*

BL 5–6 mm. **ID** 16 modest-sized black spots; elytra with cream margin. **Hab** Coniferous woodland. **Dist** First recorded in Ireland 2017. **SS** None.

Subcoccinella 1 sp.

24-spot Ladybird *Subcoccinella vigintiquattuorpunctata*

BL 3–4 mm. **ID** Reddish, hairy, with 24 black spots. **Hab** Grassland. **Dist** Also 1 old record Ireland. **SS** None.

Henosepilachna 1 sp.

Bryony Ladybird *Henosepilachna argus*

BL 5–7 mm. **ID** Red with 11 black spots. **Hab** Gardens, on White Bryony. **SS** None.

LADYBIRDS – ORANGE TO REDDISH-BROWN WITH WHITE OR CREAM SPOTS

Myrrha 1 sp.

18-spot Ladybird
Myrrha octodecimguttata

BL 4–5 mm. **ID** Reddish-brown with 18 cream spots. **Hab** Coniferous woodland. **SS** Cream-spot Ladybird.

Calvia 1 sp.

Cream-spot Ladybird
Calvia quattuordecimguttata

BL 4–5 mm. **ID** Reddish-brown with 14 cream spots. **Ha** Coniferous woodland. **SS** Orange & 18-spot Ladybirds

Halyzia 1 sp.

Orange Ladybird *Halyzia sedecimguttata*

BL 4 mm. **ID** 14 spots. **Hab** Grassland, woodland. **SS** 16-spot Ladybird, 22-spot Ladybird.

Myzia 1 sp.

Striped Ladybird *Myzia oblongoguttata*

BL 6–8 mm. **ID** Reddish with 13 cream spots and stripe **Hab** Coniferous woodland. **SS** None.

LADYBIRDS – YELLOW TO BUFF WITH BLACK SPOTS 1/

Propylea 1 sp.

14-spot Ladybird
Propylea quattuordecimpunctata

BL 4 mm. **ID** 14 spots. **Hab** Grassland, woodland. **SS** 16-spot Ladybird, 22-spot Ladybird.

Psyllobora 1 sp

22-spot Ladybird
Psyllobora vigintiduopunctata

BL 3–4 mm. **ID** 22 spots. **Hab** Grassland. **SS** 14-spot Ladybird, 16-spot Ladybird.

LADYBIRDS – BLACK WITH RED SPOTS

Chilocorus 2 spp

Heather Ladybird
Chilocorus bipustulatus

BL 3–4 mm. **ID** 6 red spots. **Hab** Heathland. **SS** None.

Kidney-spot Ladybird
Chilocorus renipustulatus

BL 4–5 mm. **ID** 2 large red spots. **Hab** Woodland, scrub. **SS** None.

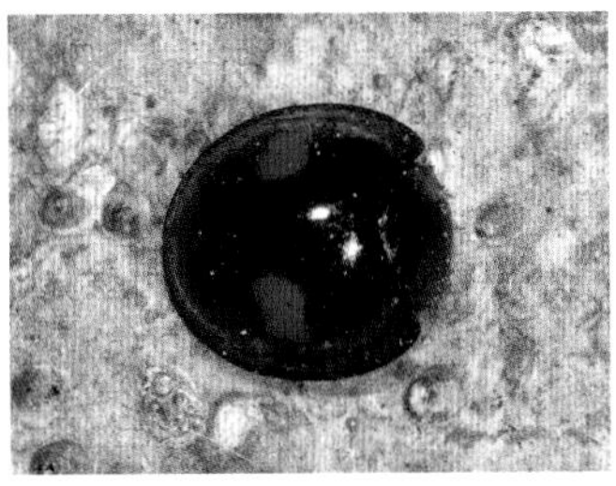

Exochomus 1 sp.

Pine Ladybird *Exochomus quadripustulatus*

BL 3–4 mm. **ID** 4 large red spots. **Hab** Woodland, gardens, parks. **SS** None.

Platynaspis 1 sp.

Ant-nest Ladybird *Platynaspis luteorubra*

BL 3 mm. **ID** Hairy, 4 large red spots. **Hab** Associated with ants. **SS** None.

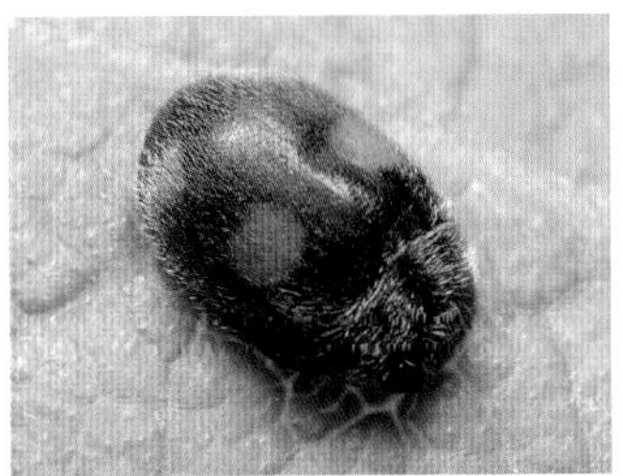

LADYBIRDS – YELLOW TO BUFF WITH BLACK SPOTS 2/2

Tytthaspis 1 sp.

16-spot Ladybird
Tytthaspis sedecimpunctata

BL 3 mm. **ID** 16 spots. **Hab** Grassland. **SS** 14-spot Ladybird, 22-spot Ladybird.

Anisosticta 1 sp.

Water Ladybird
Anisosticta novemdecimpunctata

BL 4 mm. **ID** Usually 19 black spots; elytra reddish in early season. **Hab** Wetlands. **SS** None.

LADYBIRDS – VERY SMALL BL ≤ 5 mm, including atypical species

Rhyzobius 4 spp. | 2 ILL.

Pointed-keeled Rhyzobius *Rhyzobius litura*

BL 2·5–3·0 mm. **ID** Brown, hairy; in some, elytra with large central 'U'-shaped dark mark. **Hab** Grassland. **SS** Round-keeled Rhyzobius *Rhyzobius chrysomeloides* [N/I] (**BL** *c.* 3 mm), has paler abdomen with dark marks.

Red-headed Rhyzobius *Rhyzobius lophanthae*

BL 2 mm. **ID** Black, hairy; head, thorax, legs and antennae dull orange. **Hab** Parks and gardens, often on Leyland Cypress. **Dist** First recorded in 1999; spreading north. **SS** None.

Aphidecta 1 sp.

Larch Ladybird *Aphidecta obliterata*

BL 4–5 mm. **ID** Brown, some individuals speckled with brown or black spots, or streaks [as *inset*]. **Hab** Coniferous woodland. **SS** None.

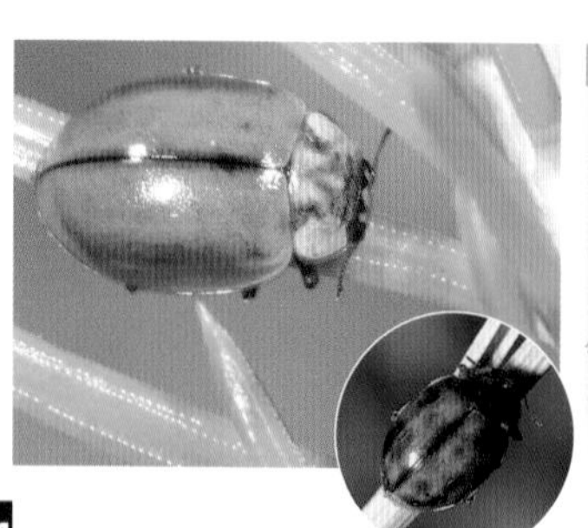

Scymnus 11 spp. | 2 ILL.

ELYTRA slightly hairy on sides towards final third. Species identified by colour/shape of elytra markings.

Pine Scymnus *Scymnus suturalis*

BL 1·5–2·0 mm. **ID** Elytra brown with dark line running along centre. **Hab** Coniferous woodland. **SS** None.

Angle-spotted Scymnus *Scymnus frontalis*

BL 3 mm. **ID** Black, elytra with two large reddish-brown spots. **Hab** Heathland, grassland, dunes. **SS** None.

Coccidula 2 spp. | 1 ILL.

Red Marsh Ladybird *Coccidula rufa*

BL 2·5–3·0 mm. **ID** Reddish-brown; elytra with dark mark on upper part. **Hab** Wetlands. **SS** Spotted Marsh Ladybird *Coccidula scutellata* [N/I] (**BL** *c.* 3 mm) has 5 black spots on elytra.

FAMILY **Chrysomelidae** (Leaf beetles) **63 GEN. | 279 spp. | 41 ILL.**

Small to large; body length 1–20 mm), broad or oval leaf-eating beetles, often very colourful and sometimes very variable in colour within species; antennae long; legs short. **They are easy enough to separate into groups by characteristic shape; about half are characteristic leaf beetles.**

FORM small (BL 2–5 mm); BODY brown or black; ELYTRA quadrate, mottled; PRONOTUM strongly narrowed in front; ANTENNAE long, 11-segmented, usually toothed or feathery.

Red-legged Seed Beetle *Bruchus rufipes* (BL 2–3 mm).

SEED BEETLES

[4 genera | 17 species, 1 illustrated]

FORM mostly small, some medium-sized (BL 1–6 mm), oval to rounded; LEGS hind femora broadened (able to jump well); ANTENNAE moderate length, with 10 to 11 thread-like segments.

Thistle Leaf-miner Beetle *Sphaeroderma testaceum* (BL 4 mm)

FLEA BEETLES

[21 genera | 127 species, 3 illustrated]

p. 316

FORM small to medium-sized (BL 4–10 mm), flattened, shell-like shape (head covered by pronotum); ANTENNAE smaller than other Chrysomelidae, segments broadened and darkened towards tip.

Pale Tortoise Beetle *Cassida flaveola* (BL 4–6 mm)

TORTOISE BEETLES

[3 genera | 14 species, 4 illustrated]

p. 316

FORM small to large (BL 2–18 mm), round to elongate; BODY often brightly coloured; ANTENNAE moderately long to long, 11-segmented, thread-like.

Dead-nettle Leaf Beetle

LEAF BEETLES

[32 genera | 100 species, 29 illustrated]

p. 317

FORM medium-sized to large (BL 5–13 mm), elongate; BODY usually metallic; PRONOTUM narrow; ANTENNAE long, 11-segmented, thread-like. HABITAT aquatic habitats, usually on reeds.

Common Reed Beetle

REED BEETLES

[3 genera | 21 species, 6 illustrated]

p. 322

Guide to Chrysomelidae p. 3

TORTOISE BEETLES

Cassida 12 spp. | 4 ILL. (one on p. 315)

FORM Small to medium, rounded body. Species identified by coloration and/or underside characters.

Green Tortoise Beetle *Cassida viridis*

BL 7–10 mm. **ID** Body green; legs pale; antenna tip dark. Underside black, except abdomen segments with yellow margin [see *inset*]. **Hab** Wetlands. Often on mints. **SS** Other plain *Cassida* species do not have rounded hind side margins to pronotum.

Thistle Tortoise Beetle *Cassida rubiginosa*

BL 6–8 mm. **ID** Body green, elytra with brown markings on and around scutellum. Legs reddish-orange, except femora; antenna tip black. **Hab** Grassland. **SS** Other *Cassida* species, but elytra markings unique.

Fleabane Tortoise Beetle *Cassida murraea*

BL 6–9 mm. **ID** Body reddish-orange [green when teneral], elytra with black spots. **Hab** Wetlands. **SS** None.

FLEA BEETLES

Phyllotreta 15 spp. | 1 ILL

ELYTRA oblong; pattern of yellow stripes helps to identify species. **HABITAT** on cabbages. Species identified by extent of yellow marks on elytra.

Small Striped Flea Beetle *Phyllotreta undulat*

BL 2–3 mm. **ID** Body black, elytra with yellow stripe, narrowed in centre. **Hab** Grassland, urban and agricultural areas. **SS** Other *Phyllotreta* species.

Crepidodera 5 spp. | 1 ILL

FORM small, metallic; **PRONOTUM** with triangular fronta tubercles, reddish, some coppery-bronze or blue; **ELYTRA** metallic green, some brownish.

Willow Flea Beetle *Crepidodera aurata*

BL 2·0–3·5 mm. **ID** Pronotum usually reddish, unevenly punctured; elytra generally metallic green. **Hab** Woodland, hedgerows, parks. **SS** Other *Crepidode* species.

Teneral Fleabane Tortoise Beetle

.EAF BEETLES – Round–elongated body (distinctive selected genera) 1/3

Oulema 4 spp. | 1 ILL.

PRONOTUM + LEGS reddish (rarely black); **HEAD + ANTENNAE** black and elongated; **ELYTRA** metallic blue.

:ereal Leaf Beetle *Oulema melanopus*

:L 4–5 mm. **ID** Pronotum and legs red; elytra blue or reenish-blue. **Hab** Grassland and cereal crops. **P** Grasses. **SS** Other *Oulema* species (confirmation of lentification requires examination of genitalia).

Agelastica 1 sp.

.lder Leaf Beetle *Agelastica alni* DD

L 6–7 mm. **ID** Dark metallic blue. **Hab** Wetlands and rban areas. **FP** Alder. **Dist** Abundant where it occurs, nd spreading. **SS** None.

Lilioceris 1 sp.

ily Beetle *Lilioceris lilii*

L 6–8 mm. **ID** Head, antennae and legs black; pronotum ıd elytra bright red. **Hab** Gardens, nurseries. **Fp** Lilies ıd fritillaries. **SS** None.

Chrysomela 4 spp. | 1 ILL.

FORM medium–large, broad, brightly coloured; **PRONOTUM** narrower than elytra base; **ELYTRA** metallic or non-metallic, with one or two rows of punctures.

Red Poplar Leaf Beetle *Chrysomela populi*

BL 10–12 mm. **ID** Head and pronotum bluish-black or dark green; elytra red. **Hab** Grassland. **FP** Willows and poplars. **SS** Other rarer *Chrysomela* species, but distinguished by larger size (others up to 9 mm).

Crioceris 1 sp.

Asparagus Beetle *Crioceris asparagi*

BL 5–6 mm. **ID** Head and elytra bluish-black, elytra with red border and 6 yellow spots; pronotum narrow, red. **Hab** Grassland, gardens. **Fp** Asparagus. **SS** None.

Lily Beetle and leaf damage.

LEAF BEETLES – Round–elongated body (distinctive selected genera) 2/

Chrysolina 19 spp. | 8 ILL.

FORM mostly medium-sized, usually metallic; **ELYTRA** with slightly hairy fringe beneath border. Species identified by body colour, shape (particularly pronotum), or e.g. punctuation on pronotum and elytra.

Banks's Leaf Beetle *Chrysolina banksii*

BL 6–12 mm. **ID** Body bronze to dark brown; underside paler; pronotum with distinctive ridge, narrowed from base to tip. Elytra coarsely punctured. **Hab** Grassland, mainly coastal. **SS** Brown Leaf Beetle *C. staphylaea* [N/I] (**BL** 5–9 mm) has almost parallel-sided pronotum.

Knotgrass Leaf Beetle *Chrysolina polita*

BL 5–9 mm. **ID** Head and pronotum metallic green, elytra non-metallic red or reddish-brown.
Hab Grassland. **SS** None.

Dead-nettle Leaf Beetle *Chrysolina fastuosa*

BL 5–7 mm. **ID** Body metallic green or coppery, with broad blue or reddish bands. **Hab** Grassland, mainly coastal. **SS** None.

Rosemary Beetle *Chrysolina americana*

BL 6–8 mm. **ID** Elytra metallic red, with green punctur in longitudinal lines. Pronotum metallic green and red **Hab** Parks, gardens. **FP** Rosemary and Lavender. **Dist** First recorded 1963. **SS** Rainbow Leaf Beetle, which is more colourful and range-restricted.

Cow Parsley Leaf Beetle *Chrysolina oricalci*

BL 6–9 mm. **ID** Body usually dark blue (some green, bronze, black). Elytra with 9 rows of punctures; pronotu with groove on side margins. **Hab** Hedgerows, road verges, woodland rides. **Dist** Old records Ireland.
SS Other *Chrysolina* species.

Hypericum Leaf Beetle *Chrysolina hyperici*

BL 5–7 mm. **ID** Body dark metallic bronze (some gree reddish or black); elytra with characteristic pits.
Hab Grassland, woodland clearings. **SS** Other *Chrysolina* species, which lack pits on pronotum/elytr

Rainbow Leaf Beetle *Chrysolina cerealis*

BL 6–10 mm. **ID** Body with longitudinal bands of metallic green, blue and red, or gold. **Hab** Mountain grassland. **SS** **Rosemary Beetle** (less distinct bands).

Rare

J F M A M J J A S O N D

Violet Leaf Beetle *Chrysolina sturmi*

BL 6–9 mm. **ID** Body metallic purple; tarsi and palps orange-brown. **Hab** Chalk grassland. **FP** Ground-ivy. **SS** None (but superficially **Small Bloody-nosed Beetle**).

Nationally Scarce

J F M A M J J A S O N D

Timarcha 2 spp.

FORM medium-sized to large, (bluish)-black, rounded; **TARSUS** S1–3 broadened; **FOODPLANT** bedstraws.

Bloody-nosed Beetle *Timarcha tenebricosa*

BL 10–20 mm. **ID** Large; pronotum front margin widened laterally. **Hab** Grassland. **Beh** Emits drop of red fluid from mouthparts in defence. **SS** **Small Bloody-nosed Beetle**.

Common

J F M A M J J A S O N D

Small Bloody-nosed Beetle *Timarcha goettingensis*

BL 8–11 mm. **ID** relatively small; pronotum front margin not widened laterally. **Hab** Chalk grassland. **SS** **Bloody-nosed Beetle**, superficially **Violet Leaf Beetle**.

Local

J F M A M J J A S O N D

Galeruca 2 spp. | 1 ILL.

Black-punctured Leaf Beetle *Galeruca tanaceti*

BL 6–12 mm. **ID** Body black, heavily punctured; elytra broad. **Hab** Grassland. **SS** **Rare Galeruca** *Galeruca laticollis* CR [N/I] (**BL** 6–9 mm) has less distinct elytral punctures and yellowish spines on tip of hind tibiae.

Common

J F M A M J J A S O N D

Gastrophysa 2 spp. | 1 ILL.

Green Dock Beetle *Gastrophysa viridula*

BL 4–6 mm. **ID** Body metallic green; legs dark. **Hab** Grassland, gardens, heathland, woodland edges. **Fp** Broad-leaved Dock. **Beh** Impregnated ♀s often seen with very swollen abdomen. **SS** None on docks.

Common

J F M A M J J A S O N D

LEAF BEETLES – Round–elongated body (distinctive selected genera) 3/

Pyrrhalta 1 sp.

Viburnum Beetle *Pyrrhalta viburni*

BL 5–7 mm. **ID** Body yellowish-brown to darker brown; head, pronotum and elytra with black central marks. **Hab** Woodland, hedgerows, gardens, parks. **Fp** Viburnums. **SS** None.

Phyllobrotica 1 sp.

Skullcap Leaf Beetle *Phyllobrotica quadrimaculata*

BL 5–8 mm. **ID** Body tawny or ocheous, elytra with larg black spots. **Hab** Grassland. **Fp** Skullcaps. **SS** None.

Cryptocephalus 20 spp. | 9 ILL.

FORM broad-bodied, pronotum broad and rounded; wide range of colours, most distinct (several species extremely rare); larvae live in protective cases ('pots') made from their own excrement.

Black-and-red Pot Beetle *Cryptocephalus bipunctatus*

BL 4–6 mm. **ID** Head, pronotum and scutellum black; elytra red with large black streaks. **Hab** Grassland. **SS** None.

Hazel Pot Beetle *Cryptocephalus coryli*

BL 6–7 mm. **ID** Body red except head and legs black (pronotum also in ♂ only); elytra with black suture margin. **Hab** Woodland. **Dist** Known from only four sites in recent years. **SS** None.

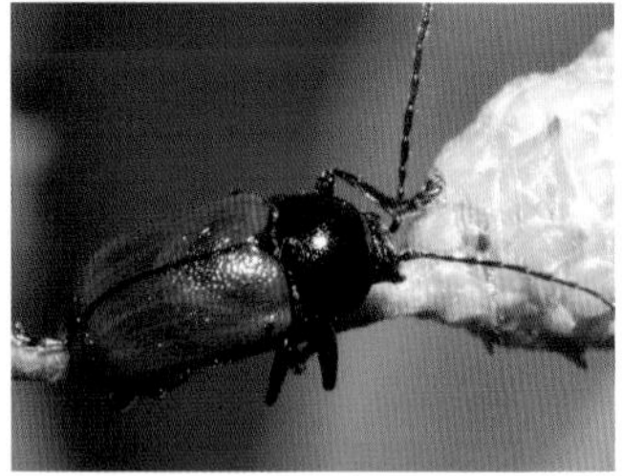

Yellow Pot Beetle *Cryptocephalus fulvus*

BL 2–3 mm. **ID** Body yellow, with some dark spots. **Hab** Grassland. **Fp** Low-growing plants. **SS** **Common Small Pot Beetle**, but antenna S4–10 shorter.

Common Small Pot Beetle *Cryptocephalus pusillus*

BL 2–3 mm. **ID** Body usually yellow, with some dark spots and bands. **Hab** Woodland. **Fp** Tree leaves. **SS** **Yellow Pot Beetle**, which occurs in different habitat

Six-spotted Pot Beetle *Cryptocephalus sexpunctatus*

EN

BL 5–6 mm. **ID** Elytra yellowish to red, with six black spots. Pronotum variably black and orange. **Hab** Woodland, chalk grassland. **Dist** Known from only three sites in recent years. **SS** None.

Hypericum Pot Beetle *Cryptocephalus moraei*

BL 3–4 mm. **ID** Body black; pronotum with yellow spots at hind angles; elytra with partly yellow sides and spots. **Hab** Grassland. **SS** Other dark *Cryptocephalus* species lack distinct yellow spots on side and tip of elytra.

Common Black Pot Beetle *Cryptocephalus labiatus*

BL 2–3 mm. **ID** Body shiny black; mouthparts yellow; legs yellowish, partly dark brown or black. **Hab** Woodland. **SS** Other small black *Cryptocephalus* species, which have entirely yellowish legs.

Green Pot Beetle *Cryptocephalus aureolus*

BL 6–8 mm. **ID** Metallic green, some individuals bluish. **Hab** Grassland. **Dist** Formerly also recorded in Ireland. **SS** The smaller **Small Green Pot Beetle** *Cryptocephalus hypochaeridis* [N/I] (**BL** 5–6 mm) has shorter scutellum (quadrate in Green Pot Beetle).

Blue Pot Beetle *Cryptocephalus parvulus*

BL 3–4 mm. **ID** Pronotum and elytra uniformly metallic blue or purple; legs dark. **Hab** Woodland. **SS** Other rare, small, dark *Cryptocephalus* species have pale legs and pronotum and elytra are different colours.

Lochmaea 3 spp. | 1 ILL.

FORM small to medium-sized, smooth-bodied, brown or reddish-brown; **PRONOTUM** with angled sides; **ELYTRA** partly rounded.

Heather Beetle *Lochmaea suturalis*

BL 4–6 mm. **ID** Head black; pronotum shining and, as elytra, reddish-brown to black; antennae black with reddish basal segments and S2 ± length of S4. **Hab** Heathland. **Fp** Heathers. **SS** Other *Lochmaea* species.

Guide to Chrysomelidae *p. 31*

REED BEETLES – Elongated body (note colour patterns)

Plateumaris — 4 spp. | 2 ILL.

FORM slightly more robust appearance than related Donacia species; **LEGS** stout; **ELYTRA** curved at tip. Species identified by body colour, pronotum shape and punctuation.

Bog Reed Beetle *Plateumaris discolor*

BL 6–9 mm. **ID** Metallic green, copper, blue, mauve, red or black; pronotum punctured, stronger around mid-line. **Hab** Bogs and mires. **SS** Jewel Reed Beetle.

Jewel Reed Beetle *Plateumaris sericea*

BL 7–10 mm. **ID** Metallic green, copper, blue, mauve, red or black; pronotum evenly punctured. **Hab** Non-acidic wetlands. **SS** Bog Reed Beetle.

Donacia — 15 spp. | 4 ILL.

LEGS mainly slender; **ELYTRA** flattened, tip often truncate. Species identified by differences in body colour, shape of elytra tip, punctuation on pronotum and presence or absence of spines on femora and tibiae.

Margined Reed Beetle *Donacia marginata*

BL 8–11 mm. **ID** Elytra bronze-brown with broad purple longitudinal band and central basal mark. Hind femora with one spine. **Hab** Wetlands. **SS** Other *Donacia* spp.

Plain Reed Beetle *Donacia simplex*

BL 7–9 mm. **ID** Elytra coppery, with silver underside; central groove on head; underside of hind tibiae not toothed. **Hab** Wetlands. **SS** Other *Donacia* species.

Common Reed Beetle *Donacia vulgaris*

BL 6–9 mm. **ID** Elytra coppery, with broad purple, red, blue or brass-coloured longitudinal band and near suture. Hind femora not toothed. **Hab** Wetlands. **SS** Other *Donacia* species; told by elytra bands.

Pondweed Reed Beetle *Donacia versicolorea*

BL 6–10 mm. **ID** Elytra punctured (appear as tiny pale spots); pronotum centre coarsely punctured; hindlegs elongated, femora broadened. **Hab** Wetlands. **SS** Other *Donacia* species; told by pronotum punctuation.

FAMILY Curculionidae (Weevils) 164 GEN. | 494 spp. | 16 ILL.

The second largest family after rove beetles, this group of small to large species (body length 1–17 mm) are broad-bodied, the antennae often elbowed (1st segment much longer than next two segments combined), also clubbed. The snout or rostrum is long and taken to an extreme in some species. However, some bark beetles within this group lack these characters. Many of the species in this family are challenging to identify, but some can be identified in the field or from photographs.

HEAD with rostrum, in some species elongated; **BODY** usually broad and robust (**BL** 1–17 mm); **ANTENNAE** often long; **TIBIAE** lacking a pair of teeth.

Nut Weevil

WEEVILS

[131 genera | 426 species, 16 illustrated]

HEAD lacking rostrum, or very short; **BODY** cylindrical (**BL** 1–9 mm); **ANTENNAE** often short; **TIBIAE** with at least one pair of teeth.

Pinhole Borer

BARK BEETLES

[33 genera | 68 species, 2 illustrated]

p. 326

WEEVILS 1/3

Hylobius 2 spp. | 1 ILL.

FORM medium-sized to large (8–13 mm), robust; **BODY** black or brown, elytra spotted; **ROSTRUM** as long as pronotum, which is narrowed at front. Species identified by body colour.

Pine Weevil *Hylobius abietis*

BL 8–13 mm. **ID** Body black, some individuals dark brown, elytra spotted orange or cream. **Hab** Coniferous woodland. **SS** None.

Cionus 6 spp. | 1 ILL.

FORM small (3–5 mm) and stocky; **PRONOTUM** narrow at front, base distinctly narrower than elytra. **ELYTRA** oblong, pale to dark, rather mottled due to pubescence (short hairs) and with bold, large black central (or just off-centre) spot. Species identified by markings on elytra, including position of black spot; in some cases extent of pronotum hairs.

Garden Figwort Weevil *Cionus hortulanus*

BL 4–5 mm. **ID** Body rather pale, elytra with short chequered greyish, black and white patches. **Hab** Gardens, parks, woodland, often on figworts. **SS** Other *Cionus* species, from which told by central spot, combined with characters mentioned in ID above.

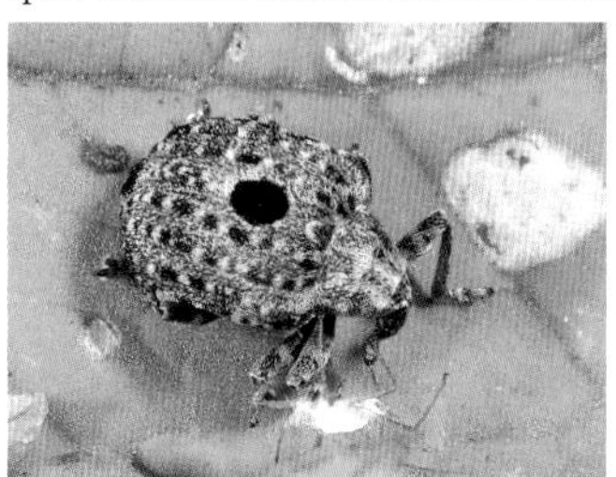

Guide to Curculionidae p. 32

WEEVILS

Otiorhynchus 25 spp. | 4 ILL.

FORM small to large (4–15 mm), brown, often mottled; **ANTENNAE** long; **WINGS** wingless.

Vine Weevil *Otiorhynchus sulcatus*

Parthenogenetic. **BL** 7–10 mm. **ID** Pronotum with many large tubercles; elytra black mottled with small brown marks. **Hab** Parks, gardens woodland.

Armadillo Weevil *Otiorhynchus armadillo*

BL 7–12 mm. **ID** Robust, elytra oval. Pronotum tuberculate; elytra with brownish-yellow mottled pattern in fresh individuals. **Hab** Parks and gardens.

Black Marram Weevil *Otiorhynchus atroapterus*

BL 6–9 mm. **ID** Body, legs and antennae black. **Hab** Sand dunes and sandy sites, mainly coastal. **SS** Other black *Otiorhynchus* species, notably **Red-legged Weevil**.

Red-legged Weevil *Otiorhynchus tenebricosus*

BL 9–13 mm. **ID** Body black, legs black/dark reddish-brown; fore tibiae tip broadened on inner and outer margins. **Hab** Grassland. **SS** Black *Otiorhynchus* spp.

Curculio 6 spp. | 2 ILL.

BODY brown; **ROSTRUM** elongated, downcurved; **PRONOTUM** narrow at front; **ELYTRA** mottled, very narrow at tip.

Acorn Weevil *Curculio glandium*

BL 4–7 mm. **ID** Antenna club narrow (3× as long as wide). **Hab** Woodland, on oaks. **SS** Other *Curculio* species, particularly **Nut Weevil**.

tip narrow

Nut Weevil *Curculio nucum*

BL 6–8 mm. **ID** Antenna club broad (2× as long as wide segments bristly. **Hab** Woodland, hedgerows, on Hazel. **SS** Other *Curculio* species, esp. **Acorn Weevil**.

tip broad

Phyllobius 10 spp. | 2 ILL.

FORM small to medium-sized (3–10 mm), often covered in metallic green or brown scales [NOTE: can be missing, so some individuals very difficult to identify].

Common Leaf Weevil *Phyllobius pyri*

BL 5–8 mm. **ID** Coppery or brown-scaled. **Hab** Woodland, scrub, hedgerows. **SS** Other *Phyllobius* spp.

Green Nettle Weevil *Phyllobius pomaceus*

BL 6–9 mm. **ID** Green-scaled (shades variable). **Hab** Grassland, sites with Common Nettle. **SS** *Phyllobius* spp.

Lixus 5 spp. | 1 ILL.

FORM small to large (4–17 mm), narrow-bodied, cylindrical.

Sea Beet Weevil *Lixus scabricollis*

BL 5–6 mm. **ID** Robust; pronotum narrowed at front; antennae short and straight. **Hab** Grassland, woodland clearings, including coastal sites; on thistles. **SS** Other *Lixus* species, which have longer antennae.

Liophloeus 1 sp.

Chequered Weevil *Liophloeus tessulatus*

BL 7–12 mm. **ID** Rounded; elytra with variable light and dark marks; short-winged and flightless. **Hab** Woodland, hedgerows, on Ivy. **SS** None.

Rhinocyllus 1 sp.

Thistle Head Weevil *Rhinocyllus conicus*

BL 4–7 mm. **ID** Robust; pronotum narrowed at front; antennae short and straight. **Hab** Grassland, woodland clearings; including coastal sites, on thistles. **SS** None.

Mogulones 3 spp. | 1 ILL.

FORM small (only one species ≥3 mm), brown; **ELYTRA** with whitish, map-like marks.

Map Weevil *Mogulones geographicus*

BL 3–5 mm. **ID** Largest species; pronotum very narrowed at front, rather collar-like. **Hab** Grassland, on Viper's-bugloss. **SS** None.

Guide to Curculionidae *p. 32*

WEEVILS

Strophosoma 6 spp. | 2 ILL.

FORM small to medium-sized (3–6 mm), stocky, often covered with scales; **ROSTRUM** short and broad; **PRONOTUM** sides rounded; **ELYTRA** oval. Species identified by characters on head, elytra, legs or shape of rostrum.

Heather Weevil *Strophosoma sus*

BL 4–5 mm. **ID** Body black, elytra with pale border; rostrum lacks a central ridge and pronotum has central, longitudinal groove. **Hab** Heathland, on heathers. **SS** Other black *Strophosoma* species.

Nut Leaf Weevil *Strophosoma melanogrammum*

Parthenogenetic. **BL** 4–6 mm. **ID** Body brown, rather mottled darker; elytra with central black line on upper half. **Hab** Woodland, hedgerows, gardens, parks. **SS** Other black *Strophosoma* species; told by scale pattern on elytra.

BARK BEETLES

Platypus 1 sp.

Pinhole Borer *Platypus cylindrus*

BL 5 mm. **ID** Head broad, same width as elongated pronotum; body black or brownish-black; elytra with orange-haired tip; antennae orange. **Hab** Woodland. **SS** None.

Scolytus 8 spp. | 1 ILL

FORM small to medium-sized (2–6 mm) [includes major forestry pests, conveying the fungus that causes Dutch Elm Disease]. Species identified by examination of abdomen segment shape, *etc.*

Small Elm Bark Beetle *Scolytus multistriatus*

BL 2–3 mm. **ID** Body black; elytra and legs brown; pronotum and elytra with punctures. **Hab** Woodland, beneath bark of dead and decaying elms. **SS** Other *Scolytus* species; this the only one with underside of abdomen S2 lacking a central 'process'.

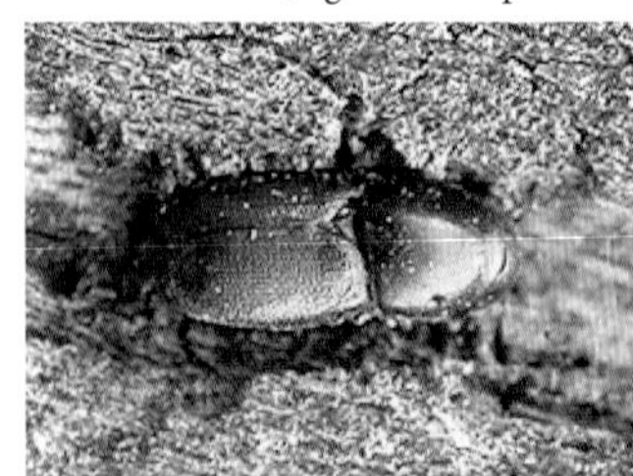

Patterning (beneath bark) made by a *Scolytus* species.

ORDER RAPHIDIOPTERA

Snakeflies

[Greek: *rhaphis* = needle; *pteron* = wing]

BI | 1 family [4 genera, **4 species**]

W | 2 families [**271 species**]

The only insects with a characteristic, elongated, 'snake'-like pronotum, raised above the body, at rest. Closely related to Neuroptera (Lacewings and Antlions (*p. 230*)), the snakeflies, along with Megaloptera (Alderflies (*p. 237*)) (whose larvae are aquatic), form a combined group 'Neuropterida'. British species (body length 7–15 mm) are blackish with two pairs of black-veined wings, used in a clumsy flight. Females use their long, needle-like ovipositor (hence the name of the order) to lay eggs in cracks of bark of trees, with the whole life-cycle taking some two years. Larvae feed on insects beneath the bark; adults prey on aphids and other small insects. Snakeflies overwinter as larvae or pupae.

Teneral newly-hatched specimens are very pale and may, like this example, be found on an oak trunk, prior to dispersal. Once hardened, the normal colour shows on the body and wing veins.

FINDING SNAKEFLIES Occasionally seen on large oak or pine trunks and low vegetation in sunny weather, but may often be out of sight, high up. Worth looking for after strong winds or low down on tree trunks when newly hatched.

Structure of a snakefly

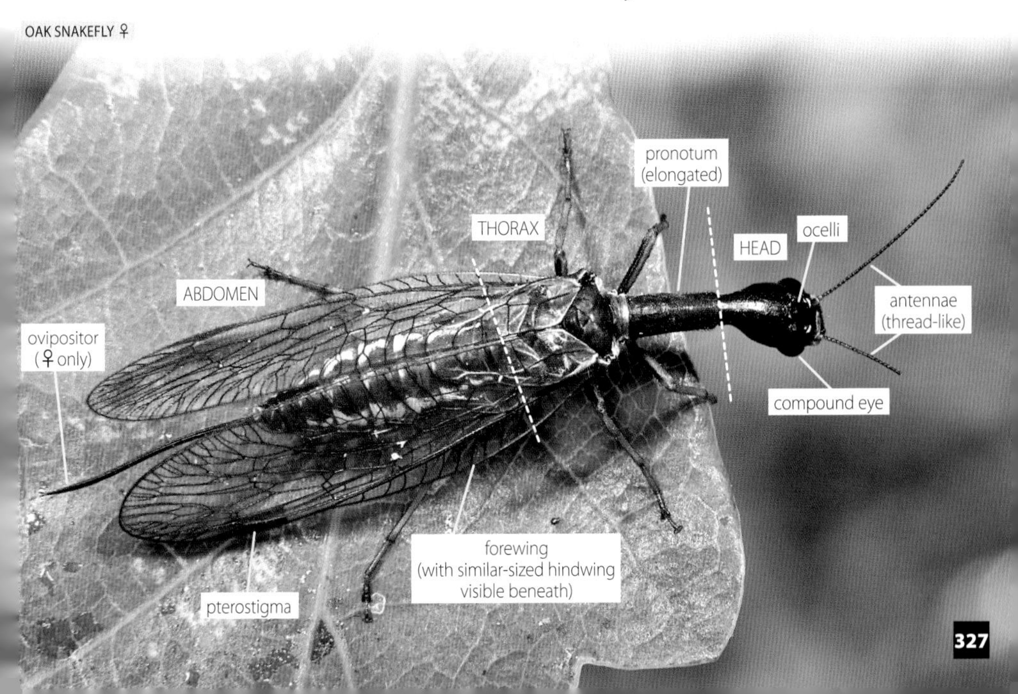

FAMILY Raphidiidae (Snakeflies) 4 GEN. | 4 spp. | 4 ILL.

Although belonging to different genera, identification to species level needs care. However, whilst some researchers will only use a microscope identifying for snakeflies, the species can be separated by checking good quality images, or in the field, by examining the forewing tips and pterostigma with a 15–20× hand lens. **Note that there may be slight variation in wing structure within species.**

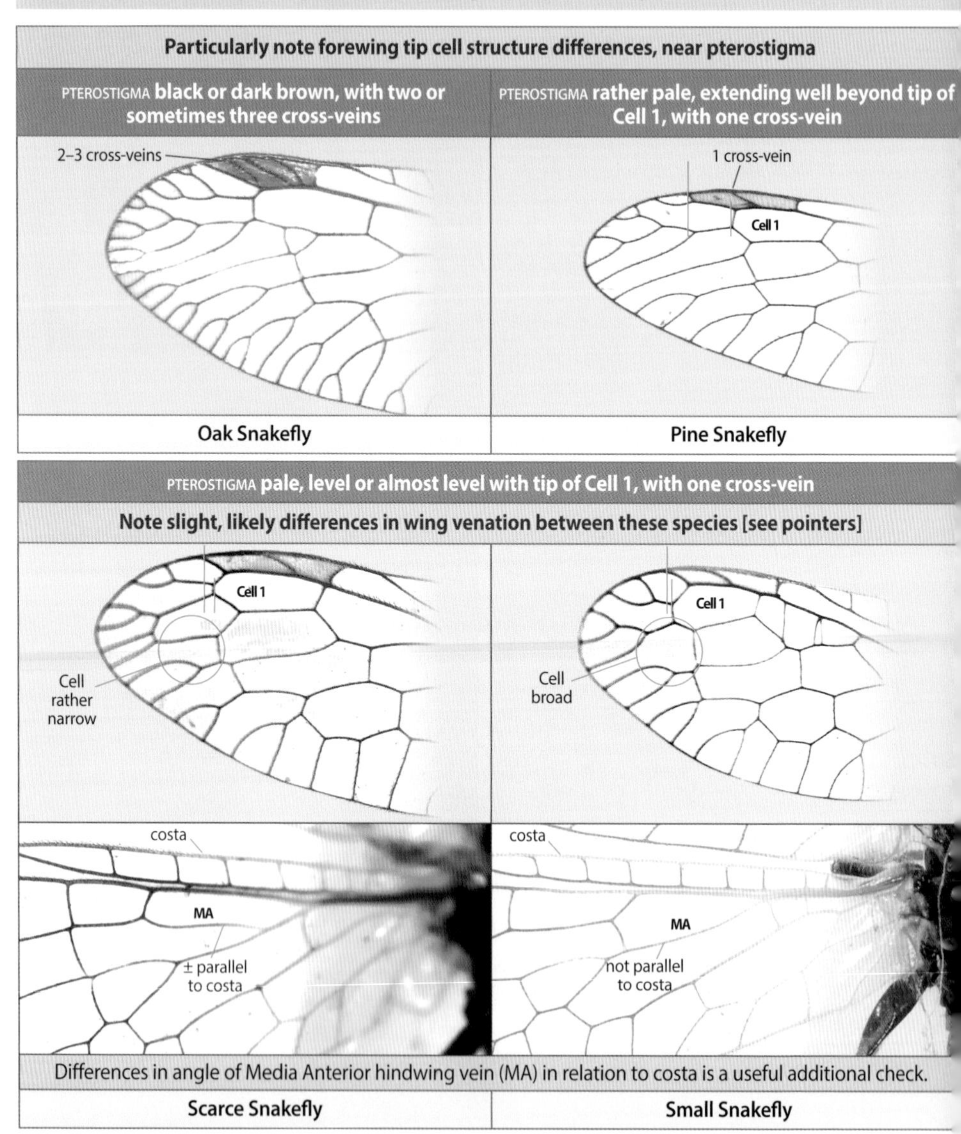

Particularly note forewing tip cell structure differences, near pterostigma
PTEROSTIGMA black or dark brown, with two or sometimes three cross-veins
2–3 cross-veins
Oak Snakefly
PTEROSTIGMA rather pale, extending well beyond tip of Cell 1, with one cross-vein
1 cross-vein
Cell 1
Pine Snakefly
PTEROSTIGMA pale, level or almost level with tip of Cell 1, with one cross-vein
Note slight, likely differences in wing venation between these species [see pointers]
Cell 1
Cell rather narrow
costa
MA
± parallel to costa
Cell 1
Cell broad
costa
MA
not parallel to costa
Differences in angle of Media Anterior hindwing vein (MA) in relation to costa is a useful additional check.
Scarce Snakefly
Small Snakefly

FURTHER READING AND USEFUL WEBSITES

Plant, C.W. 1997. ***A key to the adults of British lacewings and their allies (Neuroptera, Megaloptera, Raphidioptera and Mecoptera)***. Field Studies Council (AIDGAP).

http://lacewings.myspecies.info/ Lacewings and Allies Recording Scheme

Atlantoraphidia 1 sp.

Pine Snakefly
Atlantoraphidia maculicollis

Common

BL 9–10 mm, **WS** approx. 20 mm. **ID** Black with reddish-brown legs and antennae; also reddish marks on pronotum and abdomen, the latter with yellow marks and spots. Pterostigma brown, with one cross-vein. Femora much darker in some individuals. **Hab** Coniferous and deciduous woodland; associated with pines, larches and oaks. **SS** Other snakeflies [see *opposite*].

Phaeostigma 1 sp.

Oak Snakefly
Phaeostigma notata

Common

BL 10–15 mm, **WS** 20–30 mm. **ID** Largest British snakefly. Black; tibiae and part of tarsi reddish-brown; abdomen with yellow marks and spots. Pterostigma dark brown or black, with two or three cross-veins. Wing veins black; forewing with a number of narrow veins which are forked at tips. The outer wing vein (costa) has 12–15 small veins reaching the next vein (subcostal). **Hab** Woodland and gardens; associated with oaks. **SS** Other snakeflies [see *opposite*].

Xanthostigma 1 sp.

Small Snakefly
Xanthostigma xanthostigma

Local

BL 7–9 mm, **WS** 15–20 mm. **ID** Black. Pterostigma yellowish-brown, with one cross-vein; pterostigma level or almost level with tip of Cell 1. Wing veins brownish in part. **Hab** Oak woodland. **SS** Other snakeflies [see *opposite*].

Subilla 1 sp.

Scarce Snakefly
Subilla confinis

Local

BL 7–10 mm, **WS** 15–20 mm. **ID** Black. Pterostigma yellowish, with one cross-vein; pterostigma level with tip of Cell 1. Wing veins black. **Hab** Woodland; tree associations require research. **SS** Other snakeflies [see *opposite*].

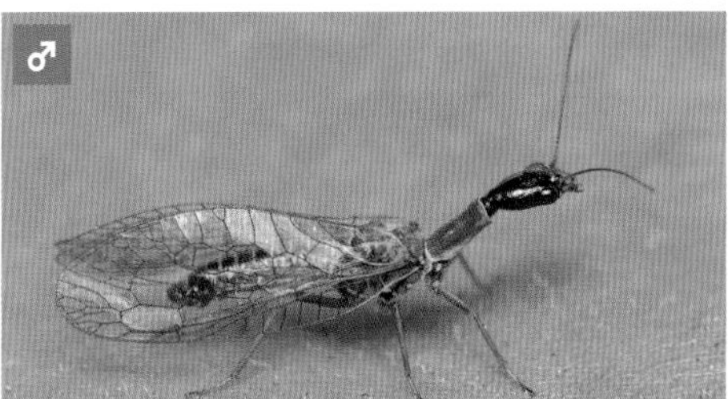

ORDER DIPTERA Flies

[Greek: *dis* = twice; *pteron* = wing]

BI | 109 families [*c.* 1,490 genera, ***c.* 7,100 species**] **W** | 188 families [***c.* 160,600 species**]

Some flies are tiny (less than 1 mm), but larger species of cranefly have a wingspan up to 60 mm. This successful group is the second largest order in Great Britain and Ireland and easily recognized as they only have one pair of wings (the name of the order literally means 'two wings'); the hindwings are modified into pin-shaped halteres, used for balance when in flight. Only a few flies are wingless. The eyes are large and compound (the space between them often broader in females), with three ocelli in between. The antennae are often rather short, with an arista (a bare or hairy bristle-like part) on the third segment. The mouthparts are designed to suck moisture or nectar, sometimes for piercing surfaces to suck blood (some flies have a bad reputation for biting or spreading disease). Legless larvae, known as maggots, are still used in some hospitals to clean wounds. Some dipterists (people who study flies) specialise in particular families, the most popular among which are larger species such as the biting horseflies, soldierflies, bee-flies, robberflies and hoverflies. Some other groups are rather neglected, in part because species may be difficult to identify.

Due to space constraints and the complexity of studying flies (a powerful microscope is needed for many families), this summary highlights some of the commoner, larger, or more interesting flies, beginning with a photo section covering selected families in taxonomic order but without describing the intricate differences in wing venation pattern or other characters that dipterists use for specific identification. Even so, many larger species are easy to place to family or even species level. Most of the families that have been omitted are small less popular insects, mainly of interest to specialists.

FINDING FLIES Plentiful in many habitats, including wetlands. Grassland and gardens full of flowers will attract a range of flies.

FURTHER READING AND USEFUL WEBSITES

Ball, S., & Morris, R. 2015. ***Britain's Hoverflies – A field guide***. Princeton University Press (Princeton WILD*Guides*). *Complements this field guide and covers identification of adult hoverflies in a particularly user-friendly format.*

Chandler, P. J. (ed.) 2010. ***A Dipterist's Handbook***. The Amateur Entomologist Vol. 15. The Amateur Entomologists' Society.

Stubbs, A. E. & Drake, M. 2014. ***British Soldierflies and their Allies***. British Entomological and Natural History Society.

Stubbs, A. E. & Falk, S. 2002. ***British Hoverflies***. British Entomological and Natural History Society.

dipterists.org.uk The Society for the study of flies (Diptera)

A BLUEBOTTLE AND GREENBOTTLES ON A DEAD BIRD

Structure of a fly

ABDOMEN: number of segments varies between genera – for example, craneflies (see *p. 333*) have 7 segments + an ovipositor in the female.

ORDER Diptera VISUAL GUIDE TO SELECTED FAMILIES | 52 of 109 BI

This guide omits minor families with little-known, often tiny British or Irish species that are either unlikely to be seen or need microscopic examination of features to confirm identification. Shape and a range of characters are important in identification, even for larger species. Families depicted by more than one species are listed alphabetically below. Length bars in the species accounts indicate forewing length (≈ ½ wingspan).

MEASUREMENTS

NOTE: most dipterists use forewing length, but some use body length; references to size in the literature can be misleading as it sometimes refers to the European fauna. In the following section, forewing length (**FWL**) is used to indicate size in the following categories:

small 1–5 mm
medium-sized 5–10 mm
large 10–15 mm
very large >15 mm

Family		Suborder	No. of genera	No. of species	Covered	Family Guide	Main accounts
Asilidae	**Robberflies**	Brachycera	16	29	6	*337*	***360***
Bibionidae	**St Mark's flies**	Nematocera	2	18	2	*335*	***347***
Bombyliidae	**Bee-flies**	Brachycera	5	10	6	*336*	***349***
Calliphoridae	**Blowflies**	Brachycera	12	29	4	*342*	***363***
Muscidae	**Houseflies and allies**	Brachycera	39	291	3	*342*	***362***
Ptychopteridae	**Fold-winged craneflies**	Nematocera	1	7	2	*334*	***347***
Stratiomyidae	**Soldierflies**	Brachycera	16	48	8	*336*	***351***
Syrphidae	**Hoverflies**	Brachycera	68	282	27	*338*	***353***
Tabanidae	**Horseflies**	Brachycera	5	30	4	*336*	***348***
Tachinidae	**Parasitic flies**	Brachycera	145	267	5	*342*	***364***
Tipulidae	**Long-palped craneflies**	Nematocera	6	87	9	*333*	***343***

Black-horned Gem
Microchrysa polita
(FWL 4–5 mm)
[Stratiomyidae (*p. 351*)]

Flecked General
Stratiomys singularior
(FWL 9–12 mm)
[Stratiomyidae (*p. 351*)]

Small Black Tachinid
Phania funesta **(FWL 3 mm)**
[Tachinidae (*p. 364*)]

Yellow-tailed Hoverfly
Xylota sylvarum
(FWL 7–12 mm)
[Syrphidae (*p. 353*)]

Giant Cranefly *Tipula maxima* **(FWL 22–30 mm)**
[Tipulidae (*p. 343*)]

The images above, all shown at life-size, highlight the range in variation of size and shape among flies, even within the same family.

ORDER Diptera — GUIDE TO SUBORDERS

SUBORDER Nematocera (Lower Diptera)

See *p. 334* for example families

Long palps, elongated body and long antennae (≥6-segmented). A total of 25 families, including the well-known craneflies or 'daddy-long-legs'.

Example shown from Tipulidae: note forewing anal veins are separated. In some other families there are fewer veins.

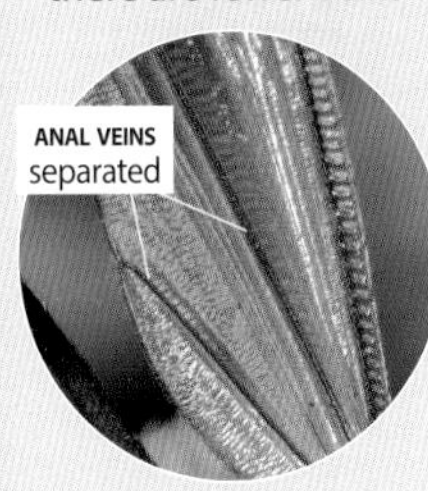

SUBORDER Brachycera

See *pp. 336–343* for example families

Short palps, usually plump body and short antennae (usually <5 segments) bearing an arista. A total of 84 families, including the well-known houseflies. Traditionally ordered with the larger, better-studied flies in various sub-groups of closely related families (superfamilies), followed by typically smaller species in a sub-group known as Schizophora, which itself is sub-divided into Acalyptrates and Calyptrates by minor characters. Identification may require microscopic evaluation until the user is familiar with the families (these divisions are highly technical, *e.g.* differences in wing venation and genitalia and beyond the scope of this book).

Example shown from Syrphidae: note forewings with anal veins merged to form a cell (a useful check on flies with ≥5 antenna segments). Forewings have several veins, which vary in number from family to family.

MIGRANT COROLLAE
Eupeodes corollae (hoverfly)

SUBORDER Nematocera (Lower Diptera)

Craneflies as a group

LIFESTYLE larvae are terrestrial, aquatic or semiaquatic, developing in rotten vegetable matter or feeding on small particles. Some *Tipula* (*p. 344*) feed on grasses and young plants and can be a pest, occurring in large numbers.

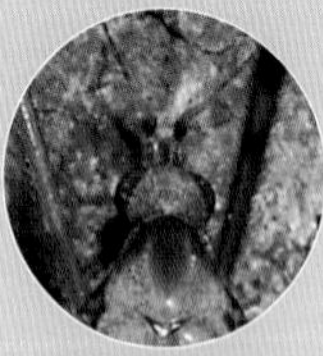

short-palped

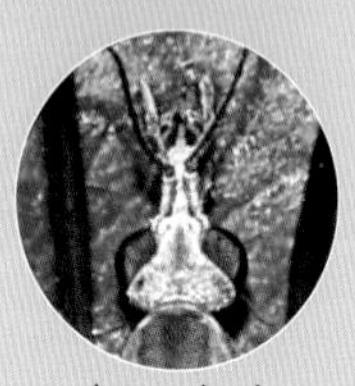

long-palped

FORM small to large (**FWL** 2–21 mm). **BODY** elongated, **PALPS** shorter than length of head; **ANTENNAE** usually 14–16-segmented. **WINGS** of some species with characteristic markings. **EYES** bare.

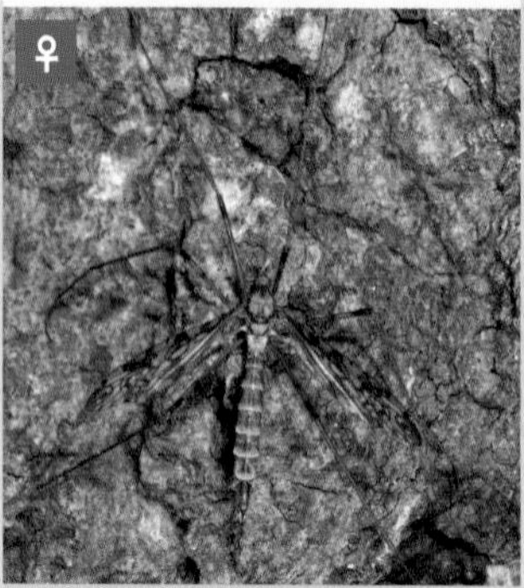

Eyed Cranefly *Epiphragma ocellare* (FWL approx. 12 mm)

Short-palped craneflies Limoniidae

[50 genera | 223 species]

FORM large to very large (**FWL** 7–30 mm). (NB ♀ short-winged in a few species.) **BODY** elongated; **ROSTRUM + PALPS** long (latter may be curved); **ANTENNAE** 13–14-segmented. **WINGS** usually plain or with few markings. **EYES** bare.

September Cranefly (FWL 13–23 mm)

Long-palped craneflies Tipulidae

[6 genera | 87 species]

p. 343

FORM medium-sized to very large (**FWL** 5–25 mm, **BL** up to 30 mm). **BODY** elongated, **PALPS** shorter than length of head; **ANTENNAE** mainly 12–17-segmented. **WINGS** often with characteristic pattern. **EYES** **hairy** (best seen under a microscope, might just be seen with good 20× hand lens).

Painted Cranefly *Pedicia rivosa* (FWL 20–24 mm)

Hairy-eyed craneflies Pediciidae

[4 genera | 20 species]

FORM medium-sized to large (**FWL** 6–12 mm). **BODY** elongated (plump in females), black with some coloured markings; **ANTENNAE** mainly 15-segmented; **WINGS** often with black markings.

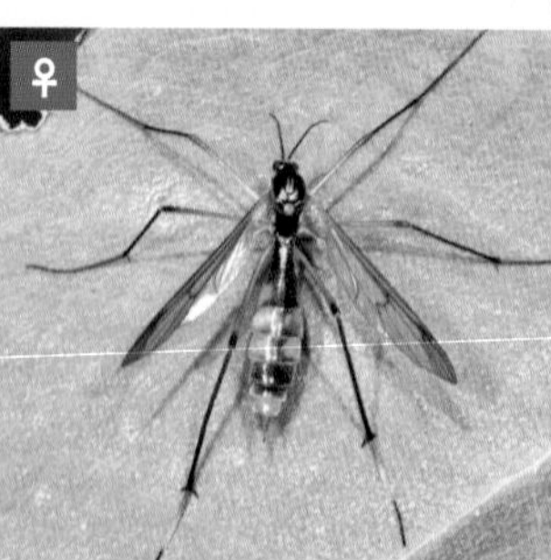

White-legged Cranefly (BL 11 mm)

Fold-winged craneflies Ptychopteridae

[1 genus | 7 species]

p. 347

FORM small to medium-sized, stout (**FWL** 2–6 mm). **BODY + WINGS** generally hairy; **ANTENNAE** usually 12–16-segmented. **LEGS** short to long. **WINGS** plain or marked; **held as shown or roof-like at rest**. **LIFESTYLE** larvae are terrestrial or semiaquatic, developing in various habitats, even drains.

A moth fly *Pericoma* **sp.**

Moth flies Psychodidae

[22 genera | 100 species]

FORM small to medium-sized (FWL 3–9 mm) **robust-looking, but legs long and delicate.** MOUTHPARTS long, piercing in ♀; ANTENNAE 15-segmented; WINGS, LEGS & ABDOMEN (PART) **covered in scales.** LIFESTYLE larvae frequent stagnant water, feeding mainly on detritus. Adult ♀s feed on blood of humans and livestock, and are vectors of diseases.

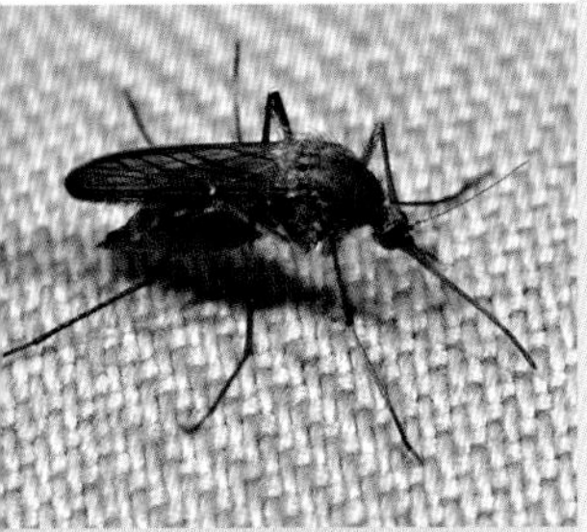

Common House Mosquito
Culex pipiens (BL 6 mm)

Mosquitoes Culicidae

[8 genera | 35 species]

FORM small, slender to broad (FWL 1–5 mm). MOUTHPARTS able to pierce in most species; ANTENNAE 13–15-segmented. WINGS patterned in some species. LIFESTYLE larvae are aquatic, developing in detritus or as predators. Adults feed on blood of mammals and can be a nuisance, none more so than the Highland Midge in Scotland.

Highland Midge
Culicoides impunctatus (BL 1 mm)

Biting midges Ceratopogonidae

[20 genera | 171 species]

FORM small to medium-sized, slender (FWL 1–10 mm). MOUTHPARTS **not elongated, or able to pierce**; ANTENNAE 13–17-segmented, **very feathery in ♂.** WINGS shorter than body, mainly clear. LIFESTYLE larvae are terrestrial or semiaquatic, developing in various habitats, even in domestic drains.

Pale Midge
Chironomus luridus (BL 10 mm)

Non-biting midges Chironomidae

[145 genera | 628 species]

FORM small to medium-sized (FWL 4–9 mm), slender and delicate to stout; ANTENNAE 16–17- segmented; WINGS variable between species, clear or marked. LIFESTYLE larvae are found in moist woodland.

Orange-banded Fungus Gnat
Platyura marginata
(BL 10–12 mm, FWL 7–9 mm)

Fungus gnats Keroplatidae

[15 genera | 51 species]

FORM small to medium-sized (FWL 1–6 mm). BODY black or in a few cases yellowish. ANTENNAE long, 16-segmented; WINGS clear or tinged. LIFESTYLE larvae are mainly terrestrial in wet habitats, where they feed on mycelium.

Yellow-bodied Black Fungus Gnat
Sciara hemerobioides
(BL 5–6 mm)

Black fungus gnats Sciaridae

[22 genera | 267 species]

FORM small to large (FWL 2–12 mm). BODY **stout, hairy; brown/black (some partly red or yellow).** EYES large in ♂; ANTENNAE **short, thickened**, 6–12-segmented, inserted low down; LEGS **fore tibiae swollen, with strong spurs, or spines.** LIFESTYLE larvae develop in decaying wood; they feed on rotting plant material.

Common Fever Fly
(BL 7–10 mm)

St Mark's flies Bibionidae

[2 genera | 18 species]

p. 347

SUBORDER Brachycera

FORM medium-sized to large (**BL** 8–17 mm, **FWL** 7–11 mm). **BODY** elongated, often black with yellow areas; non-bristly. **LIFESTYLE** larvae live under tree bark or in rot holes, where they prey on invertebrates.

Common Awl-fly
Xylophagus ater (BL 8–15 mm)

Awl-flies
Xylophagidae
[1 genus | 3 species]

FORM small to large (**BL** 2–16 mm, **FWL** 2–14 mm). **BODY** elongated, yellow to black; non-bristly. **WINGS** **with a dark stigma spot, some species with other distinct black marks**. **LIFESTYLE** larvae develop in moist soil, moss, or in decaying wood (depending on species), where they prey on invertebrates.

Downlooker Snipefly
Rhagio scolopaceus (BL 8–16 mm)

Snipeflies
Rhagionidae
[5 genera | 15 species]

FORM medium-sized to large (**BL** 7–24 mm, **FWL** 6–20 mm). **BODY** broad, lacking bristles; often yellow, brown or black; in a few cases metallic. **EYES** **very large with 1 or 2 coloured lines in most species**. **WINGS** clear or tinged. **LIFESTYLE** adults feed on mammal blood, including humans and can transmit diseases; larvae develop in soil, on land or underwater, where they prey on invertebrates.

Dark Giant Horsefly
(BL 21–24 mm)

Horseflies
Tabanidae
[5 genera | 30 species]

p. 34

FORM medium-sized (**BL** 6–9 mm, **FWL** 5–9 mm). **BODY** **robust**, often dark with paler areas, non-bristly. **LIFESTYLE** larvae live under tree bark, feeding on invertebrates.

Drab Wood-soldierfly
Solva marginata (BL 6–7 mm)

Wood-soldierflies
Xylomyidae
[2 genera | 3 species]

FORM small to large (**BL** 2–15 mm, **FWL** 2–12 mm). **BODY** **slender to broad**, often strikingly coloured, some spotted or metallic, non-bristly; **WINGS** mainly clear. **LIFESTYLE** larvae are aquatic or terrestrial, feeding on algae or decaying vegetable matter.

Four-barred Major
Oxycera rara (BL 7mm)

Soldierflies
Stratiomyidae
[16 genera | 48 species]

p. 351

FORM small to large (**BL** and **FWL** 3–14 mm). **BODY** **broad and bee-like**, brown, greyish or black. **HEAD** **with long proboscis in some**. **EYES** large; **ANTENNAE** **3–4-segmented**. **THORAX & ABDOMEN** hairy or with scales; **WINGS** usually with dark patterning. **LIFESTYLE** adults often seen nectaring; larvae are parasitoids of solitary bees and wasps.

Dark-edged Bee-fly
(♀ egglaying)

Bee-flies
Bombyliidae
[5 genera | 10 species]

p. 34

FORM **small to very large** (**BL** 8–28 mm, **FWL** 6–18 mm). **BODY** slender to robust, **often greyish to dark, some with paler legs or abdomen, very bristly** (some species hairier than others); **THORAX** **often with distinctive darker marks**; **LEGS** elongated, very bristly. **LIFESTYLE** adults often seen eating large insect prey; larvae develop in soil and dead wood, feeding on invertebrates.

Dune Robberfly (BL 13–18 mm)

Robberflies
Asilidae

[16 genera | 29 species]

p. 360

FORM **small to large** (**BL** 7–13 mm, **FWL** 5–9 mm). **BODY** slender to more robust, **yellow, silvery to dark, some covered in a fine down**; **THORAX & LEGS** bristly. **LIFESTYLE** larvae develop in sandy soil and dead wood, feeding on invertebrates.

Coastal Silver-stiletto
Acrosathe annulata (BL 8–11 mm)

Stiletto flies
Therevidae

[6 genera | 14 species]

FORM **small to medium-sized** (**FWL** 1–7 mm). **BODY** slender, often greenish and metallic; **WINGS** clear or tinged, in some species with bold spots or marks. **THORAX & LEGS** bristly. **LIFESTYLE** larvae develop in moist soil and under tree bark, feeding on invertebrates.

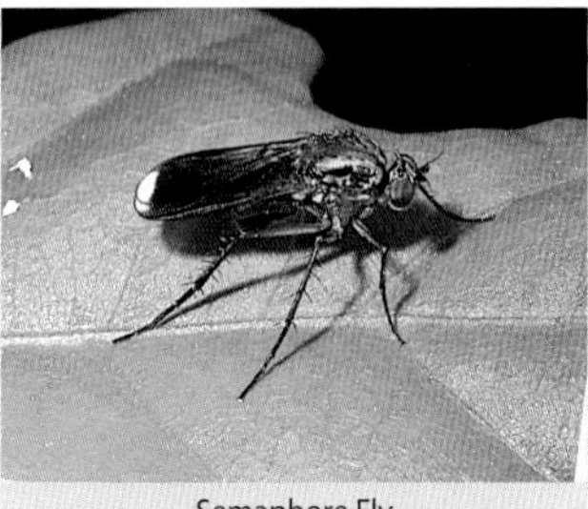

Semaphore Fly
Poecilobothrus nobilitatus (BL 7 mm)

Long-legged flies
Dolichopodidae

[46 genera | 305 species]

FORM **small to large** (**FWL** 1–14 mm). **BODY** slender to robust, light to dark; **THORAX & LEGS** bristly, some species with striped thorax; **WINGS** clear or tinged, some with stigma spot or pattern. **LIFESTYLE** larvae develop in a wide range of habitats, including aquatic, feeding mainly on Diptera larvae.

Common Dance Fly
Empis tessellata ♀ (BL 9–13 mm)

Dance flies
Empididae

[15 genera | 208 species]

FORM **small to medium-sized** (**FWL** 2–6 mm). **BODY** slender to robust, with sexual dimorphism frequent, males often dark; some species bright, including yellow and orange; **LEGS** short and stout. **LIFESTYLE** larvae develop in fungi.

Yellow Flat-footed Fly
Agathomyia wankowiczii (BL 3 mm)

Flat-footed flies
Platypezidae

[10 genera | 34 species]

FORM **small to medium-sized** (**FWL** 2–10 mm). **BODY** slender (**♀ with lance-like ovipositor**); **WINGS** long; **EYES** **large**. **LIFESTYLE** larvae are parasitoids of small Hemiptera.

A big-headed fly
Jassidophaga sp. (BL 4 mm)

Big-headed flies
Pipunculidae

[14 genera | 99 species]

SUBORDER Brachycera

FORM small to large (**FWL** 3–20 mm). **BODY** slender to robust, **often dark but many species with bright spots or marks; some pale or metallic**; **HEAD large, lacking bristles**. Although most have a distinctive appearance, check the wings for a spurious vein ('vena spuria') [can be faint] and 'false margin', which distinguishes them from other flies.
LIFESTYLE larvae develop in various habitats from gardens and parks, to grassland and woodland.

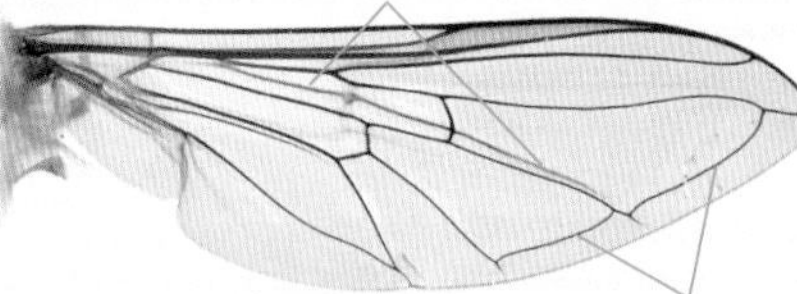

Wing of *Syrphus ribesii* showing the vena spuria and false margi[n] diagnostic of British hoverflies.

Marmalade Hoverfly (BL 6–10 mm)

Small-headed Hoverfly (BL 11–13 mm)

Common Twist-tailed Hoverfly (BL 5–7 mm)

Hoverflies
Syrphidae
[68 genera | 283 species]

p. 35

Acalyptratae 'Acalyptrates'

1/

FORM small to large (**FWL** 4–14 mm); slender to robust. **BODY** often yellow-and-black (rather wasp-like) or reddish-brown, not bristly, hairs often short; **HEAD & MOUTHPARTS large**.
LIFESTYLE larvae are parasitoids of Hymenoptera.

Common Conopid *Sicus ferrugineus* (mating pair) (BL 8–13 mm)

Thick-headed flies
Conopidae
[7 genera | 24 species]

FORM small (**FWL** 1–4 mm); slender to robust. **BODY** often yellow- or green-and-black, with few bristles; **HEAD large, with large triangular area surrounding ocelli**.
LIFESTYLE larvae include gall-forming agricultural pests of cereals.

Ladder Grass Fly *Chlorops scalaris* (BL 2·5 mm)

Grass flies
Chloropidae
[40 genera | 178 species]

FORM small to medium-sized (**FWL** 1–5 mm); robust, dark, with some bristles. **WINGS** mainly clear, in some species dark or spotted; **LEGS** in some species **forelegs broadened and adapted for catching prey** (*e.g.* Mantis Fly).
LIFESTYLE aquatic/semiaquatic.

Mantis Fly *Ochthera mantis* (BL 5 mm)

Shore flies
Ephydridae
[23 genera | 151 species]

Acalyptratae 'Acalyptrates' 2/3

FORM small to medium-sized (**FWL** 1–5 mm). **BODY** robust, yellow to dark, thorax and occasionally abdomen with some markings, including stripes and spots; **EYES** usually red, usually fairly bristly. **LIFESTYLE** larvae develop in rotting fruit and fungi; often kept in the laboratory for scientific research.

EN **Variegated Fruit Fly** *Phortica variegata* ♀ **(BL 3 mm)** [RDB1]; disease vector of *Thelazia* species (eye worm); feeds on tears or liquids around the eyes of dogs, sometimes humans.

Fruit flies
Drosophilidae

[13 genera | 65 species]

FORM small (**FWL** 3–4 mm). **BODY** robust, dark, often with small greyish hairs, usually bristly; **WINGS** clear or with dark marks. **LIFESTYLE** the larvae are thought to be associated with rotting wood.

Patched Diastatid
Diastata nebulosa
(BL 3 mm)

Diastatids
Diastatidae

[1 genus | 6 species]

FORM small to medium-sized (**FWL** 3–6 mm). **BODY** plump, dull or partly shining, yellow to black, in some species two-coloured, bristly; **WINGS** clear in almost all species. **LIFESTYLE** mainly associated with fungi.

Orange Lauxaniid
Meiosimyza rorida
(BL 3–4 mm)

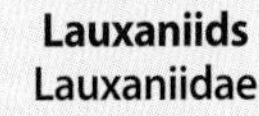

Lauxaniids
Lauxaniidae

[13 genera | 56 species]

FORM small to large (**FWL** 4–11 mm). **BODY** elongated, yellow to black, in some species two-coloured, with few bristles; **LEGS** long, stilt-like; **WINGS** clear or cloudy. **LIFESTYLE** larvae of some species develop in rotting wood, others in plant roots.

EN **Beech Échasseur** *Rainieria calceata* **(BL *c.* 10 mm)** [RDB1]; mainly Windsor Forest, Berkshire

Stilt-legged flies
Micropezidae

[5 genera | 10 species]

FORM small to medium-sized (**FWL** 2–6 mm). **BODY** slender, dull or partly shining, yellow or orange to black, bristly; **WINGS** clear to brownish. **LIFESTYLE** larvae of at least some species develop in rotting wood, in mines made by other insects.

Yellow Clusiid
Clusia flava
(BL 4–7 mm)

Clusiids
Clusiidae

[3 genera | 10 species]

FORM medium-sized (**FWL** 6–8 mm). **BODY** elongated, slightly shining black; **WINGS** pale brown; **LEGS** reddish, hind femora swollen; **HEAD** with silvery hairs but lacks bristles as elsewhere. **HABITAT** ancient oak woodland. **LIFESTYLE** larvae live under dead tree bark, feeding on invertebrates.

Bearded Fool
Megamerina dolium **(mating pair)**
(BL 6–8 mm) [Nationally Scarce]

Megamerinids
Megamerinidae

[1 genus | 1 species]

SUBORDER Brachycera

Acalyptratae 'Acalyptrates'

FORM small (**FWL** 2–5 mm). **BODY** slender, pale to black with some bristles; **WINGS** tips spotted, other spots often present. **LIFESTYLE** larvae develop in stems of grasses and a few species are agricultural pests.

Three-spotted Opomyzid
Geomyza tripunctata
(BL 4 mm)

Opomzyid flies
Opomyzidae
[2 genera | 16 species]

FORM small to medium-sized (**FWL** 2–7 mm). **BODY** slender, pale to black, with some bristles; **WINGS** much longer than abdomen, often spotted, rarely clear. **LIFESTYLE** larvae develop mainly in plants, although some species feed on mycelium or live under bark and prey on beetle larvae.

Four-spotted Fly
Palloptera umbellatarum
(BL 3 mm)

Long-winged flies
Pallopteridae
[2 genera | 13 species]

FORM medium-sized to large (**FWL** 5–10 mm). **BODY** yellow, orange or brown, bristly; **WINGS** clear or tinged, in some species with dark bands. **LIFESTYLE** larvae develop in fungi and carrion.

Orange Fungi Fly
Dryomyza analis
(BL 5 mm)

Dryomyzid flies
Dryomyzidae
[2 genera | 3 species]

FORM small to large (**FWL** 2–10 mm). **BODY** often greyish or dark, bristly; **WINGS** clear or with dark markings, including extensive spotting. **LIFESTYLE** larvae mainly feed on snails and slugs.

Mosaic Snail-killing Fly
Coremacera marginata
(BL 7–10 mm)

Snail-killing flies
Sciomyzidae
[23 genera | 72 species]

FORM small to medium-sized (**FWL** 2–6 mm). **BODY** generally black, ant-like and glossy, in some species with short silvery hairs; **WINGS** clear but often black-tipped. **LIFESTYLE** larvae develop in dung and fungi. **Most likely on leaves, waving their wings**.

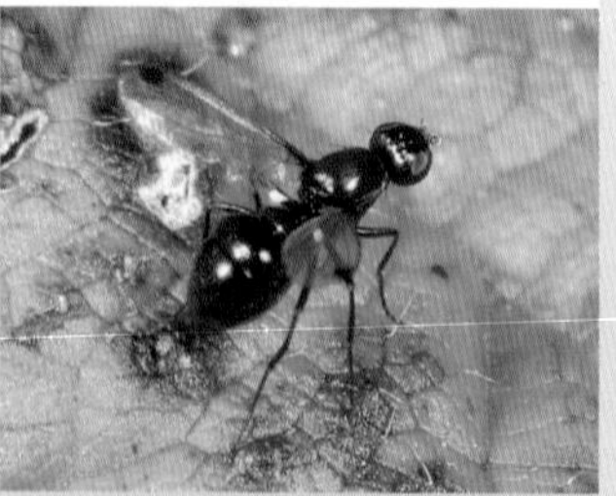

Common Black Scavenger
Sepsis cynipsea
(BL 2–3 mm)

Black scavenger flies
Sepsidae
[6 genera | 29 species]

FORM small to large (**FWL** 2–10 mm). **BODY** slender or robust, yellow to brown or black, with some bristles; **WINGS** clear or slightly tinged. **LIFESTYLE** larvae mainly feed in roots and stems of plants.

Black-faced Rust Fly
Loxocera aristata
(BL 8–13 mm)

Rust flies
Psilidae
[6 genera | 27 species]

FORM small to medium-sized (**FWL** 3–6 mm). **BODY** stout, hairy, mainly metallic bluish-black (female with lance-shaped ovipositor); **WINGS** usually clear. **LIFESTYLE** larvae develop in dead wood or some species on vegetable matter.

A lance-fly
Lonchaea sp.
(BL 3 mm)

Lance-flies
Lonchaeidae
[5 genera | 48 species]

FORM small to medium-sized (**FWL** 2–9 mm). **BODY** pale to black (female with oviscape). **WINGS** pictured in some species (helps identification to species level). **Technical characters to separate from Tephritidae require microscopic examination.** **LIFESTYLE** larvae develop in decaying vegetable matter.

Waved Picture-winged Fly
Herina frondescentiae
(BL 3 mm)

Picture-winged flies
Ulidiidae
[11 genera | 20 species]

FORM medium-sized (**FWL** 5–8 mm). **BODY** heavily mottled (female with oviscape); **WINGS** mottled or pictured; **EYES** reddish-brown. **LIFESTYLE** larvae develop in decaying vegetable matter.

Speckled Signal Fly
Platystoma seminationis
(BL 6–8 mm)

Signal flies
Platystomatidae
[2 genera | 2 species]

FORM small to medium-sized (**FWL** 2–7 mm). **BODY** often colourful (female with oviscape); **WINGS** pictured with characteristic markings (helps identification to species level – in the field, identification of ≥25% of species possible). **LIFESTYLE** larvae develop in various parts of plants.

Spear Tephritid
Urophora stylata (mating pair)
(BL 5–7 mm, FWL 4–5 mm)

Tephritids
Tephritidae
[37 genera | 77 species]

(Known by some as 'Picture-winged Flies', although the name clashes with Ulidiidae. The name 'Fruit-flies' is used outside Britain and Ireland, but that clashes with Drosophilidae.)

SUBORDER Brachycera

Calyptratae 'Calyptrates' 1/2

FORM small to medium-sized (**FWL** 2–9 mm). **BODY** grey to black, some brighter (such as species illustrated below), with few bristles. **LIFESTYLE** larvae feed on decaying organic materials and some species are a nuisance in houses.

Orange-legged Lesser Housefly *Fannia lustrator* (BL 6–9 mm)

Lesser houseflies Fanniidae

[2 genera | 61 species]

FORM small to large (**FWL** 2–12 mm). **BODY** grey to black, some paler; **THORAX** often strongly marked, with some bristles. **LIFESTYLE** larvae feed on decaying organic materials (some are aquatic); several species are a nuisance in houses and around livestock, and may be vectors of diseases.

Stable Fly (BL 8 mm)

Houseflies and allies Muscidae

[39 genera | 293 species]

p. 362

FORM medium-sized to large (**FWL** 9–18 mm). **BODY** broad, rather bee-like; **THORAX** covered in fine, soft hair (lacking bristles). **LIFESTYLE** larvae are well known for developing internally in the nostrils or guts of mammals, attacking specific animal species.

♀

Horse Bot Fly *Gasterophilus intestinalis* (BL 10–15 mm) [Nationally Scarce]

Warble flies and bot flies Oestridae

[5 genera | 11 species]

FORM small to large (**FWL** 4–14 mm); robust. **BODY** often metallic blue or green and bristly – the familiar bluebottles and greenbottles; **WINGS** usually clear, but in some species dark-veined or marked. **LIFESTYLE** larvae develop in carrion and are important in forensic entomology; adults are often a nuisance in houses and buildings, but some just use them for overwintering sites.

Yellow-faced Blowfly (BL 7–15 mm)

Blowflies Calliphoridae

[12 genera | 29 species]

p. 363

FORM small to large (**FWL** 3–15 mm); rather stout and bristly. **BODY** most species are grey to greyish-black (**eyes red**); **THORAX** black-striped; **ABDOMEN** chequered (**but never metallic blue or black – see Calliphoridae**). **LIFESTYLE** larvae of some species are parasitoids of Hymenoptera and others.

Flesh Fly *Sarcophaga carnaria* (BL 10–15 mm)

Flesh flies Sarcophagidae

[16 genera | 62 species]

FORM small to large (**FWL** 3–20 mm); stout or elongate, often bristly, some with fine soft hairs or bare. **BODY** most species are grey to greyish-black but some are lighter coloured, occasionally with spots on abdomen. **ANTENNAE** 3-segmented with hair-like arista. **LIFESTYLE** mainly internal parasites of insects, larvae commonly developing in Lepidoptera and Hemiptera larvae.

♂

Blue-winged Tachinid (BL 8–12 mm)

Parasitic flies Tachinidae

[145 genera | 268 species]

p. 36

Calyptratae 'Calyptrates' 2/2

FORM medium-sized (**FWL** 2–10 mm); atypical flies, usually with a flat, broad body. **WINGS** short in some species or may be shed once attached to a host; **TARSI** strong, hooked. **LIFESTYLE** adults are ectoparasites, feeding on blood of mostly birds but also mammals; larvae develop in the female fly.

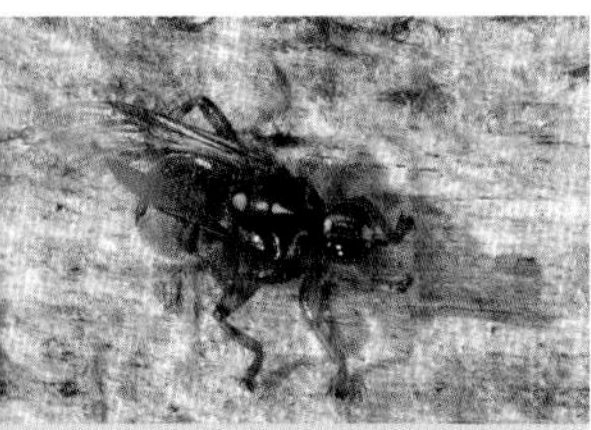

Forest Fly *Hippobosca equina* (BL 6–9 mm) [RDBk]

Flat flies
Hippoboscidae
[10 genera | 14 species]

FORM small to medium-sized (**FWL** 4–8 mm). **BODY** brown, grey or black, some with thorax and abdomen distinctly marked; **WINGS** usually clear, rarely marked. **LIFESTYLE** depending upon species, the larvae develop in dung, fungi or birds' nests, whilst others mine leaves.

Handsome Anthomyiid *Anthomyia procellaris* (BL 5–7 mm)

Anthomyiids
Anthomyiidae
[26 genera | 246 species]

FORM small to medium-sized (**FWL** 3–9 mm). **BODY** slender, yellow, yellowish-brown, or black, occasionally two-coloured, bristly; **LEGS** bristly; **WINGS** usually clear, but dark-veined or marked in some species. **LIFESTYLE** larvae of some species mine leaves, others develop in plants, or feed on prey in dung.

Yellow Dungfly *Scathophaga stercoraria* (BL 8–10 mm)

Dungflies
Scathophagidae
[23 genera | 55 species]

SUBORDER Nematocera [Lower Diptera]

FAMILY Tipulidae (Long-palped Craneflies) 8 GEN. | 87 spp. | 9 ILL.

Very distinctive-looking flies with long palps and legs, abdomen elongated and wings (usually) long, plain or patterned. Known to many people as 'daddy long legs'.

♂ ANTENNAE large, feathery	**BODY** black, orange and yellow; ♂ **ANTENNAE** 4 appendages on each segment.	*Ctenophora* [3 spp.]	p. 346
	BODY black-and-yellow; **OVIPOSITOR** (♀s) particularly elongated; ♂ **ANTENNAE** 3 appendages on each segment.	*Tanyptera* [2 spp.]	p. 344
	BODY black or black-and-orange; ♂ **ANTENNAE** 2 appendages on each segment.	*Dictenidia* [1 sp.]	p. 346
BODY brown; **ABDOMEN** often blackish; **TARSI** white.		*Dolichopeza* [1 sp.]	N/I
BODY and **WINGS** black.		*Nigrotipula* [1 sp.]	N/I
ABDOMEN banded/marked with yellow and black.		*Nephrotoma* [15 spp.]	p. 346
ANTENNAE segments swollen (unique).		*Prionocera* [3 spp.]	N/I
BODY often plain brown or grey, or **WINGS** with some colouring; **OVIPOSITOR** (♀s) normal (not elongated).		*Tipula* [61 spp.]	p. 344

Guide to Tipulidae p. 34

Tanyptera 2 spp.

Giant Sabre Comb-horn *Tanyptera atrata*

Nationally Scarce

FWL 15–19 mm. **ID** Black, abdomen with large red band (base of abdomen always red). ♂ abdomen often with red sides, but some individuals almost all-black. ♀ ovipositor sabre-like. Legs orange; femora in ♂ black-tipped (in ♀ hind femora only). Wings clear with brown stigma. **Hab** Woodland and heathland. **SS** The rare (Red Listed) **Lesser Sabre Comb-horn** *Tanyptera nigricornis* (**FWL** 13–19 mm) [*bottom*] has black (not red) trochanters; ♀ all femora black-tipped.

LESSER SABRE COMB-HORN

Tipula 61 spp. | 5 ILL

Giant Cranefly *Tipula maxima*

Common

FWL 22–30 mm. **ID** Largest cranefly in GB and Ireland. Body greyish-brown, weak stripes on thorax; wings with diagnostic dark brown patches. **Hab** Damp woodland, streamsides, grassland. **SS** None.

Black-striped Cranefly *Tipula vernalis*

Common

FWL 12–17 mm. **ID** Eyes green; body grey, thorax with darker marks; abdomen orange with central bold, black stripe and black marks on side. Legs orange, femora with black tip; wings with darkened veins and whitish lunule. ♀ ovipositor reduced. **Hab** Grassland, woodland. **SS** None.

ommon Cranefly
ipula oleracea

WL 18–28 mm (2nd brood smaller han 1st). **ID** ♂ grey, ♀ greyish-brown; ings, with brown front margin and tigma, reaching to end of abdomen t least. Antennae 13-segmented. egs orange, femora with black tip. **ab** Wetlands, damp grassland. **eh** Most likely to be seen in May. **S** September Cranefly.

Short-winged Cranefly
Tipula pagana

FWL 11–16 mm (♂), 4–5 mm (♀). **ID** Body of ♂ greyish, ♀ abdomen grey or greyish-brown; wings very short in ♀, well short of half length of abdomen (incapable of flight). Legs orange-brown, darker from middle of femora downwards. **Hab** Damp grassland, woodland, gardens. **SS** The Nationally Scarce **Bog Cranefly** *Tipula holoptera* [N/I] (**FWL** 13–15 mm), which is restricted to bogs in England and Wales, has wings longer than tip of abdomen in ♀.

eptember Cranefly *Tipula paludosa*

WL 13–23 mm **ID** ♂ grey, ♀ greyish-brown; wings do not quite reach tip of abdomen, front margin of wings and gma brown. Antennae 14-segmented. Legs orange, femora with black tip. **Hab** Damp grassland, gardens in late mmer. **Beh** Main agricultural pest among craneflies, the larvae well-known as 'leatherjackets'. **SS** Common Cranefly.

Guide to Tipulidae *p. 343* | Guide to families *pp. 332–3*

Ctenophora 3 spp. | 1 ILL.

Orange-sided Comb-horn
Ctenophora pectinicornis

Nationally Scarce

FWL 15–19 mm. **ID** Black with yellow-spotted orange sides to abdomen; ♀ less orange on sides and larger yellow spots. ♂ antennae comb-like. Legs orange (♂ femora and tibiae at least black-tipped). Wings clear with **black stigma**. **Hab** Ancient woodland. **Beh** Often on or near fallen Beech trunks. **Dist** Recorded in Ireland in 1932. **SS Twin-mark Comb-horn** (♀ only). The other, rarer, *Ctenophora* species have a yellow collar and abdominal bands.

J F M A M J J A S O N D

♀

Dictenidia 1 sp.

Twin-mark Comb-horn
Dictenidia bimaculata

Local

FWL 10–15 mm. **ID** Black, ♀ thorax and abdomen sides with orange marks, extensive in some individuals. ♂ antennae comb-like. Legs orange; femora and tibiae black-tipped, tarsi black. Wings clear with **dark stripe below stigma and brown wingtips**. **Hab** Fens, damp woodland. **SS Orange-sided Comb-horn** (♀).

♂
♀

Nephrotoma 15 spp. | 1 ILL.

Four-spotted Cranefly
Nephrotoma quadrifaria

Common

FWL 12–15 mm. **ID** Thorax and abdomen yellow-and-black, the latter with narrow central markings in ♂; ♀ ovipositor small, cerci short and blunt. Wings darkened below stigma and at wingtips (diagnostic). **Hab** Woodland and gardens. **SS** Other *Nephrotoma* species are similar and examination of genitalia is needed to identify most species.

♂
♀

FAMILY **Ptychopteridae** (Fold-winged craneflies) 1 GEN. | 7 spp. | 2 ILL.

Resemble craneflies in general appearance but more closely related to moth flies. Rather compact, forewing with only one anal vein reaching wing edge (two in true craneflies). Larvae are aquatic.

Ptychoptera 7 spp. | 2 ILL.

White-legged Cranefly
Ptychoptera albimana

BL 11 mm. **ID** Body black with some orange, notably on abdomen; hind tarsi pale. **Hab** Wetlands and woodland. **SS** Other *Ptychoptera* spp., but this is the only one with pale hind tarsi.

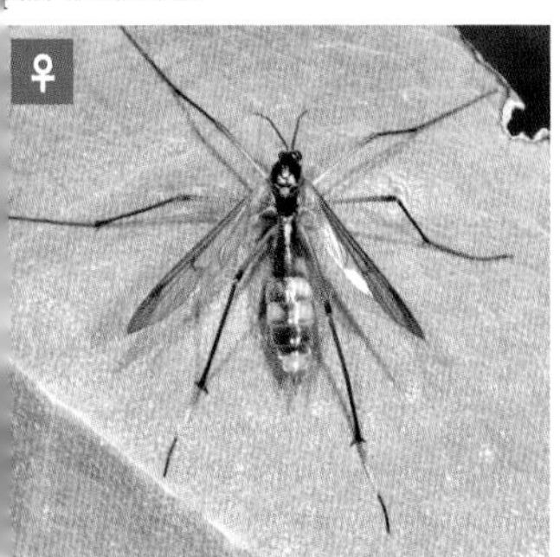

Orange-marked Cranefly
Ptychoptera contaminata

BL 12 mm. **ID** Body black, abdomen with orange marks; wings heavily mottled with dark markings; hind tarsi black. **Hab** Wetlands and woodland. **SS** Other *Ptychoptera* spp. (may require examination of genitalia).

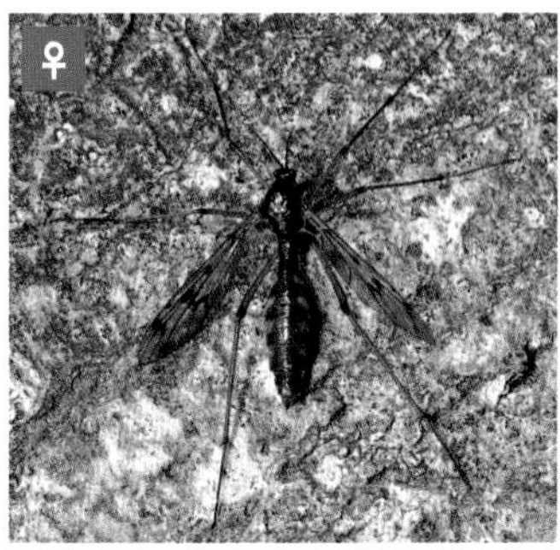

FAMILY **Bibionidae** (St Mark's flies) 2 GEN. | 18 spp. | 2 ILL.

The size of the common species shown readily distinguishes them. The sheer numbers of St Mark's Fly makes it one of the commonest insects in Britain from late April to early May.

FWL 6–13 mm. FORE TIBIAE tips with two large pointed spurs	*Bibio*
FWL 4–7 mm. FORE TIBIAE tips with several blunt spines	*Diolphus*

Bibio 14 spp. | 1 ILL.

t Mark's Fly *Bibio marci*

L 10–15 mm, **FWL** 8–12 mm. **ID** Body black, hairy; rger ♀ also with black wings; ♂ with large, bulbous yes and clear wings, held open at rest. **Hab** Woodland, rmland, hedgerows, often in damp areas. **SS** Other *ibio* species, but this is the largest species.

Dilophus 4 spp. | 1 ILL.

Common Fever Fly *Dilophus febrilis*

BL 7–10 mm, **FWL** 4–7 mm. **ID** Wings darkened with whitish tips in ♀; in ♂ clearer with black stigma. **Hab** Grassland, farmland. **SS** Other *Dilophus* species, but this is the largest and ♀ is easily distinguished by wing pattern.

Guide to families *pp.* 332–34

SUBORDER BRACHYCERA

FAMILY Tabanidae (Horseflies) 5 GEN. | 30 spp. | 5 ILL.

Stout, medium-sized to large flies, lacking bristles. Eyes often with bright metallic spots or bands. Wing patterns help to distinguish genera, but some species require examination of abdominal markings, legs (colour), eyes (presence of hairs and number of bands) to confirm identification.

Tabanus 8 spp. | 2 ILL.

Band-eyed Brown Horsefly
Tabanus bromius

BL 13–15 mm, **FWL** 11 mm. **ID** Body greyish-brown (thorax with 5 indistinct pale longitudinal lines), abdomen with 3 longitudinal rows of paler greyish or yellowish-brown (rarely reddish) spots, with central triangles. Eyes with single stripe. **Hab** Woodland, gardens, wetlands, around livestock. **SS** Other *Tabanus* species, particularly the rare **Plain-eyed Brown Horsefly** *Tabanus miki* [N/I] (**BL** 15 mm, **FWL** 12 mm) (♀ lacks eye band).

Dark Giant Horsefly
Tabanus sudeticus

BL 21–24 mm, **FWL** 18–20 mm. **ID** Very large; body dark, each abdomen segment with a wide orange band that has a central yellow triangle. **Hab** Woodland clearings, wetlands such as bogs, around livestock. **SS** Other *Tabanus* species, which are much smaller.

Haematopota 5 spp. | 2 ILL.

Notch-horned Cleg
Haematopota pluvialis

BL 8–12 mm, **FWL** 7 mm. **ID** Drab brown with very mottled wings, abdomen with series of pale, indistinct spots. Eyes patterned with stripes, slightly hairy; antenna S3 faintly reddish. **Hab** Woodland, wetlands, around livestock. **SS Black-horned Cleg** *Haematopota crassicornis* (**BL** 8–12 mm, **FWL** 7–9 mm) [*inset*], which is darker on body and wings.

Chrysops 4 spp. | 1 ILL.

Splayed Deerfly *Chrysops caecutiens*

BL 9–10 mm, **FWL** 8 mm. **ID** Body mostly black: ♀ abdomen S2 yellow with splayed black inverted 'V'-shaped mark, wings black with conspicuous clear patches; ♂ abdomen and most of wings black. **Hab** Wetlands. **SS** Other *Chrysops* species although only Splayed Deerfly has mid-tibiae with at least lower half black.

♀

Common

J F M A M J J A S O N D

FAMILY Bombyliidae (Bee-flies) 5 GEN. | 10 spp. | 6 ILL.

Head with extended proboscis (which in some species is long and an obvious character in flight and at rest). Ectoparasitoids of various bees and wasps, also Lepidoptera larvae and pupae, the species varying from species to species. The species most likely to be seen are covered briefly below.

PROBOSCIS short	WINGS with bold dark brown patches		Mottled Bee-fly	p. 349
	WINGS clear except for narrow brown front margin		Dune Villa	p. 349
PROBOSCIS long	WINGS with dark pattern (no spots)		Dark-edged Bee-fly	p. 350
	WINGS spotted		Dotted Bee-fly	p. 350
	WINGS clear	Area behind eyes with long black hairs at sides	Western Bee-fly	p. 350
		Area behind eyes with only short pale brown hairs	Heath Bee-fly	p. 350

BEE-FLIES – short proboscis

Thyridanthrax 1 sp.

Mottled Bee-fly
Thyridanthrax fenestratus

Nationally Scarce

BL 12–15 mm, **FWL** 10–12 mm. **ID** Bold dark brown mottled wings. **Hab** Heathland. **Host** Heath Sand Wasp (*p. 523*). **SS** None.

J F M A M J J A S O N D

Villa 3 spp. | 1 ILL.

Dune Villa
Villa modesta

Nationally Scarce

BL 10–14 mm, **FWL** 10–14 mm. **ID** wings with narrow front brown margin. **Hab** Mainly coastal sand dunes. **Hosts** Noctuid moth larvae and pupae (*p. 454*). **SS** Downland Villa *Villa cingulata* [N/I] (**BL** 9–11 mm) occurs on downland and in woodland.

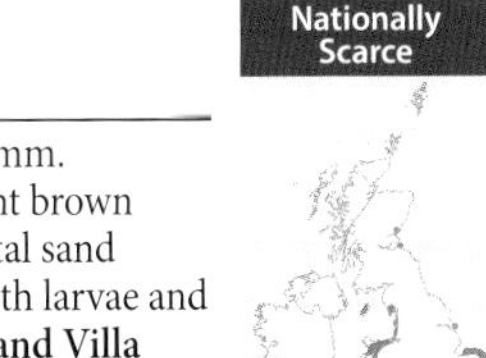
J F M A M J J A S O N D

♀

♀

Guide to Bombyliidae p. 34

BEE-FLIES – long proboscis

Bombylius 4 spp.

Dark-edged Bee-fly
Bombylius major

Common

BL 6–13 mm, **FWL** 9–14 mm.
ID Body brown; wings with broad dark front margin (difficult to see when hovering). **Hab** Woodland rides, coastal sites, parks, gardens. **Hosts** Mining bees *Andrena* spp. (*p. 562*).
SS Dotted Bee-fly [see table on *p. 349*].

Dotted Bee-fly
Bombylius discolor

Local

BL 8–12 mm, **FWL** 9–13 mm.
ID Body dark brown; wings with several dark spots (difficult to see when hovering). **Hab** Grassland, woodland, gardens. **Hosts** Mining bees *Andrena* spp. (*p. 562*). **SS** Dark-edged Bee-fly [see table on *p. 349*].

Western Bee-fly
Bombylius canescens

Nationally Scarce

BL 7 mm, **FWL** 6–8 mm. **ID** Wings clear; area behind eyes with long black hairs at sides. **Hab** Woodland. **Hosts** Base-banded furrow bees *Lasioglossum* spp. and end-banded furrow bees *Halictus* spp. (*p. 557*). **SS** Heath Bee-fly [see table on *p. 349*].

Heath Bee-fly
Bombylius minor

VU

Rare

BL 7–9 mm, **FWL** 7–9 mm.
ID Wings clear; area behind eyes with short pale brown hairs only.
Hab Heathland. **Hosts** Plasterer bees *Colletes* spp. (*p. 558*). **SS** Western Bee-fly [see table on *p. 349*].

FAMILY Stratiomyidae (Soldierflies) 16 GEN. | 48 spp. | 10 ILL. (also pp. 332, 336)

The large, broad *Stratiomys* species are easily distinguished from smaller, often more slender species in other genera. The common names relate to brightly coloured uniforms of soldiers.

Stratiomys 4 spp.

Banded General

Stratiomys potamida

Common

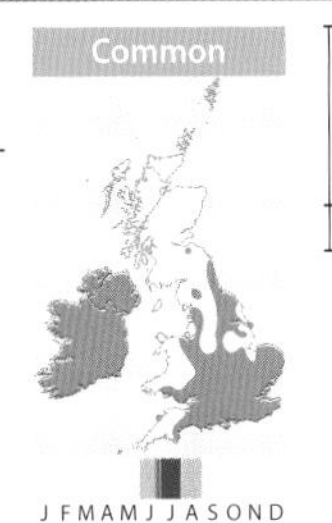

BL 12–13 mm, **FWL** 9–11 mm. **ID** Body black, abdomen segments with bright yellow marks, S3 with narrow yellow marks (< ½ length of segment). Sides of face yellow. Scutellum yellow, straight or slightly curved. **Hab** Wetlands. **Beh** Often on umbellifers. **SS** Clubbed General.

Flecked General

Stratiomys singularior

Local

BL 12–15 mm, **FWL** 9–12 mm. **ID** Body black, abdomen segments with small cream marks. **Hab** Mainly brackish coastal marshes along ditches. **SS** None, although superficially like other *Stratiomys* species.

Clubbed General

Stratiomys chamaeleon

EN Rare

BL 12–15 mm, **FWL** 10–12 mm. **ID** Abdomen black with yellow or whitish marks, segment S3 with broad yellow marks (= ½ length of segment); scutellum yellow, almost 'V'-shaped. **Hab** Fens with seepages. **SS** Banded General.

Long-horned General

Stratiomys longicornis

Nationally Scarce

BL 12–13 mm, **FWL** 10–11 mm. **ID** Abdomen plain black; thorax densely brown-haired; eyes hairy. **Hab** Brackish shallow ditches, often coastal. **SS** None, although superficially like other *Stratiomys* species

Nemotelus 4 spp. | 1 ILL.

Barred Snout
Nemotelus uliginosus

BL 5–6 mm, **FWL** 4–5 mm. **ID** Black with white marks on large head (♀ with long snout), thorax and abdomen: ♂ abdomen mainly white with black bars towards tip, ♀ with central row of white spots. **Hab** Coastal saltmarsh and brackish sites. **SS** Other *Nemotelus* species, which have different patterning on head, thorax and abdomen.

Oplodontha 1 sp.

Common Green Colonel
Oplodontha viridula

BL 6–8 mm, **FWL** 5–7 mm. **ID** Body black or brown; legs pale. Abdomen colour varies between individuals, but black above and often with apple-green or straw sides. **Hab** Ditches and ponds. **SS** None.

♀

Chloromyia 1 sp.

Broad Centurion
Chloromyia formosa

BL 8–9 mm, **FWL** 6–7 mm. **ID** Eyes with dense, short black hairs. Thorax bluish-green; abdomen of ♂ bronze; in ♀ bluish or green and wider than thorax. **Hab** Wet and dry grassland, road verges, woodland, parks, gardens. **SS** Twin-spot Centurion (♂).

Sargus 4 spp. | 1 ILL.

Twin-spot Centurion
Sargus bipunctatus

BL 12–13 mm, **FWL** 10 mm. **ID** Thorax green; abdomen narrow, bronze in ♂, black in ♀ with reddish patches at side near base. **Hab** Grassland, hedgerows, woodland margins, gardens. **Beh** On/near fresh cow dung or compost heaps. **SS** Other *Sargus* species, which are smaller and differ in colour combination; Broad Centurion (♂).

FAMILY Syrphidae (Hoverflies) 68 GEN. | 282 spp. | 32 ILL. (also *pp. 331–333*)

Many species of hoverfly can be identified in the field, as they have distinctive markings; however, others need to be checked under a microscope for confirmation.

Further details and fuller descriptions can be found in the companion WILD*Guides* volume *Britain's Hoverflies* (see *page 330* for details).

Chrysotoxum 8 spp. | 1 ILL.

FORM medium-sized to large, robust-looking wasp mimics; ABDOMEN yellow bands/bars; ANTENNAE long. Species identified by shape of abdominal bands.

Two-banded Wasp Hoverfly
Chrysotoxum bicinctum

FWL 7–10 mm. **ID** Abdomen broad with two broad yellow bands; wings black-marked. **Hab** Grassland, woodland rides. **SS** None.

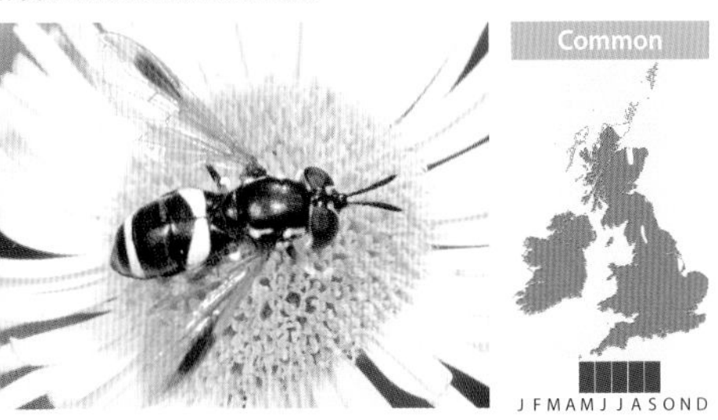

Platycheirus 25 spp. | 1 ILL.

FORM small to medium-sized, long and slender-bodied. Species identified by abdominal markings and, in ♂s, foreleg differences – but some species are difficult to distinguish.

Wetland Hoverfly
Platycheirus rosarum

FWL 5–8 mm. **ID** Abdomen black, segment 3 with pairs of yellow spots. **Hab** Wetlands. **SS** None.

Sphaerophoria 11 spp. | 1 ILL.

FORM small to medium-sized, elongate; ABDOMEN black-and-yellow banded. FACE yellow; ♂s have twisted genital capsules. Species identified by details of ♂ genitalia.

Common Twist-tailed Hoverfly
Sphaerophoria scripta

FWL 5–7 mm. **ID** Abdomen noticeably longer than wings, with broad yellow bands. **Hab** Grasslands. **SS** Some other *Sphaerophoria* species also have abdomen longer than the wings; to be certain, examination of genitalia (♂) needed, ♀s cannot be identified in the field with 100% certainty.

Xanthogramma 3 spp. | 1 ILL.

FORM small to medium-sized, robust, yellow-faced; THORAX & ABDOMEN with yellow marks; ANTENNAE short. Species identified by extent of yellow markings.

Superb Ant-hill Hoverfly
Xanthogramma pedissequum

FWL 7–10 mm. **ID** Thorax with yellow sides; abdomen segments with narrow yellow markings. **Hab** Grassland. **SS** Other *Xanthogramma* species, from which recognized by its broader triangular yellow markings on abdomen S2.

Scaeva 5 spp. | 1 ILL.

FORM medium-sized to large; **ABDOMEN** black with white or yellow curved bars. Species identified by colour and shape of abdominal spots.

White-clubbed Hoverfly *Scaeva pyrastri*

FWL 9–12 mm. **ID** Black, abdomen with six comma-shaped white marks. **Hab** Grassland, gardens. **SS** Other *Scaeva* species, but white, comma-shaped spots are distinctive (yellow in related species).

Episyrphus 1 sp.

Marmalade Hoverfly *Episyrphus balteatus*

FWL 6–10 mm. **ID** Usually very orange with two black bands on abdomen separated by two orange bands (a darker form is not uncommon). **Hab** Grassland, woodland, parks, gardens. **SS** None.

Myathropa 1 sp.

Batman Hoverfly *Myathropa florea*

FWL 7–12 mm. **ID** Body black-and-yellow, often appears very yellowish; distinctive 'batman'-shaped black marking on top of thorax. **Hab** Woodland. **SS** None.

Caliprobola 1 sp.

Orange-winged Hoverfly NT
Caliprobola speciosa

FWL 8–10 mm. **ID** Narrow-waisted; head and body black with narrow yellow bands; wings with front edge black. **Hab** Scrubby calcareous grassland. **SS** None.

Doros 1 sp.

Phantom Hoverfly *Doros profuges* NT

FWL 8–10 mm. **ID** Narrow-waisted; head and body black with narrow yellow bands; wings with front edge black. **Hab** Scrubby calcareous grassland. **SS** None.

Sericomyia 3 spp. | 1 ILL.

FORM large, stout build; **BODY** black; **ABDOMEN** with narrow yellow or cream bars, the shape of which, together with the colour of the body, scutellum and legs distinguishes the species.

ellow-barred Peat Hoverfly
ericomyia silentis

WL 10–14 mm. **ID** Abdomen segments with wedge-haped yellow bars. **Hab** Heathlands, wetlands, voodland edge. **SS** White-barred Peat Hoverfly *ericomyia lappona* [N/I] (**FWL** 10–14 mm) is somewhat imilar, but has narrower whitish abdominal bars, a eddish scutellum and reddish legs.

Helophilus 5 spp. | 1 ILL.

FORM medium-sized to large; **THORAX** distinctive bright, striped; **ABDOMEN** with bold large yellow or orange marks. Species identified by differences in colour of face and details of hind tibiae.

Tiger Hoverfly *Helophilus pendulus*

FWL 8–11 mm. **ID** Thorax with bold longitudinal stripes; abdomen yellow or orange with black bands and marks. **Hab** Grassland, woodland, parks, gardens. **SS** Other *Helophilus* species, but the black band separating abdomen 2 & 3 differentiates ♂. In ♀ only a third of hind tibiae is black (a half in other species).

Syrphus 5 spp. | 1 ILL.

FORM medium-sized to large; **ABDOMEN** broadly banded black and yellow; **FACE** yellow. Species identified by whether eyes meet on top of head or are separated, along with colour of hind femora. (In one species, **Hairy-eyed Syrphus** *Syrphus torvus* [N/I] (FWL 9–12 mm), the eyes are hairy.)

lumming Syrphus *Syrphus ribesii*

WL 7–11 mm. **ID** Abdomen with broad black and ellow bands; hind femora yellow in ♀. **Hab** Grassland, edgerows, gardens. **SS** Other *Syrphus* species and some elated genera, requiring careful examination of bands. lind femora are much darker in ♀s of closest relatives.

Ferdinandea 2 spp. | 1 ILL.

FORM small to medium-sized, robust, yellow-faced; **FORM** medium-sized to large; **THORAX** with two broad, grey, longitudinal stripes; **ABDOMEN** metallic brassy or black; **WINGS** black-spotted; **BEHAVIOUR** most likely on or near tree sap runs. Species identified by colour of abdomen and arista.

Bronze Sap Hoverfly *Ferdinandea cuprea*

FWL 7–11 mm. **ID** Thorax grey-striped; abdomen bronzy, metallic. **Hab** Woodland. **SS** The Nationally Scarce Dark Sap Hoverfly *Ferdinandea ruficornis* [N/I] (**FWL** 6–9 mm), which has a black abdomen; the arista is orange in that species (black in Bronze Sap Hoverfly).

Cheilosia — 38 spp. | 2 ILL.

FORM medium-sized to large; BODY most species black, some brown; wings normally plain. One atypical species (*Cheilosia illustrata*) [BELOW]. Species identified by extent of hairiness of eyes, and leg and antennae colour, but some require microscopic examination.

Bumblebee Cheilosia *Cheilosia illustrata*

FWL 8–10 mm. **ID** Abdomen base white-haired, along with tip; wings with a dark cloud in centre. **Hab** Woodland edge, road verges. **SS** None.

Large Spring Cheilosia *Cheilosia grossa*

FWL 9–12 mm. **ID** Body hairy, also femora. Antennae black; tibiae orange. **Hab** Grassland. **SS** None.

Rhingia — 2 spp.

FORM medium-sized, stout build; FACE obvious snout-like rostrum; ABDOMEN orange.

Common Snout Hoverfly *Rhingia campestris*

FWL 6–9 mm. **ID** Snout long. Abdomen orange with dark border; legs also dark. **Hab** Woodland, grassland, gardens. **SS** **Grey-backed Snout Hoverfly**.

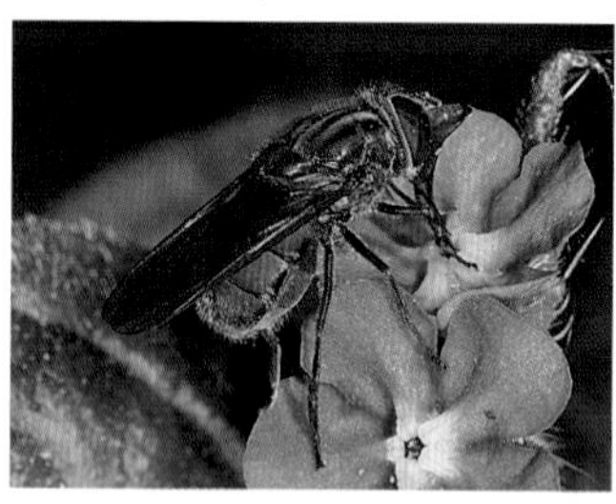

Grey-backed Snout Hoverfly *Rhingia rostrata*

FWL 8–9 mm. **ID** Snout modest length. Thorax greyish. Abdomen orange with no dark border; legs orange. **Hab** Woodland. **SS** **Common Snout Hoverfly**.

Common Snout Hoverfly extending its mouthparts.

Brachyopa 4 spp. | 1 ILL.

FORM medium-sized, stout build; **THORAX** greyish; **ABDOMEN** orange; **BEHAVIOUR** most likely on tree sap runs. Species identified by colour of scutellum and microscopic examination of shape of antennal pit.

'ale-shouldered Brachyopa
3rachyopa scutellaris

WL 6–8 mm. **ID** Thorax greyish, abdomen orange, :cutellum yellow; kidney-shaped pit on inner side of ntennae. **Hab** Woodland. **SS** Other *Brachyopa* species, rom which told by yellow scutellum (at least dusted grey 1 other species) and kidney-shaped antennal pit.

Brachypalpoides 1 sp.

łed-belted Hoverfly *Brachypalpoides lentus*

WL 10–12 mm. **ID** Robust-looking; black, except red pper half to broad abdomen; legs black. **łab** Woodland. **SS** None.

Xylota 7 spp. | 2 ILL. (see also *p. 332*)

FORM medium-sized to large, elongate; **ABDOMEN** orange-belted or spotted. Species identified by abdomen colours.

'range-belted Hoverfly *Xylota segnis*

WL 8–10 mm. **ID** Body elongate, abdomen with broad range band. **Hab** Woodland. **SS** Other *Xylota* species, ıt none of these have broad orange band on abdomen; in ›me species, the abdomen tip has yellowish hairs.

Leucozona 3 spp. | 1 ILL.

FORM medium-sized to large; **ABDOMEN** cream-and-black ɔr with bluish-grey bars, depending on the species (colouring unlike other hoverfly genera). Species dentified by body colour and markings.

łotch-winged Hoverfly *Leucozona lucorum*

WL 8–10 mm. **ID** Abdomen S2 cream; wings with black ouds; scutellum yellow. **Hab** Woodland. **SS** None, but ıperficially resembles similar-coloured, mainly unrelated ›ecies.

Volucella 5 spp. | 3 ILL.

FORM medium-sized to large (mostly), broad hoverflies; ANTENNAE with feathery arista; WINGS with black or dark clouding. Species identified by body colour and markings.

Bumblebee Hoverfly *Volucella bombylans*

FWL 8–14 mm. **ID** Usual forms black-and-yellow or black with abdomen tip orange. **Hab** Woodland, grassland. **SS** None, although has a superficial resemblance to bumble bee mimics from other genera.

Great Pied Hoverfly *Volucella pellucens*

FWL 10–15 mm. **ID** Body black, abdomen S2 white. **Hab** Woodland. **SS** Cossus Hoverfly *Volucella inflata* [N/I] (**FWL** 11–13 mm) has abdomen S2 yellow.

Hornet Hoverfly *Volucella zonaria*

FWL 15–20 mm. **ID** Thorax brown; abdomen yellow with black bands, S2 brown. **Hab** Parks, gardens and wasteland. **SS** Lesser Hornet Hoverfly *Volucella inanis* [N/I] (**FWL** 12–14 mm) thorax yellowish-grey and abdomen S2 yellow.

Merodon 1 sp.

Large Narcissus Fly *Merodon equestris*

FWL 8–10 mm. **ID** Body broad, bumble bee-like; various colour forms, ranging from brown to black-and-whitish, ea believed to mimic different bee species [3 ILLUSTRATED]. **Hab** Gardens and urban areas, grassland; larvae develop in bulbs, including daffodils. **SS** None.

Eristalis 10 spp. | 1 ILL.

FORM medium-sized to large, robust; **ABDOMEN** mostly with large yellow or orange marks; **LIFESTYLE** larvae are aquatic 'rat-tailed' maggots. Among the commonest hoverflies. Species identified by abdomen markings and leg colour.

Dronefly *Eristalis tenax*

FWL 10–13 mm. **ID** Large and robust; dark, abdomen ranging from black to orange-marked. Head with broad black facial stripe; eyes with vertical stripe of darker hairs. Hind tibiae broadened. **Hab** Grassland, gardens. **SS** Other *Eristalis* species, although three characters readily identify the large Dronefly: vertical eye stripe; black facial stripe and broad, curved hind tibiae.

Eristalinus 2 spp. | 1 ILL.

FORM medium-sized, stout build; **BODY** dark; **EYES** **heavily spotted.** **LIFESTYLE** larvae are aquatic 'rat-tailed' maggots, living among decaying vegetable matter. Species identified by details of the eyes, which are also hairy.

Large Spotted-eyed Dronefly *Eristalinus aeneus*

FWL 6–9 mm. **ID** Abdomen black or greenish black; eye spots conspicuous, lower part hairless. **Hab** Mainly coastal grassland and around rocky pools and saltmarsh. **SS** Small Spotted-eyed Dronefly *Eristalinus sepulchralis* [N/I] (**FWL** 6–8 mm) is generally smaller and has hairs covering the eyes.

Criorhina 4 spp. | 1 ILL.

FORM medium-sized to large bee mimics; **HABITAT** woodland, often around the base of tree trunks. Species identified by body colour.

Common Criorhina *Criorhina berberina*

FWL 8–12 mm. **ID** Two colour forms: thorax dark with buff stripe and abdomen with white 'tail'; also buff. **Hab** Woodland. **SS** None, but superficially resembles similar-coloured, mainly unrelated species.

Pocota 1 sp.

Small-headed Hoverfly *Pocota personata*

FWL 11–13 mm. **ID** Classic bumble bee mimic (adults even emit a loud 'buzz'), but clearly a fly as it has short antennae. Front half of thorax yellow, hind part black; abdomen black with patch of yellow hairs and white 'tail'. **Hab** Woodland, often around tree trunks with rot holes, where larvae develop. **SS** Resembles bumble bees (*p. 546*).

FAMILY **Asilidae** (Robberflies) 16 GEN. | 29 spp. | 7 ILL.

Small to very large (**BL** 8–28 mm, FWL 6–18 mm) slender to robust; thorax often distinctively marked; legs elongated, very bristly. Species identified by colour/markings on body/legs, or extent of bristles.

Dysmachus 1 sp.

Fan-bristled Robberfly
Dysmachus trigonus

BL 12–17 mm, **FWL** 8–11 mm.
ID Body stout; grey, abdomen with faint dark triangles (hind margins of S2–6 with long white bristles); thorax also very bristly; legs black. **Hab** Heathland, sand dunes. **SS** Superficially resembles other robberfly species, from which told by robust, very hairy body.

Philonicus 1 sp.

Dune Robberfly
Philonicus albiceps

BL 13–18 mm, **FWL** 10–13 mm.
ID Body slender, elongated; greyish-brown, thorax with few bristles; legs black with white bristles. **Hab** Coastal sand dunes. **SS** Superficially resembles other robberfly species, from which told by elongate shape and plain abdomen (often spotted in other larger species).

Tolmerus 3 spp. | 2 ILL.

Kite-tailed Robberfly
Tolmerus atricapillus

BL 12–15 mm; **FWL** 9–11 mm.
ID Body greyish-brown; femora black with red marks, some with red ring before tip; tibiae mainly red. ♂ has tuft black hairs on hind margin of abdomen S8. **Hab** Heathland, grassland. **SS Brown Heath Robberfly** *T. cingulatus* [N/I] (see description under Irish Robberfly); **Irish Robberfly**.

Irish Robberfly EN
Tolmerus cowini

BL 10–20 mm; **FWL** 8–12 mm.
ID Body grey or brown; femora black with red ring near tip. **Hab** Coastal sand dunes, scrublands. **SS Brown Heath Robberfly** *T. cingulatus* [N/I] (**BL** 10–13 mm; **FWL** 6–9 mm) is usually smaller, brown, and has mainly orange legs; **Kite-tailed Robberfly** (both in Britain only).

Dioctria 6 spp. | 1 ILL.

BODY elongated (very slender to stout); **THORAX** with short hairs (not bristles), sides with silver stripes or bar; **ANTENNAE** long (often longer than head); **GENITALIA** inconspicuous and lacking elaborate structures, such as ovipositor (♀) or forceps (♂). Species identified by body, wing and leg colour.

ommon Red-legged Robberfly
ioctria rufipes

L 11–13 mm, **FWL** 8–9 mm. **D** Body black; fore and mid femora nd tibiae orange-red, hind legs black. **ab** Grassland. **SS** Other *Dioctria* pecies, but the leg colour combination escribed above is unique within the enus for species with clear, not dark vings.

Laphria 1 sp.

umblebee Robberfly
aphria flava

L 15–24 mm; **FWL** 12–19 mm. **)** Hairy, broad, bee-mimic; abdomen overed in yellow or brownish furry air. **Hab** Ancient pine woodland. **S** None.

Asilus 1 sp.

ornet Robberfly
silus crabroniformis

L 18–28 mm (in Europe up to 35 mm), **WL** 15–18 mm. **ID** Head, thorax and gs plain brown with bristles and hairs; domen with front half black, back alf yellow; antennae shorter than ead; wings orange-brown with dark tches. **Hab** Grassland, including alk, heathland. **Beh** Largest robberfly Britain and able to tackle large prey, cluding grasshoppers (see *p. 32*). **SS** None.

Calyptratae 'Calyptrates'

FAMILY Muscidae (Houseflies and allies) 39 GEN. | 293 spp. | 3 ILL.

Identification for many species is difficult and usually involves microscopic work; however, the species illustrated here can usually be recognized quite readily.

Mesembrina 2 spp. | 1 ILL.

Noon Fly
Mesembrina meridiana

BL approx. 12 mm, **FWL** 11–12 mm. **ID** Body black; wing bases orange. **Hab** Grassland and woodland, around livestock. **SS** None.

NOTE: the other species in this genus, **Moustached Fly** *Mesembrina mystacea* [N/I], has been recorded in Ireland but not in Britain; it looks completely different from Noon Fly, having a brown-and-black thorax and black abdomen with a whitish tip.

Musca 3 spp. | 1 ILL.

Housefly
Musca domestica

BL 7–8 mm, **FWL** 5–8 mm. **ID** Eyes red. Thorax grey with four narrow longitudinal black stripes; abdomen with dark midline and brown patches. **Hab** Around humans, farmlands (less common in urban areas, including houses, than in the past). **Beh** Worldwide vector of diseases. **SS** Superficially like other Muscids, but ID features mentioned should help with identification.

Stomoxys 1 sp

Stable Fly *Stomoxys calcitrans*

BL approx. 8 mm, **FWL** 6–8 mm. **ID** Body greyish; thorax with four longitudinal black stripes; abdomen S2–3 with dark spots. **Hab** Farmlands, around livestock. **Beh** Biting fly, sucking blood from mammals and humans near livestock. **SS** Superficially like other Muscids, but ID features mentioned and habitat should help with identification.

FAMILY Calliphoridae (Blowflies) 12 GEN. | 29 spp. | 4 ILL.

Although this family includes the well-known metallic bluebottles and greenbottles, identification is difficult for many species and usually involves microscopic work.

Calliphora 6 spp. | 1 ILL.

Orange-bearded Bluebottle
Calliphora vomitoria

Common

FWL 8–14 mm. **ID** Body metallic blue; distinctive orange 'beard', distinguishes it from close relatives. **Hab** Grassland, woodland, gardens and urban areas, around livestock. **Dist** GB and Ireland. **SS** Other *Calliphora* species, from which told by distinctive orange beard.

Lucilia 7 spp. | 1 ILL.

Common Greenbottle
Lucilia caesar

Common

FWL 5–10 mm. **ID** Body metallic green (commonest of several greenbottles). **Hab** Grassland, woodland, gardens and urban areas, around livestock. **Beh** Often enters buildings and lays eggs on meat and dairy products. **SS** Other *Lucilia* and similar species can only be told with certainty by examination of bristles under a microscope.

J F M A M J J A S O N D

Protocalliphora 1 sp.

Bird Blowfly
Protocalliphora azurea

Common

FWL 6–10 mm. **ID** ♂ body bluish-black; ♀ more greenish. Head and antennae black in both sexes. **Hab** Woodland. **Beh** External parasites, maggots sucking the blood of bird nestlings. Adults feed on bird droppings, also human meat. **SS** None (the distinct shade of blue helps identify this species).

♀

Cynomya 1 sp.

Yellow-faced Blowfly
Cynomya mortuorum

Common

BL 7–15 mm; **FWL** 7–13 mm. **ID** Body metallic blue or turquoise; much of face yellow. **Hab** Open areas, where larvae develop in carrion. **SS** None.

♀

Guide to families *pp.* 332–34

FAMILY **Tachinidae** (Parasitic flies) **145 GEN. | 267 spp. | 4 ILL.**

Range of colours/forms; examples are shown from three genera of commonly encountered species.

Tachina 4 spp. | 2 ILL.

BODY broad; **WINGS** with orange base. Species identified by details of body colour.

Common Tachinid

Tachina fera

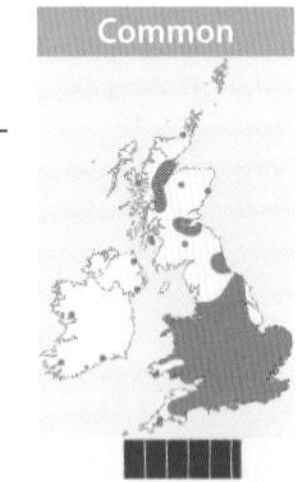

BL 10–14 mm. **ID** Abdomen orange, except for broad black central line and whitish tip. **Hab** Grassland.
SS None.

Giant Tachinid

Tachina grossa

BL 13–20 mm. **ID** Body black, except for yellowish head; largest tachinid in Britain and Ireland. **Hab** Grassland.
SS None.

Phasia 4 spp. | 1 ILL.

BODY broad; **WINGS** partly blue in common species. Species identified by wing colour and size.

Blue-winged Tachinid

Phasia hemiptera

BL 8–12 mm. **ID** ♀ thorax has orange hair on sides (lacking in ♂), wings curved and patterned, in part blue.
Hab Grassland and hedgerows.
Beh A parasitoid of shieldbugs.
SS None.

♂

Sturmia 1 sp

Beautiful Tachinid

Sturmia bella

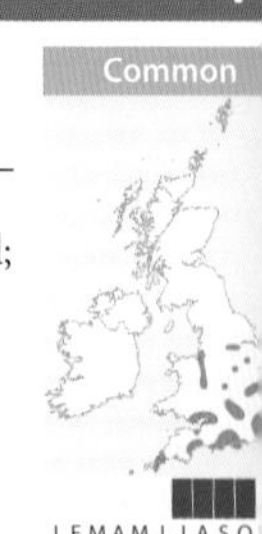

BL 10 mm. **ID** Body bluish-black, abdomen with black markings; eyes red; wings plain. **Hab** Grassland.
Beh Known to attack Small Tortoiseshell (*p.* 358) and related butterfly species. **SS** Species in various genera; ideally an association with the nymphalid butterfly host is needed to confirm identification.

ORDER MECOPTERA Scorpionflies

[Greek: *meco* = long; *pteron* = wing]

BI | 2 families [2 genera, **4 species**]

W | 9 families [**769 species**]

Elongate, small to medium-sized winged insects with a head that extends downwards into a beak. The forewings and hindwings are similar in size and enable a fluttering flight (except in the flightless Snow Flea in which the wings are reduced). The males have modified, swollen reproductive organs at the tip of the abdomen that are upturned at rest giving a scorpion-like pose. Although the structure looks menacing, it cannot sting. In females the abdomen is tapered to the tip (genitalia differs in Snow Flea, which many scientists believe are more closely related to fleas than scorpionflies). Scorpionflies mainly live on dead insects and carrion. Absent from Ireland.

FINDING SCORPIONFLIES Most likely along hedgerows or shady places, resting on vegetation such as Bramble and nettles.

Structure of a scorpionfly

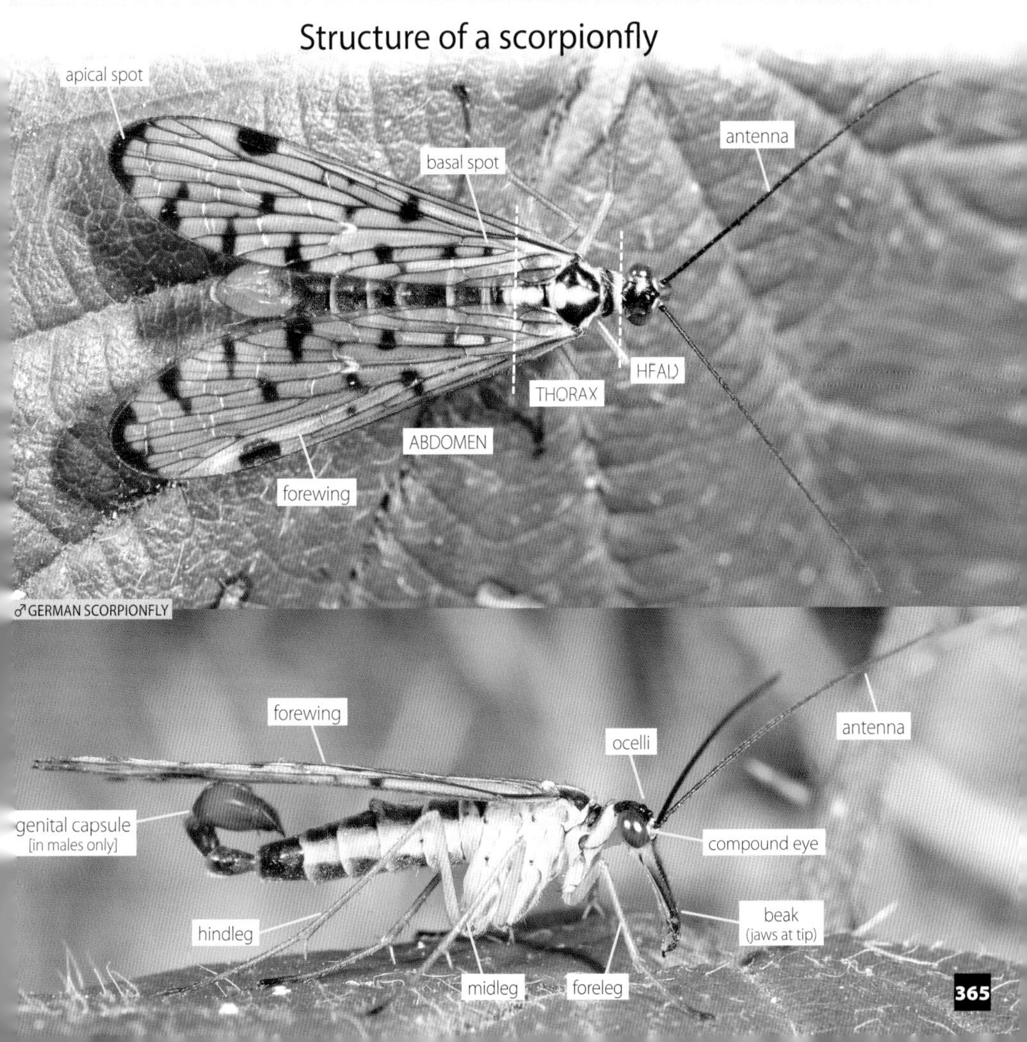

♂ GERMAN SCORPIONFLY

FAMILY **Panorpidae** (Scorpionflies) — 1 GEN. | 3 spp.

All three British species are elongate and belong to the genus *Panorpa*. Identification to species level is straightforward for both sexes from examination of wing patterning and veins, as shown in the key. If still in doubt, identification is easily confirmed in males by examining the shape of callipers within the genital capsule, either in the field with a hand lens, or via a good quality photograph taken from above the insect. **Some researchers kill and dissect females but, with experience, this is not necessary.**

Forewing and hindwing markings are diagnostic and similar in both sexes, although there is a little variation. Each species also has a difference in cell shape and size near the tip, where indicated

Scarce Scorpionfly	Common Scorpionfly	German Scorpionfly
WINGTIPS partially black (extreme margin transparent); relatively few black spots, with central spots merged and **reaching half width of wing**	WINGTIPS black, bold and broad; large central black spots merged, **forming a more-or-less continuous band across the wing**	WINGTIPS black, bold but **narrow**; medium-sized central black spots (larger in ♀), not forming a band. More heavily spotted on inner half of forewing*
		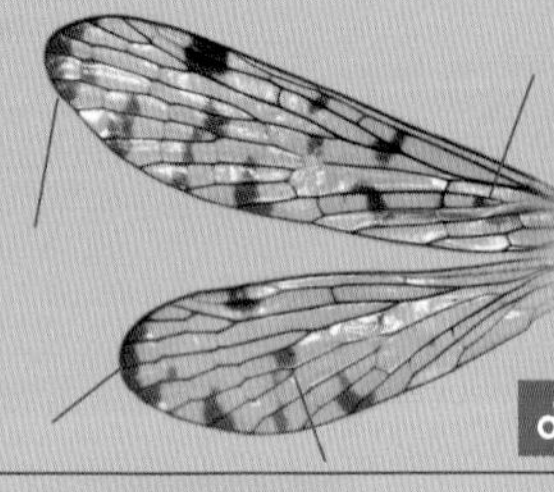
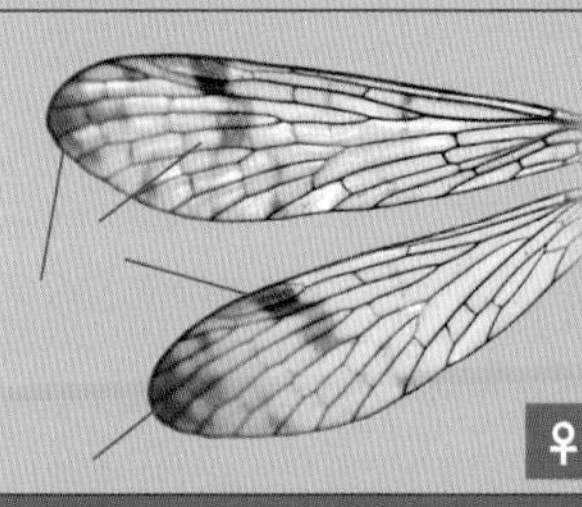		

If in doubt, male genital capsules are distinctive in each species, as shown below:

Scarce Scorpionfly	Common Scorpionfly	German Scorpionfly
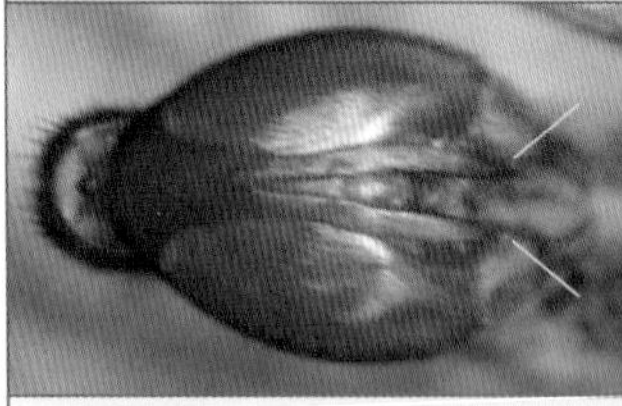	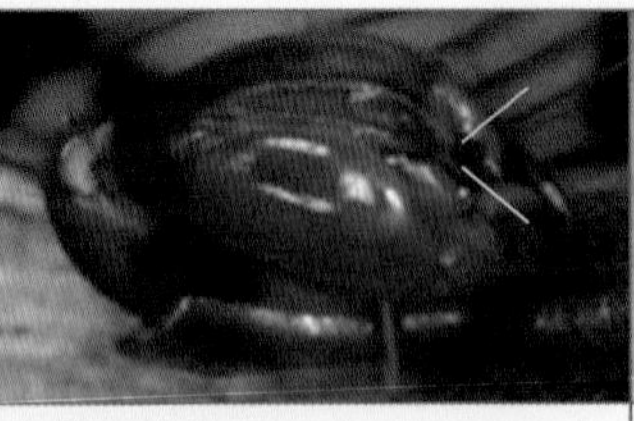	
Callipers long and slender, tips diverging	Callipers long and slender; curved outwards at middle, then almost meeting at tips	Callipers short, broad and almost straight; tips noticeably expanded

* except a form in Scotland (*borealis*), which has completely transparent wings or a reduced number of spots

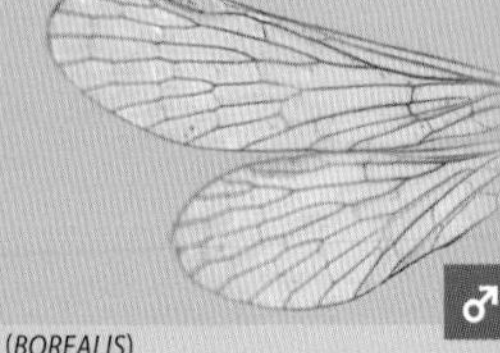

GERMAN SCORPIONFLY (*BOREALIS*)

Common Scorpionfly
Panorpa communis

BL 10–15 mm, **WS** 27–33 mm.
ID Black, with yellow sides, marks on head (which is brown in some individuals) and pronotum, also yellow abdominal rings; genital capsule, and tip of abdomen in ♀, red. All wings transparent, wingtips bold black and central spots large, merging to form a band; other spots on inner half of wings much smaller. Legs yellowish-brown. **Hab** Hedgerows, woodland; often on Bramble or nettles. **SS** Other scorpionflies.

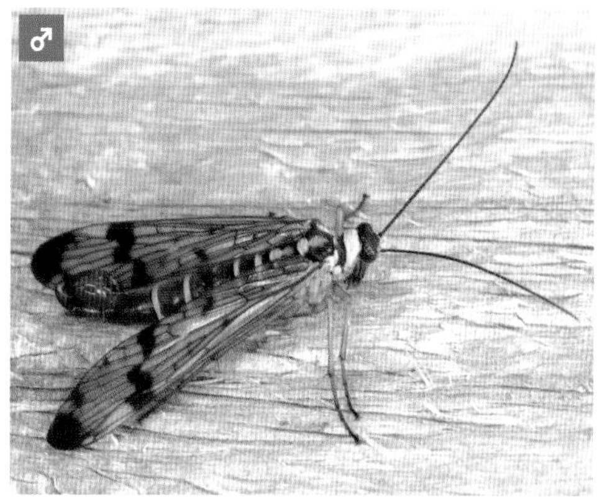

German Scorpionfly
Panorpa germanica

BL 10–12 mm, **WS** 22–28 mm.
ID Black, with yellow sides, marks on head (which is brown in some individuals) and pronotum, also yellow abdominal rings; genital capsule, and tip of abdomen in ♀, red. All wings transparent, with small to medium-sized black spots, although form *borealis* in Scotland lacks wing-spots or has much reduced spots. Otherwise, wingtips are usually bold black, but markings narrow. Legs yellowish-brown. **Hab** Hedgerows, woodland; often on Bramble or nettles. **Dist** Records from Ireland (if data labels are accurate) are from the early 1900s. **SS** Other scorpionflies.

Scarce Scorpionfly
Panorpa cognata

BL 10–12 mm, **WS** 25–28 mm.
ID Black, with yellow sides, marks on head (which is brown in some individuals) and pronotum, also yellow abdominal rings; genital capsule, and tip of abdomen in ♀, red. Brownish central marks on pronotum. All wings transparent with few black spots or marks; dark spot near tip often suffused but extreme tip more or less transparent; main black central spots merge and extend halfway across wing. Legs yellowish-brown. **Hab** Calcareous sites, damp woodland and scrub; often on low vegetation, including small birch saplings. **SS** Other scorpionflies.

FAMILY **Boreidae** (Snow fleas) 1 sp.

The single British species is known as the Snow Flea; in some countries the family is known as Snow Scorpionflies but many researchers consider they are closer to fleas than scorpionflies and a change in classification is likely in future. Wings are modified: in female very short; in male rather spine-like. The female has an obvious ovipositor. Ocelli are absent. Since they occur during the winter, these insects are sometimes seen on snow and can jump short distances, hence the vernacular name.

Snow Flea *Boreus hyemalis*

BL 3–5 mm. **ID** Yellowish-brown, with a rather dark head, antennae, pronotum and abdomen. Female ovipositor pale; male genital capsule swollen but not as noticeably as in scorpionflies and the callipers are curved when viewed from the side. **Hab** Heathland and woodland; usually seen crawling over moss. **SS** None.

Photographing Snow Fleas, Wyre Forest, Worcestershire

FURTHER READING AND USEFUL WEBSITES

Plant, C.W. 1997. ***A key to the adults of British lacewings and their allies (Neuroptera, Megaloptera, Raphidioptera and Mecoptera)***. Field Studies Council (AIDGAP).

lacewings.myspecies.info Lacewings and Allies Recording Scheme

ORDER TRICHOPTERA

Caddisflies

[Greek: *thrichos* (*thrix*) = hair; *pteron* = wing]

BI | 19 families [75 genera, **200 species** (3 endemic to Ireland)] **W** | 49 families [**approx. 15,200 species**]

The fluttering flight of caddisflies can recall some micro-moths, but the hairy wings help to separate them from these and other Lepidoptera. Another important characteristic is the wings, which are folded over their body like a roof (forewing length is 3–28 mm), and long antennae that are usually held out in front (other freshwater insects rest with their wings in different positions). In some species three ocelli are present as well as compound eyes, which can help identification to family level, in addition to differences in wing venation and the number of palps. The spurs present on the fore, mid and hind tibiae (not to be confused with spines that may also be present, but are usually a different colour) provide the 'tibia spur formula', which is important for separating some families. Specialist keys that require the use of a microscope are needed to identify families and species. Although many species lack distinctive wing patterning, the following photo guide to selected families should help with some commoner, more distinctive species. Adults do not feed and usually live for just a few days; there is little sexual dimorphism. The larvae are aquatic (except the Land Caddis), often building portable cases, the style of which can help to identify them to species level.

FINDING CADDISFLIES Day-flying species are frequently encountered flying or resting on or under waterside vegetation, including reeds and sedges. However, most are night-flying, attracted to moth traps and other lights.

Structure of a caddisfly

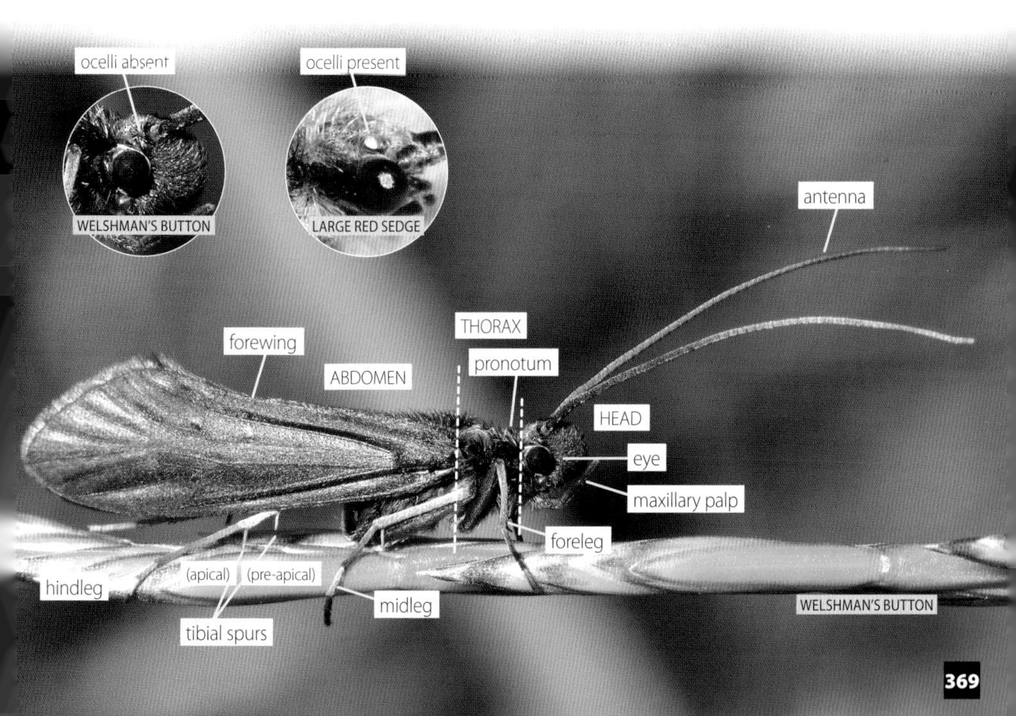

ORDER Trichoptera — GUIDE TO SELECTED FAMILIES | 9 of 19 BI

FWL = forewing length; **TSF** = tibia spur formula

OCELLI present

FWL 5–13 mm; **TSF** typically fore 2, mid 4, hind 4. **Flight** Apr–Sep.

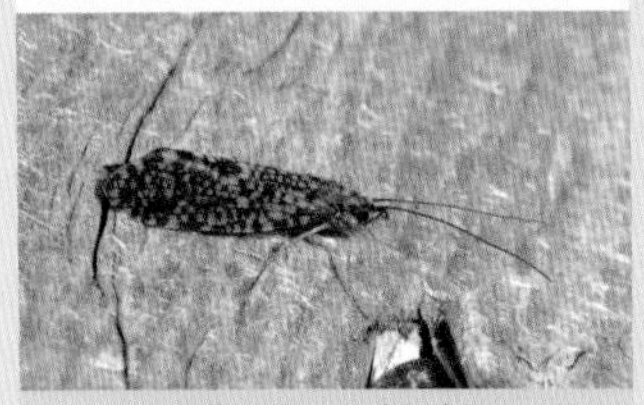

Yellow-spotted Sedge

Philopotamidae
[3 genera | 5 species]
p. 371

FWL 9–28 mm; **TSF** typically fore 2, mid 4, hind 4. **Flight** May–July.

Large Red Sedge ♀

Phryganeidae
[6 genera | 10 species]
p. 371

FWL 5–26 mm; **TSF** wide range. **Flight** May–Nov.

Bristly Sedge mating pair

Limnephilidae
[19 genera | 55 species]
p. 371

OCELLI absent

FWL 6–12 mm; **TSF** fore 2, mid 4, hind 4. **Flight** May–Aug.

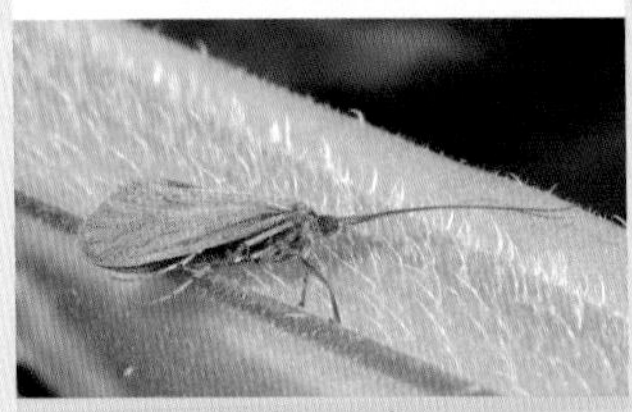

Medium Sedge

Goeridae
[2 genera | 3 species]
p. 374

FWL 13–17 mm. FORM very elongate; ANTENNAE exceeding body length; **TSF** fore 2, mid 4, hind 4. **Flight** Jun–Sep.

Silver Sedge

Odontoceridae
[1 species]
p. 376

FWL 9–16 mm. **ID** ♂ modified palps form a mask-like structure in front of the face; **TSF** fore 2, mid 4, hind 4. **Flight** Jun–Aug.

Welshman's Button mating pair

Sericostomatidae
[2 genera | 2 species]
p. 374

FWL 7–14 mm; **TSF** fore 2, mid 3, hind 3 **(unique)**. **Flight** Mar–Jun, often very early in spring.

Grannom

Brachycentridae
[1 species]
p. 374

FWL 4–6 mm; **TSF** fore 2, mid 2, hind 4. **Flight** May–Aug.

Dark Beraea

Beraeidae
[3 genera | 4 species]
p. 376

FWL 6–12 mm. FORM very elongate; ANTENNAE exceeding (often ≥2×) body length, typically banded black and white; **TSF** wide range, but 0–2 spurs on each tibia. **Flight** Jun–Sep.

White-headed Brown Silverhorn

Leptoceridae
[10 genera | 32 species]
p. 37

CADDISFLIES – OCELLI PRESENT 1/2

FAMILY Philopotamidae
3 GEN. | 5 spp. | 1 ILL.

Philopotamus 1 sp.

Yellow-spotted Sedge
Philopotamus montanus

Common

FWL 8–13 mm. **ID** Forewings spotted brown and yellow (various colour forms). **Hab** Fast-flowing rivers and streams. **SS** Superficially similar to other spotted caddisflies in the family Polycentropodidae [N/I], which have tibiae with a spur formula of fore 3, mid 4, hind 4, and lack ocelli.

FAMILY Phryganeidae
6 GEN. | 10 spp. | 1 ILL.

Phryganea 2 spp. | 1 ILL.

Large Red Sedge
Phryganea grandis

Common

FWL 18–28 mm. **ID** Very large (GB and Ireland's largest caddisfly); forewings brown, ♀ with darker central bar (less defined in ♂). **Hab** Still waters, including ponds, lakes, canals and slow rivers. **SS** Two-spotted Red Sedge *Phryganea bipunctata* [N/I] (**FWL** 18–26 mm), which is generally more uniform brown.

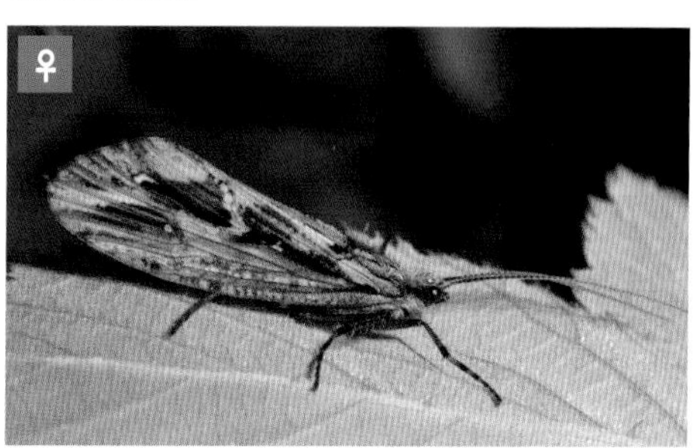

FAMILY Limnephilidae (Limnephilids)
19 GEN. | 55 spp. | 9 ILL.

Enoicyla 1 sp.

Land Caddis *Enoicyla pusilla*

FWL ♂ 5–6 mm. **BL** ♀ 4 mm. **ID** Small, ♂ forewings plain greyish-brown; ♀ wingless, brown (Britain's only wingless caddisfly). **Hab** Woodland. **Ls** Larvae develop on land in leaf-litter. **Dist** Only known from Wyre Forest, W Worcestershire and neighbouring counties. **SS** None, but ♂ superficially resembles other caddisfly species; late flight period and habitat should help.

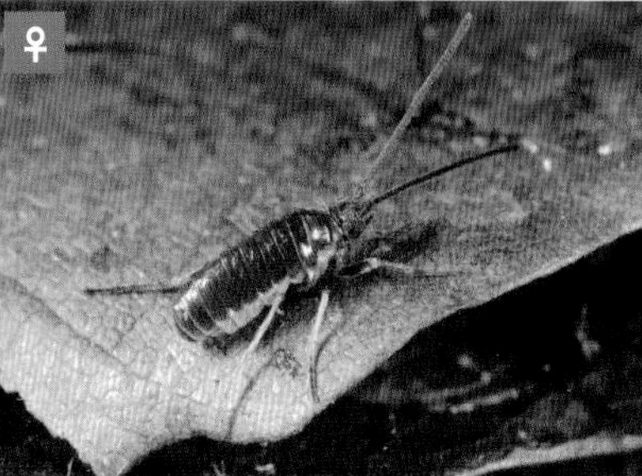

CADDISFLIES – OCELLI PRESENT 2/2

Limnephilus 29 spp. | 4 ILL.

Crescent Cinnamon Sedge
Limnephilus lunatus

Common

FWL 10–15 mm. **ID** Forewings brown with whitish markings; diagnostic pale crescent mark on wingtips, bordered with black. **Hab** Streams, lakes, ponds, rivers, wetlands. **SS** Cinnamon Sedge, Rhomboid Cinnamon Sedge and other *Limnephilus* species, from which told by the crescent mark on forewings.

Cinnamon Sedge
Limnephilus marmoratus

Common

FWL 12–17 mm. **ID** Forewings brown with several whitish markings, including larger blotches. **Hab** Lakes, ponds, rivers, pools (including temporary). **SS** Crescent Cinnamon Sedge, Rhomboid Cinnamon Sedge and other *Limnephilus* species, from which told by strongly marked forewings.

Rhomboid Cinnamon Sedge
Limnephilus rhombicus

Common

FWL 14–19 mm. **ID** Forewings very elongated, yellowish-brown with diagnostic large, pale, central rhomboid marking. **Hab** Lakes, ponds, rivers, marshes. **SS** Crescent Cinnamon Sedge, Cinnamon Sedge and other *Limnephilus* species, from which told by forewing shape and rhomboid central mark.

White-spotted Cinnamon Sedge *Limnephilus sparsus*

Common

FWL 10–13 mm. **ID** Forewings dark brown with whitish spots. **Hab** Marshes, pools, temporary puddles. **SS** Other *Limnephilus* species, from which told by white spots on forewings.

Chaetopteryx 1 sp.

Bristly Sedge
Chaetopteryx villosa

Common

FWL 6–12 mm. **ID** Forewings broad, brown, with dense, long bristly hairs; wingtips rounded. **Hab** Streams, rivers and stony-bottomed lakes. **SS** Broad Brown Sedge *Anabolia brevipennis* [N/I] (**FWL** 8–10 mm), which has narrower and less rounded tips to forewings.

Halesus 2 spp. | 1 ILL.

Streaked Caperer
Halesus radiatus

Common

FWL 17–23 mm. **ID** Forewings whitish with dark speckling and larger spots, and long, dark streaks between veins radiating towards wingtip. **Hab** Streams, rivers, lakes. **SS** Speckled Caperer *Halesus digitatus* [N/I] (**FWL** 17–23 mm), which has less well-defined markings.

Anabolia 1 sp.

Brown Sedge
Anabolia nervosa

Common

FWL 10–15 mm. **ID** Forewings dark brown, with two small white spots on each wing. **Hab** Large streams and ponds, also rivers and lakes. **SS** None.

Glyphotaelius 1 sp.

Mottled Sedge
Glyphotaelius pellucidus

Common

FWL 12–17 mm. **ID** Forewings brown, ♂ with light and darker markings (patterning mostly lacking in ♀). A notch in the outer margin of the forewing is diagnostic. **Hab** Woodland, using temporary streams and pools. **SS** None.

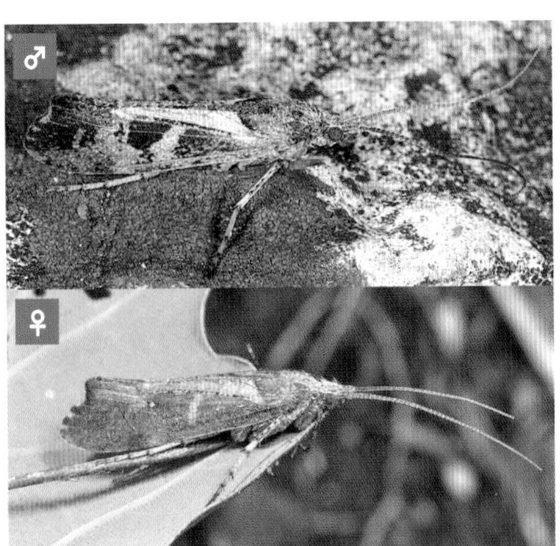

Guide to families p. 3

CADDISFLIES – OCELLI ABSENT

1/

FAMILY Brachycentridae 1 SP.

Brachycentrus 1 sp.

Grannom
Brachycentrus subnubilus

FWL 7–14 mm. **ID** Palps very hairy. Forewings greyish, with yellowish spots. **Hab** Fast rivers and streams. **SS** None.

FAMILY Goeridae 2 GEN. | 3 spp. | 1 ILL.

Goera 1 sp.

Medium Sedge *Goera pilosa*

FWL 8–12 mm. **ID** Forewings brownish-yellow, with yellow hairs. **Hab** Stony streams, rivers and lakes (commonest on chalk in SE England). **SS** Other plain brown caddisflies, from which told by careful examination of wing venation under a microscope, or from genitalia.

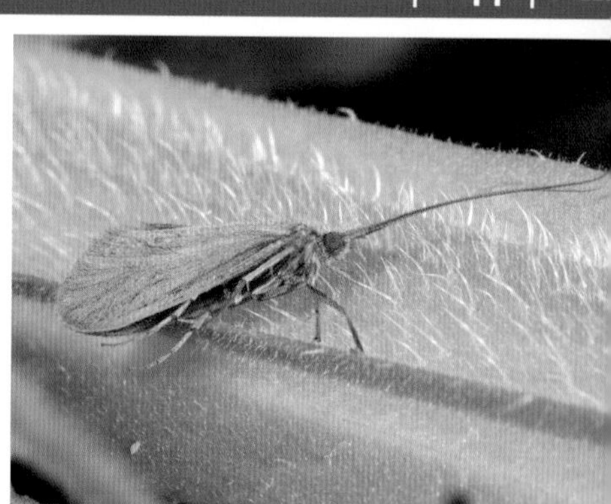

FAMILY Sericostomatidae 2 GEN. | 2 spp. | 1 ILL.

Sericostoma 1 sp.

Welshman's Button
Sericostoma personatum

FWL 9–16 mm. **ID** Head large. Modified palps in ♂ form a mask-like structure in front of the face. Forewings reddish-brown. **Hab** Streams, rivers, stony lakes. **SS** None, except superficially other brown-winged caddisflies, from which told by mask-like face structure.

FAMILY **Leptoceridae** (Long-horned caddis) **10 GEN. | 32 spp. | 4 ILL.**

Athripsodes **4 spp. | 2 ILL.**

Vhite-headed Brown ilverhorn

thripsodes albifrons

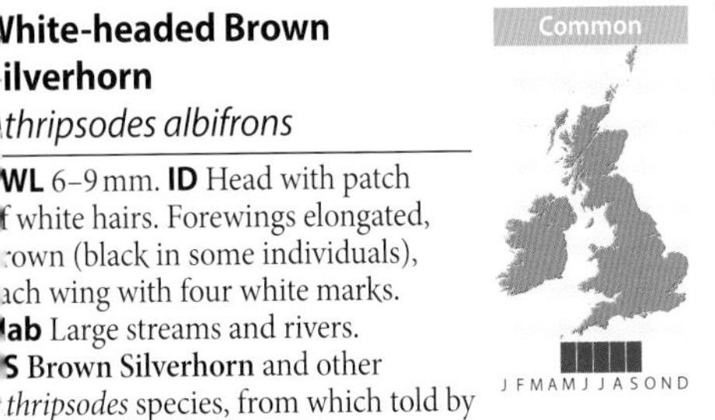

WL 6–9 mm. **ID** Head with patch f white hairs. Forewings elongated, rown (black in some individuals), ach wing with four white marks. **ab** Large streams and rivers. **S** **Brown Silverhorn** and other *thripsodes* species, from which told by aler colour or genitalia differences.

Brown Silverhorn

Athripsodes bilineatus

FWL 7–10 mm. **ID** Head black. Forewings elongated, black, each wing with up to four white marks. **Hab** Streams, lakes, rivers. **SS** **White-headed Brown Silverhorn** and other *Athripsodes* species, from which told by darker colour or genitalia differences.

Mystacides **3 spp. | 2 ILL.**

luish-black Silverhorn

lystacides azurea

WL 7–10 mm. **ID** Eyes red. Antennae ng, banded black and white (♂) or horter and white (♀). Maxillary palps rge, elbowed. Forewings elongated, lded, giving an angular shape at rest; lack with bluish sheen. **Hab** Streams, anals, stony-bottomed lakes, large onds, rivers. **SS** **Black Silverhorn** *lystacides nigra* [N/I] (**FWL** 7–10 mm) matt black (lacks sheen).

Grouse Wing

Mystacides longicornis

FWL 6–9 mm. **ID** Eyes red. Forewings elongated, usually banded yellow and brown, but a pale yellowish form is not uncommon. **Hab** Ponds, lakes, slow rivers. **SS** None (typical colour form).

Guide to families p. 3

CADDISFLIES – OCELLI ABSENT

FAMILY Beraeidae — 3 GEN. | 4 spp. | 1 ILL.

Beraea — 2 spp. | 1 ILL.

Dark Beraea *Beraea pullata*

FWL 4–6 mm. **ID** Small; forewings black; legs pale. **Hab** Flowing marshes, margins of streams. **SS** Other Beraeidae, separated by microscopic details of wing venation and genitalia.

FAMILY Odontoceridae — 1 sp.

Odontocerum — 1 sp.

Silver Sedge
Odontocerum albicorne

FWL 12–18 mm. **ID** Forewings elongated, pale with black markings and lines, particularly towards tip, that vary between individuals. Antennae distinctly toothed when viewed from the side. **Hab** Stony streams and rivers. **SS** None.

Typical caddisfly habitat.

FURTHER READING AND USEFUL WEBSITES

Barnard, P. & Ross, E. 2012. ***The adult Trichoptera (caddisflies) of Britain and Ireland. Handbooks for the identification of British insects***, Vol. 1, Part 17. Royal Entomological Society.

Crofts, S. M. 2019. ***Caddisfly Adults (Trichoptera) of Britain and Ireland. Family level keys and introductory guide.*** Freshwater Biological Association, Scientific Publication No. 70.

Wallace, I. 2006. ***Simple Key to Caddis Larvae.*** Field Studies Council (AIDGAP).

riverflies.org The Riverfly Partnership

RDER **STREPSIPTERA** Stylops

reek: *strepsis* = twisted; *pteron* = wing]

I | 4 families [7 genera, **10 species**] **W** | 9 families [**617 species**]

Mating pair of *Stylops ater* on abdomen of a solitary bee

ittle-known small endoparasitoids (also known as Twisted-winged Parasites) f other insects: in Great Britain and Ireland, *e.g. Andrena* bees (*p. 562*). A tylopised' bee is less likely to breed. The non-feeding, short-lived males ave large eyes and fan-shaped or branched antennae and look somewhat e a tiny fly – although only the few-veined hindwings function well. The rewings are the equivalent of a fly's elongated haltere. The adult female as rudimentary eyes, antennae and mouthparts, lacks wings and legs and oks like a larva. When inside a bee the female's head and thorax protrude om the bee's abdomen. After mating, eggs hatch in the female stylops' body and the resulting larvae isperse, using simple legs, possibly onto flowers so they can hitch a ride to a bees' nest. As this is a very ecialist, seldom-seen group, only one example is illustrated. Species are grouped by host: **Elenchids** species) and **Halictophagids** (1 genus, 2 species) on Hemiptera; **Stylopids** (3 genera, 5 species) n bees; and **Xenids** (2 genera, 2 species) on aculeate wasps – Vespidae and possibly Sphecidae. ifferences between the families include the structure of the antennae.

FINDING STYLOPS Examining bees closely may result in a female stylops being found, but very unlikely a male. Checking mining bee nests in April or May might be productive: look for a fly-like insect showing an interest in a bee's abdomen. Fewer than 20% of stylops are male, which only live for a few hours, mating on the host.

AMILY **Stylopidae** (Stylopids) 3 GEN. | 5 spp. | 1 ILL.

Stylops 1 sp.

ommon Stylops *Stylops* spp.

L 3–4 mm. **ID** ♀ brown, broad, flat, larva-like (head seen sticking out of the host); ♂ black with ller, narrowed abdomen and antennae broadened for ⅔rds of length from base. **Hab** Grassland, rdens, hedgerows and woodland; wherever the host mining bees, such as Chocolate Mining Bee . *564*), occur. **SS** Other stylopids, but identification is problematical as identification of the host is eded and some species use more than one host.

TE Images shown below may be of species with similar appearance, as *S. melittae* is at present wrongly ted as the only species in Britain and Ireland as based on DNA-barcoding studies in the Netherlands in 20 – which now has at least 12 *Stylops* species.

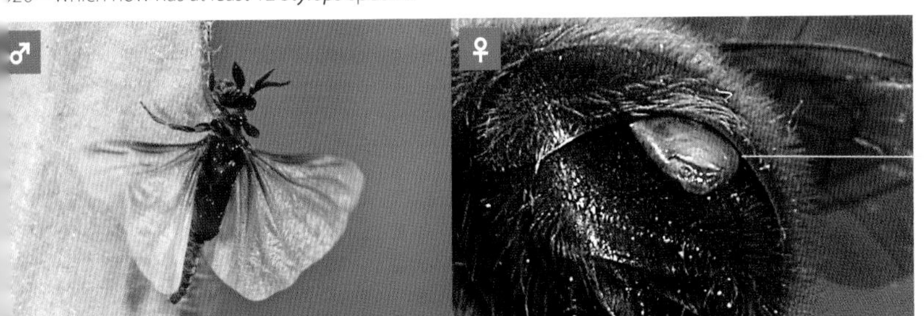

Common Stylops ♀ (in abdomen of Chocolate Mining Bee – the host indicates *S. aterrimus* or *S. nassonowi*)

ORDER SIPHONAPTERA

Flea

[Greek: *siphon* = tube; *aptera* = wingless]

BI | 7 families [28 genera, **62 species**] **W** | 18 families [**2,086 speci**

Small (<6 mm long), laterally compressed, wingless insects, light yellow to black in colour, adapted to moving between hair shafts and feathers. The antennae are very short. Often shiny, fleas have a number of bristles, most easily seen well under a microscope (which can help in identification; knowing the host also helps). Well known for their propensity to bite their hosts, fleas can jump as fa as 33 cm, or 100 times their body length, mostly due to long hindlegs with large coxae. Most species are associated with mammals (including humans), but in Great Britain and Ireland 16 species are birc parasites. Two common species are illustrated in this little-studied order, along with the important Human Flea, which has become rarer in recent generations.

FINDING FLEAS Most likely to be seen on dogs and cats, or by those who clean out birds' nest boxes.

Structure of a flea

FURTHER READING AND USEFUL WEBSITES

George, R.S. (Harding, P. [Ed].) 2008. ***Atlas of the Fleas (Siphonaptera) of Britain and Ireland***. Centre for Ecology and Hydrology.

Whitaker, A.P. 2007. ***Fleas (Siphonaptera)***. Handbooks for the identification of British insects, Part 1, vol. 17. Royal Entomological Society.

FAMILY Pulicidae (Mammal fleas) — 7 GEN. | 9 spp. | 2 ILL.

Includes well-known fleas associated with humans, cats and dogs; each species has a specific mammal (occasionally bird) host.

Ctenocephalides — 2 SPP. | 1 ILL.

Cat Flea
Ctenocephalides felis

BL 2–3 mm. **ID** Brown with paler legs; genal comb with 8 pointed spines; pronotal comb with 14–16 spines. **Hab** Urban areas. **Hosts** Cat (domestic), dog, occasionally other mammals (including humans) and birds. **SS** Many other fleas.

Common

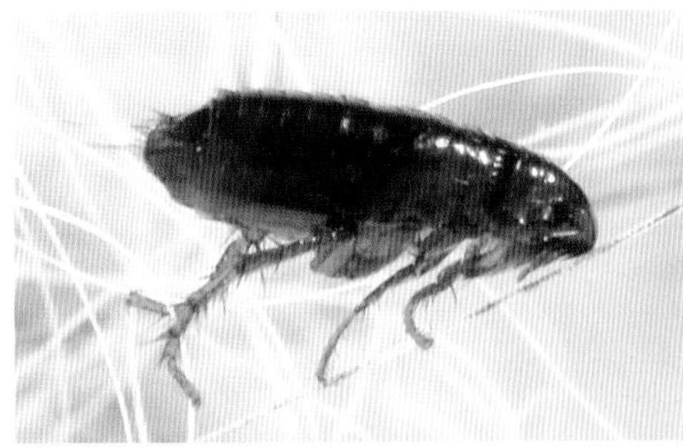

Pulex — 2 spp. | 1 ILL.

Human Flea *Pulex irritans*

BL 2–4 mm. **ID** Brown; lacking a genal or pronotal comb. **Hab** Urban areas, usually in houses. **Hosts** Humans; bites cause an itching sensation; can carry diseases (well known for being a vector of plague during the Middle Ages); often on hair, they bite anywhere on the body. Primary host possibly Badger, Fox, domesticated animals. **Dist** Although widespread, rare due to improved hygiene. **SS** None.

Rare (not Red Listed)

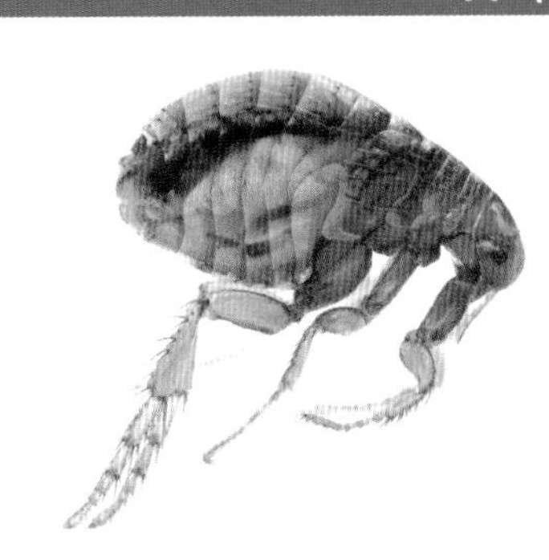

FAMILY Ceratophyllidae — 3 GEN. | 27 spp. | 1 ILL.

This family is divided into two groups: those associated with mammals (pronotal combs with <24 spines) and birds (pronotal combs with ≥24 spines). It includes Hen Flea, the commonest bird flea in Great Britain and Ireland.

Hen Flea
Ceratophyllus gallinae

BL 2–3 mm. **ID** Brown; genal comb absent; pronotal comb with 24–28 spines. **Hab** Birds' nests in trees, shrubs or nest boxes; hen-houses. **Hosts** Birds (>60 species) including poultry; occasionally mammals (including humans). **SS** Many other fleas.

Common

ORDER LEPIDOPTERA Butterflies & moth[s]

[Greek: *lepis* = scale; *pteron* = wing]

BI | 71 families [941 genera, **2,525 species**] **W** | 120 families [**158,570 specie[s]**

Popular insects with wingspans ranging from 2–135 mm in Great Britain and Ireland. As the name of the order implies, the two pairs of wings are usually covered in scales – those in the similar-looking caddisflies (*p. 369*) are hairy. Adults have a proboscis for feeding on nectar. The antennae are long: clubbed in day-flying butterflies, which normally rest with wings closed; various shapes in the mainly nocturnal moths, which rest in a range of different positions.

FINDING BUTTERFLIES & MOTHS: Butterflies Butterfly watchers tend to visit nature reserves with plentiful flowers, where the less common species tend to be more concentrated (moths are also frequent at such sites, although day-flying species are most likely to be seen).

FURTHER READING AND USEFUL WEBSITES

Newland, D., Still, R., Swash, A. & Tomlinson, D. 2020. ***Britain's Butterflies – A field guide to the butterflies of Great Britain and Ireland***. 4th Edition. Princeton University Press (Princeton WILD*Guides*).

Newland, D., Still, R. & Swash, A. 2019. ***Britain's Day-flying Moths – A field guide to the day-flying moths of Britain and Ireland.*** 2nd Edition. Princeton University Press (Princeton WILD*Guides*).

Both the books above complement this field guide and cover identification of adults in a particularly user-friendly format.

Sterling, P., Parsons M. & Lewington, R. 2018. ***Field Guide to the Micro-moths of Great Britain and Ireland***. Bloomsbury Wildlife Guides.

Thomas, J. & Lewington, R. 2019. ***The Butterflies of Britain & Ireland***. Revised Edition. British Wildlife Publishing.

Waring, P., Townsend, M. & Lewington, R. 2017. ***Field guide to the moths of Great Britain and Ireland***. 3rd Edition. Bloomsbury Wildlife Guides.
[Also a Concise Guide, 2nd Edition published in 2019.]

butterfly-conservation.org Butterfly Conservation – including recording details

ukleps.org Eggs, larvae, pupae and adult Butterflies and Moths

irishbutterflies.com Irish Butterflies

ukbutterflies.co.uk UK Butterflies

ukmoths.org.uk UK Moths

THE FOLLOWING POPULAR GUIDES ARE ALSO USEFUL:

Manley, C. 2021. ***British Moths – A Photographic Guide to the Moths of Britain and Ireland.*** 3rd Edition. Bloomsbury Wildlife Guides.
Photographic guide to macro moths and many micro moths.

Henwood, B., Sterling, P. & Lewington, R. 2020. ***Field Guide to the Caterpillars of Great Britain and Ireland***. Bloomsbury Wildlife Guides.

GREEN-VEINED WHITE (BUTTERFLY)

Structure of a butterfly

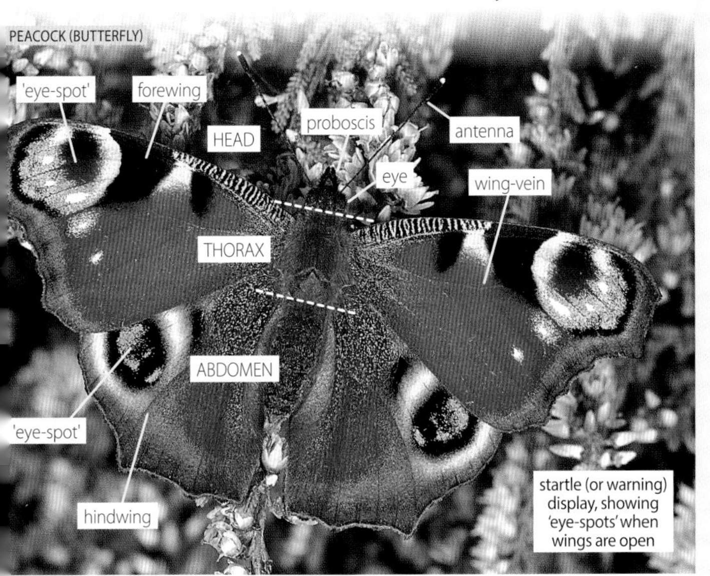

LEPIDOPTERA AT REST Although both butterflies and moths will rest with wings open, if closed most butterfly species hold their wings upright above the body; most moths hold their wings tent-like over the body. Exceptions are some skipper butterflies (see *p. 384*) and some moths (*e.g.* thorns (*p. 440*)).

Structure of a moth

LEPIDOPTERA ANTENNAE

In butterflies, tip of antennae **clubbed** or **slightly hooked** (as in the skippers); in moths, tip of antennae may be a variety of shapes – **club-like** (as in the day-flying burnet moths), **saw-edged** or **feathery**.

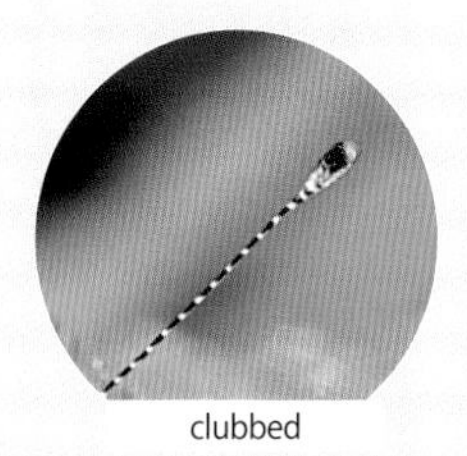

clubbed

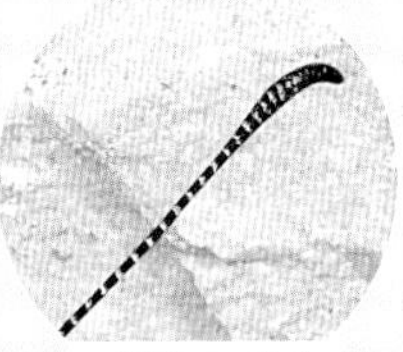

slightly hooked

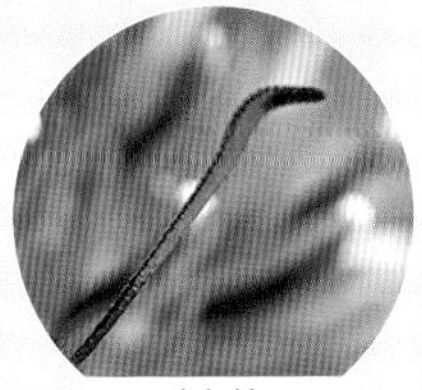

club-like

saw-edged

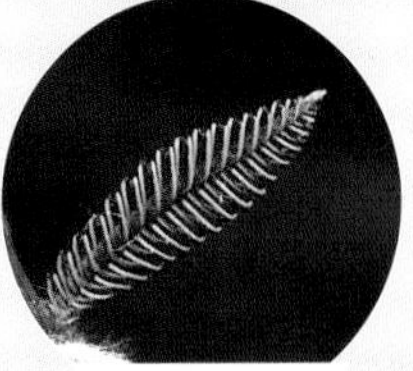

feathery

ORDER Lepidoptera (Butterflies)

This section includes all the breeding species and regularly occurring immigrants recorded in Great Britain and Ireland in recent years (a few rare vagrants (*e.g.* Berger's and Pale Clouded Yellows), 'probable vagrants' (*e.g.* Scarce Swallowtail), accidental imports or deliberate releases are omitted). **Length bars in the species accounts indicate forewing length (≈ ½ wingspan).**

For moth families see *pp. 407–413.*

Further details and fuller descriptions can be found in the companion WILD*Guides* volume *Britain's Butterflies* (see *page 380* for details).

FORM large (**WS** 76–93 mm), long-'tailed', yellow-and-black; unmistakable.

Swallowtails
Papilionidae
[Swallowtail]

p. 383

FORM small to medium-sized (**WS** 23–37 mm), rather moth-like; often orange/brown; some species rest with hindwings flat out and forewings raised. ANTENNAE slightly hooked.

Small Skipper

Dingy Skipper

Skippers
Hesperiidae
[6 genera | 8 species]

p. 38

FORM medium-sized to large (**WS** 38–76 mm), mainly white or yellow; wingtips rounded (angular in one species).

Small White

Clouded Yellow

Whites and yellows
Pieridae
[7 genera | 12 species (1†, 3 rare immigrants)]

p. 387

FORM small (**WS** 29–34 mm), orang and dark brown

Metalmarks
Riodinidae
[Duke of Burgundy]

p. 40

FORM medium-sized to large (**WS** 30–100 mm), variably coloured and patterned, including maroon, red, brown, orange and black, wingtips rounded.

Painted Lady | Small Pearl-bordered Fritillary | Gatekeeper

Nymphalids, fritillaries and browns
Nymphalidae

[20 genera | 31 species (1†, 4 rare immigrants)]

p. 390

FORM Small to medium-sized (**WS** 18–52 mm), some with a metallic sheen. Hairstreaks have short 'tails'

Purple Hairstreak | Small Copper | Common Blue

Hairstreaks, coppers and blues
Lycaenidae

[14 genera | 20 species (2†, 3 rare immigrants)]

p. 401

FAMILY Papilionidae (Swallowtails) 1 sp.

One large species, black-and-yellow, with long 'tails'; sometimes double-brooded. Occasional immigrants from Europe (the paler subspecies *gorganus*) find their way to Britain, but have not yet become established.

Papilio 1 sp.

wallowtail NT Rare

apilio machaon

/S 76–93 mm. **ID** Black-and-yellow, ndwing with blue markings and red ɔot (British subspecies, *britannicus* ıs bolder black wing veins than the re vagrant subspecies *gorganus* from ɔntinental Europe). **Hab** Open fens ıd marshes. **Beh** Sometimes seen ectaring (usually with wings flapping) ı Yellow Iris, Ragged-Robin and istles. **Fp** Milk-parsley. **SS** None.

J F M A M J J A S O N D

Guide to families *pp.* 382–38

FAMILY **Hesperiidae** (Skippers) — 6 GEN. | 8 spp.

Small to medium-sized (WS 24–37 mm) and rather moth-like, some species resting with hindwings flat out and forewings raised. The antennae are slightly hooked, unlike other butterfly families. Typical grassland species, they dart speedily from flower to flower, or to bask. The species can be identified as follows:

UPPERWINGS **golden-orange with darker edge, ♂ with black sex-brand on each forewing;** FORM **hindwings held flat with forewings raised when at rest.**			
UPPERWINGS with some paler markings; UNDERWINGS marked			
UNDERWINGS faintly mottled on underside		**Large Skipper**	***p.* 384**
UNDERWINGS distinctive silver spots on underside		**Silver-spotted Skipper**	***p.* 385**
UPPERWINGS plain; UNDERWINGS plain, with dark line inside pale edge			
UPPERWINGS male dark, almost olive-brown; female with golden circle		**Lulworth Skipper**	***p.* 385**
UPPERWINGS golden-orange	ANTENNAE tip orange underneath	**Small Skipper**	***p.* 385**
	ANTENNAE tip black underneath	**Essex Skipper**	***p.* 385**
UPPERWINGS **dark brown, strongly mottled or spotted;** FORM **wings held flat when at rest.**			
ABDOMEN noticeably broadened laterally; ANTENNAE long			
UPPERWINGS very dark with grey mottling		**Dingy Skipper**	***p.* 386**
UPPERWINGS dark with pale spotting	UPPERWINGS orange-spotted	**Chequered Skipper**	***p.* 386**
	UPPERWINGS white-spotted	**Grizzled Skipper**	***p.* 386**

SKIPPERS – Wings golden-orange with some paler markings

Ochlodes — 1 sp.

Large Skipper *Ochlodes sylvanus*

WS 29–36 mm. **ID** Upperwings bright golden with darker brown areas and paler orange spots; underwings faintly mottled. **Hab** Tall grassland, gardens, parks and woodland rides. **Fp** Various grasses, often Cock's-foot. **SS** **Silver-spotted Skipper**; also *Thymelicus* skippers.

Typical skipper habitat, Stockbridge Down, Hampshire.

Hesperia 1 sp.

;ilver-spotted Skipper NT
-lesperia comma

/VS 29–37 mm. **ID** White spots on ıpperwings and underwings. **lab** Chalk grassland with short turf. **'p** Sheep's-fescue. **SS** Large Skipper.

J F M A M J J A S O N D

;KIPPERS – Wings golden-orange; plain with dark margin

Thymelicus 3 spp.

;mall Skipper *Thymelicus sylvestris*

/VS 27–34 mm. **ID** Antennae tips orange underneath; ♂ forewings vith slightly curved black sex brands. **Hab** Tall grassland, roadside erges and woodland rides. **Fp** Grasses, especially Yorkshire-fog. 'S Other *Thymelicus* skippers, **Large Skipper**.

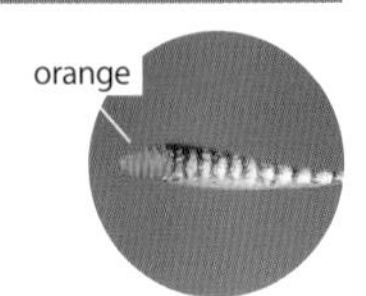

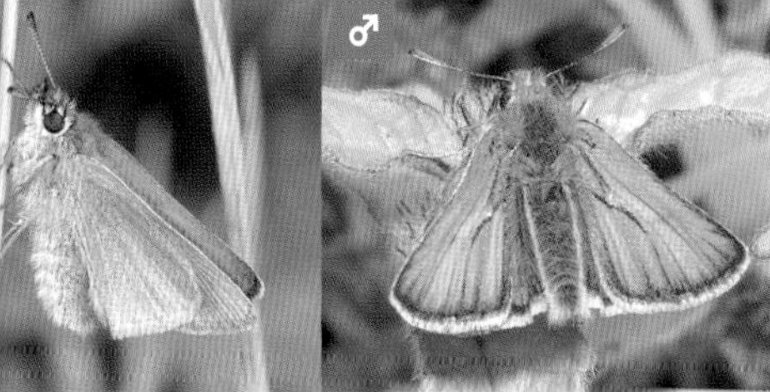

J F M A M J J A S O N D

;ssex Skipper
-hymelicus lineola

'S 26–30 mm. **ID** Antennae tips black nderneath; ♂ forewings with short, raight, black sex brands. **Hab** Tall rassland. **Fp** Grasses, usually Cock's-ot. **SS** Other *Thymelicus* skippers, arge Skipper.

Lulworth Skipper NT
Thymelicus acteon

WS 24–28 mm. **ID** ♂ upperwings dark, olive-brown; ♀ forewings with golden crescent ('sun-ray' pattern). **Hab** Chalk grassland. **Fp** Tor-grass. **SS** Other *Thymelicus* skippers, **Large Skipper**.

J F M A M J J A S O N D

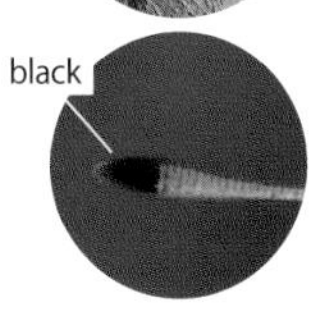

golden crescent

SKIPPERS – Wings dark; strongly patterned

Erynnis 1 sp.

Dingy Skipper *Erynnis tages* VU NT I

WS 27–34 mm. **ID** Wings dark brown with grey markings. **Hab** Open sites, such as chalk and other grasslands, heathland. **Beh** Often seen basking on bare ground. **Fp** Common Bird's-foot-trefoil. **SS** No butterflies but superficial resembles day-flying moths, such as **Burnet Companion** and **Mother Shipton** (both *p. 449*).

Pyrgus 1 sp.

Grizzled Skipper *Pyrgus malvae* VU

WS 23–29 mm. **ID** Wings dark brown (some newly emerged individuals black) with white spots. **Hab** Chalk grasslar heathland, woodland rides and waste ground. **Beh** Often seen basking on stones or bare ground, sometimes with Dingy Skipper. **Fp** Wild Strawberry, Agrimony, Creeping Cinquefoil, *etc.* **SS** None.

Carterocephalus 1 sp

Chequered Skipper *Carterocephalus palaemon* EN

WS 29–31 mm. **ID** Orange spots on upperwings; white on underwings. **Hab** Woodland edges. **Fp** Purple Moor-grass **Dist** Native in parts of western Scotland. Extinct in England since 1975 (but reintroduced at Rockingham Forest, Northamptonshire and signs that they are becoming established again). **SS** None.

FAMILY **Pieridae** (Whites and yellows)

7 GEN. | 12 spp. (including **1†** sp. and **3** rare immigrants)

Medium-sized to large (WS 36–74 mm), mainly white or yellow; wingtips rounded except in one hook-tipped species. The antennae have a clubbed tip. Includes pest species well known to gardeners, the caterpillars feeding on brassicas and nasturtiums. Species can be identified as follows:

<table>
<tr><th colspan="5">FOREWINGS with hooked tip</th></tr>
<tr><td colspan="3">WINGS sulphur-yellow in ♂; greenish-white in ♀</td><td>Brimstone</td><td>p. 387</td></tr>
<tr><th colspan="5">FOREWINGS with rounded tip</th></tr>
<tr><td colspan="5">WINGS white</td></tr>
<tr><td colspan="5">WINGS very delicate-looking, narrow, with black or grey tip</td></tr>
<tr><td colspan="3" rowspan="2">Virtually identical (underside of Cryptic Wood White typically more suffused) but genitalia differences or molecular analysis needed to confirm identification.</td><td>Wood White</td><td>p. 388</td></tr>
<tr><td>Cryptic Wood White</td><td>p. 388</td></tr>
<tr><td colspan="5">WINGS broader, ♀ forewing usually with 1 or 2 black spots</td></tr>
<tr><td rowspan="3">HINDWING underside not distinctly mottled green</td><td rowspan="2">UNDERWING plain, without bold veins</td><td>FOREWING ♂ 1 black spot; ♀ black tip (if present) extends <⅓ down front of wing</td><td>Small White</td><td>p. 389</td></tr>
<tr><td>FOREWING ♂ no black spot; ♀ black tip extends > halfway down front of wing</td><td>Large White</td><td>p. 389</td></tr>
<tr><td colspan="2">UNDERWING with bold veins</td><td>Green-veined White</td><td>p. 389</td></tr>
<tr><td rowspan="2">HINDWING underside mottled green</td><td colspan="2">FOREWING ♂ orange tip; ♀ black tip, small black spot</td><td>Orange-tip</td><td>p. 388</td></tr>
<tr><td colspan="2">FOREWING black tip with white spots, large black spot</td><td>Bath White Pontia daplidice</td><td>N/I</td></tr>
<tr><td colspan="3">WINGS with black veins</td><td>Black-veined White Aporia crataegi</td><td>N/I [†]</td></tr>
<tr><td colspan="5">WINGS yellow</td></tr>
<tr><td colspan="3">UPPERSIDE deep yellowish-orange, ♀ FOREWING yellow dots on black outer margin, HINDWING black margin broad [♀ form helice UPPERSIDE whitish]</td><td>Clouded Yellow</td><td>p. 388</td></tr>
<tr><td rowspan="2">UPPERSIDE ♂ lemon-yellow; ♀ whitish, difficult to tell apart and from ♀ form helice of Clouded Yellow.</td><td rowspan="2">HINDWING UPPERSIDE (rarely visible)</td><td>black margin very narrow or broken</td><td>Berger's Clouded Yellow</td><td>N/I</td></tr>
<tr><td>black margin narrow, unbroken</td><td>Pale Clouded Yellow</td><td>N/I</td></tr>
</table>

VHITES and YELLOWS – Forewings hooked

Gonepteryx 1 sp.

rimstone *Gonepteryx rhamni* **I**

'S 60–74 mm. **ID** ♂ wings sulphur-ellow, ♀ greenish-white; forewing ps hooked. **Hab** Woodland rides, edgerows, grassland with scrub, gardens. **o** Buckthorn and Alder Buckthorn. **eh** Overwinters as adult, sometimes es during sunny spells, but usually rst seen in spring. **SS** None if seen well, hough ♀ could be mistaken in flight for arge White (*p.* 389).

Guide to Pieridae *p. 38*

WHITES and YELLOWS – Forewings narrow

Leptidea 2 spp.

Very delicate-looking; narrow forewings with black or grey tip.

Wood White EN NT
Leptidea sinapis

WS 36–48 mm. **ID** Underwings white with suffused grey markings (in the Burren, County Clare, Ireland, underside more greenish). **Hab** Woodland rides. **Fp** Various legumes. **SS Cryptic Wood White** (Ireland).

Cryptic Wood White I
Leptidea juvernica

WS 36–48 mm. **ID & Fp** Virtually identical to Wood White, although underside typically more suffused. Identified by genitalia differences or by molecular analysis. **Hab** Open habitats, including grassland and sand dunes. **SS Wood White** (in Ireland) – see *above*.

WHITES and YELLOWS – Forewings rounded, broad

Colias 3 spp. | 1 ILL.

Clouded Yellow *Colias croceus*

WS 52–58 mm. **ID** Deep yellow (almost orange), upperwings with broad black margin (yellow-spotted in ♀). Approx. 10% of ♀s are whitish or pale lemon above and pale below (form *helice*). **Hab** Chalk and other grassland, with plentiful clovers. **Fp** Clovers. **SS Pale Clouded Yellow** *Colias hyale* [N/I] (**WS** 52–62 mm), which has unbroken narrow black border to upper side of hindwing; **Berger's Clouded Yellow** *Colias alfacariensis* [N/I] (**WS** 50–60 mm), which has very narrow or broken black border to upper side of hindwing.

Anthocharis 1 sp.

Orange-tip
Anthocharis cardamines

WS 40–52 mm. **ID** ♂ forewings orange-tipped; ♀ forewings black-tipped and have small black spot; underside mottled green in both sexes. **Hab** Damp grassland, roadside verges and woodland rides. **Fp** Crucifers, including Cuckooflower, Garlic Mustard. **SS** ♂: none; ♀: **Bath White** *Pontia daplidice* [N/I] (**WS** 48–52 mm), which has white spots in black wingtip.

WHITES and YELLOWS – Forewings rounded, wings broad and white

Pieris 3 spp.

Large White *Pieris brassicae*

WS 53–70 mm. **ID** Forewings with extensive black tip; ♀ with 2 large black spots and streak (plain in ♂).
Hab Anywhere, but often noticed in gardens and allotments, or along the coast during immigration.
Fp Crucifers, particularly cabbage (hence regarded as a pest).

Small White *Pieris rapae*

WS 38–57 mm. **ID** Forewings with grey to black tip on top edge only; ♀ with 2 large black spots and streak (one black spot in ♂). **Hab** Anywhere, but often noticed in gardens and allotments, or along the coast during immigration.
Fp Crucifers, particularly cabbage (hence regarded as a pest).

Green-veined White *Pieris napi*

WS 40–52 mm. **ID** Forewings with restricted black or greyish tip; ♀ with 2 black spots (1 spot in ♂); underside with bold veins. **Hab** Wetlands, hedgerows, woodland edges. **Fp** Wild crucifers, including Garlic Mustard and Cuckooflower.

Guide to families *pp. 382–38*

FAMILY **Nymphalidae** (Nymphalids, fritillaries and browns)

20 GEN. | 31 spp. (including **1†** sp. and **4** rare immigrants)

Medium-sized to large (WS 30–100 mm), wide range of colours and patterns, including browns and orange-and-black fritillaries; wingtips rounded. The antennae have a clubbed tip. Includes familiar species often seen nectaring on flowers, including Buddleja, the 'butterfly bush'. The species can be identified as follows:

WINGS various colours with distinctive patterns (1 sp. orange, spotted black, with ragged wing edges)		**NYMPHALIDS**	
UPPERWINGS blackish with white band and markings; UNDERWINGS orange-brown with white band and markings		**White Admiral**	*p. 391*
UPPERWINGS ♂ with purple reflection; ♀ dark brown with white bands		**Purple Emperor**	*p. 391*
UPPERWINGS upperside maroon with large blue and black 'eye-spot' on each wing; UNDERWINGS dark brown with narrow pale lines (see *p. 393*)		**Peacock**	*p. 392*
UPPERWINGS dark brown with a row of blue spots and a cream border		**Camberwell Beauty**	*p. 393*
UPPERWINGS black, scarlet and white; UNDERWINGS hindwing dark (see *p. 393*)		**Red Admiral**	*p. 392*
WINGS pinkish-orange with black marks; wingtips dark with white spots; UNDERWINGS pale, subtly patterned darker (see *p. 393*)		**Painted Lady**	*p. 392*
UPPERWINGS orange with black and white markings near tip and dark marginal borders; UNDERWINGS hindwings dark brown at base, outer part and forewings paler (see *p. 393*)	FOREWINGS upperwing with white spot near tip; narrow dark border; 3 black spots in centre	**Small Tortoiseshell**	*p. 392*
	FOREWINGS upperwing with no white spot near tip; narrow dark border; 4 black spots in centre	**Large Tortoiseshell**	*p. 393*
	FOREWINGS upperwing with white spot near tip; broad black border; 4 black spots in centre	**Scarce Tortoiseshell**	*p. 393*
WINGS ragged-edged; UNDERWINGS brown, hindwing with white 'comma' mark (see *p. 393*)		**Comma**	*p. 393*
FORM large; WINGS orange, bordered black, with black veins; unmistakable		**Monarch**	*p. 393*

WINGS orange, spotted black (wing edges not ragged); underwings patterned		**FRITILLARIES**	
FORM **large (WS 55–80 mm); powerful flight**			
UNDERWINGS hindwing with red-ringed, silver spots		**High Brown Fritillary**	*p. 394*
UNDERWINGS hindwing greenish with silver wash and whitish bars		**Silver-washed Fritillary**	*p. 394*
UNDERWINGS hindwing with green wash; lacks red-ringed silver spots		**Dark Green Fritillary**	*p. 394*
FORM **smaller (WS 30–52 mm); weaker flight**			
UNDERWINGS hindwing with border of 'pearl'-like spots	UNDERWINGS hindwing with two whitish blotches; seven silver 'pearls' along edge with **orange** chevrons	**Pearl-bordered Fritillary**	*p. 395*
	UNDERWINGS hindwing with several whitish blotches; seven silver 'pearls' along edge with **black** chevrons	**Small Pearl-bordered Fritillary**	*p. 395*
UNDERWINGS hindwing with many white areas	UNDERWINGS hindwing with black spots in white areas	**Glanville Fritillary**	*p. 396*
	UNDERWINGS hindwing without black spots	**Heath Fritillary**	*p. 396*
UPPERWINGS orange, brown and yellow; UNDERWINGS hindwing orange-and-yellow with white spots and border		**Marsh Fritillary**	*p. 39*
FORM hindwings 'squarish'; UNDERWINGS hindwing with large silver patches		**Queen of Spain Fritillary**	*p. 39*

WINGS brown and/or orange (1 sp. black-and-white) with 1 or more black 'eye-spot(s)' on both sides of forewings		BROWNS	
WINGS black-and-white (or black-and-cream)		Marbled White	*p.397*
WINGS dark brown with cream spots		Speckled Wood	*p.397*
UPPERWINGS golden-and-brown with distinctive 'eye-spots'; UNDERWINGS with distinct zigzag pattern		Wall	*p.400*
UPPERWINGS orange-and-brown; UNDERWINGS brown, mottled grey, some grey zigzag		Grayling	*p.400*
UPPERWINGS orange with broad brown margin; two white 'pupils' in black 'eye-spots'		Gatekeeper	*p.399*
UPPERWINGS pale orange	UNDERWINGS hindwing lacks 'eye-spots'	Small Heath	*p.398*
	UNDERWINGS hindwing with several 'eye-spots' [except in form *scotica* in Scotland, but in this form the long, whitish band is distinctive]	Large Heath	*p.398*
UPPERWINGS mostly dark brown (lowlands)	UPPERWINGS 'eye-spots' small and indistinct; UNDERWINGS 'eye-spots' large and yellow-ringed	Ringlet	*p.399*
	WINGS 'eye-spots' distinct, each with one white 'pupil'; UPPERWINGS ♀ with orange patches	Meadow Brown	*p.399*
UPPERWINGS dark brown; (mountains)	UPPERWINGS distinctive bright, broad orange bands; 'eye-spots' black with a central white 'pupil'	Scotch Argus	*p.398*
	UPPERWINGS orange bands (narrow in some forms, particularly ♂s); 'eye-spots' small, black, lacking white central dot	Mountain Ringlet	*p.398*

NYMPHALIDS 1/2

Limenitis 1 sp.

White Admiral VU Local

Limenitis camilla

WS 56–66 mm. **ID** Wings blackish with white band and markings; underwings orange-brown with grey and white markings. **Hab** Woodland rides and clearings. **Fp** Honeysuckle. **SS** Purple Emperor (underwings).

Apatura 1 sp.

Purple Emperor NT Local

Apatura iris

WS 70–92 mm. **ID** ♂ wings white-banded with purple reflection, ♀ dark brown with white bands. **Hab** Woodland rides and clearings. **Beh** Attracted to animal excrement and puddles. **Fp** Willows. **SS** White Admiral (underwings).

Guide to Nymphalidae *pp. 390–39*

NYMPHALIDS

Vanessa 3 spp. | 2 ILL.

Painted Lady *Vanessa cardui*

WS 58–74 mm. **ID** Wings pinkish-orange, forewing tips black with white spots; underside has a wide range of subtle markings, but is always pale. **Hab** Grassland, gardens. **Fp** Thistles. **Beh** Reliant on immigrants as unable to survive the winter in GB and Ireland. Numbers depend on arrivals from North Africa, via Europe; in a bumper year, can be seen everywhere. **SS** None.

Red Admiral *Vanessa atalanta*

WS 64–78 mm. **ID** Wings black, scarlet and white; underside of hindwing dark. **Hab** Woodland rides, meadows, gardens. **Fp** Common Nettle. **Beh** Overwinters as adult; regular migrant, population in GB/Ireland bolstered by continental immigrants, some of which make the return journey. **SS** Only possible confusion is with other nymphalids at distance, easily recognized close up.

Aglais 2 spp.

Small Tortoiseshell *Aglais urticae*

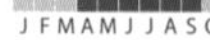

WS 45–62 mm. **ID** Wings bright orange, with large black, yellow and whitish markings and marginal blue spots; underside black and whitish. **Hab** Grassland, wastelands, gardens. **Fp** Common Nettle. **Beh** Overwinters as adult. **SS** Large Tortoiseshell.

Peacock *Aglais io*

WS 63–75 mm. **ID** Wings rich reddish-brown, each with large 'eye-spot', coloured black, blue and yellow; underside blackish-brown. **Hab** Woodland edges and clearings; gardens. **Fp** Common Nettle. **Beh** Overwinters as adult. **SS** None.

Polygonia 1 sp.

Comma *Polygonia c-album*

WS 50–64 mm. **ID** Wings bright orange, with dark brown to black markings and diagnostic **ragged-edged wings**; underside light (spring) or dark (summer); hindwing with a central white 'comma' mark. **Hab** Woodlands, hedgerows, gardens. **Fp** Common Nettle, elms, currants. **Beh** Overwinters as adult. **SS** Fritillaries (*p. 394*) at a distance, but easily recognized close up.

NYMPHALID UNDERSIDES

Nymphalids often rest with their wings held together; the patterning on the underside of the hindwings, and particularly the forewings is distinctive.

SMALL TORTOISESHELL

COMMA

PEACOCK

PAINTED LADY

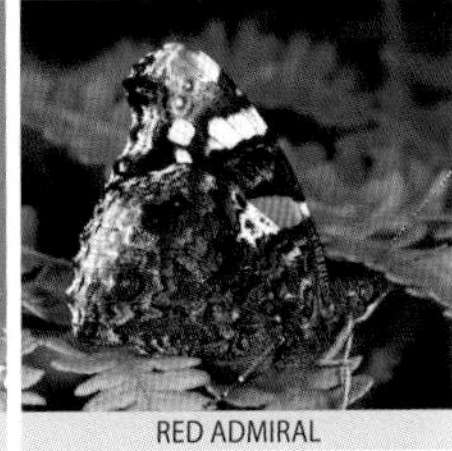
RED ADMIRAL

RARE NYMPHALIDS and FRITILLARIES

These species are rare immigrants from continental Europe, although recent records of Large Tortoiseshell (a former resident) and Queen of Spain Fritillary indicate possible evidence of breeding. Some records are regarded as suspect, as these may relate to escapees from culture stocks.

Monarch *Danaus plexippus*
WS 95–100 mm. Rare immigrant from USA and/or Macaronesia

Queen of Spain Fritillary *Issoria lathonia*
WS 34–56 mm. Rare immigrant from continental Europe

Camberwell Beauty *Nymphalis antiopa*
WS 76–88 mm. Rare immigrant from continental Europe

Large Tortoiseshell *Nymphalis polychloros*
WS 68–75 mm. Rare immigrant from Europe; occasional breeder

Scarce Tortoiseshell *Nymphalis xanthomelas*
WS 56–68 mm. Rare immigrant from continental Europe

Guide to Nymphalidae pp. 390–39

FRITILLARIES – LARGE

Speyeria 1 sp.

Dark Green Fritillary VU
Speyeria aglaja

WS 56–68 mm. **ID** Wings orange with black spots; ♀ darker than ♂ (those from Scotland and Ireland most heavily marked); underside of hindwings with white spots and green wash. **Hab** Calcareous and sometimes other grasslands, also coastal dunes; rarely woodland rides. **Fp** Various violets. **SS High Brown Fritillary**; in flight, **Silver-washed Fritillary**.

Fabriciana 1 sp.

High Brown Fritillary CR
Fabriciana adippe

WS 55–69 mm. **ID** Wings orange with black spots; 3rd spot on upperside of forewing indented compared with outer two; underside of hindwings with red-ringed, silver spots. **Hab** Bracken-covered, south-facing slopes or grass/Bracken mosaics and limestone outcrops following woodland/scrub clearance. **Fp** Various violets. **SS** Other large fritillaries.

Argynnis 1 sp.

Silver-washed Fritillary
Argynnis paphia

WS 69–80 mm. **ID** Wings orange with black spots, ♂ forewings each with four sex brands on veins, ♀ more heavily marked, appearing darker; wings with greenish tinge, particularly hindwings; underside of hindwings with whitish bars and silver wash. **Hab** Woodland rides, sometimes wooded hedgerows, churchyards. **Fp** Common Dog-violet. **SS** Other large fritillaries.

LARGE FRITILLARY UNDERSIDES

SILVER-WASHED FRITILLARY
silver wash, whitish bars (no white spots)

DARK GREEN FRITILLARY
green wash, white spots (no reddish)

HIGH BROWN FRITILLARY
red-ringed silver spots

FRITILLARIES – SMALL 1/2

Boloria 2 spp.

Small Pearl-bordered Fritillary *Boloria selene* NT

WS 35–44 mm. **ID** Underside of hindwings with **several white blotches** and seven silver 'pearls' edged with **black chevrons**. **Hab** Open woodland and clearings, damp grassland, moorlands, with plentiful foodplants. **Fp** Violets. **SS** Pearl-bordered Fritillary.

Pearl-bordered Fritillary *Boloria euphrosyne* EN EN

WS 35–47 mm. **ID** Underside of hindwings with **two white blotches** and seven silver 'pearls' edged with **orange chevrons**. **Hab** Coppiced woodland, woodland edges, grassland, with plentiful foodplants. **Fp** Violets. **SS** Small Pearl-bordered Fritillary.

Typical Pearl-bordered Fritillary habitat in the New Forest, Hampshire.

Guide to Nymphalidae *pp. 390–3*

FRITILLARIES – SMALL

Melitaea — 2 spp.

Heath Fritillary *Melitaea athalia* EN

WS 39–47 mm. **ID** Wings dark brown, chequered orange; underside of hindwings with much white but no black spots. **Hab** Woodland clearings, sheltered valleys. **Fp** Common Cow-wheat, Ribwort Plantain, speedwells, *etc.* **SS Marsh, Glanville Fritillaries.**

Glanville Fritillary *Melitaea cinxia* EN

WS 38–52 mm. **ID** Wings dark brown and orange with white rear border; underside of hindwings with much white and many black spots. **Hab** Undercliffs. **Fp** Ribwort Plantain. **SS Heath Fritillary.**

Euphydryas — 1 sp

Marsh Fritillary *Euphydryas aurinia*

WS 30–50 mm. **ID** Wings bright orange, brown and yellowish; underside of hindwings orange with yellowish blotches and white border. **Hab** Wetlands. **Fp** Devil's-bit Scabious, occasionally Field and Small Scabiouses. **SS** Heath Fritillary.

ʒROWNS 1/3

Parage 1 sp.

;peckled Wood *Parage aegeria*

VS 46–56 mm. **ID** Wings dark brown with cream spots. Underwings mottled brown. **Hab** Woodland, churchyards, ardens and hedgerows. **Fp** Grasses. **Beh** Likes dappled sunlight. **SS** None.

Melanargia 1 sp.

ˈlarbled White *Melanargia galathea*

/S 50–60 mm. **ID** Wings black-and-white or black-and-cream. **Hab** Grassland (often chalk or limestone). **p** Grasses. **SS** None, but superficially, from a distance, in flight could be confused with white butterflies.

BROWNS

2/

Coenonympha 2 spp.

Small Heath *Coenonympha pamphilus*

NT NT

WS 33–37 mm. **ID** Wings pale orange; underside of hindwings with short pale band; no 'eye-spots'. **Hab** Grassland. **Fp** Fine-leaved grasses. **SS** Large Heath.

Large Heath *Coenonympha tullia*

VU VU I

WS 35–40 mm. **ID** Wings pale orange; underside of hindwings with long white band and several 'eye-spots' (absent in form *scotica* in Scotland). **Hab** Bogs. **Fp** Mainly Hare's-tail Cottongrass. **SS** Small Heath.

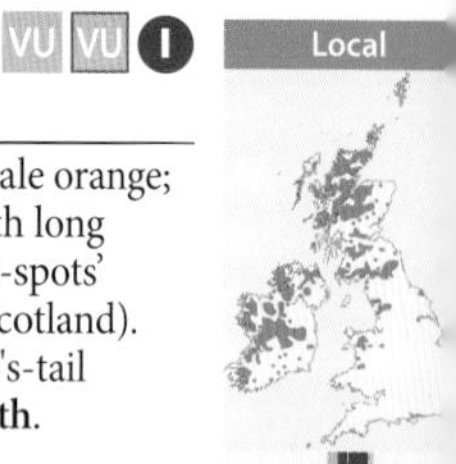

Erebia 3 spp. | 2 ILL.

Scotch Argus *Erebia aethiops*

WS 44–52 mm. **ID** Wings dark brown to blackish; orange band with black 'eye-spots' that have a white 'pupil'. Underwings with distinct pale band. **Hab** Damp sites, *e.g.* grassland and bog edges. **Fp** Purple and Blue Moor-grasses. **SS** Meadow Brown, Mountain Ringlet.

Mountain Ringlet *Erebia epiphron*

NT

WS 29–37 mm. **ID** Wings dark brown; orange band (restricted in some forms, particularly in ♂ (*e.g.* as shown below)) with small, black 'eye-spots' that lack a white central 'pupil'. **Hab** Damp mountainous grassland. **Fp** mainly Mat-grass. **SS** Scotch Argus.

Pyronia 1 sp.

Gatekeeper NT
Pyronia tithonus

WS 37–48 mm. **ID** Wings orange with broad, dark brown margins, ♂ also with dark brown band on forewings; 'eye-spot' on forewing with two 'pupils'. **Hab** Woodland rides and hedgerows. **Fp** Grasses. **SS** Meadow Brown.

Common

Maniola 1 sp.

Meadow Brown
Maniola jurtina

WS 40–60 mm. **ID** Wings dark brown, ♀ with more conspicuous orange patches; 'eye-spot' on forewing with one 'pupil'. **Hab** Grassland, including along woodland rides, and hedgerows. **Fp** Grasses. **SS** Ringlet, Gatekeeper, Scotch Argus.

Common

Aphantopus 1 sp.

Ringlet *Aphantopus hyperantus*

WS 42–52 mm. **ID** Wings very dark brown, fringed white; several yellow-ringed black spots (may be tiny or absent on upperwings). **Hab** Grassland, including along woodland rides and hedgerows. **Fp** Grasses. **SS** Meadow Brown.

Common

Guide to Nymphalidae pp. 390–39

BROWNS 2/

Lasiommata 1 sp.

Wall *Lasiommata megera* NT EN

WS 45–53 mm. **ID** Wings chequered golden-and-brown, with distinctive 'eye-spots'; underwing with zigzag pattern. **Hab** Short, open stony grassland, dunes. **Fp** Grasses. **SS** Superficially like other nymphalids.

Hipparchia 1 sp.

Grayling VU NT
Hipparchia semele

WS 51–62 mm. **ID** Wings orange-and-brown; hindwing underside brown, mottled grey, some with grey zigzag. **Hab** Dry grassland/heathland, mainly coastal areas but inland heathland in S. **Fp** Grasses. **SS** None.

FAMILY Riodinidae (Metalmarks) 1 SP

Small representative of colourful butterflies frequent in the neotropics. The antennae have a clubbed tip.

Hamearis 1 sp

Duke of Burgundy *Hamearis lucina* EN

WS 29–34 mm. **ID** Wings orange and dark brown with broken white rear border; underside of hindwings with two rows of large white spots. **Hab** Chalk and limestone grassland, woodland. **Fp** Cowslip, Primrose. **SS** None.

FAMILY **Lycaenidae** (Hairstreaks, coppers and blues)

14 GEN. | 20 spp. (including **2†** spp. and **4** rare immigrants)

Small to medium-sized (WS 18–52 mm) butterflies; variably coloured, some with a metallic sheen. All hairstreaks have short tails on the hindwings. The antennae have a clubbed tip. Some blues have well-known associations with ants.

Features		Species	Page
HINDWINGS with short tail; UPPERWINGS without blue		**HAIRSTREAKS *etc.***	
UPPERWINGS brown; **UNDERWINGS** metallic green, usually with white dotted line		Green Hairstreak	*p.* 402
UPPERWINGS ♂ purple, ♀ part-purple (around base of forewings); **UNDERWINGS** silvery-grey		Purple Hairstreak	*p.* 402
UPPERWINGS ♂ brown, ♀ part-orange (on forewings); **UNDERWINGS** orange		Brown Hairstreak	*p.* 402
UPPERWINGS brown; **UNDERWINGS** with orange margin	**UNDERWINGS** hindwings with broad orange band (also orange on forewings) and black-and-white spots	Black Hairstreak	*p.* 402
	UNDERWINGS hindwings with restricted orange band bordered by black line; white 'W'-shaped marking	White-letter Hairstreak	*p.* 402
FORM small (**WS** 15–27 mm); **UPPERWINGS** brown, bordered with small white dots; **UNDERWINGS** pale grey with bold dark grey patterning		Geranium Bronze	*p.* 406
HINDWINGS with short tail; UPPERWINGS with blue		**BLUES ('Tailed' blues)**	
FORM small (**WS** 20–30 mm); **UNDERWINGS** greyish with small black spots		Short-tailed Blue	*p.* 406
FORM medium-sized (**WS** 32–42 mm); **UNDERWINGS** brownish with white stripes		Long-tailed Blue	*p.* 406
HINDWINGS without tail; UPPERWINGS with at least some blue		**BLUES (Blues)**	
UPPERWINGS blue, ♂ forewings with narrow black outer submargin, ♀ broad black submargin; **UNDERWINGS** pale blue with small black spots (no orange)		Holly Blue	*p.* 405
FORM very small; **UPPERWINGS** ♂ greyish, with hint of blue scales, ♀ brown; **UNDERWINGS** pale grey with silver-ringed black spots (no orange)		Small Blue	*p.* 405
FORM large (**WS** 38–52 mm); **UPPERWINGS** blue, forewings with black spots (diagnostic); **UNDERWINGS** silvery-blue grey with black spots circled in white		Large Blue	*p.* 406
WINGS continuous white fringes; **UPPERWINGS** ♂ blue with black submarginal border, ♀ brown or blue with orange spots; **UNDERWINGS** submarginal black spots with shiny blue centre, no black spot on forewing close to body		Silver-studded Blue	*p.* 405
WINGS continuous white fringes; **UPPERWINGS** ♂ violet-blue with narrow black submarginal border, ♀ brown or blue with orange spots; **UNDERWINGS** with black spot on forewing near body		Common Blue	*p.* 404
WINGS white fringes broken by black veins; **UPPERWINGS** ♂ cobalt-blue with narrow black submarginal border, ♀ brown with orange spots with blue crescents below; **UNDERWINGS** with black spot on forewing near body		Adonis Blue	*p.* 404
WINGS white fringes broken by black veins; **UPPERWINGS** ♂ wings silvery blue with broad dark submarginal border; ♀ brown with orange spots with white crescents below; **UNDERWINGS** with black spot on forewing near body		Chalk Hill Blue	*p.* 404
HINDWINGS without tail; UPPERWINGS brown		**BLUES (Arguses)**	
UPPERWINGS brown with orange spots around edges; best told by range	**UPPERWINGS** forewing without central whitish spot	Brown Argus	*p.* 403
	UPPERWINGS forewing with central whitish spot in some populations, particularly in Scotland; some ♂s lack orange spots	Northern Brown Argus	*p.* 403
UPPERWINGS bright orange		**COPPERS**	
UPPERWINGS forewing orange with brown markings; hindwing brown with orange band		Small Copper	*p.* 403

Guide to families pp. 382–38

HAIRSTREAKS

Favonius 1 sp.

Purple Hairstreak *Favonius quercus* 

WS 31–40 mm. **ID** Wings: purple sheen in ♂, purple restricted to base of forewings in ♀; underwings silvery-grey with white lines. **Hab** Woodland rides, parks, large gardens. **Fp** Oaks. **SS** None.

Satyrium 2 spp.

White-letter Hairstreak *Satyrium w-album* EN

WS 25–35 mm. **ID** Underwings brown with orange band (on hindwings only), bordered by a black line; white 'W'-shaped marking. **Hab** Woodland edges and parks. **Fp** Elms. **SS** 'Brown' hairstreaks.

Black Hairstreak *Satyrium pruni*

WS 34–40 mm. **ID** Underwings brown with broad orange band, row of black-and-white spots and white line. **Hab** Woodland edges and hedgerows. **Fp** Blackthorn. **SS** 'Brown' hairstreaks.

Callophrys 1 sp.

Green Hairstreak *Callophrys rubi*

WS 27–34 mm. **ID** Wings brown; underwings metallic green, often with line of white spots. **Hab** Open grassland and heathland, with some scrub. **Fp** Common Rock-rose, Common Bird's-foot-trefoil, Gorse, Bilberry, *etc*. **SS** None.

Thecla 1 sp.

Brown Hairstreak *Thecla betulae*

WS 36–45 mm. **ID** Wings dark brown, forewings with orange patch in ♀; underwings orange with white lines **Hab** Woodland edges, hedgerows and scrub. **Fp** Blackthorn. **SS** **Black Hairstreak** and **White-letter Hairstreak**.

BLUES (Arguses)

Aricia 2 spp.

Wings brown with row of orange spots near edges; underwings greyish-brown with orange marginal spots and white-ringed black spots. No black spot close to the body on underside of forewing (unlike similar ♀ Common Blue (*p. 404*), the only similar species).

Brown Argus *Aricia agestis*

WS 25–31 mm. **ID** Forewings lack white central spot. **Hab** Chalk and limestone and sometimes other grasslands. **Fp** Common Rock-rose. **SS** *Polyommatus* blues (♀) (*p. 404*), **Northern Brown Argus.**

Northern Brown Argus *Aricia artaxerxes* VU

WS 25–31 mm. **ID** Forewings in some populations (particularly in Scotland) with central whitish spot. **Hab** Unimproved grasslands. **Fp** Common Rock-rose. **SS** **Common Blue** (♀) (*p. 404*), **Brown Argus.**

COPPERS

Lycaena 2 spp. (1†)

Small Copper *Lycaena phlaeas*

WS 26–36 mm. **ID** Forewings orange with black spots and dark brown border (tip more pointed in ♂); hindwings dark brown with orange band (blue-spotted on some individuals). **Hab** Unimproved grassland, heathland, woodland clearings. **Fp** Common Sorrel, Sheep's Sorrel. **SS** None.

Guide to Lycaenidae p. 40

BLUES (Blues) 1/

Polyommatus 3 spp.

Common Blue *Polyommatus icarus*

WS 29–36 mm. **ID** ♂ wings violet-blue with narrow black submarginal border and continuous white fringes; ♀ wing blue or dark brown with a hint of blue, with orange spots; underwings grey or brown with marginal orange spots and white-ringed black spots, including one on forewing near body. **Hab** Grassland, gardens. **Fp** Common Bird's-foot-trefoil, White Clover, *etc.* **SS** Other blues, **Brown Argus** (*p. 403*).

Adonis Blue *Polyommatus bellargus* NT

WS 30–40 mm. **ID** ♂ wings cobalt-blue with thin black submarginal border and white fringes broken by black veins (chequered effect); ♀ wings dark brown with chequered white fringes and orange spots with blue crescents on lower half; underwings brown with marginal orange spots and white-ringed black spots. **Hab** South-facing chalk grassland **Fp** Horseshoe Vetch. **SS** Other blues, **Brown Argus** (*p. 403*).

Chalk Hill Blue *Polyommatus coridon* NT

WS 33–40 mm. **ID** ♂ wings silvery blue with broad dark submarginal border and white fringes broken by black veins ♀ wings brown with chequered white fringes and orange spots with whitish crescents on lower half; underwings brown or grey with small marginal orange spots and white-ringed black spots. **Hab** South-facing chalk and limeston grassland. **Fp** Horseshoe Vetch. **SS** Other blues, **Brown Argus** (*p. 403*).

Plebejus 1 sp.

Silver-studded Blue *Plebejus argus* VU

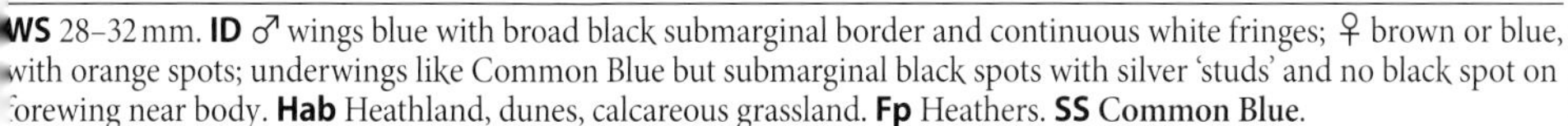

WS 28–32 mm. **ID** ♂ wings blue with broad black submarginal border and continuous white fringes; ♀ brown or blue, with orange spots; underwings like Common Blue but submarginal black spots with silver 'studs' and no black spot on forewing near body. **Hab** Heathland, dunes, calcareous grassland. **Fp** Heathers. **SS** Common Blue.

Celastrina 1 sp.

Holly Blue *Celastrina argiolus* I

WS 25–31 mm. **ID** Wings blue, forewings with black submarginal border, broad in ♀, narrow in ♂; underwings silvery-blue with black spots. **Hab** Gardens, churchyards, hedgerows, woodland rides, *etc.* **Fp** Holly (1st brood), Ivy (2nd brood). **SS** Common Blue.

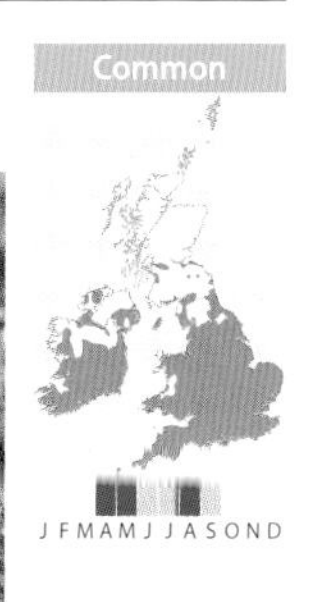

Cupido 2 spp.

Small Blue *Cupido minimus* NT EN I

WS 18–27 mm. **ID** ♂ wings uniformly greyish with hint of blue scales, but white fringe; ♀ wings brown with white fringe; underside silvery-blue with silver-ringed black spots. **Hab** Chalk, limestone and coastal grassland, dunes, quarries, *etc.* **Fp** Kidney Vetch. **SS** None.

Guide to Lycaenidae p. 40

BLUES (Blues) 2/

Maculinea 1 sp.

Large Blue CR
Maculinea arion

WS 38–52 mm. **ID** Wings blue, black spots on forewings, black submarginal border broad in ♀, narrow in ♂; underwings silvery-blue with white-ringed black spots. **Hab** Unimproved grasslands. **Fp** Wild Thyme (later Sabulet's Red Ant (*p. 499*) grubs). **SS** None.

EXTINCT SPECIES

Large Copper
Lycaena dispar

WS 33–41 mm. Extinct in Britain since 1864. Stock was released in Fens in Ireland and Cambridgeshire in the 1900s and survived with the help of further reintroduced stock - the reintroduction programme was suspended in the 1990s.

Mazarine Blue
Cyaniris semiargus

WS 22–33 mm. Extinct in Britain since 1877. Rarely recorded since (and those may be releases).

RARE 'BLUES'

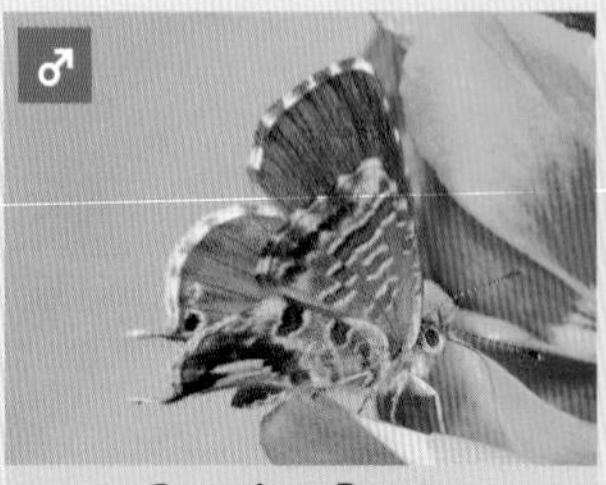

Geranium Bronze
Cacyreus marshalli

WS 15–27 mm. Accidental on pelargoniums and geraniums.

Long-tailed Blue
Lampides boeticus

WS 32–42 mm. Rare immigrant from continental Europe.

Short-tailed Blue
Cupido argiades

WS 20–30 mm. Rare immigrant from continental Europe.

ORDER Lepidoptera (Moths) SELECTED FAMILIES | 33 of 65 BI

This guide covers the moth families most likely to be found in Britain or Ireland, which can be identified without microscopic examination. The informal 'division' of moths into 'micro-moths' and 'macro-moths' has been used for many years and is followed here. For identification of families, focus on wing shape and the characters summarized.

For butterfly families see *pp. 382–383.*

Further details of diurnal species can be found in the companion WILD*Guides* volume *Britain's Day-flying Moths* (see *page 380* for details).

'MICRO-MOTHS' – guide to selected families 18 of 50 BI

Typically smaller rather primitive moths. Many have mouthparts that are not equipped for feeding. The following is a small selection of species, likely to be found throughout Britain and Ireland, mainly by day. **Length bars in the species accounts indicate forewing length (≈ ½ wingspan).**

FORM very small species, wings held roof-like at rest. Forewings glossy, bronze or with purple areas; hindwings as broad as forewings (**WS** 5–12 mm).

Yellow-barred Gold
Micropterix aureatella (**ws 9–11 mm**)

Gold moths
Micropterigidae
[1 genus | 5 species]

FORM rest with wings in a steep, roof-like position (**WS** 7–23 mm); some species with metallic wings; **ANTENNAE** very long, up to 4× forewing length; **PALPS** long.

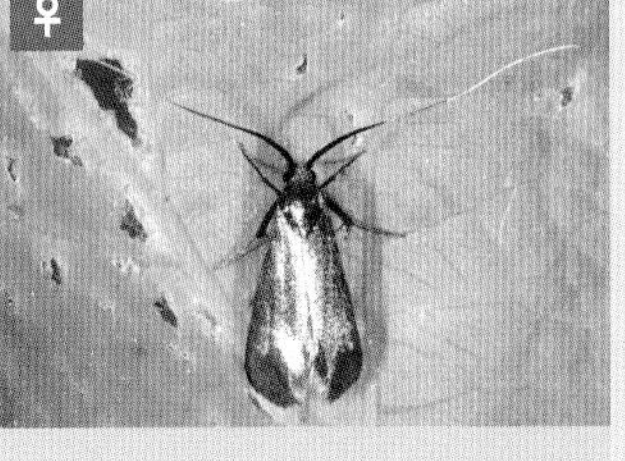

Green Long-horn

Long-horn moths
Adelidae
[4 genera | 15 species]

p. 415

FORM wings held roof-like at rest (**WS** 6–17 mm); **ANTENNAE** half to two-thirds length of forewing, tooth-like (appearing feathery) in ♂ of two species.

Feathered Leaf-cutter

Leaf-cutter moths
Incurvariidae
[2 genera | 5 species]

p. 416

FORM often brown species, wings held steeply roof-like at rest (**WS** 6–33 mm). Several species of clothes moths and allies are regarded as pests.

Four-spotted Clothes Moth

Clothes moths
Tineidae
[23 genera | 48 species]

p. 418

FORM wings held steeply roof-like at rest, forewings elongate, hindwings narrower (**WS** 6–18 mm); **LEGS** front legs sometimes raised.

Horse-chestnut Leaf-miner

Gracillariid moths
Gracillariidae
[15 genera | 93 species]

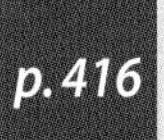
p. 416

FORM most rest with wings in a steep, roof-like position, a few with wings wrapped around abdomen (**WS** 8–27 mm). Includes **Small Emine** group (wings white with black spots).

Spindle Ermine

Ermine moths
Yponomeutidae
[9 genera | 24 species]

p. 417

FORM rest with elongated wings in a steep, roof-like position (**WS** 6–23 mm); **ANTENNAE** pointing forward; **PALPS** long.

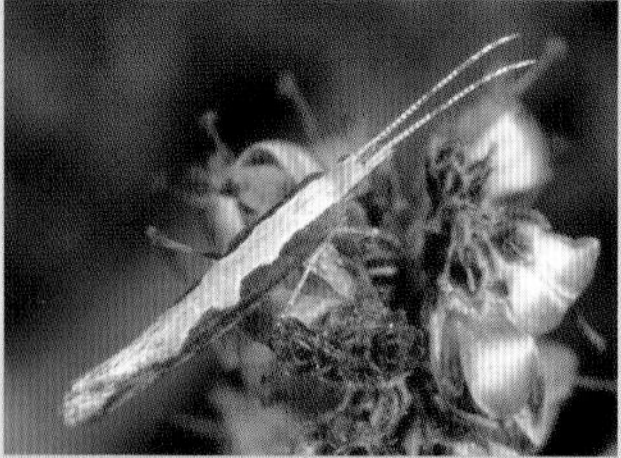

Diamond-back Moth

Diamond-back moths
Plutellidae

[3 genera | 7 species]

p. 415

FORM mostly rest with wings held flat or in a steep, roof-like position, body horizontal (**WS** 6–23 mm).

White-shouldered House-moth

Tubic moths
Oecophoridae

[19 genera | 28 species]

p. 418

FORM wings black-dotted or black-streaked on white or brown background, held roof-like at rest (**WS** 15–30 mm).

Dotted Ermel

Ermel moths
Ethmiidae

[1 genus | 6 species]

p. 417

FORM variable, wings held flat to angled, rolled, or roof-like; hindwings narrower or wider than forewings (**WS** 6–23 mm).

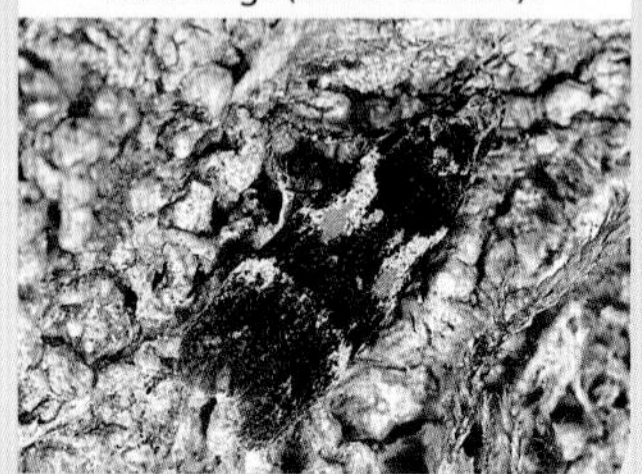

Crescent Groundling

Gelechiid moths
Gelechiidae

[54 genera | 161 species]

p. 417

FORM elongate forewings overlapping and wrapped around abdomen (**WS** 6–23 mm). Mostly accidental introductions.

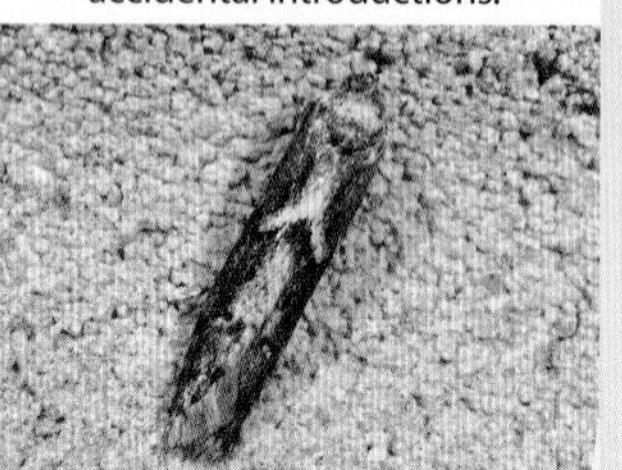

Dingy Dowd
Blastobasis adustella (ws 15–20 mm)

Dowd moths
Blastobasidae

[2 genera | 6 species]

FORM elongate wings held roof-like (**WS** 4–9 mm). Larvae develop in silk tubes on or near the foodplant.

Ling Owlet
Scythris empetrella (ws 9 mm)

Owlet moths
Scythrididae

[1 genus | 11 species]

FORM single species in family; easy to tell as each wing is separated into 6 plumes, held open at rest (**WS** 14–16 mm).

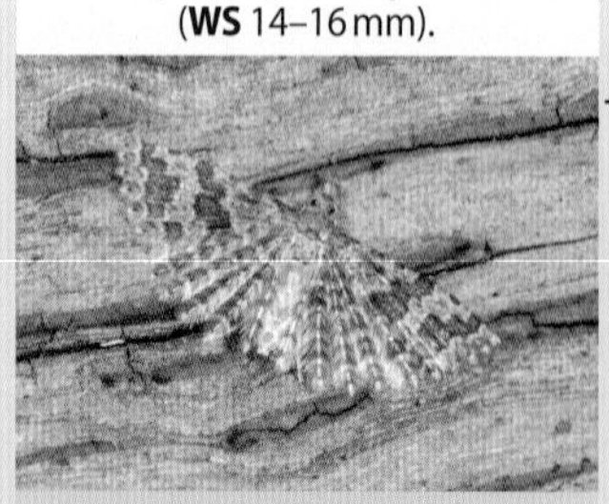

Twenty-plume Moth
Alucita hexadactyla (ws 14–16 mm)

Many-plume moths
Alucitidae

[1 genus | 1 species]

FORM plume-like wings (**WS** 10–33 mm); **LEGS** long, extending well beyond tip of abdomen.

White Plume

Plume moths
Pterophoridae

[20 genera | 45 species]

p. 421

FORM mostly rest with wings held flat, body horizontal (**WS** 8–18 mm).

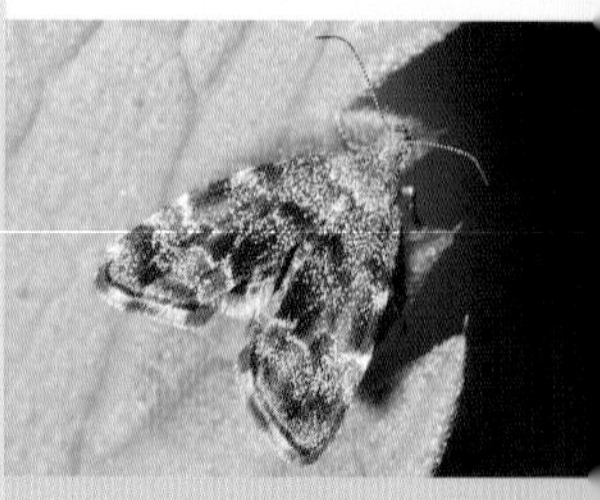

Common Nettle-tap

Metalmarks
Choreutidae

[4 genera | 7 species]

p. 416

FORM rest with wings in various positions, but always held horizontally, tent-like or wrapped around the body. Forewings are distinctly broad (**WS** 7–30 mm).

Green Oak Tortrix

Tortrix moths
Tortricidae

p. 419

[94 genera | 382 species]

FORM rest with wings in various positions, mostly roof-like, some species with end of abdomen raised. Forewings elongate to triangular (**WS** 10–45 mm).

Bee Moth

Pyralid moths
Pyralidae

[48 genera | 77 species]

FORM rest with wings in roof-like position, or slightly rolled around body (the latter known as grass moths) (**WS** 9–45 mm). Some aquatic or sub-aquatic species.

Brown China-mark

Crambid moths
Crambidae

p. 420

[54 genera | 122 species]

Micro-moth larvae

Larvae of micro-moths include well-known species feeding on woollen carpets and clothing in houses, including Common Clothes Moth *Tineola bisselliella* [Tineidae] and Black Bagworm *Pachythelia villosella* [Psychidae (Bagworms)], one of several species of which the larvae develop in a silk bag or case, with plant matter and debris attached. Adults (males winged, females wingless) are seldom seen, unlike the larvae. Some micro-moth larvae are considered to be pests; the larvae of Spindle Ermine *Yponomeuta cagnagella* [Yponomeutidae] live gregariously in a silken web and on occasion defoliate rows of Spindle bushes, particularly on chalk grasslands. Dingy Flat-body *Depressaria daucella* [Depressariidae (Flat-body moths)] is a more typical leaf-eating species

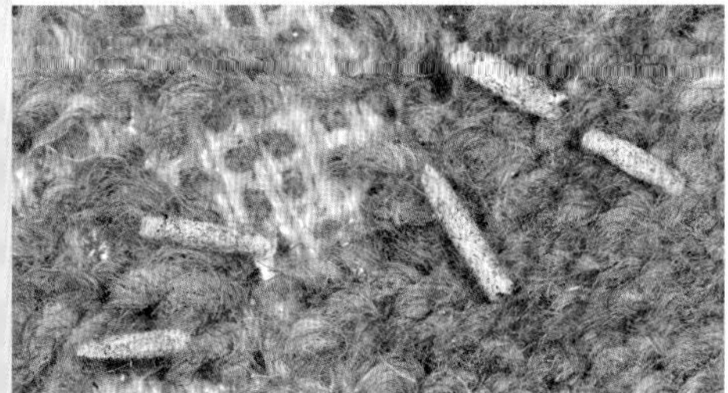

Common Clothes Moth *Tineola bisselliella* [Tineidae (*p. 418*)] – larvae showing damage to woollen carpet.

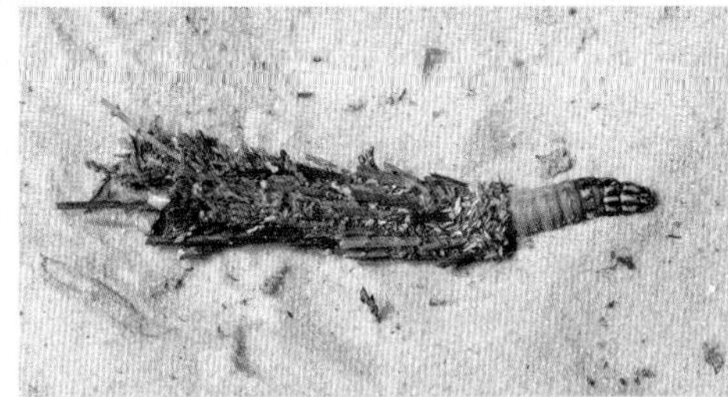

Black Bagworm *Pachythelia villosella* VU [Psychidae [N/I]] (RDB2) – larvae with plant matter and debris attached.

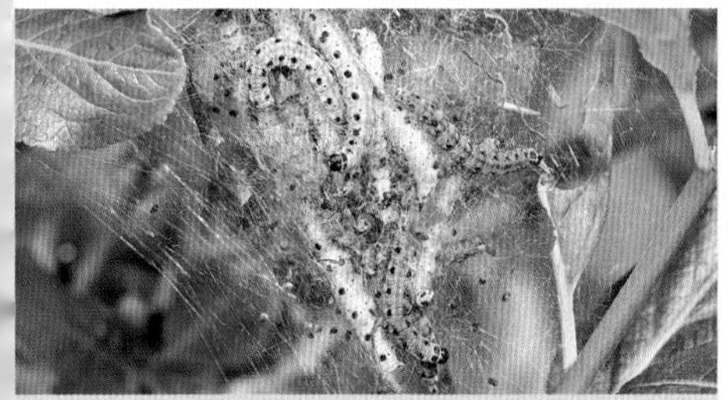

Spindle Ermine *Yponomeuta cagnagella* [Yponomeutidae (*p. 417*)] larvae in silken web.

Dingy Flat-body *Depressaria daucella* [Depressariidae [N/I]] typical leaf-eating larvae.

'MACRO-MOTHS' – guide to families

Typically, but not always, larger moths; mouthparts are usually modified for feeding; some of the 'micro-moth' species, a guide to the families of which precedes this guide [see *pages 407–409*], are often confused with 'macro-moth' species.

LARGE, BULKY SPECIES with distinctive wing shapes and/or markings

FORM large (**WS** 55–89 mm), stout; **WINGS** brown-and-white. **HABITAT** associated with birch woods.

Kentish Glory
Endromidae
[Kentish Glory]

p. 427

FORM large (**WS** 55–85 mm); **WINGS** large central 'eye-spot'; ♂ **ANTENNAE** very feathery. **LIFESTYLE** larvae spin silk cocoons.

Emperor moths
Saturniidae
[Emperor Moth]

p. 427

FORM medium-sized to large (**WS** 41–135 mm), fast-flying, streamlined; **WINGS** hindwings often brightly coloured, used to startle predators. **LIFESTYLE** larvae are fairly easy to find; most have a 'horn' at the tail end.

Poplar Hawk-moth

Elephant Hawk-moth

Hawk-moths
Sphingidae
[13 genera | 18 species (9 immigrant)]

p. 43

BULKY SPECIES with very distinctive wing shapes and/or markings

FORM medium-sized to large (**WS** 18–90 mm), bulky; **WINGS** brown or yellowish (some with white spot); some rest with wings closed; ♂ **ANTENNAE** feathery.

Oak Eggars

Eggar and lappet moths
Lasiocampidae

[10 genera | 12 species (1†)]

p. 429

FORM medium-sized to large (**WS** 24–80 mm), rather thick-bodied, furry. **BEHAVIOUR** larvae of processionary moths (pests) form processions when they move.

Puss Moths

Prominents & allies
Notodontidae

[17 genera | 29 species (6 immigrant)]

p. 443

FOREWINGS ELONGATE, held tightly against body

FORM medium-sized (**WS** 25–50 mm); wings held tight against body at rest; **ANTENNAE** short.

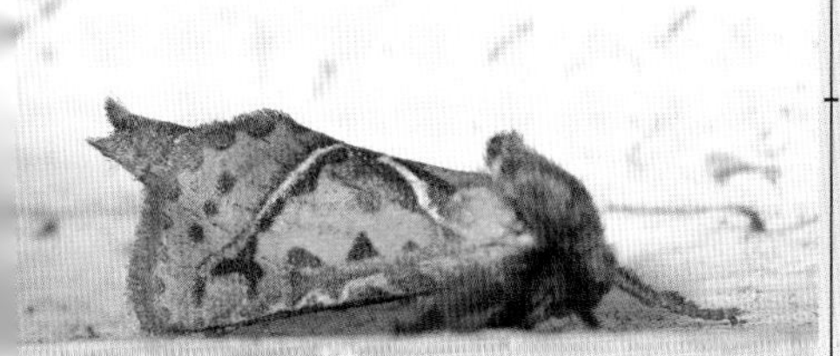

Orange Swift

Swift moths
Hepialidae

[4 genera | 5 species]

p. 423

FORM large (**WS** 30–96 mm), bulky; **ABDOMEN** elongate, extending beyond closed wings in two species.

Goat Moth

Leopard and goat moths
Cossidae

[3 genera | 3 species]

p. 422

SMALLER SPECIES with very distinctive wing shapes and/or markings

FORM medium-sized (**WS** 22–40 mm); **ABDOMEN** slender in some; **WINGS** broad and spread out at rest in some, close to the body in others; hook-tips have hooked forewing tips.

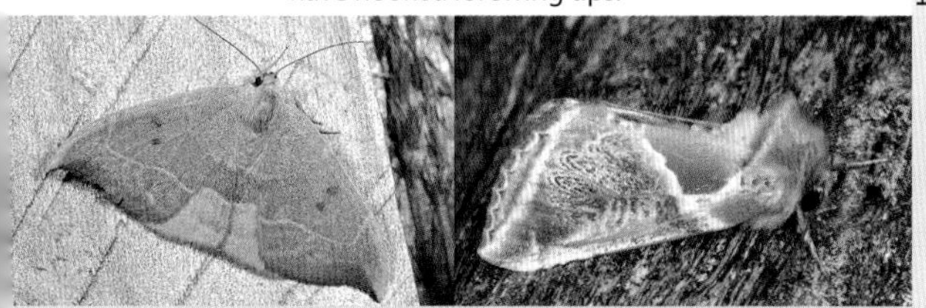

HOOK-TIPS (see also Geometrids (*p. 434*))
Oak Hook-tip

ALLIES
Buff Arches

Hook-tips, lutestrings & allies
Drepanidae

[13 genera | 16 species]

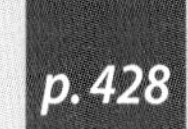

p. 428

FORM small (**WS** 16–18 mm); forewings broad and rounded, rests in a tent-like position.

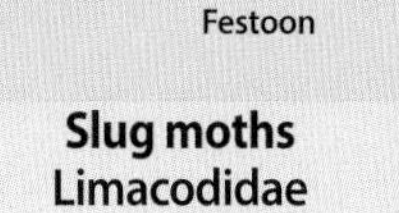

Festoon

Slug moths
Limacodidae

[2 genera | 2 species]

p. 422

DISTINCTIVE WING FEATURES

FORM medium-sized (**WS** 16–50 mm), wasp-like, day-flying; **WINGS** with large transparent areas, forewings with dark bars or marks. **ANTENNAE** clubbed. **NOTE:** males attracted to pheromone lures.

Yellow-legged Clearwing

Clearwing moths
Sesiidae
[6 genera | 16 species (1†)]

p. 424

FORM medium-sized (**WS** 20–46 mm), day-flying; **WINGS** narrow, round-tipped; black with red spots (burnets) or green (foresters); **ANTENNAE** clubbed (burnets). **NOTE:** adults and larvae are toxic.

Six-spot Burnet   Forester

Burnet & forester moths
Zygaenidae
[3 genera | 10 species]

p. 426

'TYPICAL' MOTHS 1/2

FORM varied – the second largest British macro-moth family (**WS** 16–68 mm); **WINGS** broad, often held flat when resting; **ABDOMEN** slender; **ANTENNAE** feathery in some ♂s, slender in ♀, never clubbed.

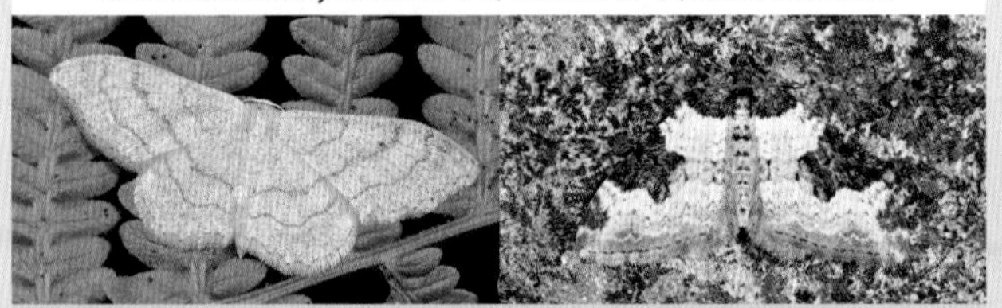

MOCHAS & WAVES (Sterrhinae)
Riband Wave

CARPETS (Larentiinae)
Garden Carpet

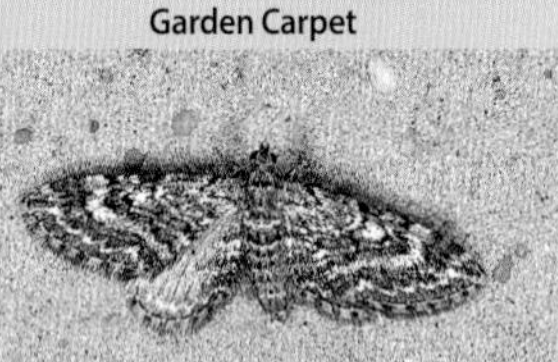

EMERALDS (Geometrinae)
Large Emerald

PUGS (Larentiinae)
Narrow-winged Pug

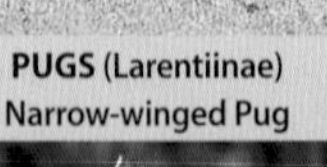

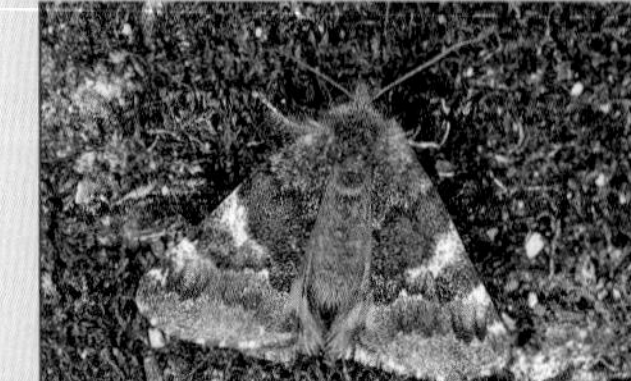

ORANGE UNDERWINGS (Archiearinae)
Orange Underwing

THORNS, BEAUTIES, UMBERS, *etc.*
(Ennominae) Purple Thorn

Geometrids
Geometridae
[141 genera | 307 species (5†)]

p. 434

FORM small to medium-sized (**WS** 16–48 mm), whitish/brown with dark bars; some green.

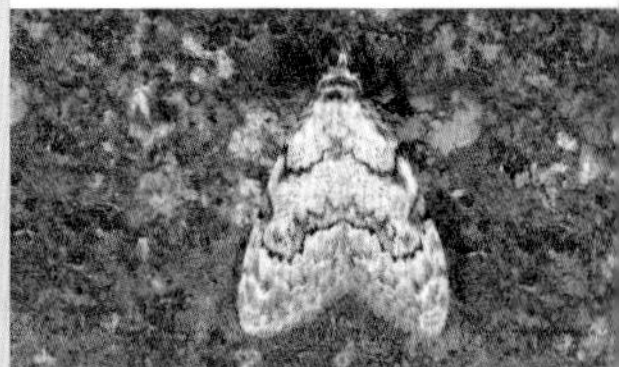

BLACK ARCHES
Least Black Arches

SILVER-LINES
Green Silver-lines

Black arches, silver-lines & allies
Nolidae
[6 genera | 12 species]

p. 463

'TYPICAL' MOTHS 2/2

FORM varied, including groups notable for the large, brightly coloured underwing moths and boldly striped tiger moths; some rest with forelegs outstretched (**WS** 14–106 mm); **WINGS** often broad.
NOTE: many species were formerly placed in other families prior to reclassification.

SNOUTS
Snout

FOOTMAN MOTHS
Common Footman

Cinnabar (see also burnets (*p. 426*))

UNDERWINGS
Red Underwing

TUSSOCKS/VAPOURERS
Pale Tussock

TIGERS
Wood Tiger

Tussocks, ermines, tigers, footman moths, snouts, fan-foots, marbleds, blacknecks and underwings

Erebidae
[53 genera | 88 species (1†)]

p. 447

FORM varied – the largest British macro-moth family; mostly medium-sized (**WS** 19–74 mm), stout-bodied, mainly brown species; **WINGS** much longer than broad, equipped for powerful flight (many are immigrant species). **NOTE:** recent taxonomic changes have resulted in some species being re-classified in Erebidae.

Angle Shades

Setaceous Hebrew Character

Silver Y

Common Wainscot
Mythimna pallens (ws 32–40 mm)

Large Yellow Underwing

Grey Dagger

Noctuids

Noctuidae
[175 genera | 368 species (6†)]

p. 454

Guide to families *pp. 407–40*

'MICRO-MOTHS'

Generally smaller than 'macro-moths', 'micro-moths' are mainly primitive species, found virtually everywhere, with some species associated with humans (feeding on carpets and clothes), as well as larvae on vegetation, algae and fungi; some larvae are notable miners of leaves. The wide range of resting postures of adults and the often considerable variation in form within some families can be confusing and make identification difficult. Indeed, many species are impossible to identify without microscopic examination of genitalia. The species included in this section have been selected to show the wide variation within the many families within the somewhat artificial grouping of 'micro-moths'. All are readily identified in the field and occasionally encountered.

FAMILY Pyralidae (Pyralid moths) — 48 GEN. | 77 spp. (1†) | 2 ILL.

Aphomia — 2 spp. | 1 ILL.

Bee Moth *Aphomia sociella*

Common

WS 25–38 mm. **ID** Elongated forewings buff in basal third, otherwise greyish-brown with dark lines (♂) or brown with greenish areas (♀). **Hab** Woodland, hedgerows, parks, gardens: in nests of bumble bees and wasps. **Fp** Larvae feed on debris. **SS** None.

Pyralis — 4 spp. | 1 ILL.

Meal Moth *Pyralis farinalis*

Common

WS 22–29 mm. **ID** Forewings dark brown, with yellowish brown between whitish cross-lines. Abdomen curved upwards. **Hab** Often in urban and agricultural areas. **Fp** Stored cereals and cereal refuse. **SS** Scarce Tabby *Pyralis lienigialis* [N/I] (**WS** 22–26 mm) is smaller and darker, lacking pale area between cross-lines.

FAMILY **Adelidae** (Long-horn moths) 4 GEN. | 15 spp. | 2 ILL.

Adela 3 spp. | 1 ILL.

ìreen Long-horn
ıdela reaumurella

/S 14–18 mm. **ID** Wings metallic dark reen; antennae white with black base, × length of forewing (♂), 1·5× (♀ (see . *407*)). **Hab** Woodland, heathland, etlands. **Beh** Flies in hundreds around aks in sunny weather. **Fp** In late larval age on dead fallen leaves. **SS** Early ong-horn *Adela cuprella* [N/I] **WS** 15 mm), which is not as green.

Nemophora 5 spp. | 1 ILL.

Yellow-barred Long-horn
Nemophora degeerella

WS 22–29 mm. **ID** Head orange, with antennae almost 4× length of forewing (♂), only slightly longer than forewing and white-tipped (♀). Forewings brownish with central yellow band; yellow streaks elsewhere. **Hab** Woodland, wetlands, hedgerows. **Fp** Dead leaves on ground. **SS** None.

FAMILY **Plutellidae** (Diamond-back moths) 3 GEN. | 7 spp. | 1 ILL.

Plutella 3 spp. | 1 ILL.

iamond-back Moth
'utella xylostella

/S 12–17 mm. **ID** Forewings greyish-rown with paler whitish-brown iamond-back' band. **Hab** Cabbage-owing sites, hence often in ricultural and urban areas. **p** Cabbage family. **SS** None.

FAMILY Gracillariidae (Gracillariid moths) 15 GEN. | 93 spp. | 1 ILL.

Cameraria 1 sp.

Horse-chestnut Leaf-miner
Cameraria ohridella

WS 8 mm. **ID** Forewings orange-brown with white cross-lines and basal streak. **Hab** Woodland, parks, gardens. **Beh** Larva mines leaves, the telltale damage obvious on host trees [see *inset*]. **Fp** Horse-chestnut. **SS** None with same angled bands on the forewings.

FAMILY Choreutidae (Metalmarks) 4 GEN. | 7 spp. | 1 ILL.

Anthophila 1 sp.

Common Nettle-tap
Anthophila fabriciana

WS 11–15 mm. **ID** Forewings dark brown, speckled with grey and some whitish markings. **Hab** Hedgerows, woodland edges, gardens. **Fp** Common Nettle. **SS** None.

FAMILY Incurvariidae (Leaf-cutter moths) 2 GEN. | 5 spp.

Incurvaria 4 spp. | 1 ILL.

Feathered Leaf-cutter
Incurvaria masculella

WS 12–17 mm. **ID** Head yellow, with feathery antennae. Upper margin of forewings dark brown with white spots about halfway along and towards hind part. **Hab** Woodland, gardens. **Fp** Hawthorn. **SS Common Leaf-cutter** *Incurvaria oehlmanniella* [N/I] (**WS** 14 mm), which lacks feathery antennae, and has a third white spot on lower margin of forewing towards wingtip.

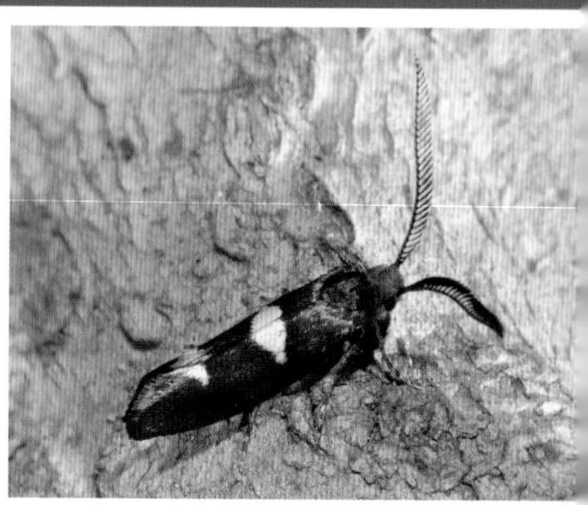

FAMILY Yponomeutidae (Ermine moths) 9 GEN. | 24 spp. (2†) | 1 ILL.

Yponomeuta 8 spp. | 1 ILL.

;pindle Ermine
'ponomeuta cagnagella

/S 19–26 mm. **ID** Forewings white with umerous black dots. **Hab** Hedgerows, :rub, gardens. **Beh** Defoliates large reas, with many webs visible. **p** Spindle. **SS** Other *Yponomeuta* oecies, but these have different pattern f spots and greyish areas.

FAMILY Ethmiidae (Ermel moths) 1 GEN. | 6 spp. | 1 ILL.

Ethmia 6 spp. | 1 ILL.

otted Ermel
thmia dodecea

'S 11–12 mm. **ID** Forewings whitish ith about 10 black dots, some large. **ab** Woodland rides, grassland. **)** Common Gromwell. **SS** None.

FAMILY Gelechiidae (Gelechiid moths) 64 GEN. | 161 spp. (11†) | 1 ILL.

Teleiodes 5 spp. | 1 ILL.

rescent Groundling
leiodes luculella

S 11–12 mm. **ID** Forewings dark ey to black with large semicircular iite blotches that have a yellow or ange mark which reaches the costa. **ab** Woodland, parks. **Fp** Oaks. **; Chestnut Groundling** *Teleoides vimaculella* [N/I] (**WS** 12 mm), which the mark near the costa is ange, but does not reach the costa.

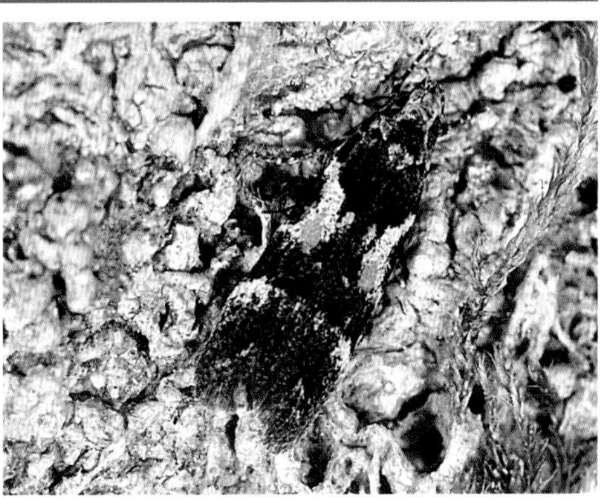

FAMILY Oecophoridae (Tubic moths) 19 GEN. | 28 spp. (2†) | 2 ILL

Endrosis 1 sp.

White-shouldered House-moth *Endrosis sarcitrella*

Common

WS 13–20 mm. **ID** Forewings dark greyish-brown, with small paler marks and several black spots or marks. Head and thorax white. **Hab** Urban areas and outdoor buildings. **Fp** Cereals, animal and vegetable matter. **SS** None.

Esperia 1 sp

Sulphur Tubic
Esperia sulphurella

Common

WS 12–17 mm. **ID** Head partly orange; rests with antennae held forward, which have whitish mark two-thirds along their length. Forewings dark greyish brown with yellow scales, ♂ with yellow streaks from the base reaching just over half length. **Hab** Woodland, hedgerows, heathland, gardens.
Fp Decaying wood and fungi. **SS** None.

FAMILY Tineidae (Clothes moths) 23 GEN. | 48 spp. (1†) | 2 ILL

Nemapogon 10 spp. | 1 ILL.

Cork Moth
Nemapogon cloacella

Common

WS 11–18 mm. **ID** Forewings whitish, with reddish and pale brown areas. **Hab** Woodland, parks. **Fp** Bracket fungi. **SS** Other *Nemapogon* species, which require detailed inspection of markings.

Triaxomera 1 sp

Four-spotted Clothes Moth
Triaxomera fulvimitrella

Local

WS 13–22 mm. **ID** Head yellow; forewings blackish with four large white spots. **Hab** Woodland.
Fp Bracket fungi on oaks and Beech.
SS None.

FAMILY **Tortricidae** (Tortrix moths) 94 GEN. | 382 spp. (7†) | 4 ILL.

Tortrix 1 sp.

;reen Oak Tortrix
'ortrix viridana

VS 17–24 mm. **ID** Forewings light ;reen; hindwings grey. **Hab** Woodland, ıedgerows, scrub. **Fp** Oaks. **SS** Cream-ıordered Green Pea *Earias clorana* ɴ/ɪ] (**WS** 20–24 mm) (a 'macro-moth', amily Nolidae (see *p. 463*)), which is ess flat at rest and has white hindwings.

Olethreutes 1 sp.

Arched Marble
Olethreutes arcuella

WS 14–17 mm. **ID** Distinctive: forewing orange brown with silver lines and markings. **Hab** Woodland; flies by day. **Fp** Dead leaves on ground. **SS** None.

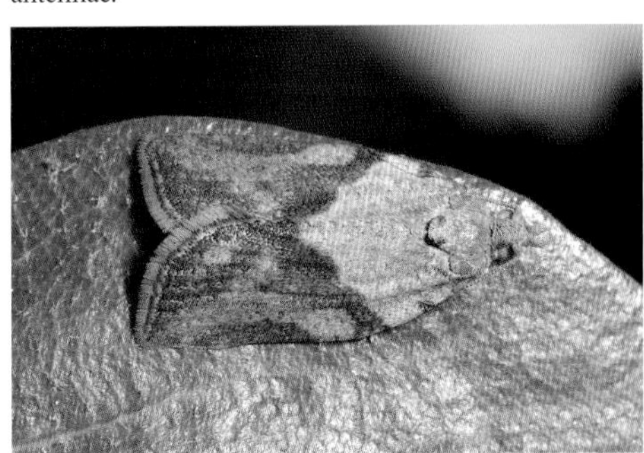

Agapeta 2 spp. | 1 ILL.

:ommon Yellow Conch
ıgapeta hamana

VS 15–24 mm. **ID** Forewing whitish ellow to deep yellow, with orange or rown marks and partial bold line. **Iab** Grassland. **Fp** Thistles. **S** Knapweed Conch *Agapeta zoegana* ɴ/ɪ] (**WS** 15–25 mm), which has a ear-circular brown ring on outer half f forewings.

Epiphyas 1 sp.

Light Brown Apple Moth
Epiphyas postvittana

WS 16–25 mm. **ID** Forewing basal half pale yellowish in ♂, more brown in ♀, rest of wing reddish-brown with darker border. **Hab** Gardens. **Fp** Many trees and shrubs. **SS** Cinquefoil Tortrix *Philedone gerningana* [ɴ/ɪ] ♂ (**WS** 13–19 mm), which is smaller and has a more arched costa and thread-like antennae.

Guide to families *pp. 407–40*

FAMILY **Crambidae** (Crambid moths) — 54 GEN. | 122 spp. | 7 ILL.

Anania — 10 spp. | 1 ILL.

Small Magpie
Anania hortulata

WS 33–35 mm. **ID** Wings white with dark greyish-black spots and markings; much of head and thorax yellow-orange; abdomen black with yellow rings. **Hab** Hedgerows, gardens. **Fp** Common Nettle. **SS** None.

Pleuroptya — 1 sp.

Mother of Pearl
Pleuroptya ruralis

WS 33–37 mm. **ID** Wings pale yellowish-brown with mottled greyish areas and cross-lines; overall with a pearly sheen, spots and markings. Much of head and thorax yellow, abdomen black with yellow rings. **Hab** Hedgerows, woodland, grassland, waste ground, gardens. **Fp** Common Nettle. **SS** None.

Agriphila — 7 spp. | 1 ILL.

Common Grass-veneer
Agriphila tristella

WS 25–30 mm. **ID** Forewings pale to dark brown with pale cream or yellowish streak, branched towards wingtip. **Hab** Grassland. **Fp** Grasses. **SS** Various grass-veneers, including **Pale-streak Grass-veneer** *Agriphila selasella* [N/I] (**WS** 22–30 mm), which has forewings with a white longitudinal streak.

Elophila — 5 spp. | 1 ILL.

Brown China-mark
Elophila nymphaeata

WS 25–33 mm. **ID** Wings with brown markings and large white blotches. **Hab** Wetlands, in ponds, canals and bogs. **Fp** Water plants; the larvae live underwater in a floating case of partial leaves. **SS** None.

Pyrausta 7 spp. | 3 ILL.

:ommon Purple & Gold
'yrausta purpuralis

Common

VS 15–22 mm. **ID** Forewings dark
urple with yellowish broken cross-
and with variable number of spots.
Iindwings with yellow spot as well
s band, but not always visible when
t rest. **Hab** Grassland (mainly
alcareous). **Fp** Corn Mint, thymes.
S Small Purple & Gold; also the
calized Scarce Purple & Gold *Pyrausta*
strinalis [*inset*] (see under Small Purple & Gold).

Small Purple & Gold
Pyrausta aurata

Common

WS 15–18 mm. **ID** Forewings dark purple with orange-yellow spot and variable, but usually small yellow marks. Hindwings black with single yellow central band. **Hab** Grassland, gardens. **Fp** Mints, Marjoram, thymes. **SS** Common Purple & Gold; Scarce Purple & Gold *Pyrausta ostrinalis* [*inset*] (**WS** 15–21 mm), which has narrower forewings and paler yellow markings.

Scarce Purple & Gold

FAMILY Pterophoridae (Plume moths) 20 GEN. | 45 spp. (2†) | 2 ILL.

Pterophorus 1 sp.

/hite Plume
terophorus pentadactyla

Common

'S 24–35 mm. **ID** Wings pure white
umes. **Hab** Grassland, gardens.
ɔ Bindweeds. **SS** None.

Emmelina 2 spp. | 1 ILL.

Common Plume
Emmelina monodactyla

Common

WS 18–27 mm. **ID** Wings pale brown (some dark), tightly rolled, forewing tips curved. **Hab** hedgerows, woodland edges, scrub, gardens. **Fp** Bindweeds. **SS** Superficially other brown plume moths; examination of genitalia required.

 Guide to families *pp. 410–41*

'MACRO-MOTHS'

FAMILY Cossidae (Leopard & goat moths) 3 GEN. | 3 spp. | 2 ILL.

Cossus 1 sp.

Goat Moth *Cossus cossus*

Nationally Scarce

WS 68–96 mm. **ID** Large, robust; wings dark brown with black lines. **Hab** Open woodland and also wetlands, parks and gardens. **Fp** Under bark and heartwood of trees, including oaks and others. **Beh** Larva may take about five years to mature; said to smell of 'goat', hence the name. Tell-tale sap runs exude from Goat Moth-infected trees. **SS** None.

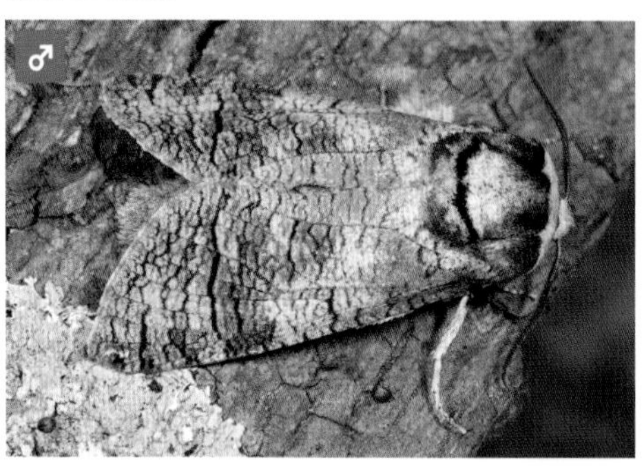
♂

Zeuzera 1 sp

Leopard Moth *Zeuzera pyrina*

Common

WS 45–78 mm. **ID** Wings whitish with numerous black spots; thorax with six larger black spots. ♂ antennae black, comb-like. **Hab** Open woodland and scrub; also parks and gardens. **Fp** Under bark and heartwood of trees, including fruit trees. **SS** None, but superficially resembles **Puss Moth** (*p. 443*).

♂

FAMILY Limacodidae (Slug moths) 2 GEN. | 2 spp. | 1 ILL.

Apoda 1 sp

Festoon *Apoda limacodes*

Nationally Scarce

WS 28–32 mm. **ID** Wings orange-brown, rests tent-like; ♂ abdomen tip upturned. Forewings with curved cross-lines. **Hab** Woodland. **Fp** Oak, Beech. **SS** None.

♂

FAMILY Hepialidae (Swift moths) 4 GEN. | 5 spp. | 4 ILL.

Hepialus 1 sp.

Ghost Moth *Hepialus humuli*

WS 44–48 mm. **ID** ♂ wings white, ♀ yellowish, marked with orange-brown. **Hab** Grassland. **Fp** Roots of grasses. **SS** None.

Phymatopus 1 sp.

Gold Swift *Phymatopus hecta*

WS 26–32 mm. **ID** ♂ forewings partly golden with **whitish markings and spots,** ♀ drabber with two pale bands. **Hab** Grassland. **Fp** Bracken roots, also grasses. **SS** Common Swift.

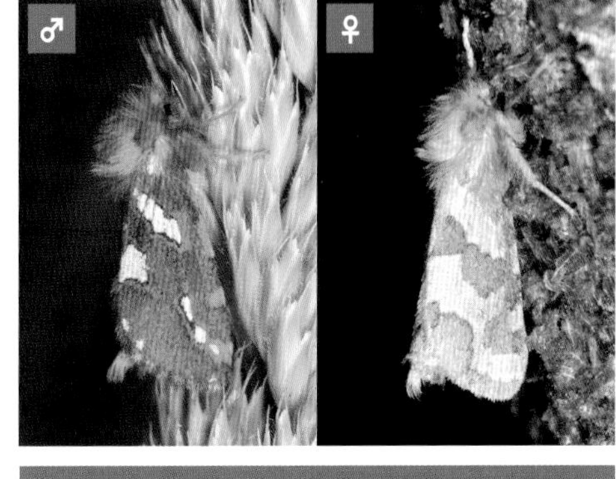

Korscheltellus 2 spp. | 1 ILL.

Common Swift
Korscheltellus lupulina

WS 25–40 mm. **ID** ♂ wings usually brown with whitish markings, but some individuals plain, ♀ drabber. **Hab** Grassland. **Fp** Roots of grasses and many other plants. **SS Orange Swift; Gold Swift; Map-winged Swift** *Korscheltellus fusconebulosa* [N/I] (**WS** 32–40 mm), which has more elaborate, map-like markings on forewings (in ♂).

Triodia 1 sp.

Orange Swift *Triodia sylvina*

WS 32–48 mm. **ID** ♂ orange-brown forewings with **two narrow diagonal lines** (not spots); ♀ drabber.
Hab Grassland. **Fp** Roots of Bracken, dandelions and other species.
SS Common Swift.

♂

Guide to families pp. 410–4

FAMILY Sesiidae (Clearwing moths) 6 GEN. | 16 spp. (1†) | 7 ILL.

Bembecia 1 sp.

Six-belted Clearwing *Bembecia ichneumoniformis*

WS 17–22 mm. **ID** Abdomen with **5** (♀) or **6** (♂) yellow bands; tail fan black with some yellow streaks; legs mainly yellow. **Hab** Chalk and limestone, also coastal grassland. **Fp** Common Bird's-foot-trefoil, Kidney Vetch. **SS Yellow-legged Clearwing, Thrift Clearwing.**

Pyropteron 2 spp. | 1 ILL.

Thrift Clearwing CR *Pyropteron muscaeformis*

WS 16–20 mm. **ID** Smallest British/Irish clearwing, somewhat drab but thorax with 3 yellowish stripes; abdomen with 3 or 4 narrow yellowish-white bands; tail fan black with some yellow streaks. **Hab** Coastal, rocky grassland. **Fp** Thrift. **SS Six-belted Clearwing, Yellow-legged Clearwing, Currant Clearwing.**

Sesia 2 spp. | 1 ILL.

Hornet Moth *Sesia apiformis*

WS 34–50 mm. **ID** Large, broad, hornet-like (*p. 511*); head, shoulder patches and most abdomen segments yellow; collar black. **Hab** Rows of poplar trees, parks, hedgerows. **Fp** Poplars. **SS Lunar Hornet Moth** *Sesia bembeciformis* [N/I] (**WS** 32–44 mm) has a yellow collar; although commoner than Hornet Moth, also seldom observed.

Synanthedon 9 spp. | 4 ILL.

‑arge Red-belted Clearwing
‑ynanthedon culiciformis

Nationally Scarce

‑**VS** 23–29 mm. **ID** Abdomen with ‑road red band; base of forewings ‑vith reddish scales. **Hab** Sparse ‑voodland, heathland. **Fp** Birches. ‑**S** The Nationally Scarce **Red-belted ‑learwing** *Synanthedon myopaeformis* ‑N/I] (**WS** 18–26 mm) has black ‑orewing markings; **Red-tipped ‑learwing**.

Red-tipped Clearwing
Synanthedon formicaeformis

Nationally Scarce

WS 18–26 mm. **ID** Abdomen with red band; tip of forewings with reddish scales. **Hab** Wetlands. **Fp** Willows. **SS Red-belted Clearwing** *Synanthedon myopaeformis* [N/I] (**WS** 18–26 mm) has black forewing markings, **Large Red-belted Clearwing**.

♀

♂

‑urrant Clearwing
‑ynanthedon tipuliformis

Nationally Scarce

‑**/S** 18–22 mm. **ID** Abdomen with ‑ (♀) or 4 (♂) yellow bands; tail fan ‑ack with yellow streaks; collar and ‑orax lines yellow. **Hab** Gardens, ‑lotments, fruit farms. **Fp** Currant ‑ushes. **SS** The Nationally Scarce **Sallow ‑learwing** *Synanthedon flaviventris* ‑/I] (**WS** 18–22 mm), which has no ‑llow on thorax; **Thrift Clearwing**.

Yellow-legged Clearwing
Synanthedon vespiformis

Nationally Scarce

WS 19–27 mm. **ID** Abdomen with 4 yellow bands. Legs and tail fan yellow. **Hab** Open woodland and parks. **Fp** Oaks. **SS Six-belted Clearwing**, **Thrift Clearwing**.

♂

‑AIL FAN black with yellow streaks

♀

TAIL FAN yellow

Guide to families pp. 410–41

FAMILY **Zygaenidae** (Burnet & forester moths) 3 GEN. | 10 spp. | 5 ILL.

Zygaena 7 spp. | 4 ILL.

Six-spot Burnet
Zygaena filipendulae

Common

WS 25–40 mm. **ID** Forewing black with **6 bright red spots**; hindwing red with narrow black border. **Hab** Grassland, woodland rides. **Fp** Common Bird's-foot-trefoil. **Beh** Day-flying. **SS** Other burnet moths, but Six-spot Burnet is the only species with six spots on forewing.

Five-spot Burnet
Zygaena trifolii

Local

WS 32–40 mm. **ID** Forewing tip rather rounded; black with 5 bright red spots, middle pair often merged; hindwing red with slightly broader black border than Narrow-bordered Five-spot Burnet. **Hab** Rough grassland, woodland rides. **Fp** Common and Greater Bird's-foot-trefoils. **Beh** Day-flying. **SS** **Narrow-bordered Five-spot Burnet.**

Narrow-bordered Five-spot Burnet *Zygaena lonicerae* VU

Common

WS 30–46 mm. **ID** Forewing long and pointed; black with **5 bright red spots; hindwing red with narrow black border**. **Hab** Rough grassland, woodland rides. **Fp** Meadow Vetchling, Red Clover. **Beh** Day-flying. **SS** **Five-spot Burnet.**

Transparent Burnet
Zygaena purpuralis

Nationally Scarce

WS 30–39 mm. **ID** Forewing black with 3 red streaks; hindwing red with narrow black border. **Hab** Grassland. **Fp** Wild Thyme. **Beh** Day-flying. **Dist** Restricted to W Scotland (islands), the Burren and Aran Islands, Ireland. **SS** None.

Adscita 2 spp. | 1 ILL.

Forester *Adscita statices* EN

Local

WS 24–29 mm. **ID** Wings green. **Hab** Chalk and limestone grassland. **Fp** Common Sorrel, Sheep's Sorrel. **Beh** Day-flying. **SS** Two other, less common forester moths (absent in Ireland): **Scarce Forester** *Jordanita globulariae* NT [N/I] (**WS** 24–30 mm) has forewing tip broader and more rounded; the Nationally Scarce **Cistus Forester** *Adscita geryon* [N/I] (**WS** 20–25 mm) is slightly smaller.

FAMILY Endromidae (Kentish Glory) 1 SP.

Endromis 1 sp.

Kentish Glory NT
Endromis versicolora

Rare

WS 55–89 mm. **ID** Large and stout, wings brown with distinctive dark lines and white markings. **Hab** Open birch woodland. **Fp** Silver Birch. **Beh** ♂ day-flying. **SS** None.

FAMILY Saturniidae (Emperor moths) 1 SP.

Saturnia 1 sp.

Emperor Moth
Saturnia pavonia

Common

WS 55–85 mm. **ID** Wings brightly coloured (♂), or paler grey or whitish (♀) (see *p. 410*), with large central 'eye-spot' on each wing. **Hab** Heathland, moorlands, wetlands, scrub. **Fp** Heathers, Bramble. **SS** None.

Guide to families pp. 410–41

FAMILY Drepanidae (Hook-tips, lutestrings & allies) 13 GEN. | 16 spp. | 6 ILL.

Cilix 1 sp.

Chinese Character
Cilix glaucata

WS 22–27 mm. **ID** Wings white with dark markings; held tight over body at rest, when resembles a bird dropping. **Hab** Woodland and hedgerows. **Fp** Blackthorn, Hawthorn and Crab Apple. **SS** None.

Habrosyne 1 sp.

Buff Arches
Habrosyne pyritoides

WS 40–44 mm. **ID** Forewings orange-brown with white lines forming an 'arch', and a grey 'saddle' – a unique pattern. **Hab** Open woodland and scrub. **Fp** Bramble, Dewberry. **SS** None.

Watsonalla 2 spp. | 1 ILL.

Oak Hook-tip VU
Watsonalla binaria

WS 28–35 mm. **ID** Wings orange-brown, forewings with two pale cross-lines. **Hab** Woodland, parks, gardens. **Fp** Oaks. **SS** **Barred Hook-tip** *Watsonalla cultraria* [N/I] (**WS** 24–33 mm) has less obviously hooked forewings and a dark central band.

Drepana 2 spp. | 1 ILL.

Pebble Hook-tip
Drepana falcataria

WS 36–40 mm. **ID** Wings pale to partly dark brown, forewings have dark central spot and, in some, purplish markings towards hooked tip. **Hab** Woodland, heathland, gardens. **Fp** Birches. **SS** Immigrant **Dusky Hook-tip** *Drepana curvatula* [N/I] (**WS** 34–42 mm) has smaller central spots on forewing and dark cross-band on hindwing; rare **Scarce Hook-tip** *Sabra harpagula* NT [N/I] (**WS** 36–42 mm) has more strongly hooked forewings marked with black.

Ochropacha 1 sp.

Common Lutestring *Ochropacha duplaris*

Common

WS 25–34 mm. **ID** Wings drab greyish; individually variable but forewings have small black spot, black dash on wingtip and pale band across centre usually visible. **Hab** Sparse woodland and scrub. **Fp** Birches. **SS** Other lutestrings (3 spp., each in its own genus [N/I]), which do not have black spot on forewing and lack the black dash on wingtip.

J F M A M J J A S O N D

Thyatria 1 sp.

Peach Blossom *Thyatira batis*

Common

WS 39–44 mm. **ID** Forewings dark brown with distinctive pink blotches. **Hab** Woodland. **Fp** Brambles. **SS** None.

J F M A M J J A S O N D

FAMILY Lasiocampidae (Eggar/lappet moths) 10 GEN. | 12 spp. (1†) | 6 ILL.

Poecilocampa 1 sp.

December Moth *Poecilocampa populi*

Common

WS 18–22 mm. **ID** Wings very dark brown, forewings with cream cross-line. **Hab** Woodland, hedgerows and gardens. **Fp** Broadleaved trees. **SS** The Nationally Scarce **Small Eggar** *Eriogaster lanestris* NT [N/I] (**WS** 36–47 mm) is reddish-brown and has white spots on forewings.

J F M A M J J A S O N D

Malacosoma 2 spp. | 1 ILL.

Lackey *Malacosoma neustria* VU

Common

WS 30–41 mm. **ID** Wings brown or yellowish, forewings with two cross-lines. **Hab** Hedgerows, parks and gardens. **Fp** Hawthorn, Blackthorn, **Beh** Larval nests often seen, with many distinctive orange, white and blue-striped larvae. **SS** Nationally Scarce **Ground Lackey** *Malacosoma castrensis* [N/I] (**WS** 31–41 mm), which has first forewing cross-line kinked and paler hindwing fringe.

J F M A M J J A S O N D

Lasiocampa 2 spp. | 1 ILL.

Oak Eggar *Lasiocampa quercus*

WS 58–90 mm. **ID** Wings dark brown (♂ (see *p. 411*)) or buff (♀); forewings with yellowish cross-line, also central white spot. **Hab** Heathland, moorlands, scrub. **Fp** Heather, Bilberry, Bramble, Hawthorn. **Beh** Two-year life-cycle in the North, one in the south. **SS** The Nationally Scarce **Grass Eggar** *Lasiocampa trifolii* [N/I] (**WS** 46–76 mm) is smaller.

Euthrix 1 sp.

Drinker *Euthrix potatoria*

WS 50–70 mm. **ID** Wings reddish-brown, with yellowish patches (♂), yellow or brown (♀); forewings with two darker cross-lines, one running to wingtip; also two central spots, including one large spot. **Hab** Wetlands and some drier grassland. **Fp** Grasses and reeds. **SS** None.

Macrothylacia 1 sp.

Fox Moth *Macrothylacia rubi*

WS 48–72 mm. **ID** Wings reddish-brown (♂) or grey (♀); forewings with two yellowish cross-lines, but no central spot. **Hab** Heathland, moorlands, coastal grassland, open woodland. **Fp** Heather, Bilberry, Bramble. **Beh** Overwinters as a fully-grown larva. **SS** None.

Gastropacha 1 sp

Lappet *Gastropacha quercifolia* EN

WS 56–88 mm. **ID** Wings purplish-brown, scallop-shaped; snout distinctive. **Hab** Open woodland, hedgerows, downlands. **Fp** Hawthorn, Blackthorn. **SS** None.

FAMILY Sphingidae (Hawk-moths) 13 GEN. | 18 spp. | 11 ILL.

Macroglossum 1 sp.

Hummingbird Hawk-moth
Macroglossum stellatarum

Regular Immigrant

WS 50–58 mm. **ID** Forewings grey, with darker markings; hindwings orange. **Hab** Woodland rides, gardens. **Fp** Lady's-bedstraw, Hedge Bedstraw. **SS** None.

Sphinx 2 spp.

Pine Hawk-moth *Sphinx pinastri*

Local

WS 72–80 mm. **ID** Forewings plain grey with three to four black streaks. Abdomen black-barred. **Hab** Coniferous woodland, gardens. **Fp** Scots Pine. **SS** Convolvulus Hawk-moth *Agrius convolvuli* (**WS** 94–120 mm), uncommon immigrant, has pink bands on abdomen.

Pine Hawk-moth Convolvulus Hawk-moth

Flight period normally September

Privet Hawk-moth *Sphinx ligustri*

Common

WS 100–120 mm. **ID** Forewings dark brown with some paler areas; hindwings and abdomen pink and black banded. **Hab** Open woodland, hedgerows, parks, gardens. **Fp** Privets, Lilac. **Beh** Hindwings flashed open and abdomen displayed in startle display. **SS** Convolvulus Hawk-moth *Agrius convolvuli* [*above right*], an uncommon immigrant, is drabber grey, rather than brown.

Laothoe 1 sp.

Poplar Hawk-moth
Laothoe populi

WS 72–92 mm. **ID** Wings grey, some lighter and darker shades; hindwings with chestnut patch. Forewings pushed up at rest. **Hab** Woodland, gardens. **Fp** Poplars, willows. **SS** None.

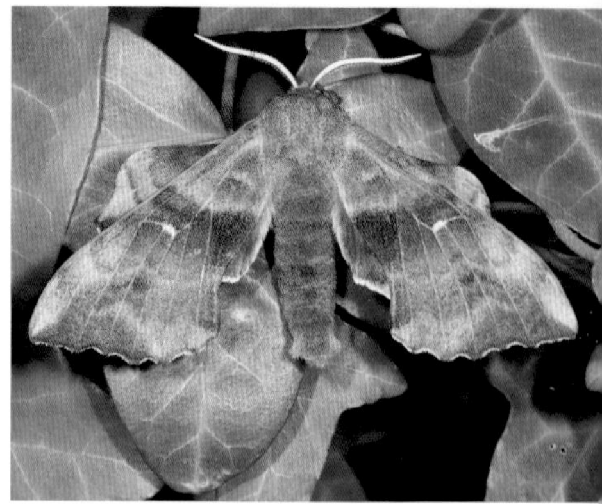

Mimas 1 sp.

Lime Hawk-moth
Mimas tiliae

WS 70–80 mm. **ID** Wings green or brownish, forewings with dark olive-green, central blotches. **Hab** Woodland, parks, gardens. **Fp** Limes. **SS** None.

Smerinthus 1 sp.

Eyed Hawk-moth
Smerinthus ocellata

WS 75–95 mm. **ID** Forewings pinkish-brown with some black areas; hindwing pink with large 'eye-spot'. Thorax with black central patch. **Hab** Woodland, scrub, parks, gardens. **Fp** Willows. **Beh** 'Eye-spots' flashed open in startle display against possible predators. **SS** None.

Deilephila 2 spp.

lephant Hawk-moth
eilephila elpenor

VS 62–72 mm. **ID** Forewings pink and live-green (same shade as abdomen); indwings pink and black. **ab** Woodland edges, rough grassland, arks, gardens. **Fp** Willowherbs, edstraws, fuchsias. **Beh** Wings flashed pen in startle display against possible redators. **SS** Small Elephant Hawk-noth.

Small Elephant Hawk-moth
Deilephila porcellus

WS 47–56 mm. **ID** Forewings pink and yellowish-brown (same shade as abdomen); hindwings pink and black. **Hab** Chalk grassland, heathland. **Fp** Bedstraws. **SS** Elephant Hawk-moth.

Hemaris 2 spp.

road-bordered Bee Hawk-noth *Hemaris fuciformis*

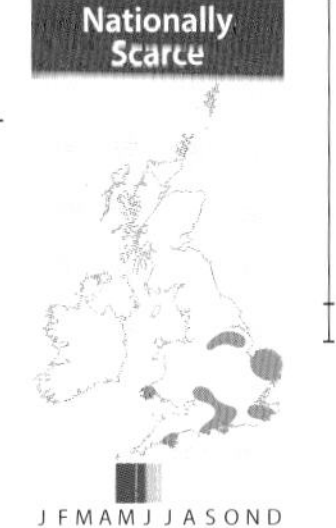

/S 46–52 mm. **ID** Wings transparent, ith broad, brown border. Thorax and art of abdomen olive-green; latter with road, brown band **Hab** Woodland des and clearings. **Fp** Honeysuckle. **S** Narrow-bordered Bee Hawk-moth.

Narrow-bordered Bee Hawk-moth *Hemaris tityus*

WS 41–46 mm. **ID** Wings transparent, with narrower, brown border. Thorax and part of abdomen olive or yellowish-brown; abdomen with broad, dark band, with yellowish beneath. **Hab** Woodland, grassland, heathland. **Fp** Devil's-bit Scabious, Small Scabious, Field Scabious. **SS** Broad-bordered Bee Hawk-moth.

FAMILY **Geometridae** (Geometrids) 141 GEN. | 307 spp. (5†) | 35 ILL.

Idaea 18 spp. | 2 ILL.

Riband Wave *Idaea aversata*

Common

WS 30–35 mm. **ID** Wings cream-coloured with three dark lines or dark band, outermost line/edge of band 'kinked inwards' at front edge of forewing. **Hab** Woodlands, chalk grasslands, wetlands, gardens. **Fp** Bedstraws, dandelions, docks. **SS** Other waves, including **Plain Wave** *Idaea straminata* [N/I] (**WS** 26–31 mm), generally smaller with silky wings and fainter cross-lines.

Purple-bordered Gold *Idaea muricata* EN

Nationally Scarce

WS 18–20 mm. **ID** Wings bright pink and yellow (including yellow fringe); some variation in amount of pink. **Hab** Wetlands including boggy heathland. **Fp** Marsh Cinquefoil. **SS** None.

Timandra 1 sp.

Blood-vein *Timandra comae*

Common

WS 30–34 mm. **ID** Wings cream-coloured, with distinctive shape and vivid pink border. **Hab** Wetlands, gardens. **Fp** Docks. **SS** None.

Camptogramma 1 sp

Yellow Shell *Camptogramma bilineata* NT

Common

WS 28–32 mm. **ID** Wings yellow, with thin brown wavy lines; often with darker yellow bands and white lines. **Hab** Woodland, grassland, hedgerows, gardens. **Fp** Cleavers, bedstraws. **SS** None.

Cyclophora 8 spp. | 2 ILL.

3irch Mocha

:yclophora albipunctata

WS 25–29 mm. **ID** Wings grey or vhitish with darker lines and spots; arge 'eye-spots'. **Hab** Woodland, ıeathland. **Fp** Downy Birch, Silver 3irch. **SS Dingy Mocha** *Cyclophora ›endularia* **NT** [N/I] (**WS** 26–29 mm) s pinkish-grey with more hooked ɔrewings and more pointed hindwings.

Maiden's Blush EN

Cyclophora punctaria

WS 25–32 mm. **ID** Forewings pale, pinkish-brown or buff, with bold cross-line and rosy flush. **Hab** Woodland, hedgerows. **Fp** Oaks. **SS** The Nationally Scarce **False Mocha** *Cyclophora porata* [N/I] (**WS** 25–32 mm) has wings with central dark-ringed 'eye-spots'.

Ourapteryx 1 sp.

;wallow-tailed Moth

)urapteryx sambucaria

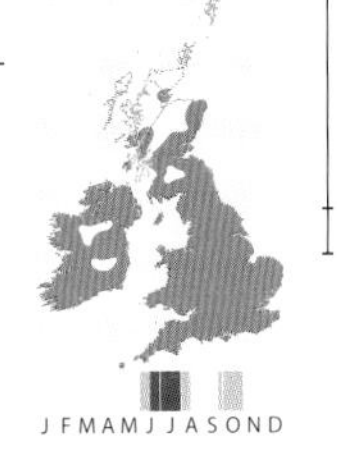

WS 50–62 mm. **ID** Wings pale yellow, ɔrewings with two darker cross-lines ınd a dash in between; hindwing with ail . **Hab** Woodland, hedgerows, crub, parks and gardens. **'p** Blackthorn, Hawthorn. **SS** None.

Menophra 1 sp.

Waved Umber

Menophra abruptaria

WS 36–42 mm. **ID** Wings flat, whitish, brown and black markings blending well. **Hab** Woodland, hedgerows, parks and gardens. **Fp** Privets, Lilac, Winter Jasmine. **Dist** S GB. **SS** None.

Odezia 1 sp.

Chimney Sweeper VU
Odezia atrata

Common

J F M A M J J A S O N D

WS 27–30 mm. **ID** Wings dark brown to black, forewings white-tipped. **Hab** Chalk and limestone grassland, hedgerows, damper areas. **Fp** Pignut. **SS** None.

♀

Abraxas 2 spp. | 1 ILL.

Magpie
Abraxas grossulariata

Common

J F M A M J J A S O N

WS 42–48 mm. **ID** Wings white with black and yellow pattern, which can vary between individuals. **Hab** Moorlands, allotments, gardens, hedgerows. **Fp** Blackthorn, Hawthorn, Bramble, currants. **SS** None.

Archiearis 1 sp.

Orange Underwing
Archiearis parthenias

Local

J F M A M J J A S O N D

WS 35–39 mm. **ID** Forewings brown, with whitish bands; hindwings orange-and-black. **Hab** Woodland, heathland. **Fp** Downy Birch, Silver Birch. **Beh** Day-flying. **SS** The Nationally Scarce **Light Orange Underwing** *Boudinotiana notha* [N/I] (**WS** 33–36 mm), which lacks whitish central cross-band on forewings.

♀

Rheumaptera 1 sp

Argent & Sable VU
Rheumaptera hastata

Nationally Scarce

J F M A M J J A S O

WS 30–34 mm. **ID** Wings black-and-white, distinctively patterned. **Hab** More open woodland. **Fp** Downy Birch and Silver Birch saplings. **Beh** Active by day. **SS** Superficially similar to **Small Argent & Sable** *Epirrhoe tristata* VU [N/I] (**WS** 24–26 mm), which is much smaller.

♀

Ematurga 1 sp.

ommon Heath
maturga atomaria

WS 22–34 mm. **ID** Wings white particularly ♀] to light brown, ellowish or grey; with dark brown ross lines (rests butterfly-like). **ab** Heathland and moorlands; rassland. **Fp** Heathers, clovers, trefoils nd vetches. **Beh** Day-flying. **S** Latticed Heath.

Chiasmia 2 spp. | 1 ILL.

Latticed Heath NT
Chiasmia clathrata

WS 26–32 mm. **ID** Wings white with dark brown cross lines (chequered effect); rests butterfly-like. **Hab** Chalk and other grassland. **Fp** Clovers and trefoils. **Beh** Day-flying. **SS** Common Heath.

Opisthograptis 1 sp.

rimstone Moth
pisthograptis luteolata

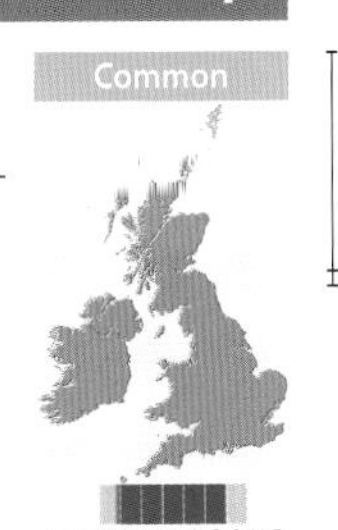

WS 33–46 mm. **ID** Wings yellow, ith brown tips and other marks long front edge. **Hab** Woodland, edgerows, gardens. **Fp** Blackthorn, lawthorn. **SS** None.

Pseudopanthera 1 sp.

Speckled Yellow
Pseudopanthera macularia

WS 28–30 mm. **ID** Wings yellow, with brown blotches. **Hab** Woodland, grassland, scrub. **Fp** Wood Sage. **Beh** Day-flying. **SS** None.

Eupithecia 47 spp. | 2 ILL.

Lime-speck Pug
Eupithecia centaureata

WS 20–24 mm. **ID** Wings whitish with brown and grey markings; forewings with large dark blotch on front edge of wing. **Hab** Open sites including gardens, hedgerows, scrub, grassland. **Fp** A wide variety of low-growing plants. **SS** None.

Narrow-winged Pug
Eupithecia nanata

WS 17–20 mm. **ID** Wings elongated, with light and dark greyish-brown bands. **Hab** Heathland and moorlands. **Fp** Heather flowers. **SS** None.

Epirrhoe 4 spp. | 1 ILL.

Common Carpet
Epirrhoe alternata

WS 27–30 mm. **ID** Wing markings vary between individuals but have dark cross-band and light grey line through narrower whitish band. **Hab** Woodland, grassland, heathland, hedgerows, gardens. **Fp** Cleavers, bedstraws. **SS** Various other carpet species, from which told by shape and colour of forewing cross-band and other markings.

Xanthorhoe 8 spp. | 1 ILL

Garden Carpet
Xanthorhoe fluctuata

WS 27–31 mm. **ID** Wings grey or whitish-grey, central band black; band faded in lower half. **Hab** Woodland, allotments, gardens. **Fp** Cabbage family. **SS** Various other carpet species, from which told by shape and colour of forewing cross-band and other markings.

Erannis 1 sp.

Mottled Umber VU Common
Erannis defoliaria

WS 40–45 mm. **ID** ♂ forewings whitish to brown, with broad bands of brown or black and usually dark spot; hindwings pale. ♀ mottled, with tiny wing stumps. **Hab** Woodland. **Fp** Oaks, birches, Hazel, Hawthorn. **SS** Other superficially similar species are generally plainer, with cross-lines on forewings and lack dark spot.

Operophtera 2 spp. | 1 ILL.

Winter Moth Common
Operophtera brumata

WS 28–33 mm. **ID** ♂ Wings brown, often drab and with a darker central forewing band (frequently indistinct); ♀ wingless. **Hab** Woodland, gardens. **Fp** Broadleaved trees, pest on apples in orchards. **SS Northern Winter Moth** *Operophtera fagata* [N/I] (**WS** 32–40 mm) is generally larger and usually greyer.

Aplocera 3 spp. | 2 ILL.

Treble-bar *Aplocera plagiata*

WS 37–43 mm. **ID** Wings with bold, grey bands. **Hab** Grassland, Heathland, moorlands, gardens. **Fp** St. John's-worts. **SS Lesser Treble-bar** *Aplocera efformata* [*below*] (**WS** 35–41 mm) (absent Ireland) is slightly smaller with the narrow basal cross-band sharply angled.

Treble-bar

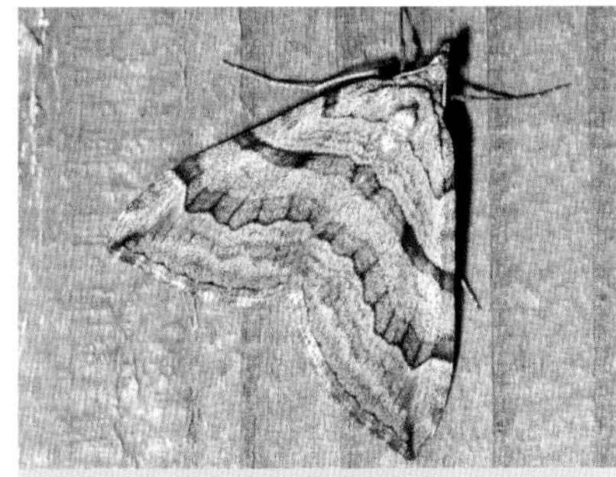
Lesser Treble-bar

Common

Lycia 3 spp. | 1 ILL.

Brindled Beauty *Lycia hirtaria*

WS 42–52 mm. **ID** Hairy. Wings grey or with a yellowish tinge; always with bold black cross-lines and bands. **Hab** Woodland, hedgerows, scrub, parks and gardens. **Fp** Birches, Hawthorn and other deciduous trees. **SS** The local **Pale Brindled Beauty** *Phigalia pilosaria* [N/I] (**WS** 45–50 mm), which is rather paler and lacks heavy black lines.

Selenia 3 spp. | 1 ILL

Purple Thorn *Selenia tetralunaria*

WS 44–52 mm. **ID** Wings dark, purplish-brown; rests with wings part-spread, rather like some butterflies. **Hab** Woodland, grassland, scrub. **Fp** Hazel, birches. **SS** The local **Lunar Thorn** *Selenia lunularia* [N/I] (**WS** 44–52 mm), which has rather paler, more scalloped hindwings.

Biston 2 spp.

Peppered Moth *Biston betularia*

WS 45–62 mm. **ID** Wings and body white, 'peppered' with black; almost all-black in some (form *carbonaria*), which is most frequent in industrial areas. **Hab** Woodland, hedgerows, scrub, parks and gardens. **Fp** Birches, Hawthorn a wide range of deciduous trees and shrubs. **SS** None.

Oak Beauty *Biston strataria*

WS 51–56 mm. **ID** Forewings whitish to grey, with two broad brown bands. ♂ antennae feathery. **Hab** Woodland. **Fp** Oaks, Hazel. **SS** None.

Scotopteryx 6 spp. | 1 ILL.

›haded Broad-bar *Scotopteryx ·henopodiata*

Common

VS 34–38 mm. **ID** Wings various hades of brown with dark central ross-band that has narrower bands vithin. Forewing tips with black streak. **Iab** Grassland, heathland, woodland ides. **Fp** Clovers, vetches. **SS** None.

Petrophora 1 sp.

Brown Silver-line *Petrophora chlorosata*

Common

WS 31–37 mm. **ID** Forewings brown with central dark dot and two straight silvery, dark-edged cross-lines. **Hab** Woodland, heathland. **Fp** Bracken. **Beh** Often disturbed by day, can be abundant. **SS** None.

Crocallis 2 spp. | 1 ILL.

·calloped Oak *'rocallis elinguaria*

Common

'S 40–46 mm. **ID** Wings pale yellowish, ›rewings slightly scalloped with darker and and central black spot. **Hab** Voodland, scrub, parks and gardens. **p** Hawthorn, Blackthorn, birches, ıks. **SS** The rare immigrant **Dusky calloped Oak** *Crocallis dardoinaria* ı/ı] (**WS** 46–52 mm) has wings less llowish. Superficially similar to **Scalloped Hazel**.

Odontopera 1 sp.

Scalloped Hazel *Odontopera bidentata*

Common

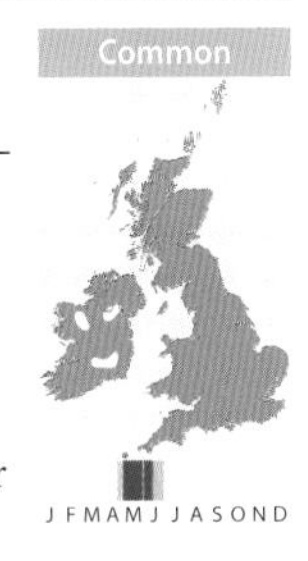

WS 46–50 mm. **ID** Wings whitish-brown to dark brown, forewings very scalloped with two cross-lines and central black ring. **Hab** Woodland, scrub, parks and gardens. **Fp** Hazel, birches, Hawthorn, Blackthorn, oaks. **SS** None, although superficially similar to **Scalloped Oak**.

Colostygia 3 spp. | 1 ILL.

Green Carpet
Colostygia pectinataria

WS 25–29 mm. **ID** Wings green with darker areas and whitish zigzag lines; background colour fades to whitish. **Hab** Woodland, heathland, gardens. **Fp** Bedstraws. **SS** Beech-green Carpet *Colostygia olivata* [N/I] (**WS** 26–35 mm), which is less vividly marked.

Bupalus 1 sp.

Bordered White
Bupalus piniaria

WS 34–40 mm. **ID** Wings whitish, yellowish or orange with broad, dark brown border (rests butterfly-like, underside brown with whitish streaks and dark markings); ♂ has feathery antennae. **Hab** Coniferous woodland. **Fp** Scots Pine. **SS** None, but needs a careful look, as upperside is not easy to examine.

Geometra 1 sp.

Large Emerald
Geometra papilionaria

WS 50–64 mm. **ID** Wings vivid green, with whitish cross-lines (rests butterfly-like). **Hab** Woodland, heathland, grassland, hedgerows. **Fp** Birches, Hazel, Alder. **SS** None.

Campaea 1 sp.

Light Emerald
Campaea margaritaria

WS 42–54 mm. **ID** Wings pale greenish-white with two pale cross-lines. **Hab** Woodland, parks and gardens. **Fp** Oaks, Hawthorn, birches, Hazel. **SS** None.

FAMILY **Notodontidae** (Prominents & allies) 17 GEN. | 28 spp. | 15 ILL.

Thaumetopoea 2 spp. | 1 ILL.

Oak Processionary
Thaumetopoea processionea

Immigrant; adventive pest

WS 31–41 mm. **ID** Wings greyish-brown with darker cross-lines; forewings with pale basal area. **Hab** Woodland. **Beh** The hairy larvae live in communal nests, in huge numbers. Although the Forestry Commission eradicate nests, where possible, still regularly recorded in London and elsewhere. **Fp** Oaks. **SS** The immigrant **Pine Processionary** *Thaumetopoea pityocampa* (**WS** 36–46 mm) [N/I], which lacks pale basal area on forewings.

J F M A M J J A S O N D

Notifiable pest: report to Forestry England (*p. 571*)

Cerura 2 spp. | 1 ILL.

Puss Moth *Cerura vinula*

Common

WS 62–80 mm. **ID** Hairy, wings whitish or greyish with numerous black lines and markings; thorax with small black spots. **Hab** Woodland, hedgerows, gardens. **Fp** Poplars, willows. **SS** Leopard Moth (*p. 422*) is superficially similar; Feline *Cerura erminea* [N/I] (**WS** 45–75 mm) is a rare immigrant with less extensive black markings.

Stauropus 1 sp.

Lobster Moth *Stauropus fagi*

Common

WS 55–70 mm. **ID** Forewings greyish-brown with hint of reddish-brown; at rest, hindwings extend sideways beyond edge of forewings. **Hab** Woodland, hedgerows. **Beh** Strange-shaped larva, with long legs. **Fp** Beech, birches, Hazel, oaks. **SS** None.

Notodonta 4 spp. | 2 ILL.

Iron Prominent
Notodonta dromedarius

WS 42–50 mm. **ID** Forewings dark greyish-brown; yellowish base and reddish-brown streaks. **Hab** Woodland, heathland, gardens. **Fp** Birches, Alder, Hazel, oaks. **SS** Only rare immigrant prominents, which lack distinctive reddish-brown streaks.

Pebble Prominent
Notodonta ziczac

WS 42–52 mm. **ID** Forewings yellowish-brown; pebble-like blotch towards tip. **Hab** Woodland, scrub, gardens. **Fp** Willows, Aspen, poplars. **SS** None.

Pheosia 2 spp.

Lesser Swallow Prominent
Pheosia gnoma

WS 46–58 mm. **ID** Forewings whitish with brown and black marks; white 'wedge' in corner short and broad. **Hab** Woodland, heathland, parks and gardens. **Fp** Downy Birch, Silver Birch. **SS** Swallow Prominent.

Swallow Prominent
Pheosia tremula

WS 50–64 mm. **ID** Forewings whitish with brown and black marks; whitish 'wedge' in corner long and narrow, extending more than half length of wing. **Hab** Woodland, scrub, parks and gardens. **Fp** Poplars, Aspen, willows. **SS** Lesser Swallow Prominent.

Leucodonta 1 sp.

White Prominent EN Rare

Leucodonta bicoloria

WS 38–42 mm. **ID** Forewings white, with orange marks edged black. **Hab** Woodland. **Fp** Birches. **Dist** Rediscovered in Co. Kerry, Ireland, in 2008. **SS** None.

Pterostoma 1 sp.

Pale Prominent Common

Pterostoma palpina

WS 42–60 mm. **ID** Forewings pale greyish-brown; appears to have a 'snout' due to the presence of large palps; also has conspicuous scale tufts; in ♂, abdomen extends well beyond wings. **Hab** Woodland, scrub, parks and gardens. **Fp** Poplars, Aspen, willows. **SS** None.

Peridea 1 sp.

Great Prominent Local

Peridea anceps

WS 52–72 mm. **ID** Forewings marbled and streaked greenish-brown; hindwings yellowish-white. **Hab** Woodland, hedgerows. **Fp** Oaks. **SS** None.

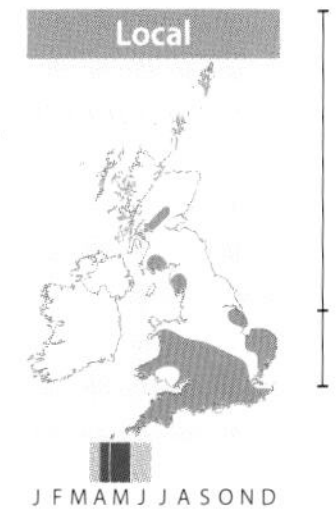

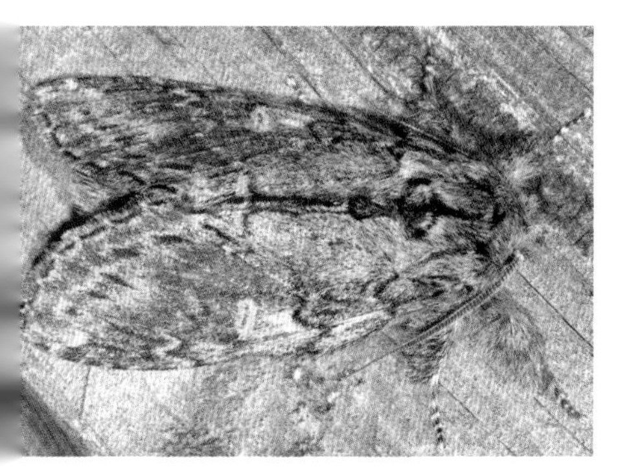

Ptilodon 2 spp. | 1 ILL.

Coxcomb Prominent Common

Ptilodon capucina

WS 40–50 mm. **ID** Forewings pale to dark brown with dark scale tuft and scalloped edges. Thorax with forward-pointing 'crest'. **Hab** Woodland, scrub, parks and gardens. **Fp** Birches, Hazel, Hawthorn. **SS** None.

Guide to families *pp.* 410–41

Furcula 3 spp. | 1 ILL.

Sallow Kitten *Furcula furcula*

Common

WS 35–42 mm. **ID** Forewings whitish with central greyish cross-band and markings, including some yellow marks and black spots. **Hab** Woodland, hedgerows. **Fp** Willows, Aspen, poplars. **SS** Other kittens, which are distinguished by subtle differences in shape of forewing bands or markings.

Phalera 1 sp.

Buff-tip *Phalera bucephala*

Common

WS 55–68 mm. **ID** Forewings greyish with broad yellow tip. Thorax with additional yellow marking. **Hab** Woodland, scrub, hedgerows and gardens. **Fp** Willows, birches, Hazel, oaks. **SS** None.

NOTE: photo of larva on *p. 137*.

Drymonia 2 spp. | 1 ILL.

Lunar Marbled Brown *Drymonia ruficornis*

Common

WS 38–46 mm. **ID** Forewings greyish-brown; central cross-band whitish, with black 'comma'. **Hab** Woodland, hedgerows, gardens, parks. **Fp** Oaks. **SS** The local **Marbled Brown** *Drymonia dodonaea* [N/I] (**WS** 39–44 mm) has forewings without a black 'comma'.

Clostera 3 spp. | 1 ILL.

Chocolate-tip *Clostera curtula*

Local

WS 36–38 mm. **ID** Forewings grey, tip chocolate-brown. Thorax with black patch on top; tip of abdomen chocolate. **Hab** Woodland, scrub, hedgerows. **Fp** Aspen, poplars, willows. **SS** **Small Chocolate-tip** *C. pigra* [N/I] (**WS** 22–27 mm) and the rare **Scarce Chocolate-tip** *C. anachoreta* [N/I] (**WS** 37 mm) are less obviously chocolate-brown.

FAMILY **Erebidae** (Tussocks, tigers, underwings & allies)

53 GEN. | 88 spp. (5†) | 27 ILL.

Euproctis — 2 spp. | 1 ILL.

Yellow-tail *Euproctis similis*

Common

WS 35–45 mm. **ID** Wings white; tip of abdomen yellow and clearly visible; forewings with grey or blackish spots in ♂. **Hab** Grassland, gardens. **Fp** Hawthorn, Blackthorn; other broadleaved trees and shrubs. **SS** The local **Brown-tail** *Euproctis chrysorrhoea* [N/I] (**WS** 36–42 mm) has a brown tip to the abdomen.

NOTE: larvae of both these species [*inset*] have irritating hairs and should not be handled.

YELLOW-TAIL LARVA

Calliteara — 1 sp.

Pale Tussock *Calliteara pudibunda*

Common

WS 50–70 mm. **ID** Wings grey (some individuals very dark) with darker lines. **Hab** Woodland, hedgerows, gardens. **Fp** Hawthorn, Blackthorn, birches; other broadleaved trees and shrubs. **SS** The local **Dark Tussock** *Dicallomera fascelina* [N/I] (**WS** 40–53 mm) is usually darker, forewings with black scales.

Orgyia — 2 spp. | 1 ILL.

Vapourer *Orgyia antiqua*

Common

WS ♂ 35–38 mm. **ID** ♂ wings orange-brown, forewings white-spotted. ♀ greyish-brown, wingless. **Hab** Scrub, hedgerows, parks, gardens. **Beh** ♂s are day-flying. **Fp** Broadleaved trees and shrubs, including willows and birches. **SS** Scarce Vapourer *Orgyia recens* EN [N/I] (**WS** 34–40 mm), which has much more extensively marked forewings, including additional white spots.

♂

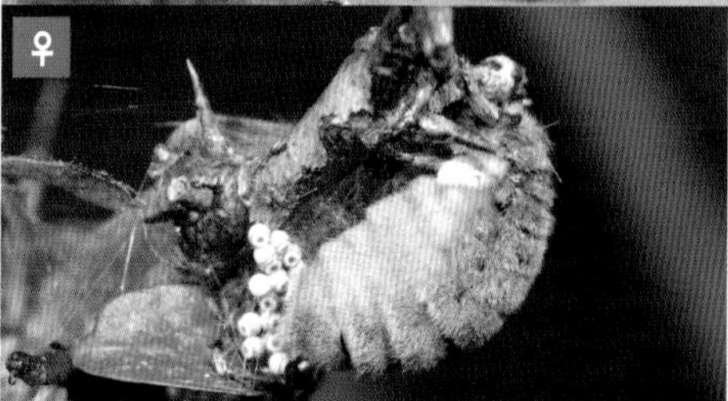

♀

Arctia 2 spp.

Garden Tiger NT
Arctia caja

WS 50–78 mm. **ID** Forewings white with bold dark brown blotches, hindwings red with large bluish spots. **Hab** Wet and dry grassland, hedgerows, sand dunes, parks, gardens. **Fp** Docks, dandelion, Common Nettle. **SS** None.

Cream-spot Tiger
Arctia villica

WS 50–66 mm. **ID** Forewings black with bold cream blotches, hindwings yellow with large black spots; abdomen yellow with bright red tip. **Hab** Open grassland, heathland and woodland; often coastal. **Fp** White Dead-nettle, ragworts and many other herbaceous plants. **SS** None.

Callimorpha 1 sp.

Scarlet Tiger
Callimorpha dominula

WS 52–58 mm. **ID** Forewings black with yellow and white blotches, hindwings red with black spots; thorax and abdomen black, thorax with two yellow spots. **Hab** Wetlands. **Beh** Day-flying. **Fp** Common Comfrey, Hemp-agrimony, Common Nettle, Bramble. **SS** None.

Euplagia 1 sp.

Jersey Tiger
Euplagia quadripunctaria

WS 52–65 mm. **ID** Forewings black with bold white stripes, hindwings red or orange with black spots; thorax black-and-white striped, abdomen matches those of hindwings. **Hab** Hedgerows. **Beh** Day and night-fliers. **Fp** Common Nettle, Hemp-agrimony, White Dead-nettle, plantains, Bramble. **SS** None.

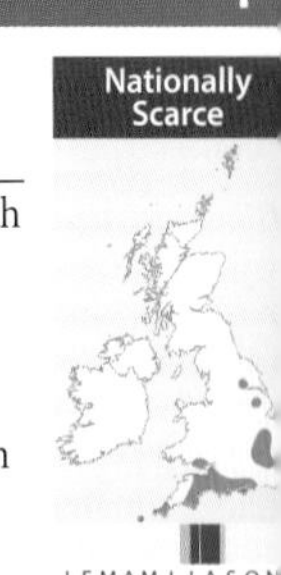

Parasemia 1 sp.

Wood Tiger NT

Parasemia plantaginis

Local

WS 34–42 mm. **ID** Forewings white with bold black stripes and patterning; hindwings orange (white in some individuals) with range of black spots and streaks. **Hab** Heathland, moorlands, grassland and scrub. **Beh** Day- and night-fliers. **Fp** Bell Heather, plantains, Common Rock-rose. **SS** None.

Phragmatobia 1 sp.

Ruby Tiger

Phragmatobia fuliginosa

Common

WS 28–38 mm. **ID** Forewings pinkish-brown; hindwings vivid pink with black markings and greyish areas (hindwings all-black with pink fringe in some individuals); abdomen vivid pink with black markings. **Hab** Chalk grassland, heathland, moorlands, gardens. **Fp** Ragworts, plantains, heathers. **SS** None.

Euclidia 2 spp.

Burnet Companion

Euclidia glyphica

Common

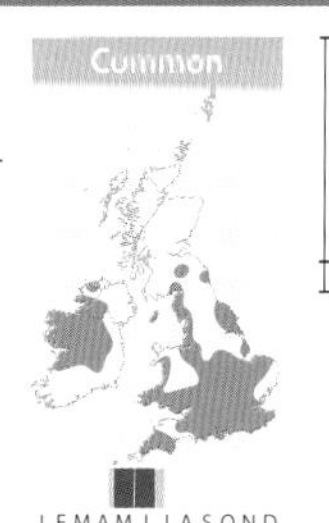

WS 28–34 mm. **ID** Forewings brown with darker brown bands and markings; hindwings orange with brown bands; often mistaken for a butterfly. **Hab** Grassland, roadside verges. **Beh** Day-flying. **Fp** Clovers, trefoils and vetches. **SS** None.

Mother Shipton *Euclidia mi*

Common

WS 30–34 mm. **ID** Forewings with large brown blotches and distinctive markings; hindwings brown with yellowish spots; often mistaken for a butterfly. **Hab** Grassland, heathland, moorlands, roadside verges. **Beh** Day-flying. **Fp** Clovers, trefoils. **SS** None.

Spilosoma 3 spp. | 2 ILL.

Buff Ermine *Spilosoma lutea*

WS 34–42 mm. **ID** Wings white (♀) or cream/buff with black spots, mainly in a diagonal line. **Hab** Woodland, hedgerows, parks, gardens. **Fp** Common Nettle, Honeysuckle and many others. **SS** superficially similar to other ermines (2), **Muslin Moth** and other relatives, but all these species lack a diagonal line of spots.

Common

White Ermine *Spilosoma lubricipeda*

WS 34–48 mm. **ID** Wings white (or cream/buff in Scotland and Ireland) with black spots, pattern varying between individuals. **Hab** Woodland, moorlands, parks, gardens. **Fp** Common Nettle, docks. **SS** superficially similar to other ermines (2), **Muslin Moth** and other relatives, but White Ermine typically has far more spots.

Common

Diaphora 1 sp.

Muslin Moth *Diaphora mendica*

WS 30–43 mm. **ID** Wings brown or grey (♂) or white (♀), each with several small black spots. **Hab** Grassland, hedgerows, gardens. **Fp** Docks, plantains. **SS** None.

Common

Diacrisia 1 sp.

Clouded Buff *Diacrisia sannio*

WS 35–50 mm. **ID** Forewings yellow with pink fringe (♂) or orange (♀), in both cases with central dark mark; hindwings whitish with black shading. **Hab** Heathland and moorlands. **Beh** ♂s are day-flying. **Fp** Heathers, Sheep's-sorrel, plantains. **SS** None.

Local

Scoliopteryx 1 sp.

Herald *Scoliopteryx libatrix*

Common

WS 44–48 mm. **ID** Forewings greyish-brown with orange blotches; hooked. **Hab** Woodland, scrub, hedgerows, wetlands, parks and gardens. **Beh** Adults hibernate in sheltered areas. **Fp** Willows, Aspen, poplars. **SS** None.

J F M A M J J A S O N D

Hypena 6 spp. | 1 ILL.

Snout *Hypena proboscidalis*

Common

WS 36–42 mm. **ID** Forewings hooked, fairly plain brown, with darker cross-lines; palps very elongated. **Hab** Grassland, gardens. **Fp** Common Nettle. **SS** Other snout species, although only Snout has hooked wingtips.

J F M A M J J A S O N D

Rivula 1 sp.

Straw Dot *Rivula sericealis*

Common

WS 19–25 mm. **ID** Forewings rather plain, yellowish-brown, with darker central kidney-shaped mark. **Hab** Grassland, heathland, gardens. **Fp** Grasses. **SS** None.

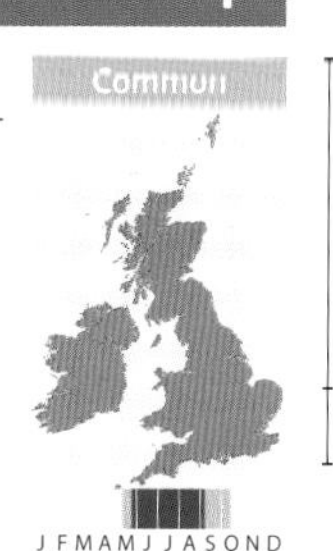

J F M A M J J A S O N D

Lymantria 2 spp. | 1 ILL.

Black Arches *Lymantria monacha*

Local

WS 44–54 mm. **ID** Forewings white with bold black lines, spots and marks; thorax with black spots; abdomen pinkish. **Hab** Woodland and parks. **Fp** Oaks. **SS** None.

J F M A M J J A S O N D

Eilema 7 spp. | 1 ILL.

Common Footman
Eilema lurideola

WS 31–38 mm. **ID** Elongated wings, grey with narrow yellow border, not reaching forewing tip; rests with forewings curled over. Thorax collar yellow, narrower in centre. **Hab** Woodland, wetlands, gardens. **Fp** Lichens on trees. **SS** Other footman species, which either lack a distinct yellow stripe on forewings, or yellow stripe reaches wingtip.

Miltochrista 1 sp.

Rosy Footman VU
Miltochrista miniata

WS 25–33 mm. **ID** Forewings pinkish-red with delicate black markings. **Hab** Woodland, heathland, hedgerows. **Fp** Lichens on oaks. **SS** None.

Atolmis 1 sp.

Red-necked Footman
Atolmis rubricollis

WS 28–36 mm. **ID** Wings uniform black, thorax collar red; abdomen yellow-and-black. **Hab** Woodland. **Fp** Lichens growing on trees. **SS** None.

Tyria 1 sp.

Cinnabar *Tyria jacobaeae*

WS 35–45 mm. **ID** Forewings greyish-black with large red streak and two large red spots; hindwings red with black border. **Hab** Grassland, heathland, gardens. **Beh** Day-flying. **Fp** Common Ragwort. **SS** None.

Catocala **7 spp. | 4 ILL.**

lifden Nonpareil
atocala fraxini

Immigrant, rare resident locally

VS 90–106 mm. **ID** Very large; orewings grey, hindwings black with road violet-blue band and broad vhite fringe. **Hab** Woodland. **p** Aspen, poplars. **Dist** Former esident, has recolonized several ounties since about 2007 and is now eing reported more frequently. **S** None, once the hindwings are seen.

Red Underwing
Catocala nupta

Common

WS 70–90 mm. **ID** Large; forewings grey with faint jagged markings; hindwings red with black bands and broad white fringe. **Hab** Woodland, parks, gardens. **Fp** Poplars, willows. **SS** Rare immigrant 'red' underwings (2), which are distinguished by forewing pattern.

ark Crimson Underwing NT
atocala sponsa

Rare

VS 58–74 mm. **ID** Large; forewings ark greyish-brown with irregular ross-lines and bands; hindwings red vith black bands (end of central band W'-shaped) and whitish-black fringe. **Hab** Woodland. **Fp** Oaks. **Dist** Stronghold New Forest, Hampshire. **S** **Light Crimson Underwing**, possibly ther 'red' underwings.

Light Crimson Underwing NT
Catocala promissa

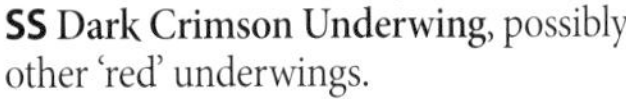

Rare

WS 50–60 mm. **ID** Large; forewings greyish-brown with irregular cross-lines and bands; some whitish marks and paler areas; hindwings red with black bands and whitish fringe, suffused with black. **Hab** Woodland. **Fp** Oaks. **SS** **Dark Crimson Underwing**, possibly other 'red' underwings.

Guide to families *pp.* 410–41

FAMILY **Noctuidae** (Noctuids) **175 GEN. | 368 spp. (6†) | 36 ILL.**

Autographa 4 spp. | 1 ILL.

Silver Y *Autographa gamma*

WS 32–52 mm. **ID** Forewings marbled brown and grey (tinged mauve in some individuals), with distinctive silver 'Y'-shaped central mark. **Hab** Grassland, wetlands, hedgerows, woodland edges. **Beh** Day-flying. **Fp** Bedstraws, clovers, Common Nettle. **SS** Several less common species, often immigrants, are superficially similar, although at the very least the central mark is a different shape.

Agrotis 13 spp. | 1 ILL.

Heart & Dart
Agrotis exclamationis

WS 35–44 mm. **ID** Forewings narrow, light to dark brown, bold central 'dart' near spot with larger blotch beneath; thorax with broad, blackish collar. **Hab** Virtually anywhere, but often agricultural areas and gardens. **Fp** Plantains and many other herbaceous plants. **SS** Heart & Club *Agrotis clavis* [N/I] (**WS** 35–40 mm), which lacks a black collar.

Xestia 11 spp. | 1 ILL.

Setaceous Hebrew Character
Xestia c-nigrum

WS 35–45 mm. **ID** Forewings dark greyish-brown with triangular 'hebrew character' mark near edge in centre. **Hab** Woodland, wetlands, parks, gardens. **Fp** Common Nettle, White Dead-nettle, willowherbs. **SS** None.

Orthosia 8 spp. | 1 ILL.

Hebrew Character
Orthosia gothica

WS 30–40 mm. **ID** Forewings greyish to reddish-brown with saddle-shaped 'hebrew character' mark near edge in centre. **Hab** Woodland, gardens. **Fp** Oaks, birches, Hawthorn, willows. **SS** None.

Ceramica 1 sp.

room Moth *Ceramica pisi*

Common

/S 33–42 mm. **ID** Forewings light to ark brown with whitish cross-line ear border, forming a corner spot or lotch. **Hab** Heathland, moorlands. **p** Heathers, Bracken, Broom, ramble. **SS** Superficially similar to arious other brown Noctuids but utermost creamy yellow cross-line orms diagnostic spot in corner.

Lycophotia 1 sp.

True Lover's Knot
Lycophotia porphyria

Common

WS 26–34 mm. **ID** Forewings reddish-brown (some individuals deep pinkish-brown) with pattern of black and whitish markings. **Hab** Heathland, moorlands, gardens. **Fp** Heathers. **SS** None.

Dypterygia 1 sp.

ird's Wing
ypterygia scabriuscula

Local

/S 34–42 mm. **ID** Forewings dark rown with paler markings on edge of nerwing widening towards corner nd forming a distinctive pattern esembling a bird's wings. **ab** Woodland, heathland, parks. **p** Docks, sorrels. **SS** None.

Cerapteryx 1 sp.

Antler Moth
Cerapteryx graminis

Common

WS 27–39 mm. **ID** Forewings light to dark brown with whitish antler-shaped marks and black streaks. **Hab** Grassland. **Fp** Grasses. **SS** None.

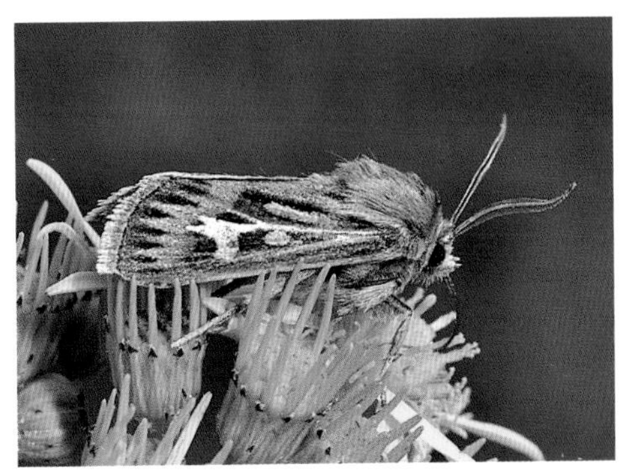

Plusia 2 spp. | 1 ILL.

Gold Spot
Plusia festucae

WS 34–36 mm. **ID** Forewings yellowish-brown with two large whitish central marks. **Hab** Wetlands, heathland. **Fp** Sedges. **SS** The local **Lempke's Gold Spot** *Plusia putnami* [N/I] (**WS** 34–46 mm) is usually smaller and more orange-brown.

Diachrysia 3 spp. | 1 ILL.

Burnished Brass
Diachrysia chrysitis

WS 34–44 mm. **ID** Forewings hooked; brassy yellow bands either split into blotches or separated. **Hab** Grassland, wetlands, hedgerows, woodland edges, gardens. **Fp** Common Nettle, White Dead-nettle. **SS** **Cryptic Burnished Brass** *D. stenochrysis* [N/I] (**WS** 34–44 mm) has central brown band across forewing broken by brassy belt.

Cirrhia 3 spp. | 1 ILL.

Sallow
Cirrhia icteritia

WS 32–40 mm. **ID** Forewings slightly hooked at tip, pale yellow with orange-brown markings and black central spot. **Hab** Woodland, heathland, moorlands and gardens. **Fp** Willows and poplars (in catkins). **SS** Other sallows look superficially similar, but have different markings and colours.

Eremobia 1 sp

Dusky Sallow
Eremobia ochroleuca

WS 34–37 mm. **ID** Forewings whitish-brown with dark bands; wing fringes brown and white. **Hab** Grassland, mainly chalk. **Fp** Grasses. **SS** None.

Panolis 1 sp.

'ine Beauty
'anolis flammea

VS 32–40 mm. **ID** Forewings orange-rown with darker band that includes wo large whitish markings. **lab** Coniferous woodland, parks. **p** Pines. **SS** None.

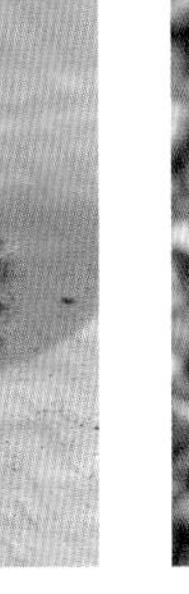

Phlogophora 1 sp.

Angle Shades
Phlogophora meticulosa

WS 45–52 mm. **ID** Forewings light brown with pinkish and olive areas; characteristic resting position, with edge of forewings creased and folded. **Hab** Woodland, hedgerows, gardens. **Fp** Common Nettle, docks, Bramble, Barberry, birches. **SS** None.

Cosmia 4 spp. | 1 ILL.

un-bar *Cosmia trapezina*

'S 28–38 mm. **ID** Forewings yellowish-own to dark brown with distinctively aped cross-veins. **Hab** Woodland, dgerows and gardens. **Fp** Elms, oaks, rches, Blackthorn, Hazel. **SS** None.

Cucullia 9 spp. | 1 ILL.

Mullein *Cucullia verbasci*

WS 44–52 mm. **ID** Forewings buff with dark band each side and whitish marks; hindwings grey. **Hab** Grassland, including on chalk, gardens, woodland. **Fp** Mulleins, figworts, buddleja. **SS** Other related 'shark' moths, some of which require examination of genitalia to confirm identification.

Noctua 7 spp. | 2 ILL

Large Yellow Underwing *Noctua pronuba*

WS 50–60 mm. **ID** Forewings narrow, light to dark brown, with darker bands and black mark towards wingtip; hindwings yellow with black band. **Hab** Virtually anywhere, but often grassland and gardens. **Fp** Docks, cabbages, Foxglove. **SS** Other yellow underwing species but black wingtip mark distinctive.

Broad-bordered Yellow Underwing *Noctua fimbriata*

WS 50–58 mm. **ID** Robust-looking, forewings light to dark brown or greenish with darker bands; hindwings yellow with broad black band. **Hab** Woodland and parks. **Fp** Common Nettle, docks. **SS** Other yellow underwing species, but the large size and distinct forewing pattern makes confusion unlikely.

Anarta 3 spp. | 1 ILL.

Beautiful Yellow Underwing *Anarta myrtilli*

WS 24–28 mm. **ID** Forewings reddish-brown with greyish-white markings; hindwings yellow, with broad black border. **Hab** Heathland, moorlands. **Beh** Day-flying. **Fp** Heathers. **SS** The Nationally Scarce **Small Dark Yellow Underwing** *Coranarta cordigera* [N/I] (**WS** 24–28 mm) has darker forewings (only in Scottish Highlands).

Panemeria 1 sp

Small Yellow Underwing *Panemeria tenebrata*

WS 19–22 mm. **ID** Forewings brown, with darker bands and grey marks; hindwings black with broad yellow central band; wing fringes whitish. **Hab** Grassland, including chalk. **Beh** Day-flying. **Fp** Common Mouse-ear, Field Mouse-ear seed capsules. **SS** None.

Amphipyra 3 spp. | 1 ILL.

:opper Underwing *Amphipyra ›yramidea*

Common

VS 47–54 mm. **ID** Forewings brown, vith darker central band and whitish nes; hindwings orange (underside part-range at margin, not extending beyond ark cross-band). **Hab** Woodland, arks, gardens. **Fp** Hawthorn, 3lackthorn, Hazel. **SS** Svensson's :opper Underwing *Amphipyra berbera* N/I] (**WS** 47–56 mm), hindwings (underside) with more xtensive orange, well beyond dark cross-band.

Mormo 1 sp.

Old Lady *Mormo maura*

Common

WS 64–74 mm. **ID** Large, wings broadened, dark brown with darker bands. **Hab** Woodland, hedgerows, scrub, gardens. **Fp** Docks, Ivy, Hawthorn. **SS** None.

Apamea 14 spp. | 1 ILL.

›ark Arches *pamea monoglypha*

Common

/S 46–54 mm. **ID** Forewings tapered, reyish to dark brown or black ith large oval and kidney marks ıd a conspicuous pale 'W'-shaped ıark. **Hab** Grassland, hedgerows, eathland, gardens. **Fp** Grasses. **S** None, although other less tapered, rown Noctuids can look superficially milar.

Colocasia 1 sp.

Nut-tree Tussock *Colocasia coryli*

Common

WS 30–38 mm. **ID** Forewings with dark brown inner half, paler outer half. **Hab** Woodland. **Fp** Beech, birches, Hazel, oaks. **SS** None.

Acronicta — 10 spp. | 4 ILL.

Alder Moth *Acronicta alni*

WS 37–43 mm. **ID** Forewings greyish-brown with black areas including 'dagger-like' mark and central pale kidney spot. **Hab** Woodland. **Fp** Birches, Alder, willows, oaks. **SS** None.

Sycamore *Acronicta aceris*

WS 40–45 mm. **ID** Forewings light grey, mottled with darker grey markings. **Hab** Parks, gardens, woodland. **Fp** Sycamore, maples, Horse-chestnut. **SS** **Poplar Grey**, which has pale area beyond kidney mark.

Grey Dagger *Acronicta psi*

WS 34–45 mm. **ID** Forewings grey with black markings, including 'daggers'. **Hab** Woodland, hedgerows, gardens. **Fp** Hawthorn, Blackthorn, apples. **SS** **Dark Dagger** *Acronicta tridens* [N/I] (**WS** 33–44 mm); the two species cannot be identified in the field (genitalia dissection required).

Knot Grass *Acronicta rumicis*

WS 34–44 mm. **ID** Forewings grey, rather mottled with black markings and dusting, and distinct curved white spots. **Hab** Grassland, marshes, heathland, parks, gardens, woodland. **Fp** Docks, Bramble, plantains. **SS** Superficially resembles several dagger species but forewings lack 'daggers'.

Subacronicta 1 sp.

'oplar Grey
ubacronicta megacephala

VS 40–44 mm. **ID** Forewings grey vith darker markings; rather mottled, vith pale area beyond kidney mark. **ab** Woodland, parks, gardens. **p** Poplars, Aspen. **SS** Sycamore, hich lacks pale area beyond kidney ark.

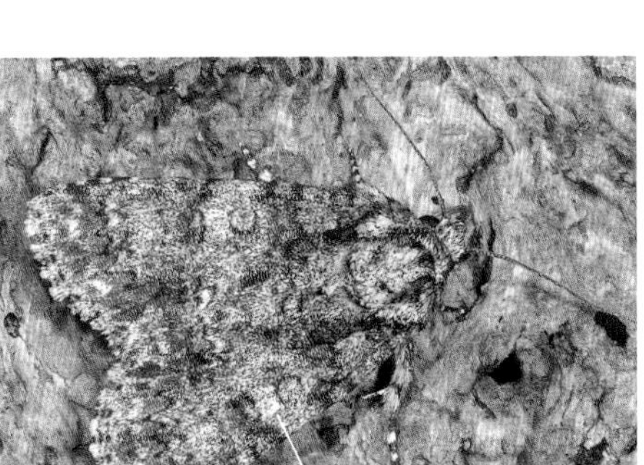

Abrostola 2 spp. | 1 ILL.

Spectacle
Abrostola tripartita

WS 32–38 mm. **ID** Forewings grey with large central brown band with blotches and markings; from the front, looks like it is wearing spectacles! **Hab** Grassland, wetlands, hedgerows, woodland edges. **Fp** Common Nettle. **SS** Dark Spectacle *Abrostola triplasia* [N/I] (**WS** 34–40 mm), common, is much darker and greyer.

Omphaloscelis 1 sp.

unar Underwing
mphaloscelis lunosa

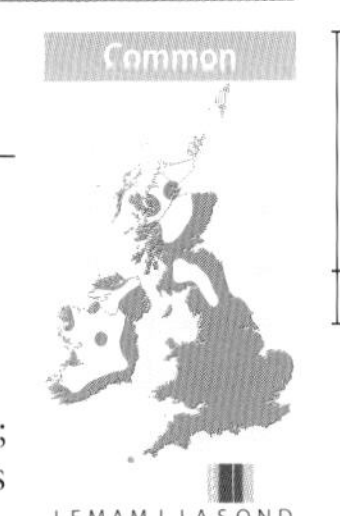

S 32–38 mm. **ID** Forewings grey or own to black, but with pale veins d black mark towards tip; hindwing le, with 'crescent moon' mark. **ab** Grassland, parks and gardens. **)** Grasses. **SS** Other brown Noctuids; en if underwings not seen, markings forewing are distinctive.

Aporophyla 4 spp. | 1 ILL.

Black Rustic
Aporophyla nigra

WS 32–39 mm. **ID** Forewings subtle shades of black with central white mark. **Hab** Grassland, heathland, moorlands, woodland rides, gardens. **Fp** Blackthorn, clovers, docks. **SS** None.

Griposia 1 sp.

Merveille du Jour
Griposia aprilina

WS 42–52 mm. **ID** Forewings pale green-and-black, edged with white; hindwings black, fringes white and blackish. **Hab** Woodland, hedgerows and parks. **Fp** Oaks. **SS** **Scarce Merveille du Jour**, which flies much earlier in the year.

Moma 1 sp.

Scarce Merveille du Jour
Moma alpium

WS 32–40 mm. **ID** Forewings distinctively patterned with pale green-and-black markings, the largest black markings being near the base and outer edge. **Hab** Woodland. **Fp** Oaks. **SS** **Merveille du Jour**, which flies much later in the year.

Craniophora 1 sp.

Coronet
Craniophora ligustri

WS 35–43 mm. **ID** Forewings very dark, with subtle greenish hues and bold white markings, particularly near wingtips. **Hab** Woodland, hedgerows. **Fp** Ash, Wild Privet. **SS** Superficially resembles dark forms of **Poplar Grey** (*p. 461*) and **Knot Grass** (*p. 460*), which have a more textured appearance.

Calamia 1 sp.

Burren Green NT
Calamia tridens

WS 37–42 mm. **ID** Forewings green with two whitish marks. **Hab** Limestone grassland. **Fp** Blue Moor-grass. **Dist** Restricted to the Burren area, Ireland, but one probable immigrant in Essex in 2014. **SS** None.

FAMILY Nolidae (Black arches, silver-lines & allies) 6 GEN. | 12 spp. | 3 ILL.

Nola 4 spp. | 1 ILL.

.east Black Arches
Vola confusalis

VS 16–24 mm. **ID** Small; narrow ɔrewings whitish with black cross-nes in distinctive pattern; hindwings ɾeyish. **Hab** Woodland, parks, ardens. **Fp** Downy Birch, Blackthorn, ɩuckthorn, limes. **SS** Various small lack arches species, which have ifferent patterning, and can in no way ɛ confused with the unrelated, much ɩrger **Black Arches** (*p. 451*).

Pseudoips 1 sp.

ɪreen Silver-lines
seudoips prasinana

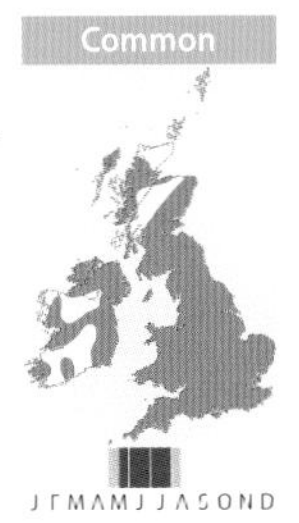

VS 32–40 mm. **ID** Forewings green ɪith three whitish cross-lines; ɪndwings white or yellowish-white. ntennae, palps, front legs pink. **ab** Woodland. **Fp** Oaks, birches. **S** Scarce Silver-lines.

Bena 1 sp.

carce Silver-lines
ɜna bicolorana

S 40–48 mm. **ID** Forewings broad, ɪeen with two yellowish-white cross-ɪes; hindwings white. Palps and front ɡs pink. **Hab** Woodland, parks. **ɔ** Oaks, birches. **SS** Green Silver-lines.

ORDER HYMENOPTERA Ants, bees, wasps & relative

[Greek: *hymen* = membrane; *pteron* = wing]

BI | 64 families [approx. 1,292 genera, **7,760 species**] **W** | 94 families [approx.**153,000 spec**

The largest order in Great Britain and Ireland, many of which are little-known parasitic wasps (some tiny; the largest, a giant Darwin wasp, is 50 mm long, plus an ovipositor of similar length). Many species have two pairs of wings, with hindwings smaller. Some females have an ovipositor that, depending on the group, is used to drill, saw or sting. This order features a range of complex relationships, with many being parasitoids and some having complex social structures. Only the 'aculeates' ('stinging Hymenoptera', *i.e.* bees, ants and wasps) are covered in recent popular guides, and wasps are yet to be fully addressed. Body length is extremely variable in many species and as such is only occasionally useful for identification. Here, 'small' is regarded as a body length of ≤5 mm 'medium-sized' 6–10 mm, and 'large' >10 mm. Note that the terms thorax and abdomen are not wid used in the Hymenoptera literature, as the first abdomen segment (S1) is associated with the thora Effectively, the thorax is divided into several sections and a petiole (narrow waist) links the propode to the metasoma (also known as gaster, mainly in older literature). The metasoma, then, can be regarded as the 'abdomen below the waist.' However, for simplicity, the terms thorax and abdomen are used in the text in this book; the term body refers to head, thorax and abdomen.

FINDING ANTS, BEES, WASPS & RELATIVES Check sandy areas, such as coastal sand dunes and sandy heathland with plenty of bare patches for nesting and hunting insects; these insects are sometimes found on flowers or resting on vegetation. Good sites include Thursley Commor in Surrey, where 227 aculeate species have been recorded since 1984; heathlands are more productive in summer; in Ireland, Ballyteige Burrow near Kilmore Quay, County Wexford has a wide range of species. Other habitats, including woodland and wetlands, can have a good faur in spring to autumn.

Structure of an ant

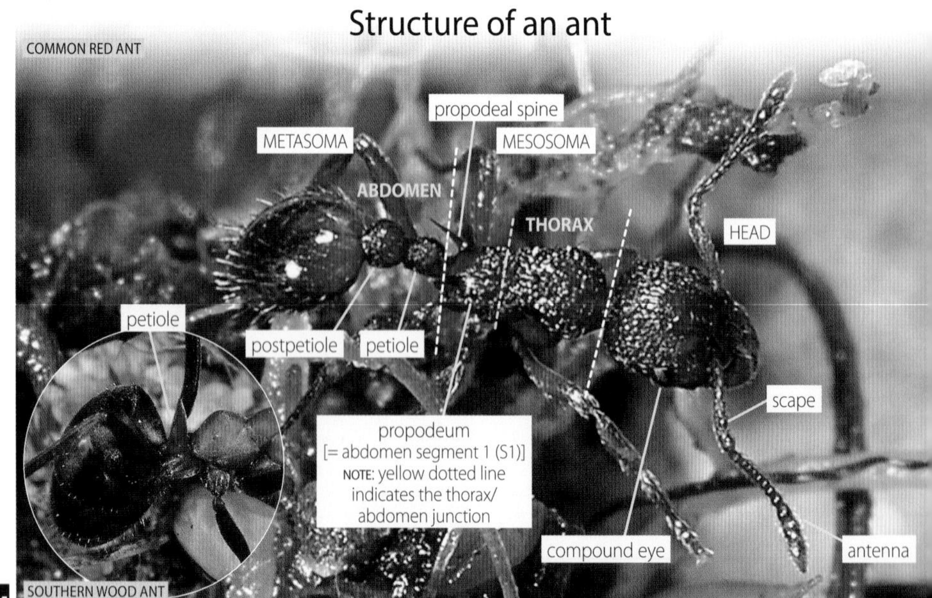

Structure of a bee

Structure of a wasp

ORDER **Hymenoptera** GUIDE TO SUBORDERS

Abdomen–thorax join with **no clear narrowing** (no 'wasp-waist')

FORM small to large (WS 2–28 mm). ANTENNAE threadlike or clubbed.

FORM medium-sized to very large (WS 6–40 mm), ANTENNAE threadlike. ♀ OVIPOSITOR sting-like.

SAWFLIES
Orange-horned Scabious Sawfly (BL 10–12 mm)

WOODWASPS
Giant Woodwasp (BL 24–40 mm)

SAWFLIES & WOODWASPS (SUBORDER Symphyta)

p. 46

'wasp-waist'

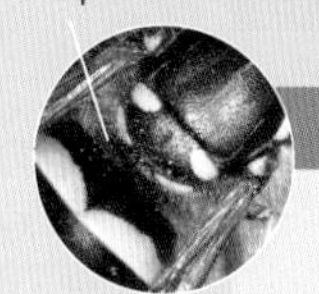

Abdomen*–thorax join **with clear narrowing** ('wasp-waist')

* The abdomen behind the constricted waist is known as the 'metasoma'. In Hymenoptera, segment S1 (propodeum) lies in front and is the first part of the abdomen, although fused to the thorax (mesosoma) (see *p. 464–465*).

♀ OVIPOSITOR **conspicuous** – adapted for the parasitic life of many species (although not all are parasitoids). ANTENNAE long in some families, **at least 16-segmented**.

Gelis sp. (BL 5 mm)

Hedgerow Darwin Wasp (BL 15 mm)

Red-eyed Ormyrid Wasp (BL 2–5 mm

PARASITIC WASPS (SUBORDER Apocrita | Parasitica)

p. 47

♀ OVIPOSITOR retractable and hidden – modified as a sting (some species parasitic on other insects). ANTENNAE 12-segmented (♀) or 13-segmented (♂).

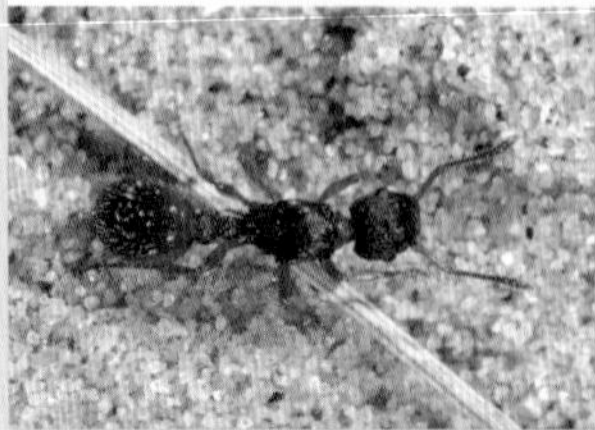

Myrmica sp. (ant) (BL approx. 5 mm)

Chocolate Mining Bee (BL 10–13 mm)

Common Wasp (BL 12–18 mm)

ANTS, BEES & WASPS (SUBORDER Apocrita | Aculeates)

p. 47

SUBORDER Symphyta (Sawflies & woodwasps)

12 FAMILIES | APPROX. 540 spp. BI

This group, which includes some colourful species, is little known but is attracting more attention from naturalists. Britain and Ireland's species are small to large-sized and sometimes mistaken for wasps; females have 'saws' used to cut open plant tissues when laying eggs. Most sawfly larvae feed on vegetation and have six or more pairs of fleshy prolegs in addition to three pairs at the front (Lepidoptera have five or fewer pairs of prolegs). Several mainly common representatives from 5 families are illustrated; identification to species level can often be difficult, requiring examination of wing vein differences, although body and wing colour is sometimes helpful, as is knowing the foodplant association. Maps are not provided as most species are so poorly recorded.

Examples of sawfly larvae

Figwort Sawfly larva

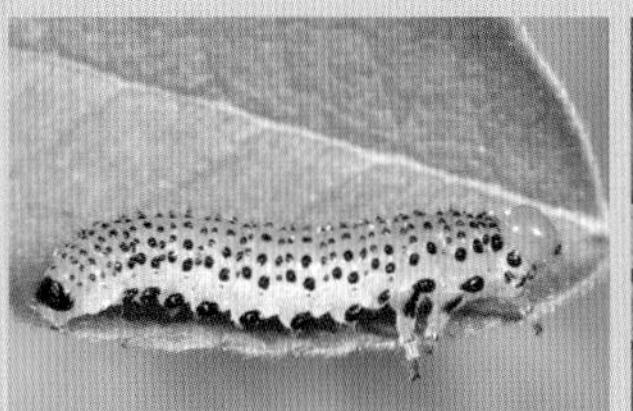
Large Rose Sawfly larva

Orange-horned Scabious Sawfly larva

FAMILY Argidae (Fuse-horned sawflies)

3 GEN. | 20 spp.

Medium-sized to large (**BL** 5–12 mm); **ANTENNA** short, 3-segmented.

Bramble Sawfly *Arge cyanocrocea*
(BL 7–8 mm)

Oak Arge *Arge rustica*
(BL 9 mm)

Rose Sawfly *Arge ochropus*
(BL 7–10 mm)

Large Rose Sawfly *Arge pagana*
(BL 7–9 mm)

FAMILY **Tenthredinidae** (Common sawflies) APPROX. **76 GEN. | 446 spp.**

Small to large (BL usually 2–15 mm); includes 30 *Tenthredo* spp., which are often colourful (BL 8–15 mm); antennae medium length and 9-segmented is normal, but a few species have 15 or 7 segments.

Wetland Sawfly
Dolerus madidus (BL 8–10 mm)

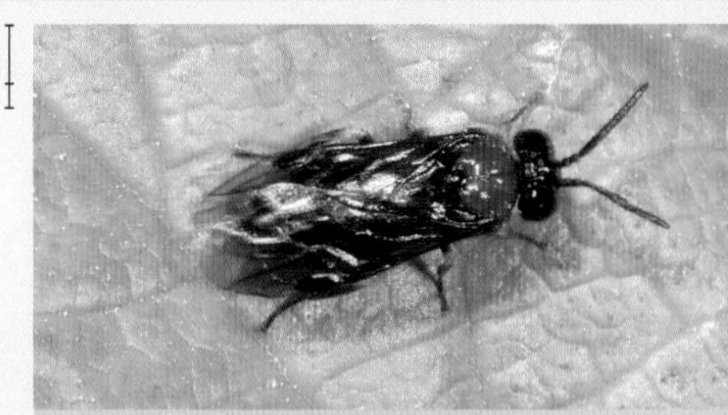

Alder Sawfly
Eriocampa ovata (BL 5–7 mm)

♀

Common Rhogogaster
Rhogogaster viridis (BL 8–12 mm)

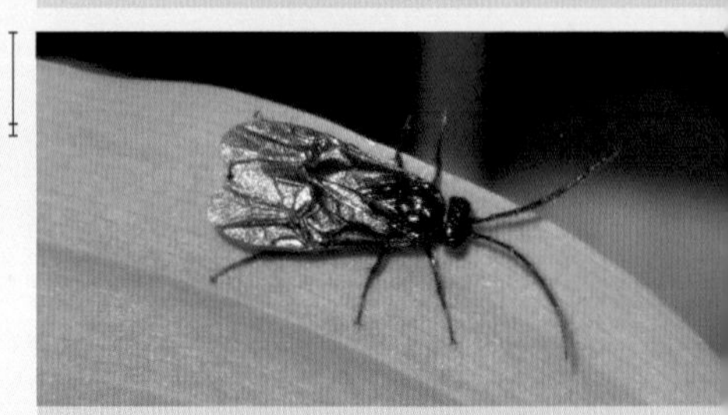

Solomon's-seal Sawfly
Phymatocera aterrima (BL 5–7 mm)

Charming Sawfly
Tenthredo amoena (BL 9–11 mm)

♀

Slate Sawfly
Tenthredo livida (BL 10–13 mm)

Figwort Sawfly
Tenthredo scrophulariae (BL 11–15 mm)

Variable Sawfly
Tenthredopsis litterata (BL 11–12 mm)

Turnip Sawfly
Athalia rosae (BL 6–8 mm)

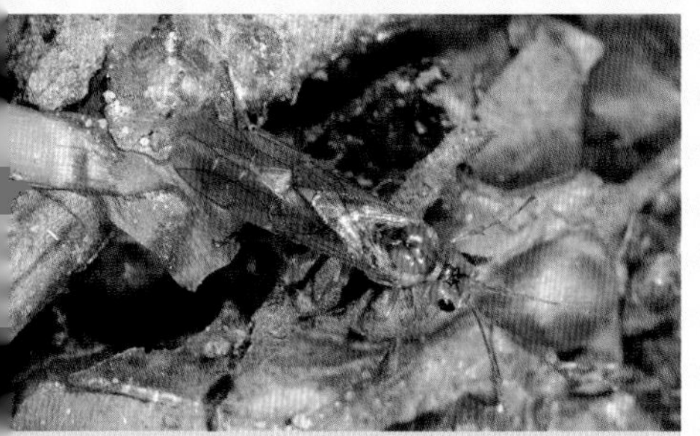

Orange Sawfly
Dineura virididorsata (BL 5–9 mm)

Black-backed Sawfly
Tenthredo mesomela (BL 10–13 mm)

Spotted Sawfly
Tenthredo maculata (BL 13–14 mm)

Tormentil Sawfly
Allantus truncatus (BL 5–8 mm)

Yellow-banded Sawfly
Tenthredo temula (BL 10–12 mm)

Bracken Sawfly
Strongylogaster multifasciata (BL 9–11 mm)

FAMILY Cimbicidae (Club-horned sawflies) 5 GEN. | 17 spp.

Medium-sized to very large (BL 9–28 mm); antennae with 6–7 segments, tip strongly clubbed; often broad-bodied species.

Scarce Honeysuckle Sawfly
Abia aenea
(**BL** approx. 10 mm)

Dusky-horned Scabious Sawfly
Abia candens
(**BL** 10–11 mm)

Birch Sawfly
Cimbex femoratus
(**BL** 20–28 mm)

Honeysuckle Sawfly
Zaraea fasciata
(**BL** approx. 10 mm)

Orange-horned Scabious Sawfly
Abia sericea
(**BL** 10–12 mm)

FAMILY Xiphydriidae 1 GEN. | 3 spp. (Necked woodwasps)

Large (**BL** 6–21 mm); **ANTENNA** 13–19 segments; **NECK** elongated.

Alder Woodwasp
Xiphydria camelus (**BL** 10–21 mm)

FAMILY Siricidae 4 GEN. | 11 spp. (Woodwasps)

Large (**BL** 10–40 mm); **ANTENNA** 14–30 segments; **HEAD** strongly swollen behind eyes.

Giant Woodwasp
Urocerus gigas (**BL** 24–40 mm)

SUBORDER Apocrita | Parasitica (Parasitic wasps)

32 FAMILIES | APPROX. 6,600 spp. BI

This group of 32 families is represented by an astonishing number (over 6,600) of species in Great Britain and Ireland. Although the parasitic wasps are little known, lacking popular guides, they are the subject of some recent publications. Most species kill the host, which includes Lepidoptera and even tiny aphids. A small selection of mainly common representatives from 8 families is illustrated, but identification to species level can sometimes be difficult, as many are small, often requiring microscopic study – and hence beyond the scope of this book. Even some larger Darwin wasps can be a challenge.

Male parasitic wasps are similar to some aculeate wasps and bees but can be differentiated by their antennae – **Parasitic wasps have ≥16 segments**; aculeate wasps and bees have 12 segments (♂ has 13).

Marble Gall (**PARASITICA**)

Short-fringed Mining Bee (**ACULEATES**)

Structure of a Darwin wasp

FAMILY **Ichneumonidae** (Darwin wasps) **490 GEN. | 2,586 spp.**

Small to very large (BL 5–50 mm); superficially like other wasps, with a narrow waist. Most easily told by long antennae (16+ segments, more than aculeates) and often long female ovipositor.

Hedgerow Darwin Wasp
Heteropelma amictum (BL 15 mm)

Yellow-and-black Darwin Wasp
Dusona falcator (BL 10 mm)

Short-winged Darwin Wasp
Agrothereutes abbreviatus (BL 4 mm)

Banded-wing Darwin Wasp
Gelis areator (BL approx. 5 mm

Yellow-tipped Darwin Wasp
Ichneumon stramentor (BL 13–18 mm)

Black Darwin Wasp
Ichneumon deliratorius (BL 10–16 mm)

Yellow-striped Darwin Wasp
Ichneumon xanthorius (BL 15 mm)

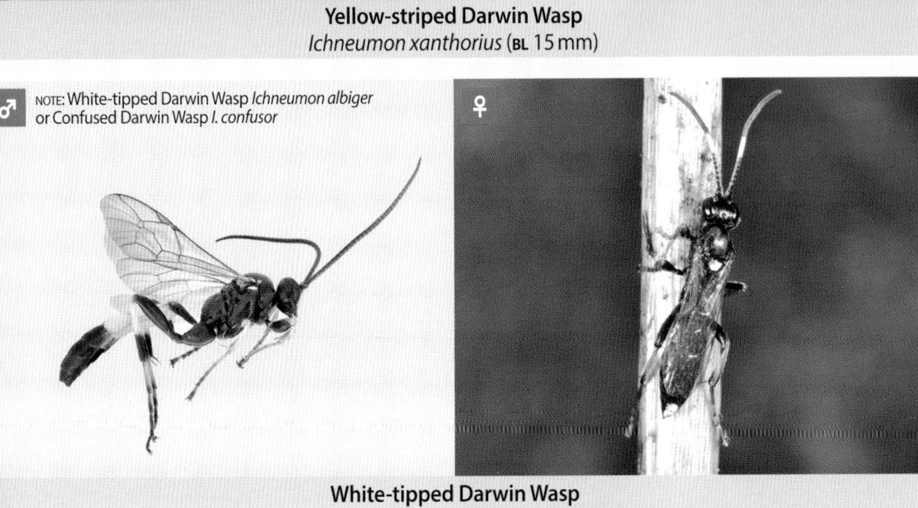

♂
NOTE: White-tipped Darwin Wasp *Ichneumon albiger* or Confused Darwin Wasp *I. confusor*
♀

White-tipped Darwin Wasp
Ichneumon albiger (BL 12–16 mm)

White-striped Darwin Wasp
Ichneumon sarcitorius (BL 10–12 mm)

Pupa-killer Darwin Wasp
Pimpla turionellae (BL 9–12 mm)

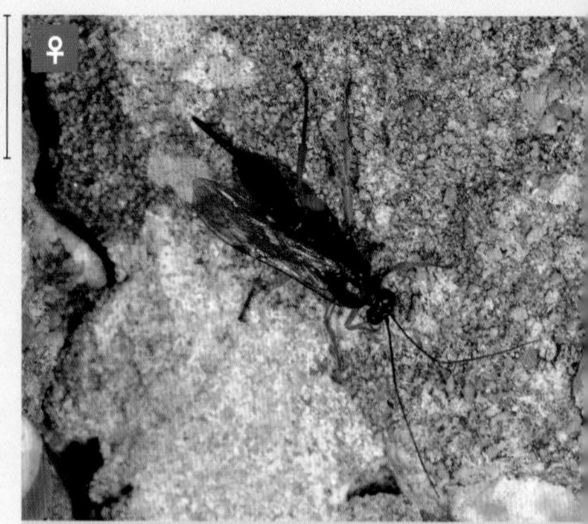

Orange-legged Darwin Wasp
Pimpla rufipes (BL 15 mm)

Lobster Darwin Wasp
Stauropoctonus bombycivorus (BL 25–30 mm)

Cream-striped Darwin Wasp
Ophion obscuratus (BL 15–22 mm)

Small Elephant Darwin Wasp
Protichneumon pisorius (BL 22–28 mm)

Woodland Darwin Wasp
Echthrus reluctator (BL 19 mm)

Orange Darwin Wasp
Netelia melanura (BL 20 mm)

Oak Eggar Darwin Wasp
Metopius dentatus (BL 22 mm)

Black-tipped Darwin Wasp
Enicospilus ramidulus (BL 13 mm)

Orange-brown Darwin Wasp
Ophion scutellaris (BL 14–20 mm)

Holly Blue Darwin Wasp
Listrodromus nycthemerus (BL 5–8 mm)

Striped Darwin Wasp
Perithous scurra (BL 7–14 mm)

Umbellifer Darwin Wasp
Amblyteles armatorius (BL 12–16 mm)

Sabre Wasp *Rhyssa persuasoria*
(BL 18–50 mm (+ 50 mm ovipositor)

Bee Darwin Wasp *Ephialtes manifestator*
(BL 20 mm (+ 50 mm ovipositor)

FAMILY Chalcididae (Chalcidid wasps)

Small (BL up to 5 mm), robust; hind femora swollen, with 1+ teeth on inner margin. Hind tibiae curved.

Black-and-yellow Chalcid Wasp
Brachymeria tibialis (BL 3–5 mm)

FAMILY Ormyridae (Ormyrid wasps)

Small (BL up to 5 mm), robust; body metallic; abdomen segments with coarse sculpturing.

Red-eyed Ormyrid Wasp
Ormyrus nitidulus (BL 2–5 mm)

FAMILY **Cynipidae** (Gall wasps) **24 GEN. | 91 spp.**

Small (BL less than 5 mm), often robust gall formers or inquilines of gall formers; the galls can be very large, such as Robin's Pincushion, with a width up to 70 mm.

Further details and additional species can be found in the companion WILD*Guides* volume *Britain's Plant Galls* (see *page 569* for details).

Marble Gall Wasp
Andricus kollari
(**BL** approx. 2 mm; **GALL** up to 25 mm diameter)

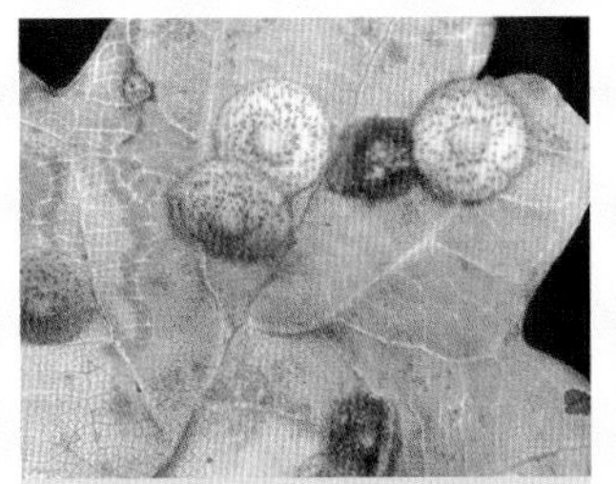

Common Spangle Gall
Neuroterus quercusbaccarum
(**GALL** up to 5 mm wide)

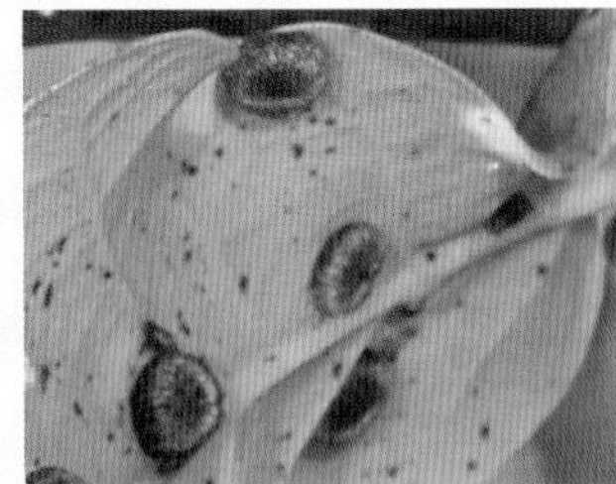

Silk Button Spangle Gall
Neuroterus numismalis
(**GALL** up to 5 mm wide)

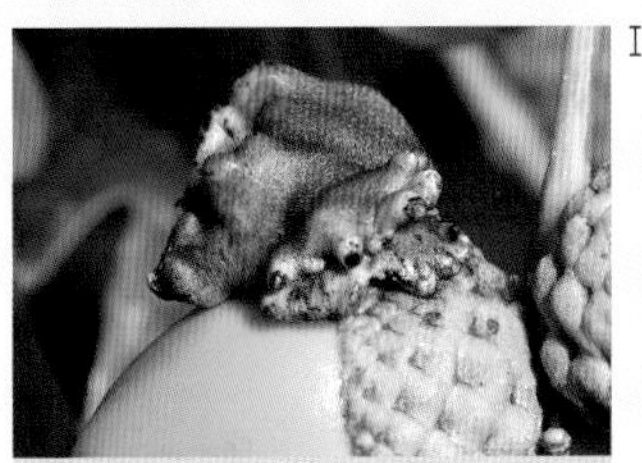

Knopper Gall
Andricus quercuscalicis
(**GALL** up to 28 mm long)

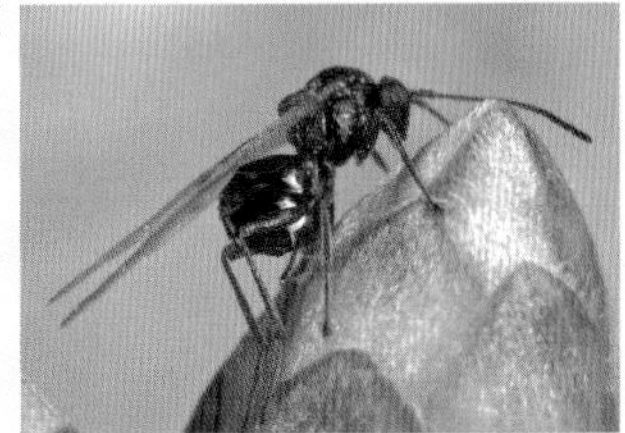

A gall wasp *Andricus* sp.
(**BL** approx. 2 mm)

A gall wasp *Synergus* sp.
(**BL** approx. 2 mm; **GALL** up to 25 mm diameter)

Robin's Pincushion
Diplolepis rosae
(**GALL** up to 70 mm wide)

FAMILY Platygastridae

(Platygastrid wasps) **20 GEN. | 362 spp.**

Tiny (mainly BL approx. 1 mm), with simple waist; many species are associated with plant galls.

♀

On Gorse Shieldbug (*p. 162*) eggs

A scelionid (subgroup of the Platygastridae) *species not determined* (BL approx. 1 mm)

FAMILY Torymidae

(Torymine wasps) **9 GEN. | 111 spp**

Small (BL up to 5 mm), metallic, hindlegs thickened. Include parasitoids of gall-forming wasps; females have a long ovipositor.

♀

Metallic Green Wasp *Torymus auratus* (BL 2–5 mm)

FAMILY Gasteruptiidae

(Gasteruptiid wasps) **1 GEN. | 5 spp.**

Medium-sized (BL ≤ 18 mm); abdomen slender; antennae fairly short; neck thick; hind tibiae broad.

♀

Javelin Wasp *Gasteruption jaculator* (BL 10–18 mm)

♀

Small Javelin Wasp *Gasteruption assectator* (BL 8–12 mm)

FAMILY Braconidae (Braconids)

approx. **190 GEN. | 1,355 spp**

Tiny (BL 2–6 mm), otherwise superficially like Darwin wasps; fewer wing veins than related groups.

♀

Pale Braconid *Aphaereta pallipes* (BL approx. 2 mm)

♀

On Fox Moth (*p. 430*) larva

A braconid *Cotesia* sp. (BL approx. 2 mm)

SUBORDER Apocrita | Aculeates (Ants, bees & wasps)

Basic division into 3 superfamilies (at this stage ignoring some obvious differences in antennae and lack of wings in some species):

SHORT-FRINGED MINING BEE (ACULEATES) MARBLE GALL (PARASITICA)

Male parasitic wasps are similar to some aculeate wasps and bees, but can be differentiated by their antennae – **Aculeate wasps and bees have 12 segments (♂ has 13); Parasitic wasps have ≥16 segments**.

ABDOMEN with 6 (♀) or 7 (♂) visible segments; BODY plain or bright in one or two colours, but not metallic; ♀ OVIPOSITOR a modified sting in some.

ABDOMEN often with 3 visible segments (or no more than 5 in a few species); BODY usually metallic green or reddish; ♀ OVIPOSITOR short (hidden in some).

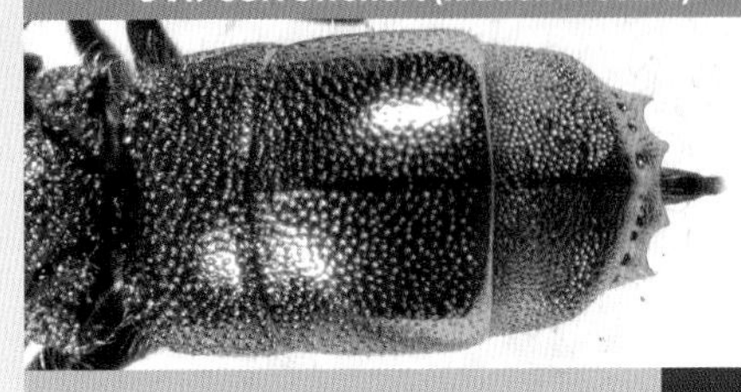

CHRYSIDOIDEA *p. 481*

PRONOTUM extended along side of thorax to tegulae; WINGS 'pleated' (*i.e.* folded) when closed, held along side of body

VESPOIDEA
POMPILOIDEA
FORMICOIDEA
TIPHIOIDEA *p. 489*

PRONOTUM short (collar-like); hind part not reaching to tegulae

APOIDEA *p. 521*

NOTE: the collar is not always easy to see in the field on hairy bees or quick-moving wasps. However, in these cases the colour and general appearance of the families helps. Female bees are often seen carrying pollen.

SUPERFAMILY **CHRYSIDOIDEA**

NOTE: the following omits Embolemidae, reddish-brown parasitoid wasps (1 little-known British species).

FORM very small (**BL** 2–5 mm). **BODY** **black; not metallic or strongly punctured**. **WINGS** winged but ♀s can be wingless or have reduced wings. **LEGS** fore tarsi (♀) **lack 'pincers'**.

FAMILY **Bethylidae** (Bethylids)

8 GEN. | 13 spp. | 1 ILL.
(in addition, there are 6 known or suspected introductions and 1 doubtful species)

Females often overwinter as adults, whilst males die off by about October. The larvae are external parasitoids on Coleoptera and Lepidoptera larvae. One example species is shown; a microscope is needed for identification to species level (minor differences in ocelli and propodeum), but genera can be distinguished by wing structure.

Bethylus — 4 spp. | 1 ILL.

Large-headed Bethylid
Bethylus cephalotes

Common

BL 4 mm. **ID** Black, forewings not reaching end of abdomen, although rarely ♀ can occur with shorter wings. **Hab** Various open areas including sand dunes; parasitoids of moth larvae. **Dist** True distribution probably far more extensive than mapped. **SS** Other *Bethylus* species, from which told by position of ocelli and marks on propodeum.

FORM very small to medium-sized (**BL** 1–7 mm). **BODY** **black, brown or whitish; not metallic or strongly punctured**. **WINGS** winged but ♀s can be wingless or have reduced wings. **LEGS** fore tarsi (♀) **with 'pincers'**.

FAMILY **Dryinidae** (Dryinids)

8 GEN. | 36 spp. | 1 ILL.

Parasitoids on bugs; males are often rare, possibly absent in some species. Probably under-recorded as these wasps are rarely observed, hence only one easily recognizable species is illustrated.

Dryinus — 2 spp. | 1 ILL.

Collared Dryinid
Dryinus collaris

Rare

J F M A M J J A S O

BL 7 mm. **ID** Black, partly orange thorax (collar), antennae and legs. Forewings with two black bands. **Hab** Hedgerows and gardens. **Hosts** Bug nymphs, including Common Issid Bug (*p. 221*). **Dist** Rare, SE England; as found few times in England (including in very recent years), this distinctive dryinid does not yet have a conservation status and is probably under-recorded. **SS** The other UK *Dryinus* species, **Black Dryinid** *Dryinus niger* [N/I], which has a black body.

GUIDE TO FAMILIES | 4 BI

FORM small to medium-sized (**BL** 4–11 mm).
BODY colourful, **often green, blue and red, usually metallic; strongly punctured.**

Linnaeus's Cuckoo Wasp *Chrysis ignita*

FAMILY Chrysididae (Cuckoo wasps) — 11 GEN. | 35 spp. | 23 ILL.

The most colourful British and Irish wasps, these mainly small to medium-sized parasitoids (body length 4–11 mm) have an armoured, often metallic-looking cuticle. Sometimes known as jewel or ruby-tailed wasps, some species can be a challenge to identify, even under a microscope. Fortunately, others are very distinctive and can be readily identified to species, making them an attractive group to study. Females have a small ovipositor. Most species have one generation a year, developing within the nests of the host. The larvae are believed to eat the eggs or larvae of the host before eating the host's food.

IOTE: the following omits the genus *Philoctetes*; one small British species (formerly in *Omalus*), only recorded ɔnce since 1910 (in 1983).

			Genus	Page
BDOMEN **with 4 (♀) or 5 (♂) segments; black-tipped.** RONOTUM **with distinctive bulbous collar**			***Cleptes*** [2 species]	*p. 488*
BDOMEN **with 3 segments; not black-tipped.** PRONOTUM **without collar**				
ORM **rather elongated**				
BDOMEN ast egment sually **vith** eeth	ABDOMEN last segment with 3 teeth; BODY all greenish-blue		***Trichrysis*** [Blue Cuckoo Wasp]	*p. 482*
	ABDOMEN last segment usually with 4 teeth (rarely lacking); BODY greenish-blue; THORAX different colour to abdomen		***Chrysis*** [15 species]	*p. 486*
BDOMEN ast egment **vithout** eeth	ABDOMEN end subtruncate; THORAX blue or bluish-green, in some individuals with yellow patches; ANTENNAE positioned beneath lower part of eyes when viewed head-on		***Chrysura*** [2 species]	*p. 482*
	ABDOMEN end rounded; THORAX bluish-green, with minimal yellow on the collar or side of thorax; ANTENNAE positioned in line with lower part of eyes when viewed head-on		***Pseudospinolia*** [Neglected Cuckoo Wasp]	*p. 482*
ORM **rather stout (particularly abdomen);** ABDOMEN **without teeth**				
HORAX hind part with **large central flange**			***Elampus*** [2 species]	*p. 482*
HORAX ind part **vithout** entral ange	THORAX **not humped**; multi-coloured, never plain		***Hedychridium*** [4 species]	*p. 484*
			Hedychrum [3 species]	*p. 485*
	THORAX **distinctly humped**; often plain	BODY bluish-green	***Omalus*** [2 species]	*p. 483*
		Combination of green or blue thorax and red or yellowish-red abdomen in one species, to all-violet body in the other species	***Pseudomalus*** [2 species]	*p. 483*

Trichrysis 1 sp.

Blue Cuckoo Wasp
Trichrysis cyanea

BL 4–8 mm. **ID** Wholly metallic blue with green patches. **Hab** Woodland edges and open areas, including gardens; often seen on fence posts, logs and stumps. **Hosts** Crabronid wasps in the genera *Pemphredon* (*p.526*) and *Trypoxylon* (*p.528*), also pompilid wasps (*p.503*). **SS** None.

Pseudospinolia 1 sp

Neglected Cuckoo Wasp
Pseudospinolia neglecta

BL 5–9 mm. **ID** Head and thorax blue, abdomen red (tip rounded). **Hab** Vertical clay or sandy banks frequented by the hosts.
Hosts Mason wasps in the genus *Odynerus*, including Spiny Mason Wasp (*p.519*). **SS** *Chrysis* species (*p.486*), **Radiant Cuckoo Wasp**.

Elampus 2 spp. | 1 ILL.

Panzer's Cuckoo Wasp
Elampus panzeri

BL 4–8 mm. **ID** Head and thorax yellowish-green with blue areas; abdomen yellowish-red. Hind part of thorax with large central flange (not easily seen in the field). **Hab** Open sandy areas in heathland for a few days only (usually in July), flying around grass or heather; also likely on the ground around nests of the hosts.
Hosts Crabronid wasps in the genus *Mimesa* (*p.529*).
SS Rare **Elampus Cuckoo Wasp** *Elampus foveatus* [N/I] (**BL** 5–8 mm), which has a central inverted 'U'-shape in centre of abdomen tip.

THORAX large central flange

Chrysura 2 spp. | 1 ILL

Radiant Cuckoo Wasp
Chrysura radians

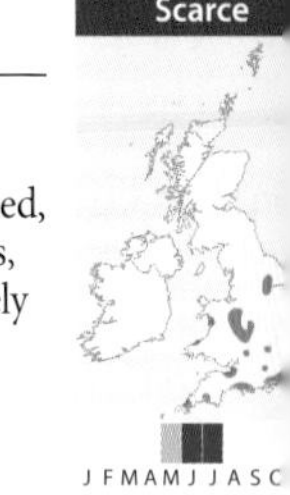

BL 7–11 mm. **ID** Head and thorax bluish-green, part-yellow in some individuals. Abdomen red, tip rounded, no teeth. **Hab** Various sunny habitats, *e.g.* woodland and gardens; most likely on wooden posts or dead wood.
Hosts Orange-vented Mason Bee (*p.540*). **SS** *Chrysis* species (*p.486*), which have teeth at tip of abdomen, **Neglected Cuckoo Wasp.** Also the rare **Northern Osmia Cuckoo Wasp** *Chrysura hirsuta* (RDB3) [N/I], which is restricted to the Scottish Highlands, where Radiant Cuckoo Wasp does not occur.

Omalus 2 spp.

THORAX central part smooth, shiny, without punctures	Shining Cuckoo Wasp
THORAX central part with punctures	Punctured Cuckoo Wasp

Shining Cuckoo Wasp
Local

Omalus aeneus

BL 3–7 mm. **ID** Thorax and abdomen
/ery shiny, bluish-green, some
ndividuals more violet or even black;
ıbdomen broad. **Hab** Open areas
vhere hosts occur; sometimes on
purge flowers. **Hosts** *Passaloecus*
p. 525) and *Pemphredon* (*p. 526*).
;S Punctured Cuckoo Wasp [see table
ıbove], Violet Cuckoo Wasp, from which told
ɔy shorter-haired thorax and smooth abdomen.

Punctured Cuckoo Wasp
Nationally Scarce

Omalus puncticollis

BL 3–6 mm. **ID** Thorax very shiny, bluish-green; head and thorax heavily punctured, including mesoscutum. **Hab** Open areas where the stem and wood nest hosts occur. **Hosts** Crabronid wasps in the genera *Passaloecus* (*p. 525*) and *Pemphredon* (*p. 526*). **SS** Shining Cuckoo Wasp [see table *above*], Violet Cuckoo Wasp, from which told by shorter-haired thorax.

Pseudomalus 2 spp.

THORAX green or blue; **METASTOMA** red or yellowish red	Golden Cuckoo Wasp
BODY violet with green; **METASTOMA** violet	Violet Cuckoo Wasp

;olden Cuckoo Wasp
Common

'seudomalus auratus

L 3–6 mm. **ID** Small, stocky; green
r blue thorax and shining red or
ellowish-red abdomen; abdomen S3
ith long hairs. **Hab** Scrubby areas
ıd gardens, where the stem and wood
ests of crabronid wasps (the hosts)
ccur. Nests in bramble and other
oody stems; often seen on fence posts
ıd on vegetation. **Hosts** Crabronid wasps in the genera
ıssaloecus (*p. 525*), *Pemphredon* (*p. 526*), *Rhopalum*
. 528) and *Trypoxylon* (*p. 528*). **SS** None.

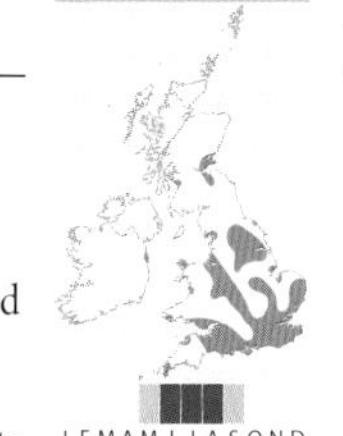

Violet Cuckoo Wasp
Nationally Scarce

Pseudomalus violaceus

BL 5–8 mm. **ID** Stocky species, dark violet to blue, tending to be greener in ♀. **Hab** Woodland edges, gardens and several other habitats, where the stem and wood nests of crabronid wasps (the hosts) occur. Likely to be active low down on leaves of trees. **Hosts** Crabronid wasps in the genera *Passaloecus* (*p. 525*) and *Pemphredon* (*p. 526*). **SS** *Omalus* species, from which told by longer-haired thorax.

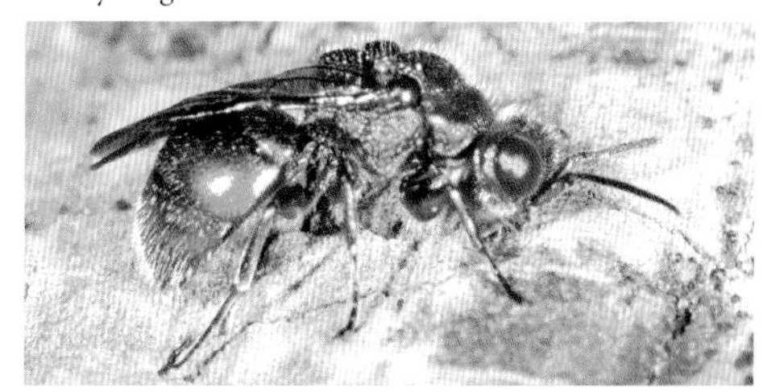

Hedychridium — 4 spp. | 3 ILL.

ABDOMEN dull orange, not metallic; **THORAX** bluish-green, with some yellow		**Dull Cuckoo Wasp**
ABDOMEN metallic	**HEAD** reddish and green; **THORAX** wholly covered with large punctures; front reddish and green, remainder rather darker blue; **ABDOMEN** red, partly yellow	**Glowing Cuckoo Wasp**
	HEAD red and green; **THORAX** upper half rather wrinkled, no large punctures; mainly red with green, hind part blue; **ABDOMEN** red	**Rugged Cuckoo Wasp**
	ABDOMEN green, partly purple	**Dryudella Cuckoo Wasp** [N/I]

The Nationally Scarce **Dryudella Cuckoo Wasp** *Hedychridium cupreum* [N/I] (**BL** 4·5–5·5 mm) is a difficult species to find but can be distinguished by being rather darker with a purple abdomen (in part) with green markings.

Dull Cuckoo Wasp
Hedychridium roseum

Local

BL 5–8 mm. **ID** Larger than closely related species; thorax bluish-green, with some yellow; abdomen dull orange or reddish, not metallic. **Hab** Sand dunes and inland heaths, often settling on the ground. **Hosts** Include Shieldbug Digger Wasp (*p. 529*), possibly also Common Tachysphex (*p. 529*). **SS** None. Superficially resembles many *Chrysis* species (*p. 486*) but abdomen is dull.

Glowing Cuckoo Wasp
Hedychridium ardens

Common

BL 4–6 mm. **ID** Small; head and front of thorax reddish and green (fairly dark in some individuals, much greener in others), rest of thorax blue; thorax with conspicuous large punctures; abdomen red, partly yellow. **Hab** Open sandy areas such as sand dunes and sandy heathland, often flying rapidly, but occasionally settling near nests of the host. **Host** Common Tachysphex (*p. 529*). **SS** Other *Hedychridium* species [see table *above*].

Rugged Cuckoo Wasp
Hedychridium coriaceum

Rare (RDB3)

BL 4–5 mm. **ID** Small; head and much of thorax mainly red and green; hind part of thorax blue, upper half rather rugged; abdomen red. **Hab** Open sandy areas near nests of the host. **Host** White-lipped Digger Wasp (*p. 531*). **SS** Other *Hedychridium* species [see table *above*]. Although very reddish, needs close examination of the thorax to distinguish it from related species, particularly in the field.

Hedychrum	**3 spp. | 2 ILL.**
BL small 4–7 mm; ♀ **HEAD** all one colour	**Niemalä's Cuckoo Wasp**
BL larger 6–10 mm; ♀ **HEAD** red spot between eyes	**Noble Cuckoo Wasp**
BL 4–10 mm; **THORAX** much less extensive red	**Bee-wolf Cuckoo Wasp** [N/I]

Bee-wolf Cuckoo Wasp *Hedychrum rutilans* was only known from historic specimens, until being rediscovered in London in 2017 and Dorset. The host is the Bee-wolf (*p. 537*)

Niemelä's Cuckoo Wasp
Hedychrum niemelai

Rare (RDB3)

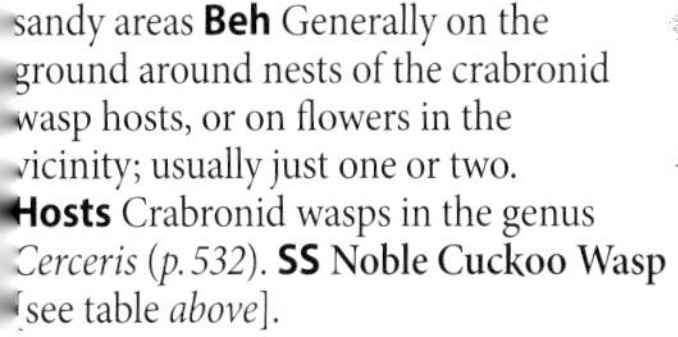

BL 4–7 mm. **ID** A small, robust-looking *Hedychrum*; head and thorax green with blue hue (broad red band in ♀); abdomen reddish. **Hab** Open, sandy areas **Beh** Generally on the ground around nests of the crabronid wasp hosts, or on flowers in the vicinity; usually just one or two. **Hosts** Crabronid wasps in the genus *Cerceris* (*p. 532*). **SS** Noble Cuckoo Wasp [see table *above*].

Noble Cuckoo Wasp
Hedychrum nobile

Local

J F M A M J J A S O N D

BL 6–10 mm. **ID** Robust-looking; head and thorax green with blue hue (broad red band in ♀, in which head usually has a reddish or yellowish streak between the eyes); abdomen reddish. **Hab** Open, sandy areas. **Beh** Around nests of the crabronid wasp hosts, or on flowers in the vicinity, particularly in sunny weather; often in numbers. **Dist** Only recognized as a British species from specimens found in Surrey in 1998. **Hosts** Crabronid wasps in the genus *Cerceris* (*p. 532*). **SS** Niemelä's Cuckoo Wasp [see table *above*].

<table>
<tr><th colspan="2">*Chrysis*</th><th colspan="2">15 spp. | 9 ILL.</th></tr>
<tr><td colspan="4">ABDOMEN **2- or-3-coloured**</td></tr>
<tr><td colspan="2">**ABDOMEN** S1 blue; S2 red (♂ with some blue); S3 red</td><td>**Aspen Cuckoo Wasp**</td><td>*p.486*</td></tr>
<tr><td colspan="2">**ABDOMEN** front S1 green or blue; rear S1 + S2 red; S3 green or blue</td><td>**Multi-coloured Cuckoo Wasp**</td><td>*p.486*</td></tr>
<tr><td colspan="4">ABDOMEN **all-red**</td></tr>
<tr><td colspan="2">**THORAX** partly red (mesoscutum); **HEAD & PRONOTUM** with red patches</td><td>**Illiger's Cuckoo Wasp**</td><td>*p.488*</td></tr>
<tr><td rowspan="6">**THORAX** blue with green; species similar in general appearance, key features vary; other species [N/I] differ marginally from Linnaeus's Cuckoo Wasp and the validity of some of these is doubted by some authors.</td><td>**BODY** particularly elongated; **ABDOMEN** S2 not bulbous</td><td>**Narrow-bodied Cuckoo Wasp**</td><td>*p.487*</td></tr>
<tr><td>**HEAD** front between eyes with four projecting tubercles; **HEAD & THORAX** rather blue</td><td>**Tuberculate Cuckoo Wasp**</td><td>*p.488*</td></tr>
<tr><td>**HEAD & THORAX** in part yellow; **ABDOMEN** S2 dorsally with smaller punctures than S1</td><td>**Rudd's Cuckoo Wasp**</td><td>*p.488*</td></tr>
<tr><td>**ABDOMEN** tip teeth shape useful in identification; 4 large, evenly spaced teeth</td><td>**Linnaeus's Cuckoo Wasp**</td><td>*p.487*</td></tr>
<tr><td>**ABDOMEN** tip teeth shape useful in identification; inner teeth usually wider apart but fine thorax structure and other characters should be checked (microscope needed and specialist keys)</td><td>**Impressive Cuckoo Wasp**</td><td>*p.487*</td></tr>
<tr><td>**BEHAVIOUR** usually seen around nests of the chimney-building Spiny Mason Wasp (*p.519*)</td><td>**Linsenmaier's Cuckoo Wasp**</td><td>*p.487*</td></tr>
</table>

Multi-coloured Cuckoo Wasp

Local

Chrysis viridula

BL 6–9 mm. **ID** Head, underside and sides of thorax, legs, abdomen S3 and base of S1 green or blue; rest of thorax and remainder of abdomen red. The meaning of the species' name in Latin 'greenish' does not do this stunning species justice. **Hab** Vertical clay or sandy banks frequented by the hosts. **Hosts** Mason wasps in the genus *Odynerus* (*p.519*). Sometimes found in the company of Neglected Cuckoo Wasp (*p.482*), as well as other *Chrysis* species. **SS** None.

Aspen Cuckoo Wasp

EN Rare (RDB1)

Chrysis fulgida

BL 7–12 mm. **ID** Head, thorax and abdomen S1 bluish; abdomen S2 and S3 red (♂ has some blue on S2). **Hab** Scrubby heathland edges. Sometimes on logs and tree trunks near the host's nests. **Host** Large Aspen Mason Wasp (*p.518*), which preys on Red Poplar Leaf Beetle (*p.317*) larvae on Aspen. **SS** None.

♀

Narrow-bodied Cuckoo Wasp

Common

Chrysis angustula

J F M A M J J A S O N D

BL 5–9 mm. **ID** Usually small and narrow-bodied. Head and thorax bluish, with green hue; abdomen red. **Hab** Heathland. Sometimes on wooden posts and dead trees, near nests of the hosts. **Hosts** Mason wasps, including the Three-banded Mason Wasp (*p. 517*) and most likely the Willow Mason Wasp (*p. 518*).
SS Linnaeus's Cuckoo Wasp and other related species. Although listed as a synonym of Linnaeus's Cuckoo Wasp by some authors, Narrow-bodied Cuckoo Wasp is more elongated.

Linnaeus's Cuckoo Wasp

Common

Chrysis ignita

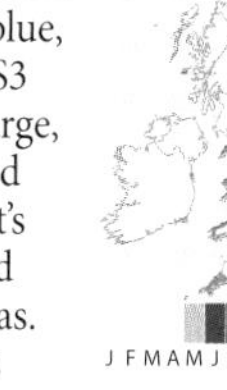

J F M A M J J A S O N D

BL 6–10 mm. **ID** Head and thorax blue, with green patches; abdomen red, S3 bulbous, tip with four distinctive large, evenly spaced teeth. **Hab** Woodland edges, walls and cliffs, near the host's nests. Likely in sunny spots on dead wood or on flowers such as angelicas.
Hosts Wood-nesting mason wasps in the genus *Ancistrocerus* (*p. 516*) and probably other similar wasps.
SS Several other *Chrysis* species, including **Narrow-bodied Cuckoo Wasp**.

ABDOMEN TIP with 4 large, evenly spaced teeth

insenmaier's Cuckoo Wasp

Common

hrysis mediata

J F M A M J J A S O N D

BL 5–11 mm. **ID** Head and thorax bluish-green; abdomen red. irst small antenna segment blue, in ome individuals also the second. **ab** Sandy and clay banks, also walls, ear the host's nests. **Host** Spiny Mason Vasp (*p. 519*). **SS** Several other *Chrysis* pecies of the *ignita* group (closely elated species), from which told by xamination of minor spine shape differences at tip of bdomen, hence many old records are not separated to pecies level.

Impressive Cuckoo Wasp

Common

Chrysis impressa

J F M A M J J A S O N D

BL 7–11 mm. **ID** Head and thorax dark blue or black; abdomen red, tip with ill-defined teeth. **Hab** Sandy cliffs, also walls, woodlands. **Host** Wood-nesting mason wasps in the genus *Ancistrocerus* (*p. 516*). **SS** Several other *Chrysis* species of the *ignita* group (closely related species), from which told by examination of tip of abdomen, fine thorax details and other minor characters (using microscope).

ABDOMEN TIP with ill-defined teeth

Rudd's Cuckoo Wasp
Chrysis ruddii

Local

BL 6–10 mm. **ID** Head and thorax bluish-green, in part yellow. Pronotum short; abdomen red, S2 dorsally with smaller punctures than on S1. **Hab** Limestone and soft rock cliffs and banks; often seen resting on cliff walls. **Dist** Stronghold in Devon and Cornwall; the few recent records are mainly coastal. **Host** Stocky Mason Wasp *Ancistrocerus oviventris* [N/I] (**BL** 6–14 mm), which has declined since 1970. **SS** Other *Chrysis* species, but these have a longer pronotum and lack yellowish markings.

J F M A M J J A S O N D

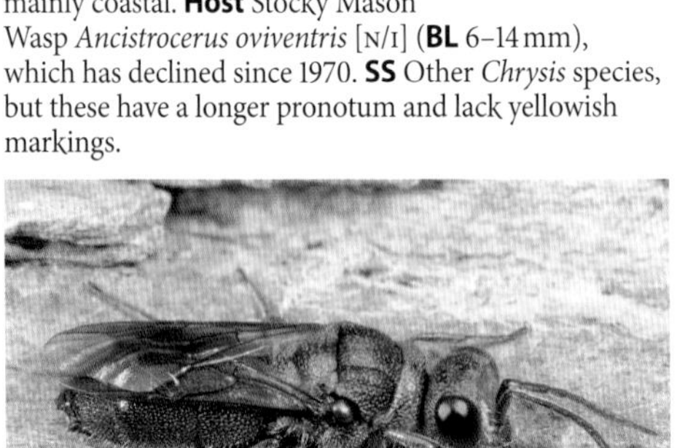

Tuberculate Cuckoo Wasp
Chrysis terminata

Local (recent colonist)

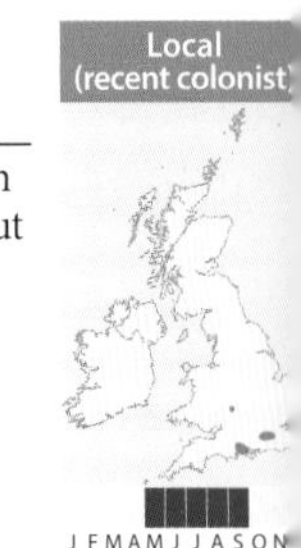

BL 5–9 mm. **ID** Front of head between eyes with four projecting tubercles, but this character needs to be looked for carefully. Head and thorax distinctly blue; abdomen red. **Hab** Woodland and other habitats, individuals often exploring dead wood. **Dist** Only added to the British list in 2016 and distribution unclear, but from scattered records is believed to be fairly widespread. **Hosts** Thought to be mason wasps, *Ancistrocerus* species (*p. 516*). **SS** Other *Chrysis* species, from which told by turbercles on head.

J F M A M J J A S O N

Illiger's Cuckoo Wasp
Chrysis illigeri

Nationally Scarce

BL 5–8 mm. **ID** Head and thorax blue with patches of green and also of red, a colour combination also present on the mesoscutum and abdomen. **Hab** Sandy areas such as dunes and heathland; sometimes found on flowers (Apiaceae). **Host** Common Tachysphex (*p. 529*). **SS** None.

J F M A M J J A S O N D

Cleptes 2 spp. | 1 ILL

Part-golden Cuckoo Wasp
Cleptes semiauratus

Nationally Scarce

BL 5–7 mm. **ID** Head and thorax bluish-black (♂), reddish or violet (♀); abdomen orange with black tip. **Hab** Gardens and sandy sites. **Host** Prepupal-stage tenthredinid sawflies (*p. 468*) in their cocoons. **SS** The Nationally Scarce **Black-headed Cuckoo Wasp** *Cleptes nitidulus* [N/I] (**BL** 5–7 mm), which has a black head and red on upper part of thorax.

J F M A M J J A S O

GUIDE TO SUPERFAMILIES AND FAMILIES | 6 BI

SUPERFAMILY **POMPILOIDEA** 3 BI

ANTENNAE **not elbowed**; BODY broad (winged or wingless); PETIOLE **densely hairy**

Large Velvet Ant

Velvet ants Mutillidae
[3 genera | 3 species]
p. 500

BODY typically narrow; ANTENNAE long, **not club-tipped**; FORM variable, often black or black-and-red

Jumping Spider Wasp

Spider-hunting Wasps
Pompilidae
[14 genera | 44 species]
p. 503

BODY narrow; ANTENNAE long, **club-tipped**
FORM abdomen black with yellow marks or front red-banded (one species black-and-yellow)

Club-horned Wasp

Club-horned wasps
Sapygidae
[2 genera | 2 species]
p. 501

SUPERFAMILY **FORMICOIDEA** 1 BI

True ants; ANTENNAE elbowed; PETIOLE not hairy

Lasius sp. (ant)

Ants Formicidae
[13 genera | 51 species]
p. 490

SUPERFAMILY **TIPHIOIDEA** 1 BI

ANTENNAE **not elbowed**; BODY narrow (winged or wingless); PETIOLE not hairy

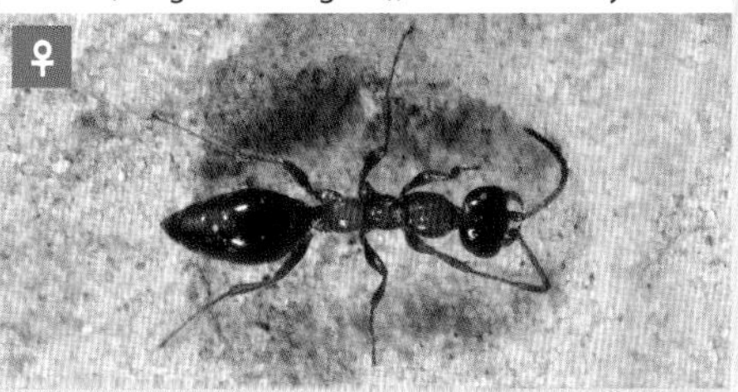

Tiger Beetle-hunting Wasp

Tiphiid wasps Tiphiidae
[2 genera | 3 species]
p. 502

SUPERFAMILY **VESPOIDEA** 1 BI

BODY **typical wasp shape** (fairly broad) and colour (black-and-yellow)

Early Mason Wasp

Social, potter and mason wasps
Vespidae
[12 genera | 33 species]
p. 508

FAMILY Formicidae (Ants) 13 GEN. | 51 spp. | 21 SPP. ILL.

(in addition there are 13 non-established introductions)

Many people are familiar with the conspicuously shaped black-and-red *Formica* species, or yellow to black *Lasius* species, particularly those in the garden; other genera tend to be little-known although may be common. Some ants eat other insects, seeds or feed on the honeydew secreted by aphids, in return for protection offered. Ants can bite or sting in defence; some also spray acid at intruders. Ants live in complex communities usually underground, the life-cycle often lasting more than a year; wood ants may build huge mounds. Some species of ants have symbiotic relationships with other insects, including lycaenid butterflies and various beetles. Workers (BL 2–6 mm) are wingless. Larger males and queens are usually winged; the females of some species are larger and plumper, shedding their wings after mating; measurements given range from the typically small workers to queens. Many species require examination under a microscope but those illustrated here can usually be identified in the field, with a good hand lens. The following is a guide to workers, the most likely form to be seen.

GUIDE TO SELECTED GENERA

'WAIST' **1-segmented**; PROPODEUM lacks spines

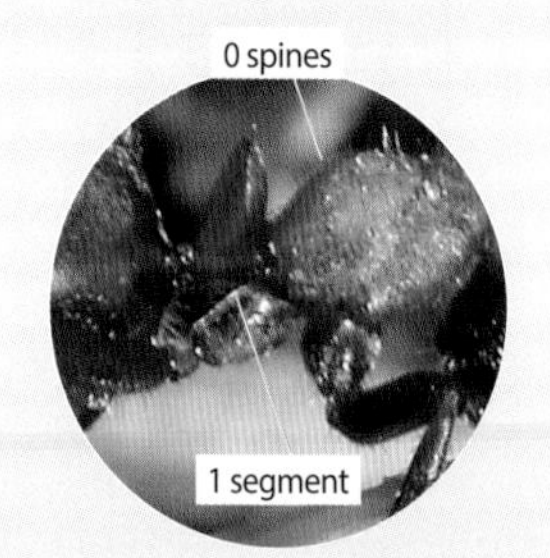

Hairy Wood Ant

SUBFAMILY **Formicinae ❯**

'WAIST' **2 bead-like segments**; PROPODEUM with a pair of spines on the hind part

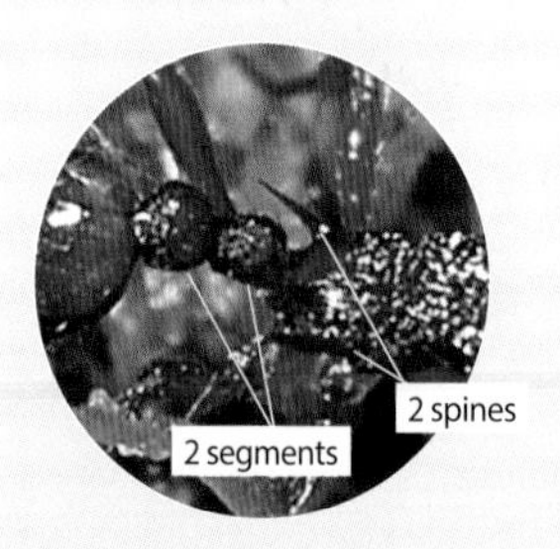

Common Red Ant

SUBFAMILY **Myrmicinae ❯❯**

› SUBFAMILY **Formicinae**

FORM medium-sized, robust-appearance (workers typically 5–7 mm); often red-and-black; **ABDOMEN** large and rounded; **HIND TIBIAE** underside with a double row of bristles, as well as at tip

Southern Wood Ant

Formica
[11 species | 9 ILL.]

p. 494

FORM small (workers typically 3–4 mm, occasionally 5 mm); mainly plain brown, black or yellow. **HIND TIBIAE** underside with bristles at tip only; lacking rows

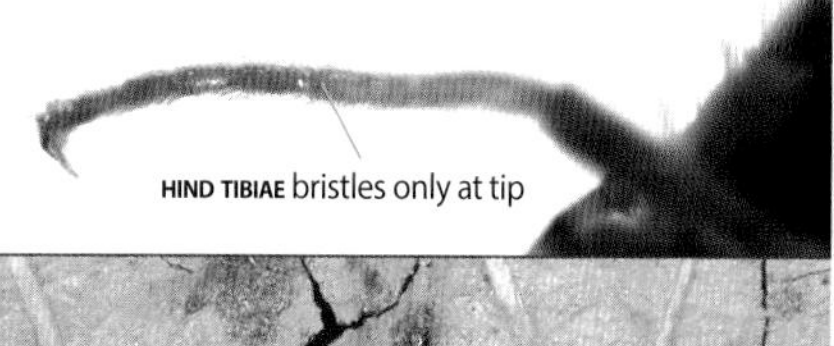

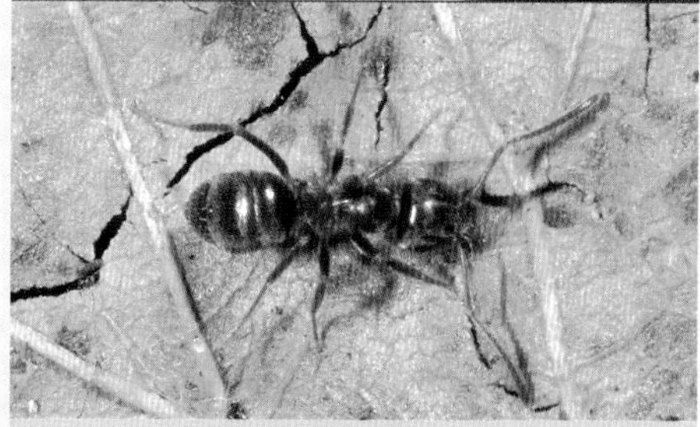

Small Black Ant

Lasius
[12 species (+ 1 non-established introduction) | 6 ILL.]

p. 492

›› SUBFAMILY **Myrmicinae**

POSTPETIOLE with spine on underside (inquiline 'guest' in wood ants' nests)

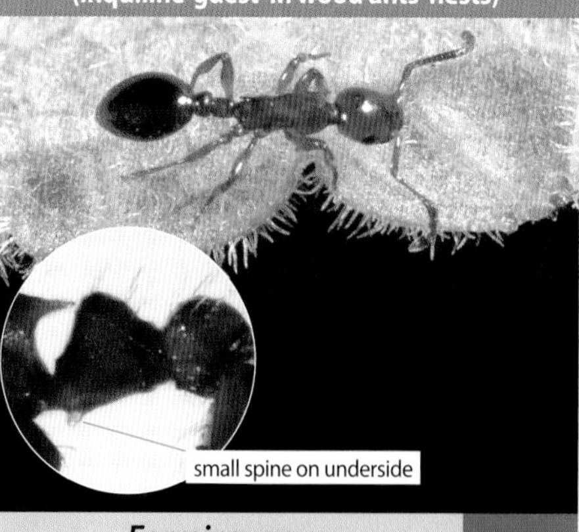

Formicoxenus
[Shining Guest Ant]

p. 497

POSTPETIOLE without spine on underside

PRONOTUM rounded; reddish

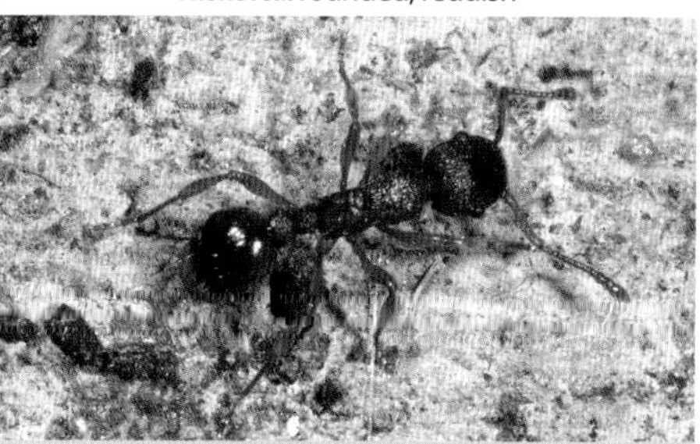

Woodland Red Ant

Myrmica
[12 species | 4 ILL.]

p. 498

POSTPETIOLE without spine on underside

PRONOTUM with angled corners; dark brown/black

Turf Ant

Tetramorium [1 species (+ 2 non-established introductions) | 1 ILL.]

p. 498

 Guide to Formicidae p. 490

Lasius 12 spp. (+1 non-established introduction) | 6 ILL.

GUIDE TO SELECTED SPECIES

<table>
<tr><td colspan="3">BODY reddish-yellow and brown; mainly North London</td><td>Wall Ant</td></tr>
<tr><td colspan="3">BODY yellow</td><td>Yellow Meadow Ant</td></tr>
<tr><td colspan="3">BODY shining black</td><td>Jet Ant</td></tr>
<tr><td rowspan="3">BODY dull black or dark brown</td><td colspan="2">ANTENNAE SCAPES + TIBIAE without long hairs</td><td>Small Sand Ant</td></tr>
<tr><td rowspan="2">ANTENNAE SCAPES + TIBIAE with long hairs</td><td>CLYPEUS with dense pubescence</td><td>Small Black Ant</td></tr>
<tr><td>CLYPEUS with sparse pubescence</td><td>Humid Ant</td></tr>
</table>

Yellow Meadow Ant
Lasius flavus

BL 3–7 mm. **ID** Yellow; body hairs long. Queen dark brown; abdomen and legs yellowish. **Hab** Open grassland, woodland and gardens. **SS** Other *Lasius* species, particularly **Mixed Yellow Ant** *L. mixtus* [N/I] (**BL** 3–5 mm), which has shorter body hairs; that species takes over the nests of other *Lasius* species.

Wall Ant
Lasius emarginatus

BL 3–7 mm. **ID** Yellowish-red with dark brown head, legs and abdomen. **Hab** Sparsely vegetated areas, nesting in old brick walls. **Dist** There have also been isolated past records from Dorset and the Isle of Wight. **SS** Other *Lasius* species.

Workers tending to the queen (much larger, dark brown with a yellowish abdomen and legs).

Nest mounds covered in vegetation are a familiar site in grassland and may be numerous.

Nests in brick walls; at this site near Chalk Farm, London, ar move from the wall shown to trees with plenty of aphids

Small Black Ant *Lasius niger*

BL 4–9 mm. **ID** Dark brown to black; clypeus with dense pubescence. **Hab** Sunny fairly dry sites, including gardens and heathland; this is usually the ant known to enter houses and is most prolific in urban areas, often found under pavements and stones. **SS** **Humid Ant** [see *inset below*]; other dark brown or black ants.

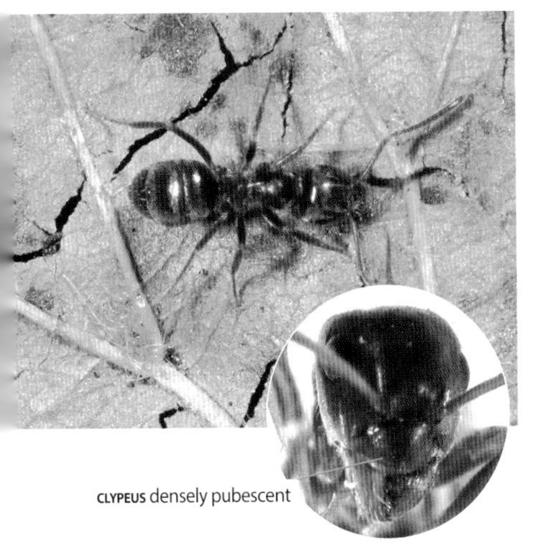

Humid Ant *Lasius platythorax*

BL 4–8 mm. **ID** Dark brown to black; clypeus with sparse pubescence. **Hab** Various, including wetlands and bogs and woodland with dead wood; more likely in rural areas. **SS** **Small Black Ant** [see *inset below*]; other black ants.

mall Sand Ant *asius psammophilus*

L 3 mm. **ID** Small, brownish. **ab** Sandy heathland and grassland capes and tibiae without long hairs. **S** Other *Lasius* species; until recently onfused with **Chalk Ant** *Lasius lienus* [N/I] (**BL** 3–4 mm), which avours calcareous grassland.

Jet Ant *Lasius fuliginosus*

BL 4–7 mm. **ID** Black; glossy, back of head heart-shaped. **Hab** Hedgerows, areas with trees and bushes. Forages above ground. **SS** Other black ants, particularly **Black Bog Ant** (*p. 495*) (which lives in a different habitat).

 Guide to Formicidae p. 49

Formica 11 spp. | 9 ILL.

GUIDE TO SELECTED SPECIES

BODY red-and-black			
CLYPEUS margin with central notch; **FORM** often redder than other *Formica*, including head and legs [does not build a nest; instead, workers of other *Formica* species, often Large Black Ant, perform nest duties]		**Slave-making Ant**	***p. 496***
EYES tiny hairs between facets or nearby	**FACE** standing hairs near eyes not reaching beneath eyes; **MESOTHORAX + PROPODEUM** with few hairs; Scotland and Ireland	**Scottish Wood Ant**	***p. 497***
	Standing hairs on face reaching beneath eyes; **MESOTHORAX + PROPODEUM** with numerous long hairs; rather darker red than some related species; N England, Scotland and Ireland	**Hairy Wood Ant**	***p. 497***
BACK OF HEAD heart-shaped; brighter red than most *Formica*; mainly Scotland		**Narrow-headed Ant**	***p. 496***
BACK OF HEAD not heart-shaped	**UNDERSIDE OF HEAD** with long hairs	**Southern Wood Ant**	***p. 496***
	UNDERSIDE OF HEAD lacking long hairs	**Red-barbed Ant**	***p. 496***
BODY black			
BODY glossy; associated with bogs		**Black Bog Ant**	***p. 495***
BODY not glossy	**MID + HIND FEMORA** without hairs on underside; **THORAX** top usually hairless, at most up to 3 short hairs present	**Large Black Ant**	***p. 495***
	MID + HIND FEMORA with numerous hairs on underside; **THORAX** top with short hairs on top of thorax; **LEGS** brownish	**Leman's Ant**	***p. 495***

Formica ants' nests

Hairy Wood Ant: a thatched mound of conifer needles and twigs, almost always much less than 1 m high.

Scottish Wood Ant: a thatched mound of conifer needle and twigs, up to 1·5 m high and 2 m diameter.

Southern Wood Ant: large mounds of vegetation.

Narrow-headed Wood Ant: a small thatched mound of grass fragments, up to 25 cm high.

FORMICINAE | *FORMICA* ANTS – Body black

Large Black Ant

Formica fusca

Common

BL 4–9 mm. **ID** Dull black; mid and hind femora without hairs on underside; top of thorax usually hairless, at most up to 3 short hairs present. **Hab** Open parts of woodland and heathland. **SS** Other black *Formica* species, very similar to and often confused with **Leman's Ant** [see comparison of features *below*].

Leman's Ant

Formica lemani

Common

BL 5–8 mm. **ID** Shining black; short hairs on top of thorax; abdomen with grey reflections. Mid and hind femora with numerous hairs on underside; brownish legs. **Hab** Various (up to 575 m in altitude). **SS** Other black *Formica* species, very similar to and often confused with **Large Black Ant** [see comparison of features *below*].

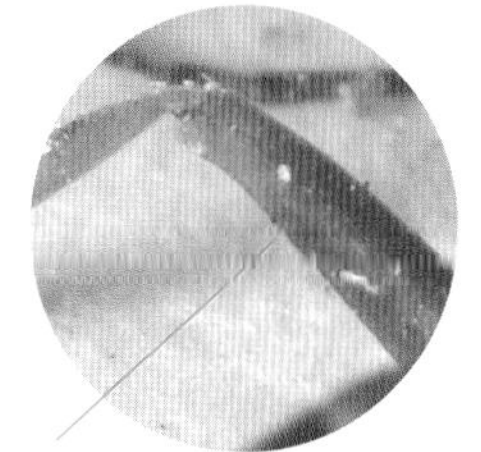

MID + HIND FEMORA underside without hairs

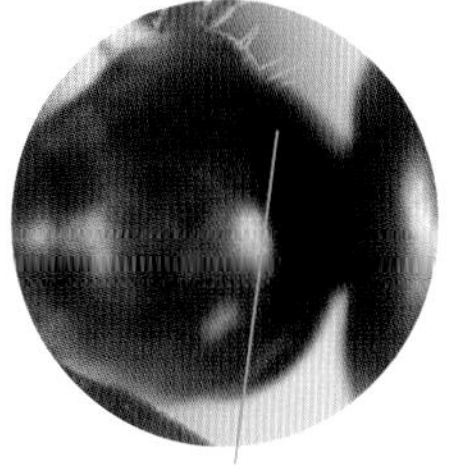

MID + HIND FEMORA underside with hairs

THORAX with hairs

Black Bog Ant

Formica picea

EN Rare (RDB1)

BL 5–9 mm. **ID** Black with glossy appearance. **Hab** Bogs and wetlands. **SS** Other black *Formica* species; **Jet Ant** (*p. 493*).

 Guide to *Formica* species p. 49

FORMICINAE | *FORMICA* ANTS – Body red-and-black

Southern Wood Ant
Formica rufa

Common

BL 5–10 mm. **ID** Red-and-black; underside of head with standing hairs. Winged ♂ black with reddish legs. **Hab** Woodland, including coniferous. **Dist** Only slight overlap with Hairy Wood Ant. **SS** Other red-and-black *Formica* species [see table on *p. 494*].

Red-barbed Ant
Formica rufibarbis

EN Rare (RDB1)

BL 5–12 mm. **ID** Red-and-black; pronotum and mesonotum with long hairs. The queen has a red thorax and dark patches. **Hab** Coastal grassland and heathland. **Dist** St Martin's, Isles of Scilly; possibly still present in Surrey, in low numbers. **SS** Other red-and-black *Formica* species [see table on *p. 494*].

Narrow-headed Ant
Formica exsecta

EN Rare (RDB1)

BL 5–8 mm. **ID** Red-and-black; back of head heart-shaped. Winged ♂ black. **Hab** Open heathland and woodland glades, up to 350 m in altitude. **Dist** Stronghold Scottish Highlands; still persists in England at Chudleigh Knighton, Devon. **SS** Other red-and-black *Formica* species.

Slave-making Ant
Formica sanguinea

Nationally Scarce

BL 6–10 mm. **ID** Red-and-black; redder than relatives, including head and legs. Margin of clypeus with central notch. **Hab** Open woodland and heathland. **SS** Other red-and-black *Formica* species.

Guide to Formicidae *p. 490*

SELECTED RED-AND-BLACK *FORMICA* ANTS COMPARED

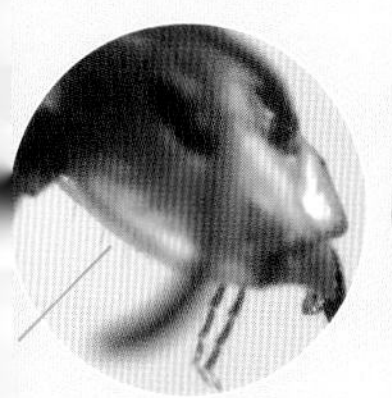

Southern Wood Ant
HEAD UNDERSIDE
standing hairs

Hairy Wood Ant
HEAD HAIRS
reach eyes

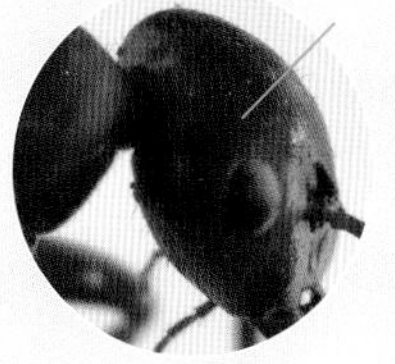

Scottish Wood Ant
HEAD HAIRS
do not reach eyes

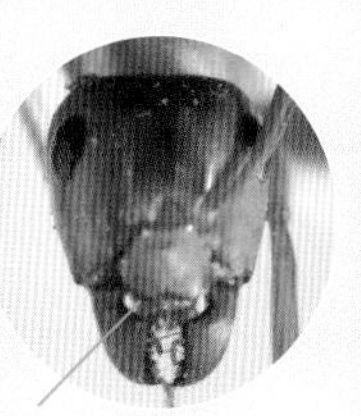

Slavemaker Ant
CLYPEUS
central notch

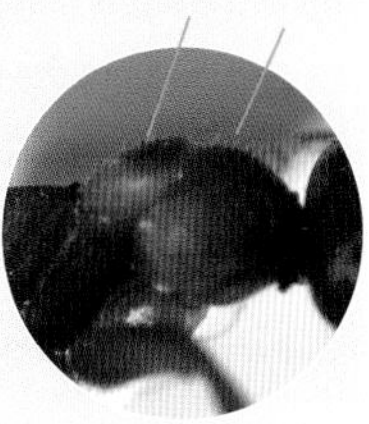

Red-barbed Ant
PRONOTUM + MESONOTUM
long hairs

Hairy Wood Ant
Formica lugubris

BL 5–10 mm. **ID** Red-and-black (darker han some relatives); head with erect iairs reaching to eyes; mesothorax and ›ropodeum with numerous long hairs. Winged ♂ black. **Hab** Open coniferous ınd deciduous woodland, to 400 m n altitude. **SS** Other red-and-black *Formica* species; very similar to and often confused with **Scottish Wood Ant**.

Scottish Wood Ant
Formica aquilonia

BL 6–10 mm. **ID** Red-and-black; mesothorax and propodeum (workers only) with few hairs. ♂ black. **Hab** Coniferous and sometime birch woodland, up to 360 m in altitude. **SS** Other red-and-black *Formica* species; in Scotland, range overlaps with very similar **Hairy Wood Ant**, with which often confused.

MESOTHORAX + PROPODEUM long hairs

MESOTHORAX + PROPODEUM few hairs

MYRMICINAE | Ants – postpetiole with spine

Formicoxenus 1 sp.

Shining Guest Ant
Formicoxenus nitidulus

BL 2–4 mm. **ID** Reddish-yellow to brown; shining; body with scattered pale hairs. **Hab** Open woodland and heathland, in wood ants' nest mounds. **Dist** Probably under-recorded due to their habitat. **SS** None.

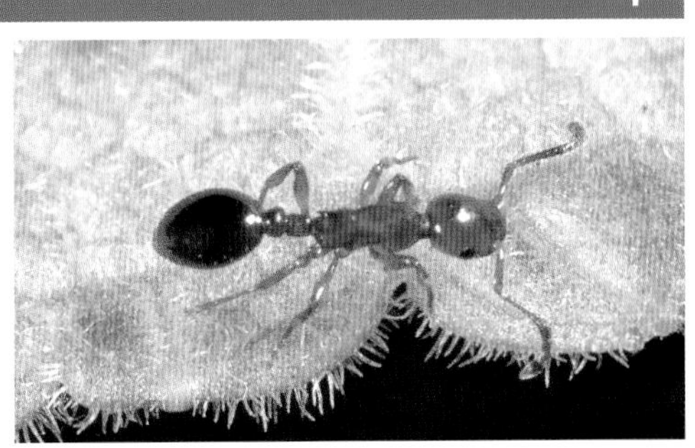

 Guide to Formicidae p. 49

MYRMICINAE | Ants – postpetiole without spine

Tetramorium 4 spp. | 1 ILL.

Turf Ant
Tetramorium caespitum

BL 3–8 mm. **ID** Dark brown to black, legs paler. Propodeal spines short. Wingless queens dark. **Hab** Rocky coastal grassland and some heathlands; usually under stones. **Beh** An aggressive species that readily stings. **SS** Dark Guest Ant *Tetramorium atratulum* [N/I] (**BL** 3 mm), is the only other *Tetramorium* sp. likely to occur in Britain now; this workerless rarity [RDB2] lives in the nest of the Turf Ant; queen dark and winged, with a small head; ♂ yellowish and wingless.

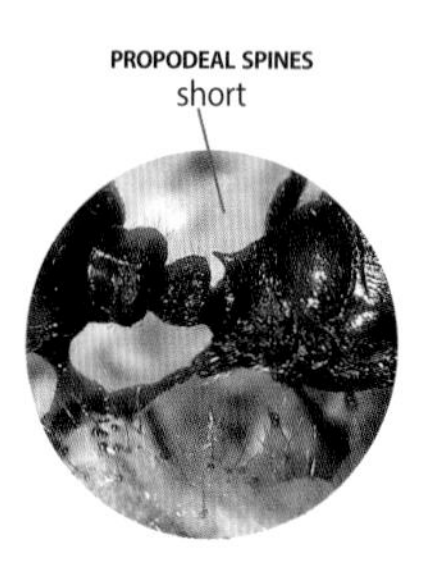

Myrmica 12 spp. | 4 ILL.

GUIDE TO WORKERS OF MORE COMMON SPECIES

Red Ant	Woodland Red Ant	Common Red Ant	Sabulet's Red Ant
ANTENNAE SCAPES base gently curved		**ANTENNAE SCAPES** base sharply bent	
PROPODEUM with a pair of short spines	**PROPODEUM** with a pair of long spines	**SCAPE** lacking flange along shaft	**SCAPE** with flange along shaft

Voodland Red Ant
Ayrmica ruginodis

BL 5–7 mm. **ID** Red; antenna scapes
lightly curved at base. Propodeal
pines longer than space between
hem. **Hab** Woodland, grassland;
ome damp habitats. Shade-tolerant.
SS Other red ants.

Red Ant
Myrmica rubra

BL 4–8 mm. **ID** Reddish. Propodeal spines shorter than width of space between them. Winged ♂ black. **Hab** Open, damper habitats. **Beh** Can readily sting humans. **SS** Other red ants.

abulet's Red Ant
Ayrmica sabuleti

L 4–6 mm. **ID** Reddish or yellowish;
ntenna scape base sharply bent with
ange along shaft. Winged ♂ black.
ueens dark. **Hab** Chalk downland
nd sandy heathland. **SS** Other red
nts.

Common Red Ant
Myrmica scabrinodis

BL 4–6 mm. **ID** Reddish; antenna scape base sharply bent. Winged ♂ black. Queens dark. **Hab** Sunny spots where not too dry, including some wetlands. **SS** Other red ants.

FAMILY Mutillidae (Velvet ants) — 3 GEN. | 3 spp. | 3 SPP. ILL.

Small to large (BL 3–14 mm), ant-like wasps with wingless females; black with reddish areas. Larvae are parasitoids on ground-nesting bees and wasps. Species can be identified as follows:

<table>
<tr><td colspan="2">BL 5–11 mm. THORAX narrow, divided into two dorsal sections; HEAD black; PRONOTUM + ABDOMEN BASE red</td><td>Black-headed Velvet Ant</td></tr>
<tr><td rowspan="2">THORAX broad, not divided; ABDOMEN BASE black</td><td>BL 9–14 mm. FORM robust appearance; BODY hairy but hairs not long; ANTENNAE black</td><td>Large Velvet Ant</td></tr>
<tr><td>BL 3–7 mm. FORM rather slender appearance; BODY with long silver hairs; ANTENNAE partly red</td><td>Small Velvet Ant</td></tr>
</table>

Mutilla — 1 sp.

Large Velvet Ant
Mutilla europaea

Nationally Scarce

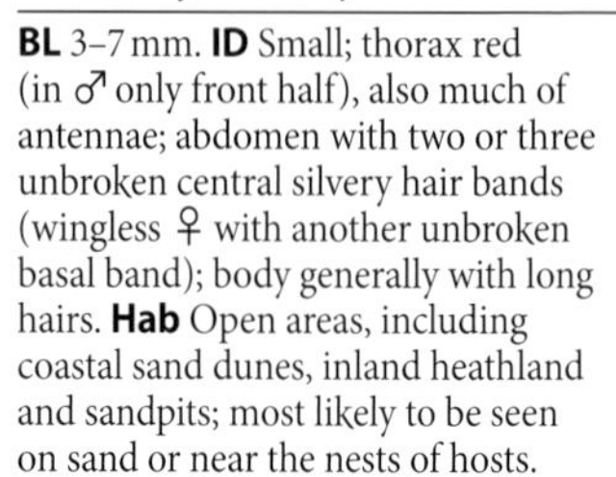

BL 9–14 mm. **ID** Large and hairy; thorax (except pronotum) red (less extensive in ♂); abdomen with two broken central bands of silvery hairs (wingless ♀ with another unbroken basal band). **Hab** Open areas, including sand dunes and chalk downland, heathland and woodland; most likely to be seen on tracks or near the nests of hosts. **Beh** Beware if handling a ♀, as the sting is painful! **Hosts** Bumble bee (*p. 546*) pupae. **SS** Small Velvet Ant.

♂

♀

Smicromyrme — 1 sp.

Small Velvet Ant
Smicromyrme rufipes

Nationally Scarce

BL 3–7 mm. **ID** Small; thorax red (in ♂ only front half), also much of antennae; abdomen with two or three unbroken central silvery hair bands (wingless ♀ with another unbroken basal band); body generally with long hairs. **Hab** Open areas, including coastal sand dunes, inland heathland and sandpits; most likely to be seen on sand or near the nests of hosts. **Hosts** Halictine bees (*p. 556*), crabronid wasps (*p. 524*) and pompilid wasps (*p. 503*). **SS** Large Velvet Ant, Black-headed Velvet Ant.

♂

♀

Myrmosa 1 sp.

›lack-headed Velvet Ant
Ʌyrmosa atra

›L 5–11 mm. **ID** Small, elongate; ↗ black, winged, ♀ wingless; body red ɿith black head and final two-thirds of ɓdomen, which has four narrow pale ands. **Hab** Open, mainly sandy sites; ɫso chalk grassland. **Hosts** Halictine ees (*p. 556*) and crabronid wasps ›. *524*). **SS** Small Velvet Ant.

ꟻAMILY **Sapygidae** (Club-horned wasps) 2 GEN. | 2 spp. | 2 SPP. ILL.

Medium-sized to large (BL 6–13 mm), elongate black solitary wasps, with red, yellow or white marks and distinctive club-tipped antennae. Larvae are cleptoparasites on solitary bees.

Sapyga 1 sp.

ive-spotted Club-horned Ꞁasp
apyga quinquepunctata

L 7–13 mm. **ID** Black. ♀ abdomen S2 S3 red, S4–5 with two pale yellow · whitish spots, S6 with a larger pale, ·ntral spot. ♂ abdomen white-spotted **ab** Woodland and gardens. **Hosts** *smia* species (*p. 552*), particularly Red ꞁason Bee (*p. 553*). **SS** None.

Monosapyga 1 sp.

Club-horned Wasp
Monosapyga clavicornis

BL 6–10 mm. **ID** Black; head, collar and legs and abdomen black with yellow spots and markings. **Hab** Woodland and gardens, wherever the nests of hosts occur. **Hosts** Sleepy Carpenter Bee (*p. 556*) and *Osmia* species (*p. 552*). **SS** None.

FAMILY Tiphiidae (Tiphiid wasps) — 2 GEN. | 3 spp. | 2 SPP. ILL.

Small to large black wasps (BL 3–14 mm) with reddish areas. Parasitoids on beetles. Species can be identified as follows:

FORM winged (both sexes); ♀ **THORAX** broad; black except for **MID + HIND TIBIAE; MID + HIND FEMORA** reddish (♀)	Large Tiphia
As Large Tiphia but usually much smaller, ♂ otherwise only told by thorax and abdomen being less distinctly punctured (microscope); ♀ **ALL FEMORA** black	Small Tiphia [N/I]
FORM winged in ♂ only; ♀ **THORAX** narrow (divided into three sections); red except for black head, abdomen and part of antennae	Tiger Beetle-hunting Wasp

Tiphia — 2 spp. | 1 ILL.

Large Tiphia
Tiphia femorata

Common

BL 5–14 mm. **ID** Elongate black wasp with spiny legs; in ♀ mid and hind tibiae red. ♂ body distinctly punctured. **Hab** Open areas, including sand dunes and chalk downland, often visiting umbellifers. **Hosts** Scarabaeid (*p. 282*) larvae, including small dung beetles. **SS** The Nationally Scarce **Small Tiphia** *Tiphia minuta* [N/I] (**BL** 4–6 mm) has all-black femora and the ♀ is clearly much smaller; ♂ body sparsely punctured.

Methocha — 1 sp

Tiger Beetle-hunting Wasp
Methocha articulata

Nationally Scarce

BL 3–12 mm. **ID** Elongate black wasp, ♂ winged with longer antennae than ♀; wingless ♀ red, except head and abdomen black. **Hab** Open sandy areas, including sand dunes and inland heathland. **Hosts** Tiger beetle (*p. 262*) larvae. **SS** None, except superficially resembles an ant (*p. 490*), but easily distinguished by the curved (not elbowed) antenna structure.

FAMILY **Pompilidae** (Spider-hunting wasps) **14 GEN. | 44 spp. | 16 SPP. ILL.**

Characteristic long-legged, ground-running, solitary wasps (BL 4–18 mm), appearing continually agitated, with antennae and wings constantly vibrating. Most of these wasps nest in the ground, provisioning each nest with a paralyzed spider. They often overwinter as full-grown larvae.

NOTE: ♂s ANTENNAE 13-segmented; ABDOMEN 7 visible segments;
♀s ANTENNAE 12-segmented; ABDOMEN 6 visible segments.

The colour and general build of the wasp are important in identification to genus level, but many species are difficult to identify, even by experts using a microscope.

		Genus	Page
BODY black			
♂ FACE yellow; ABDOMEN S7 white		*Auplopus* [Yellow-faced Spider Wasp]	p. 504
FEMORA part red; ABDOMEN with white tip		*Caliadurgus* [White-spurred Spider Wasp]	p. 505
FOREWINGS with two black bands		*Dipogon* [3 species | 1 ILL.]	p. 504
LEGS part red; STIGMA large; EYE margins with oval fleck		*Agenioideus* [3 species | 1 ILL.]	p. 505
PRONOTUM broad	♀ ABDOMEN black, part dark reddish	*Aporus* [Purseweb Spider Wasp]	p. 505
	♀ ABDOMEN black (thorax and abdomen blood red-and-black in an uncommon colour form); TIBIAL SPURS pale	*Homonotus* [Blood Spider Wasp]	p. 505
THORAX + ABDOMEN white-spotted; LEGS reddish, in part		*Episyron* [2 species | 1 ILL.]	p. 504
BODY covered in pale grey hairs		*Pompilus* [Leaden Spider Wasp]	p. 504
BODY black-and-red			
ABDOMEN ♀ tip with many thick bristles. NOTE: in one species the body is black		*Anoplius* [5 species | 3 ILL.]	p. 507
PROPODEUM wrinkled		*Cryptocheilus* [Ridge-saddled Spider Wasp]	p. 506
WINGS brownish, in some species forewings with clear white spot near tip		*Priocnemis* [13 species | 1 ILL.]	p. 506
FORM stout, some with white markings; WINGS brownish		*Arachnospila* [7 species | 1 ILL.]	p. 506
ABDOMEN ♀ tip lacking thick bristles		*Evagetes* [3 species | 1 ILL.]	p. 506
BODY black, red and white			
BODY white and red markings; ♀ OVIPOSITOR visible		*Ceropales* [2 species | 1 ILL.]	p. 507

Guide to Pompilidae p. 50

Pompilus 1 sp.

Leaden Spider Wasp
Pompilus cinereus

BL 4–10 mm. **ID** Black, with short grey hairs; forewings dark-tipped. **Hab** Sandy sites, particularly coastal areas, also inland heathland and sandpits. **SS** None.

Episyron 2 spp. | 1 ILL.

Red-legged Spider Wasp
Episyron rufipes

BL 5–14 mm. **ID** Black; hind tibiae and base of femora red; abdomen with 2–3 pairs of white spots. Head also with white markings by eyes. **Hab** Sandy sites, often coastal; also many inland. **SS** French Spider Wasp *Episyron gallicum* [N/I] (**BL** 6–12 mm), which has abdomen with one pair of large white spots; other spider wasp species, but white spots are characteristic.

Dipogon 3 spp. | 1 ILL.

Variable Spider Wasp
Dipogon variegatus

BL 4–9 mm. **ID** Black, abdomen short; forewings double-banded. **Hab** Grassland, heathland, gardens and parks; a cavity nester. **SS** Other *Dipogon* species, from which told by microscopic examination of ♂ genitalia.

Auplopus 1 sp

Yellow-faced Spider Wasp
Auplopus carbonarius

BL 6–10 mm. **ID** Black, ♂ with pale yellow face and clypeus; abdomen S7 white. ♀ forewing has two elongated submarginal cells. **Hab** Woodland, parks and gardens; around old tree stumps. **SS** Other black spider wasps, from which told by white tip to abdomen (♂); in nearly all other spider wasps, ♀s have three elongated submarginal cells.

Caliadurgus 1 sp.

Vhite-spurred Spider Wasp
:aliadurgus fasciatellus

Local

3L 6–10 mm. **ID** Black-and-red; orewings with dark patches; ♀ front f tibiae with short, curved spine; ♂ with white spurs on tibiae, also red n legs and abdomen with white-potted tip. **Hab** Heathland. **SS** Other lack-and-red spider wasps, although hese lack the white marks described bove.

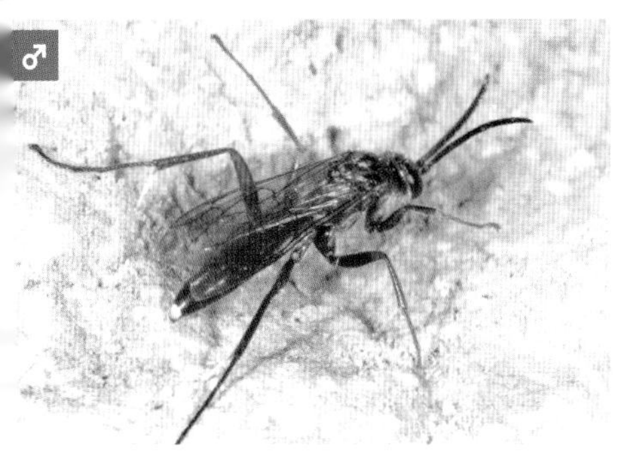
♂

Aporus 1 sp.

Purseweb Spider Wasp
Aporus unicolor

Nationally Scarce

BL 5–11 mm. **ID** Black-and-red; rather elongate; wings dark. **Hab** Coastal cliffs, heathland and chalk downland; may be found on Wild Carrot flowers. **SS** None.

♀

Homonotus 1 sp.

lood Spider Wasp
'omonotus sanguinolentus

EN Rare (RDB1)

L 7–10 mm. **ID** Black or a ♀ form ccasionally black with parts of orax blood red; robust, bullet-like pearance. **Hab** Damp heathland. **S** None.

♀

Agenioideus 3 spp. | 1 ILL.

Jumping Spider Wasp
Agenioideus cinctellus

Local

BL 4–7 mm. **ID** Black; legs red; ♀ clypeus whitish, also head and pronotum with whitish marks; ♂ black, face, base of hind tibiae and abdomen S7 with small white marks. **Hab** Sandy soils. **SS** None.

NOTE: other *Agenioideus* species are all-black, including the legs, and are rare recent additions to the UK fauna.

♀

 Guide to Pompilidae p. 50

Priocnemis 13 spp. | 1 ILL.

Perturbed Spider Wasp
Priocnemis perturbator

Common

BL 9–17 mm. **ID** One of the largest *Priocnemis* species in Britain and Ireland. Black-and-red. **Hab** Open woodland; often seen on Wood Spurge flowers. **SS** Other *Priocnemis* species, from which told by size and microscopic examination of characters such as extent of hairs on fore and mid femora; even then, very difficult to identify.

♀

Cryptocheilus 1 sp.

Ridge-saddled Spider Wasp VU
Cryptocheilus notatus

Rare (RDB2)

BL 7–15 mm. **ID** Large. Black-and-red; propodeum deeply furrowed; forewings with dark tips. **Hab** Heathland. **SS** Superficially like other (usually smaller) black-and-red spider wasps but fairly distinctive in the field once familiar with the species.

♀

Arachnospila 7 spp. | 1 ILL.

Heath Spider Wasp
Arachnospila anceps

Common

BL 5–10 mm. **ID** Black-and-red; propodeum in ♀ coarse; end of subgenital plate (on underside) of ♂ with small tuft of hairs. **Hab** Sandy sites. **SS** Other *Arachnospila* species; this is a difficult genus to separate, even for specialists using technical keys. Distinguished from *Evagetes* species by shorter and thicker antennae.

with *Trochosa* spider prey

♀

Evagetes 3 spp. | 1 ILL.

Common Evagetes Spider Wasp *Evagetes crassicornis*

Common

BL 5–9 mm. **ID** Black-and-red, red fairly dull; first abdomen segments with black borders. **Hab** Sandy sites; a cleptoparasite on other spider wasps. **SS** Other, rarer, *Evagetes* species, from which told by microscopic examination of submarginal wing cells; black-and-red spider wasps, from which told by general appearance confirmed by microscopic examination.

♂

Anoplius 5 spp. | 3 ILL.

Black-banded Spider Wasp
Anoplius viaticus

BL 8–14 mm. **ID** Black; abdomen S1–3 with distinctive red bands not reaching hind margins (feature less obvious in more slender ♂). Wings mostly dark. **Hab** Heathland and sand dunes. **Dist** Only old records from Ireland. **SS** Dusky Spider Wasp *Anoplius infuscatus* [N/I] (**BL** 5–10 mm) has abdomen S1–3 all-red.

Common

Black Spider Wasp
Anoplius nigerrimus

BL 5–8 mm. **ID** Black, ♀ abdomen with silvery pubescence; pale hairs on face. **Hab** Grassland, heathland, woodland, gardens and parks. **SS** Other *Anoplius* species, from which told by body colour or forewing venation.

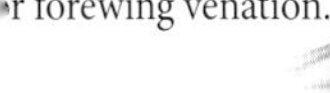

Common

Elegant Spider Wasp
Anoplius concinnus

BL 6–12 mm. **ID** Black; dark hairs on face. **Hab** Stony riversides, near shingle beaches and gravel pits. **SS** Other *Anoplius* species, from which told by body colour or forewing venation.

Local

Ceropales 2 spp. | 1 ILL.

White-streaked Spider Wasp
Ceropales maculata

BL 5–9 mm. **ID** Black, with white or yellowish markings on head, thorax and abdomen; legs reddish. **Hab** Heathland, sand pits and coastal sand dunes; might also be seen on Wild Carrot flowers. **SS** Variegated Spider Wasp *Ceropales variegata* [N/I] EN (RDB1) (**BL** 5–9 mm), which is more colourful, with red spots on abdomen S1–2.

Local

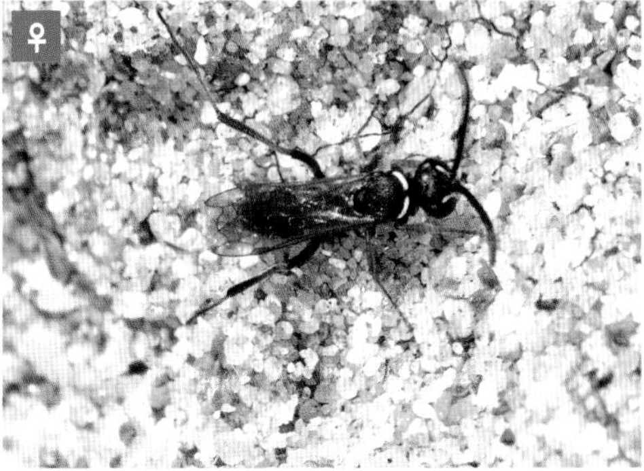

FAMILY **Vespidae** (Social, potter & mason wasps) **12 GEN. | 33 spp. | 28 SPP. ILL.**

This is a combined group of medium-sized to large wasps (BL 6–30 mm), many of which are brightly coloured. The characteristic yellow-and-black social wasps are familiar insects, living in large colonies in nests built of wood pulp. Several hundred wasps serve a queen (only the queen overwinters) and they have a reputation for stinging. Food includes fallen fruit and prey in the form of many insect orders. The potter and some mason wasps build amazing structures in a short time period.

There is a huge size range in the Vespidae – from the diminutive 5 mm long Little Mason Wasp [INSET] to the Hornet, at 30 mm in length, both shown here at life-size.

NOTE: in ♂s, the observation of some subtle characters is needed to distinguish between some genera.

BODY broad, often large (BL ≥10 up to 30 mm); ABDOMEN segments often yellow or wit broad, yellow bands; many species have a pair of black spots on most segments.

SOCIAL WASPS (3 GEN.)

BODY **broad**

OCELLI slight distance from ocelli to brow at back of head

MALAR SPACE long [= long-faced]

♀

Tree Wasp

Dolichovespula [4 species | 4 ILL.] p. 514

long space

MALAR SPACE very short

Common Wasp

Vespula [4 species | 4 ILL.] p. 512

OCELLI hind pair behind eyes

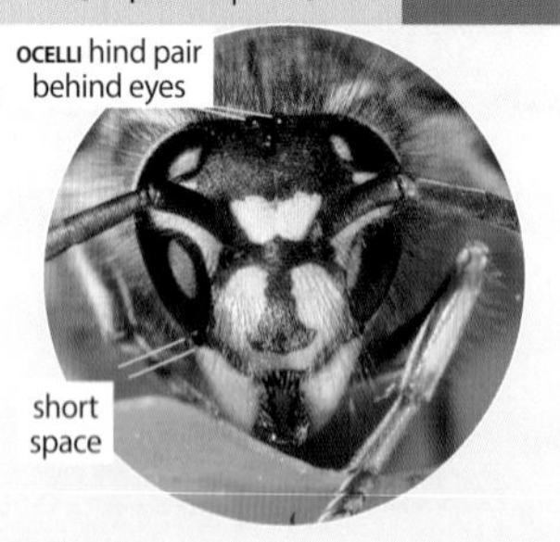

OCELLI long distance from ocelli to brow at back of head

Hornet

Vespa [2 species | 2 ILL.] p. 511

OCELLI between eyes

BODY slender, small to large (BL 5–16 mm). ABDOMEN segments usually with narrow, whitish or yellow rings, not broad bands

ANTENNAE partly orange

PAPER WASPS (1 GEN. *Polistes*)

ABDOMEN S1 ½ as broad as S2

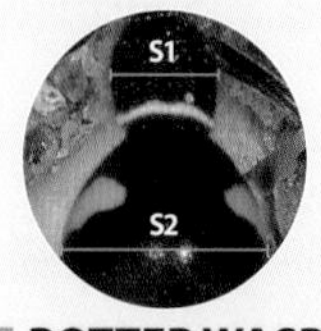

POTTER WASPS (1 GEN. *Eumenes*)

ABDOMEN S1 ≈ broad as S2

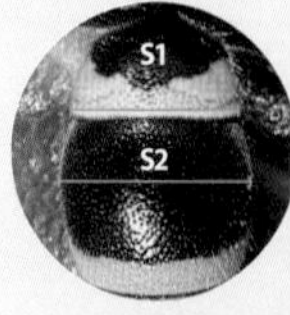

MASON WASP (7 GEN.)

MASON WASPS (7 GEN.)

ABDOMEN S1 ≈ broad as S2; S1 yellow band without distinct 'notch' 1/2

ABDOMEN S1 with red; BL 9–13 mm

Purbeck Mason Wasp

Pseudepipona [Purbeck Mason Wasp] p. 520

ABDOMEN S1 with broad yellow band; BL 6–15 mm

Wall Mason Wasp

Ancistrocerus [9 species | 6 ILL.] p. 516

ABDOMEN S1 ≈ broad as S2

ABDOMEN S1 with **distinct 'notch'** on yellow band; BL 6–15 mm

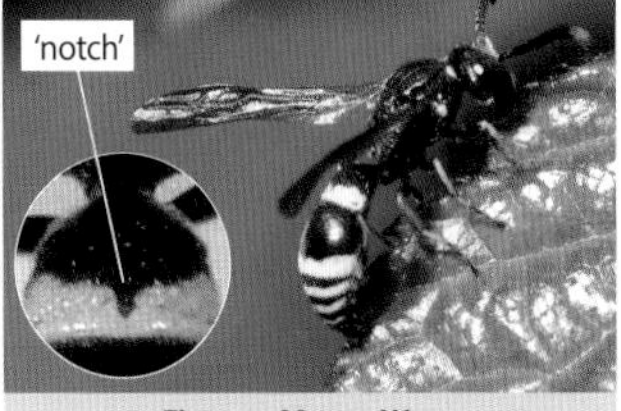

Figwort Mason Wasp

Symmorphus [4 species | 3 ILL.] p. 518

ABDOMEN S1 ≈ broad as S2; S1 yellow band without distinct 'notch' 2/2

ABDOMEN S1 not red; THORAX + PROPODEUM slightly longer than wide

BL small–medium-sized (5–8 mm)

Little Mason Wasp

Microdynerus [Little Mason Wasp] p. 520

ABDOMEN with 4–5 yellow bands; ♀ THORAX black; FORM broad, 'box'-headed; BL 9–12 mm

Box-headed Mason Wasp

Gymnomerus [Box-headed Mason Wasp] p. 520

ABDOMEN with 4 yellow bands; ♀ THORAX with yellow mark; BL 9–12 mm

Four-banded Mason Wasp

Euodynerus [Four-banded Mason Wasp *Euodynerus quadrifasciatus*]

ABDOMEN with 4–6 yellow bands; THORAX black; HEAD bulbous between eyes, but not as broad as Box-headed Mason Wasp

Spiny Mason Wasp

Odynerus [4 species (1†) | 3 ILL.] p. 519

PAPER WASPS (1 GEN.)

ANTENNAE partly orange

BL 10–17 mm; ♂ FACE yellow

European Paper Wasp

Polistes [3 species | 1 ILL.] p. 515

POTTER WASPS (1 GEN.)

ABDOMEN S1 ½ as broad as S2

MID TIBIA with one spur; BL 9–15 mm

Heath Potter Wasp

Eumenes [2 species | 1 ILL.] p. 520

SOCIAL WASPS

Social wasp castes

DISTINGUISHING WORKERS, QUEENS AND MALES OF SOCIAL WASPS

Queen	Worker	Male
Can look the same as worker, or differ in colour (as in example shown *below*), but always larger; queens are the reproducing females that develop from fertilized eggs; overwinters as an adult.	Small (at least in comparison with queen). A non-reproducing female that develops from fertilized eggs, often seen in abundance, performing chores.	Looks much the same as a worker (develop from unfertilized eggs laid by queens or sometimes workers), but antennae are longer. Males cannot sting.

Median Wasp	Median Wasp	German Wasp

German Wasp workers (BL approx. 12 mm) and queen (BL 19 mm).

Vespa	2 spp.
BODY reddish-brown and yellow; THORAX: only black on mesoscutum; HEAD yellow; ABDOMEN final segments mainly yellow	Hornet
BODY black-and-yellow; THORAX wholly black; HEAD black on top, yellow beneath; ABDOMEN only S4 yellow	Asian Hornet

Hornet *Vespa crabro*

BL 15–30 mm. **ID** Very large, reddish-brown and yellow with minimal amount of blackish on the mesoscutum (part of thorax). Head enlarged, yellow. **Hab** Woodland, parks and gardens. **Beh** Nests in hollow trees, buildings and bird boxes. **SS** **Asian Hornet**; **Median Wasp** (queen) (*p. 515*).

Asian Hornet *Vespa velutina*

BL 18–30 mm. **ID** Large, yellow-and-black; head black on top, yellow beneath. Abdomen S4 yellow. **Hab** Woodland, parks and gardens. **Beh** Usually nests in tree canopies; attacks Honey Bees (*p. 546*). **Flight** Apr–Nov, peak May–Oct. **Dist** Rare immigrant; first recorded in Britain in 2016; ten nests of this invasive species have been destroyed at several locations in England since then (to 2020), ranging from Cornwall to Kent and N to Lancashire. **SS Hornet, Median Wasp** (queen) (*p. 515*).

Sightings should be reported to the GB Non-Native Species Secretariat via the Asian Hornet Watch app or online (see *p. 571*).

Guide to Vespidae *p. 50*

SOCIAL WASPS

Vespula — 4 spp.

'Short-faced' (small gap between eye and mandible). Best identified using abdomen S1 hair colour, body pattern and face (clypeus) details as below:

Austrian Cuckoo Wasp	Red Wasp	German Wasp	Common Wasp
ABDOMEN S1 with long black hairs		**ABDOMEN** S1 with long pale hairs	
ABDOMEN yellow, S1 + S2 never reddish	**ABDOMEN** yellow, S1 + S2 with reddish patches	**ABDOMEN** yellow with large black central marks [NB some individuals have inverted triangles, similar to Common Wasp]	**ABDOMEN** yellow with broad, inverted black triangles (particularly on S1 + S2)
 ♀			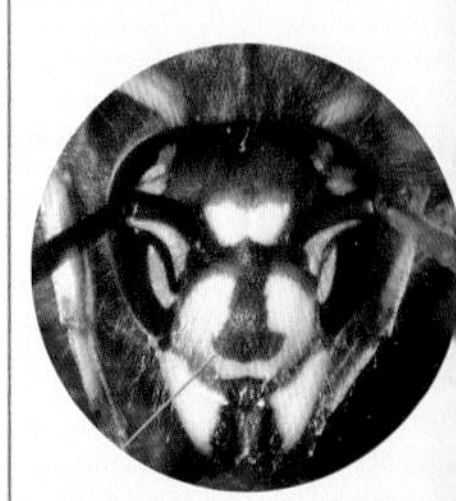
CLYPEUS plain yellow (♂) or with 3 black spots (♀)	**CLYPEUS** with broad black anchor-like mark	**CLYPEUS** with 3 black spots, or central black bar and 2 black spots; **'CHEEKS'** yellow, no black patch	**CLYPEUS** with broad black anchor-like mark; **'CHEEKS'** with black patch

VESPULA WASPS – ABDOMEN S1 WITH DARK HAIRS

Red Wasp *Vespula rufa*

BL 12–18 mm. **ID** Yellow-and-black. Clypeus with broad black anchor-like mark. Abdomen S1 + S2 yellow with reddish patches. Tibiae lack long hairs. **Hab** Open woodland, moorlands, scrubby grassland and hedgerows. **Beh** Nests underground. **Cuckoo** Austrian Cuckoo Wasp. **SS** Other social wasps in the genera *Vespula* [see *above*] and *Dolichovespula* (*p. 514*).

Common

Austrian Cuckoo Wasp
Vespula austriaca

BL 12–18 mm. **ID** Yellow-and-black. Clypeus with three black spots (♀), or plain (♂). Abdomen mainly yellow with some black markings. Tibiae with long black hairs. **Hab** Open woodland, moorlands and hedgerows. **Host** Red Wasp. **Beh** The cuckoo parasite of Red Wasp, therefore only producing queens and males. After entering the host colony, the queen immediately kills the host queen, unless it flees the nest. **SS** Other social wasps in the genera *Vespula* [see *opposite*] and *Dolichovespula* (*p. 514*); most likely to be confused with **Common Wasp**.

VESPULA WASPS – ABDOMEN S1 WITH PALE HAIRS

Common Wasp
Vespula vulgaris

BL 12–18 mm. **ID** Yellow-and-black. Clypeus with broad, black anchor-like mark. 'Cheeks' with black patch; abdomen with broad, inverted black triangles (particularly conspicuous on S1 & 2), workers with extensive yellow. **Hab** Almost anywhere; often in urban areas in parks, gardens and homes. **Beh** Nests in houses, outbuildings and underground. **SS** Other social wasps in the genera *Vespula* [see *opposite*] and *Dolichovespula* (*p. 514*); often confused with **German Wasp** and **Austrian Cuckoo Wasp**.

German Wasp
Vespula germanica

Common

J F M A M J J A S O N D

BL 12–19 mm. **ID** Yellow-and-black. Clypeus with three black spots, or a central bar and two spots. Abdomen segments with black central marks. **Hab** Grassland, heathland, hedgerows; often in urban areas in gardens and parks, sometimes in homes. **Beh** Nests underground (such as old rodent burrows), sometimes above ground. **SS** Other social wasps in the genera *Vespula* [see *opposite*] and *Dolichovespula* (*p. 514*); often confused with **Common Wasp**.

SOCIAL WASPS 3/

Dolichovespula 4 spp.

'Long-faced' (large gap between eye and mandible). Best identified using body pattern and face (clypeus) details as below:

Tree Wasp	Norwegian Wasp	Saxon Wasp	Median Wasp
ABDOMEN yellow bands broad, lacking black spots and marks present in other species long black hairs	**ABDOMEN** S1 & S2 (possibly S3) usually partly red	**ABDOMEN** never any red	**MESOSOMA** with narrow yellow '7' at front; **ABDOMEN** with narrow yellow bands
	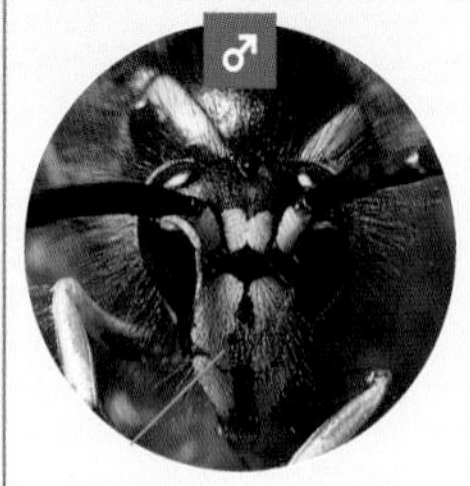	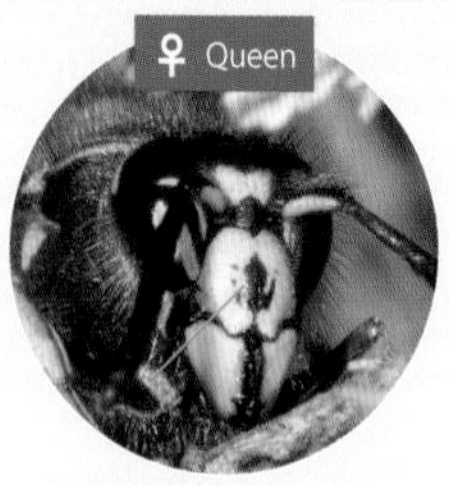	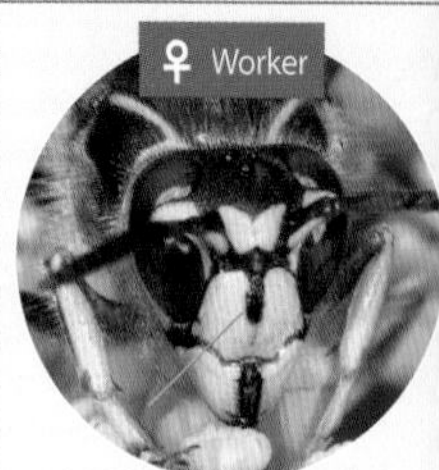
CLYPEUS plain yellow or with single black central spot	**CLYPEUS** with broad black line and mark	**CLYPEUS** with black line (partial in some individuals) and central mark, surrounded by a few small spots	**CLYPEUS** with slim black ba (often plain in queen)

Tree Wasp

Common

Dolichovespula sylvestris

BL 13–19 mm. **ID** Yellow-and-black. Clypeus yellow or with single black central spot, although the particularly hairy-faced ♂ has an incomplete black stripe. Abdomen segments with almost straight, often broad yellow bands, lacking black spots. **Hab** Wide range, including gardens. **Beh** Nests are often in trees, sometimes in shallow ground. **SS** Other social wasps in the genera *Dolichovespula* [see table *above*] and *Vespula* (*p. 512*).

J F M A M J J A S O N D

Norwegian Wasp

Local

Dolichovespula norwegica

BL 11–18 mm. **ID** Yellow-and-black; abdomen S1 + S2 (possibly S3) partly reddish. Clypeus with broad black line and mark, narrowest in ♂. **Hab** Heathland and moorland; sometimes in urban areas. **Beh** Nests attached to shrubs or trees. **SS** Other social wasps in the genera *Dolichovespula* [see table *above*] and *Vespula* (*p. 512*).

ABDOMEN yellow bands broad, lacking spots

♂

ABDOMEN at least S1 partly reddis

;axon Wasp
)olichovespula saxonica

Common

;L 11–18 mm. **ID** Yellow-and-black. :lypeus with black line (only partial in ome) and central mark, surrounded y a few, typically, four, small spots. **Iab** Heathland and woodland. **Beh** Nests are usually attached to branches r shrubs, also beehives and buildings. **)ist** First recorded in Britain in 1987, eaching Scotland in 2013. **SS** Other ocial wasps in the genera *Dolichovespula* [see table *pposite*] and *Vespula* (*p. 512*).

Median Wasp
)olichovespula media

Nationally Scarce

;L 14–22 mm. **ID** Large, yellow-and- lack (partly reddish-brown on queen, otably on thorax, including sides; the ueen also has head yellowish). Basal ntenna segments yellow or reddish- rown. Clypeus with slim black bar not eaching base (in queen clypeus often lain). Yellow stripes on 'shoulders', ather like an inverted pair of '7's. Abdomen has narrow yellow bands. **Iab** Woodland, agricultural areas, gardens and parks. **;eh** Nests typically attached to branches or shrubs. **)ist** First recorded in Britain in 1980. **SS** Other social /asps in the genera *Dolichovespula* [see table *opposite*] nd *Vespula* (*p. 512*); queen like **Asian Hornet**, **Hornet** *). 511*).

queen Median Wasp is somewhat hornet-like, having ɔme reddish-brown coloration and a plain yellow ypeus – but is smaller, has black stripes on the abdomen, nd ocelli nearer the back of the head.

'APER WASPS

Polistes — 3 spp. | 1 ILL.

uropean Paper Wasp
olistes dominula

Immigrant; rare

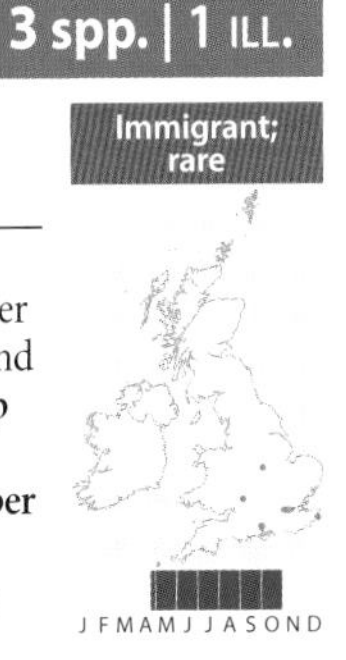

L 10–17 mm. **ID** Yellow-and-black; ntenna S1 and clypeus yellow (the latter nay have a black mark). **Hab** Woodland nd gardens. **SS** **Nymphal Paper Wasp** *olistes nimpha* [N/I] (**BL** 10–17 mm), hich has darker antennae; **French Paper** **/asp** *P. gallica* [N/I] (**BL** 10–15 mm), hich was recorded in Ireland in 1962: nicroscopic examination of various haracters is needed.

A typical paper wasp nest: a comb hanging from a pedicel.

Guide to Vespidae p. 50

Ancistrocerus 9 spp. | 6 ILL.

GUIDE TO SELECTED SPECIES

ABDOMEN S2 yellow side band about level with continuing band on underside (about level)		
ABDOMEN yellow bands on S1–6	band on S1 not wider at sides	**Garden Mason Wasp**
	band on S1 gradually widens towards sides; front of S1 with deep, central 'V'-shaped notch	**Wall Mason Wasp**
ABDOMEN yellow band on S1 expanded at sides; bands S1–4 (up to S6)		**Small-notched Mason Wasp**
ABDOMEN yellow bands on S1–**3** (some also S4)		**Three-banded Mason Wasp**
ABDOMEN S2 yellow side band extends well forward of edge of band on underside (well forward)		
LEGS yellow; **ABDOMEN** yellow bands on S1–**5 (up to S6)**		**Early Mason Wasp**
LEGS (tibiae and tarsi) reddish; **ABDOMEN** yellow bands on S1–**4** (some only S1–3)		**Maritime Mason Wasp**

S2 yellow side band extends well forward of underside band

Early Mason Wasp
Ancistrocerus nigricornis

Common

BL 6–13 mm. **ID** Black; abdomen S1–5 (up to S6 in ♂) with yellow markings. **Hab** Grassland, woodland, gardens and parks. **Dist** Also an old record from Ireland. **SS** Other *Ancistrocerus* species [see table *above*].

Maritime Mason Wasp
Ancistrocerus scoticus

Local

BL 7–12 mm. **ID** Black; abdomen S1–3 (or up to S4 in ♂) with yellow markings (whitish in Scotland and Ireland). Tibiae and tarsi reddish. **Hab** Various, but often coastal, sandy sites. **SS** Other *Ancistrocerus* species [see table *above*].

Early Mason Wasp bands on S1–6

Maritime Mason Wasp bands on S1–3

›2 yellow side band about level with underside band

›arden Mason Wasp
›ncistrocerus parietinus

Common

›L 9–15 mm. **ID** Black; abdomen S1–6 ›ith yellow markings. **Hab** Open areas ›uch as grassland, gardens and parks. ›**S** Other *Ancistrocerus* species [see ›able *opposite*].

Wall Mason Wasp
Ancistrocerus parietum

Common

BL 7–12 mm. **ID** Black; abdomen S1–6 with yellow markings, S1 yellow band widened towards sides. **Hab** Grassland, woodland, gardens and parks. **SS** Other *Ancistrocerus* species [see table *opposite*].

›mall-notched Mason Wasp
›ncistrocerus gazella

Common

›L 7–12 mm. **ID** Black; abdomen ›–4(–5) (up to S6 in ♂) with yellow ›arkings. **Hab** Grassland, woodland ›d parks, where hunts Lepidoptera ›rvae to take to the nest. As with other *›ncistrocerus* species, the parasitoids ›e *Chrysis* species (*p.* 486). ›**S** Other *Ancistrocerus* species [see ›ble *opposite*].

Three-banded Mason Wasp
Ancistrocerus trifasciatus

Common

BL 7–12 mm. **ID** Black; abdomen S1–3(–4) with yellow markings. **Hab** Grassland, heathland, gardens and parks; sometimes in damper areas. **SS** Other *Ancistrocerus* species [see table *opposite*].

Guide to Vespidae *p.* 50

Symmorphus 4 spp. | 3 ILL.

ABDOMEN ♂ S1–6 with yellow markings; ♀ S1–4 (some to S5)		
PRONOTUM lateral corners **pointed**; spines project **forwards**		Figwort Mason Wasp
PRONOTUM lateral corners **rounded**; spines project **sideways**		Large Aspen Mason Wasp
ABDOMEN S1+S2+S4 with yellow markings		
PRONOTUM lateral corners **rounded**; spines project **sideways**	**PRONOTUM + SCUTELLUM** yellow markings	Willow Mason Wasp
	PRONOTUM + SCUTELLUM no yellow markings	Small Aspen Mason Wasp [N/I]

Willow Mason Wasp
Symmorphus bifasciatus

BL 6–10 mm. **ID** Black, abdomen S1–2 + S4 with yellow markings. **Hab** Wetlands, woodlands and gardens. **Host** Blue Willow Beetle *Phratora vulgatissma* [N/I] larvae. **SS** The rare (RDB3) **Small Aspen Mason Wasp** *Symmorphus connexus* [N/I] (**BL** 6–9 mm) is very similar but pronotum and scutellum lack yellow markings.

Figwort Mason Wasp
Symmorphus gracilis

BL 7–12 mm. **ID** Black, abdomen S1–6 (S1–4 or S5 in ♂) with yellow markings. **Hab** Woodland and gardens. **Host** Red Poplar Leaf Beetle (*p. 317*), Garden Weevil *Cionus hortulanus* [N/I] on figworts. **SS Large Aspen Mason Wasp** [see table *above*].

Large Aspen Mason Wasp
Symmorphus crassicornis

BL 8–15 mm. **ID** Black, abdomen S1–6 with yellow markings. **Hab** Woodland containing Aspen. **Host** Red Poplar Leaf Beetle (*p. 317*) larvae on Aspen. **Parasitoid** Aspen Cuckoo Wasp (*p. 486*). **SS Figwort Mason Wasp** [see table *above*].

Abdomen with whitish-yellow markings

Odynerus		**4 spp. (1†) | 3 ILL.**
ABDOMEN with yellow markings; ♀ TIBIAE + TARSI in part dark brown		**Spiny Mason Wasp**
ABDOMEN with whitish-yellow markings; ♀ TIBIAE + TARSI pale	ABDOMEN S2 band broadened at sides; ♀ METANOTUM black	**Black-headed Mason Wasp**
	ABDOMEN S2 band hardly broadened at sides; ♀ METANOTUM with a whitish-yellow line	**Fen Mason Wasp**

·piny Mason Wasp
)dynerus spinipes

Common

·L 10–12 mm. **ID** Black; abdomen 1–4 (S1–5 in some ♂s) with yellow ıarkings. **Hab** Various open areas; ·urrow entrance in form of a chimney, ·hich is approx. 30 mm long and often ·urved (see *p. 32*); provisioned with up ɔ 30 weevil larvae. **Dist** Old records nly from Ireland. **SS** Other *Odynerus* ɔecies [see table *above*].

·lack-headed Mason Wasp
)dynerus melanocephalus

Nationally Scarce

·L 8–10 mm. **ID** Black; abdomen ·–4 (S1–6 in ♂) with whitish-yellow ıarkings. ♂ clypeus yellowish. ·ibiae and tarsi yellowish-orange. **·ab** Various open areas. Ground-·esters: chimney at burrow entrance ıort, approx. 10 mm; provisioned ·ith weevil larvae. **SS** Other *Odynerus* ·ecies [see table *above*].

Fen Mason Wasp
Odynerus simillimus

EN | Rare (RDB1)

BL 11–12 mm. **ID** Black; abdomen S1–5 (S1–6 in ♂) with whitish-yellow markings. ♀ metanotum also with whitish-yellow mark. ♂ clypeus white. Tibiae and tarsi yellowish-orange. **Hab** Fenlands. Ground-nesters: chimney at burrow entrance short, approx. 10 mm; provisioned with weevil larvae. **SS** Other *Odynerus* species [see table *above*].

Microdynerus 1 sp.

Little Mason Wasp
Microdynerus exilis

Nationally Scarce

BL 5–8 mm. **ID** Black, abdomen S1–2 with whitish-yellow markings. **Hab** Grassland, parks, woodland, gravel pits and gardens, where seeks weevil larvae. **SS** None.

♀

Gymnomerus 1 sp.

Box-headed Mason Wasp
Gymnomerus laevipes

Local

BL 8–11 mm. **ID** Black, abdomen S1–4 (S1–5 in ♂) with yellow markings. **Hab** Grassland, wetlands, gardens, woodland clearings and others, where seeks weevil larvae. **SS** ♂s resemble *Ancistrocerus* species (*p. 516*), all of which have a much broader yellow band on abdomen S1.

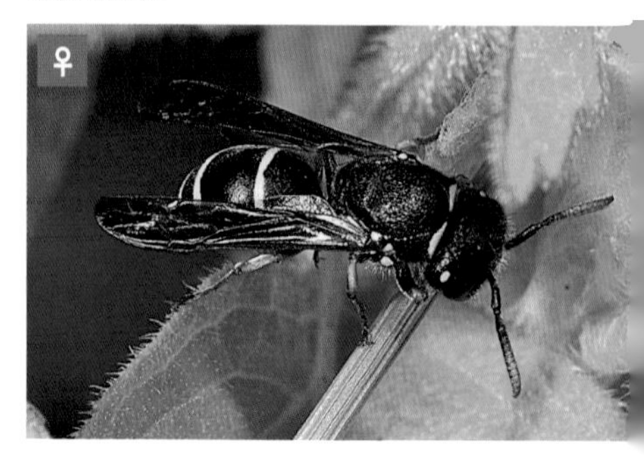

Pseudepipona 1 sp.

Purbeck Mason Wasp
Pseudepipona herrichii

EN Rare (RDB1)

BL 9–13 mm. **ID** Black, abdomen S1 with large reddish markings; otherwise S1–5 (S1–6 in ♂) and thorax with whitish markings. **Hab** Heathland; Bell Heather used for nectar and searching for tortricid moth (*p. 419*) larvae to provision underground nests. **SS** None.

Eumenes 2 spp. | 1 ILL

Heath Potter Wasp
Eumenes coarctatus

Nationally Scarce

J F M A M J J A S O

BL 9–15 mm. **ID** Black, with yellow markings on head, thorax and abdomen S1–4; S2 with two yellow spots. ♂ seldom seen (usually only early in the season). **Hab** Damp heathland, usually near patches of exposed clay soil. **Beh** Pots are constructed on vegetation, often Heather or Gorse; these are often single, but several may be built side by side. **SS Rare Potter Wasp** *Eumenes papillarius* [N/I] (**BL** 9–15 mm) is a very rare vagrant, with much more extensive yellow markin[gs] on the thorax and abdomen.

SUPERFAMILY **APOIDEA** — **GUIDE TO FAMILIES | 3 BI**

BODY shape as shown; generally very hairy – the true bees

BODY shape as shown; generally very hairy (but note some cuckoo bees can appear hairless *e.g. Nomada* spp. (*p. 543*)) – these are the true bees, with pollen-collecting hairs

Bees
Apidae
[**29 genera** (incl. 2†) | **259 species** (incl. 18†)]

p. 538

BODY shape as shown; usually smooth or slightly hairy – the 'spheciformes' wasps

ABDOMEN elongate (black-and-red);
PETIOLE long

Sand wasps
Sphecidae
[**2 genera | 4 species**]

p. 522

ABDOMEN less elongate ranging from all one colour to various colour combinations, often black-and-yellow);
PETIOLE if present, short (except a few species, including *Mimesa* (*p. 529*))

Digger wasps
Crabronidae
[**33 genera | 120 species**]

p. 524

FAMILY **Sphecidae** (Sand wasps) — 2 GEN. | 4 spp.

Elongate black wasps with red-banded abdomen, long legs and a long petiole (those in Great Britain and Ireland range in body length from 13–24 mm). Sand wasps hunt often large caterpillars of butterflies, moths and sawflies to provision nests in sandy ground; this is quite a feat as they sometimes have to travel a considerable distance. Features other than those included here are considered to be either unreliable or need to be checked under a microscope.

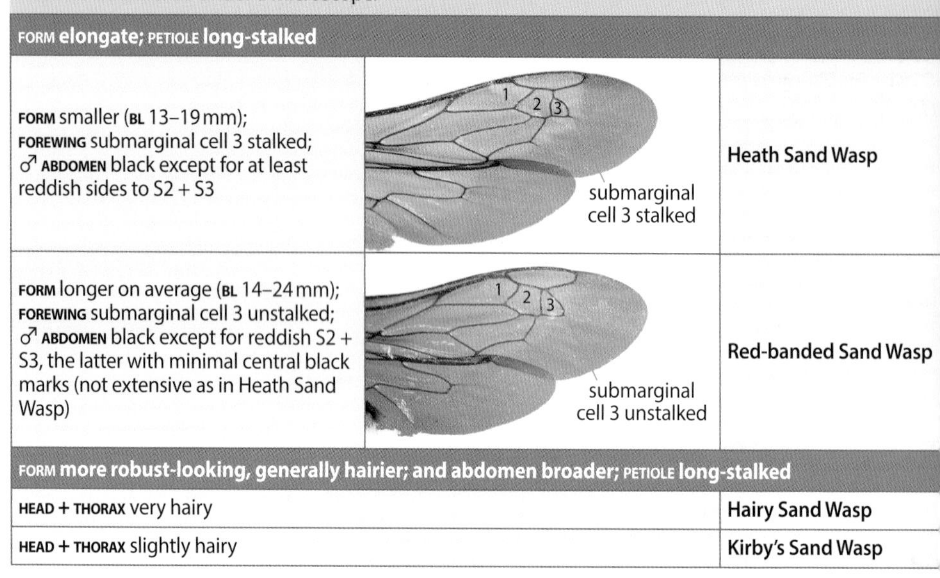

FORM elongate; PETIOLE long-stalked		
FORM smaller (**BL** 13–19 mm); **FOREWING** submarginal cell 3 stalked; ♂ **ABDOMEN** black except for at least reddish sides to S2 + S3	submarginal cell 3 stalked	**Heath Sand Wasp**
FORM longer on average (**BL** 14–24 mm); **FOREWING** submarginal cell 3 unstalked; ♂ **ABDOMEN** black except for reddish S2 + S3, the latter with minimal central black marks (not extensive as in Heath Sand Wasp)	submarginal cell 3 unstalked	**Red-banded Sand Wasp**
FORM more robust-looking, generally hairier; and abdomen broader; PETIOLE long-stalked		
HEAD + THORAX very hairy		**Hairy Sand Wasp**
HEAD + THORAX slightly hairy		**Kirby's Sand Wasp**

Kirby's Sand Wasp with moth caterpillar prey.

Hairy Sand Wasp at burrow.

Podalonia 2 spp.

Kirby's Sand Wasp
Podalonia affinis

Rare (RDB3)

BL 13–20 mm. **ID** Black with red band on upper part of abdomen; tip of abdomen black. Head and thorax only slightly hairy. **Hab** Coastal sand dunes; occasionally inland heathland. **SS** Hairy Sand Wasp [see table *opposite*].

Hairy Sand Wasp
Podalonia hirsuta

Nationally Scarce

BL 14–23 mm. **ID** Black, with red band on upper part of abdomen; tip of abdomen black. Head and thorax extensively hairy. **Hab** Coastal sandy sites. **SS** Kirby's Sand Wasp [see table *opposite*].

♀

♀

Ammophila 2 spp.

Red-banded Sand Wasp
Ammophila sabulosa

Common

BL 14–24 mm. **ID** Black, with red band on upper part of abdomen; tip of abdomen black but can appear slightly bluish. Forewing submarginal cell 3 unstalked [see *opposite*]. ♂ abdomen with minimal black on S2–3. **Hab** Dry heathland, sand dunes. **Beh** ♀s often rob the prey of other ♀s. **SS** Heath Sand Wasp.

Heath Sand Wasp
Ammophila pubescens

Local

BL 13–19 mm. **ID** Black, with red band on upper part of abdomen; tip of abdomen black but appears slightly bluish at certain angles. Forewing submarginal cell 3 stalked [see *opposite*]. ♂ abdomen black except for at least reddish sides to S2–3. **Hab** Dry heathland. **Parasitoid** Mottled Bee-fly (*p. 349*). **SS** Red-banded Sand Wasp.

♂

♀

FAMILY **Crabronidae** (Digger wasps) **33 GEN. | 120 spp. | 24 GEN., 41 SPP. ILL.**

Small to large (BL 7–13mm) solitary wasps, nesting in the ground, in hollow stems and dead wood. The genera shown in this section represent a good cross-section of those most likely to be seen.

NOTE: characters that are likely to help with identification in the field are given here. In addition, careful examination of wing venation under a microscope can help.

Characters	Genus	Page
BODY black		
HEAD large; **PETIOLE** short; **PROPODEUM** pitted	***Diodontus*** [4 species]	p. 525
HEAD short; **PETIOLE** very short	***Passaloecus*** [8 species]	p. 525
HEAD (large) + **THORAX** hairy; **WINGS** often dark; **PETIOLE** long	***Pemphedron*** [6 species]	p. 526
ANTENNAE pale; **MANDIBLES** yellow; **STIGMA** large	***Stigmus*** [2 species]	p. 526
FORM slender; **ANTENNAE** positioned low down; **STIGMA** large	***Spilomena*** [4 species]	p. 526
FORM elongate; **PETIOLE** long	***Mimumesa*** [4 species]	p. 526
FORM very elongate; **FOREWINGS** well short of abdomen tip; **CLYPEUS** silvery	***Trypoxylon*** [5 species]	p. 528
LEGS partly yellow; **PETIOLE** absent; **CLYPEUS** silvery	***Lindenius*** [2 species]	p. 531
BODY black-and-yellow		
LEGS fore tibiae ♂ with shield; black-and-yellow; **CLYPEUS** silvery	***Crabro*** [3 species]	p. 533
CLYPEUS silver or gold	***Ectemnius*** [10 species]	p. 534
ANTENNAE partly brown or yellowish	***Gorytes*** [2 species]	p. 537
STIGMA long and dark	***Lestiphorus*** [Dark-winged Digger Wasp]	p. 537
ANTENNAE black; **PROPODEUM** short	***Argogorytes*** [2 species]	p. 534
PETIOLE thick; **PROPODEUM** long	***Mellinus*** [2 species]	p. 53
HEAD large, **EYES** diverge at back; **PETIOLE** absent	***Cerceris*** [6 species (1†)]	p. 53

BODY black with extensive yellow; **HEAD + THORAX** large; **PETIOLE** absent; **FACE** with white or yellow areas	***Philanthus*** [Bee-wolf]	***p. 537***
BODY black-and-red		
HEAD + EYES large (in ♂ these meet above); **ANTENNAE** base thickened	***Astata*** [Shieldbug Digger Wasp]	***p. 529***
THORAX + METANOTUM with white spots	***Harpactus*** [White-spotted Digger Wasp]	***p. 529***
FORM very elongate; **PETIOLE** long	***Mimesa*** [4 species]	***p. 529***
FORM compact; **EYES** diverge at back	***Tachysphex*** [3 species]	***p. 529***
BODY genera with mixed colour combinations		
BODY black or black-and-red; **FORM** slender and delicate-looking; **PETIOLE** swollen	***Rhopalum*** [3 species]	***p. 528***
BODY mainly black but some black-and-yellow; **CLYPEUS** silvery	***Crossocerus*** [22 species]	***p. 527***
BODY black-and-red or black-and-yellow; **PROPODEUM** 2-spined	***Nysson*** [4 species]	***p. 536***
FORM broad; **METANOTUM** yellow- or white-spotted; **PROPODEUM** with spine	***Oxybelus*** [3 species]	***p. 530***

Diodontus 4 spp. | 1 ILL.

Minute Black Wasp

Diodontus minutus

Local

BL 3–5 mm. **ID** Small, black and large-headed; mandibles yellow; tibiae and tarsi partly brownish. **Hab** Sandy coastal and inland areas, including heathland. **Beh** Nests in bare sandy ground; predates aphids. **SS** Other *Diodontus* species and superficially like other small-sized black wasps; microscopic examination of clypeus (♀) and tarsi (♂) needed to confirm identification.

♀

Passaloecus 8 spp. | 1 ILL.

Horned Black Wasp

Passaloecus corniger

Common

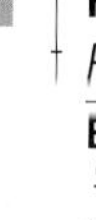

BL 5–7 mm. **ID** Elongate, black; ♀ clypeus with silver pubescence; fore and mid tibiae with narrow reddish bands at base and end, hind tibiae band only at base. **Hab** Heathland, open woodland and gardens with dead wood. **Beh** Nests in beetle holes in dead wood and fence posts, provisioned with aphids; ♀s can rob other ♀s and those of other *Passaloecus* species of aphid prey. **SS** Other *Passaloecus* species, which require microscopic examination of mouthparts and other minor characters to confirm identification.

♀

Guide to Crabronidae pp. 524–525

Pemphredon 6 spp. | 1 ILL.

Mournful Wasp *Pemphredon lugubris*

Common

BL 7–12mm. **ID** Elongate, black; wings with a brownish tinge. **Hab** Woodland, parks and gardens, with plentiful dead wood. **Beh** Nests in dead wood, provisioned with aphids. **SS** Other *Pemphredon* species, but these are all considerably shorter.

♀

Stigmus 2 spp. | 1 ILL.

Solsky's Wasp *Stigmus solskyi*

Local

BL approx. 4mm. **ID** Black; forewing with bold pterostigma. Antennae, mandibles and lower part of legs, brown. **Hab** Open woodland, parks and gardens. **Beh** Nests in dead wood in old beetle holes, provisioned with aphids. **SS Woodland Digger Wasp** *Stigmus pendulus* [N/I] (**BL** approx. 4mm) [RDBk], from which told by microscopic examination of thorax sculpturing; distinguished from other small black wasps by pterostigma.

♀

Spilomena 4 spp. | 1 ILL.

Thrips Wasp *Spilomena troglodytes*

Common

BL approx. 3mm. **ID** Body black, legs brown. **Hab** Open woodland and gardens. **Beh** Nests in dead wood in old beetle holes, provisioned with thrips nymphs. **SS** Other *Spilomena* species, from which told by microscopic examination of minor characters, including head size; other small black wasps, from which told by body form and colour.

♀

Mimumesa 4 spp. | 1 ILL.

Dahlbom's Digger Wasp *Mimumesa dahlbomi*

Common

BL approx. 7mm. **ID** Elongate, black; petiole long. **Hab** Heathland, open woodland and gardens, with dead wood. **Beh** Nests in old beetle holes in dead wood, provisioned with bugs. **SS** Other *Mimumesa* and *Mimesa* (*p. 529*) species, from which told by microscopic examination of several minor characters including ♂ genitalia and differences in antenna segments.

♀

Crossocerus — 22 spp. | 3 ILL.

FORM small to medium-sized (BL 3–11 mm), large-headed; **BODY** black, except in a few of the larger species that have yellow marks on the abdomen. Species identified by microscopic examination of various characters, including shape of mandibles, presence or absence of teeth on clypeus and subtle leg differences.

GUIDE TO SELECTED SPECIES

TIBIAE base with white band	**Ring-legged Crossocerus**
UTIBIAE of ♂ black-and-white, enlarged	**Black and white-legged Crossocerus**
FORM all-black; usually longer than related species; **HEAD + PRONOTUM** with distinct indentations	**Large-headed Crossocerus**

Ring-legged Crossocerus
Crossocerus annulipes

BL 5–8 mm. **ID** Black; tibiae with whitish basal band, fore tibiae yellow. **Hab** Woodland, hedgerows and gardens. **Beh** Nests in dead trunks or logs, provisioned with small bugs. **SS** Other *Crossocerus* species, from which told by microscopic examination (this species is identifiable by colour of tibiae) [see table *above*].

Black-and-white-legged Crossocerus
Crossocerus cetratus

BL 6–9 mm. **ID** Black; ♂ has black-and-white, enlarged, fore tibiae. **Hab** Woodland and hedgerows. **Beh** Nests in dead wood or hollow stems, provisioned with small flies. **SS** Other *Crossocerus* species, from which told by microscopic examination [see table *above*].

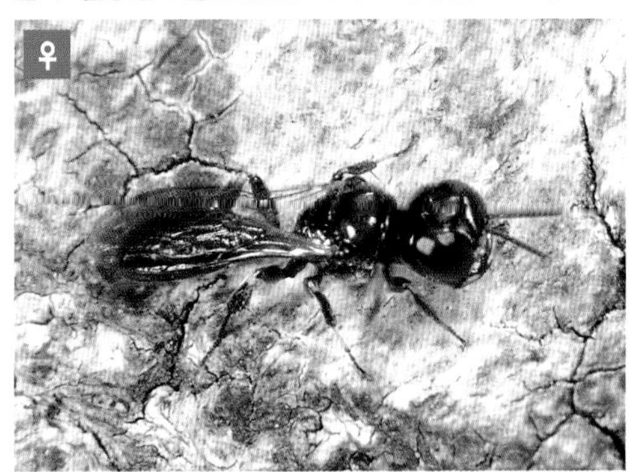

Large-headed Crossocerus
Crossocerus megacephalus

BL 6–10 mm. **ID** One of the larger *Crossocerus* species: large-headed, all-black; head and pronotum with distinct indentations. **Hab** Woodland, parks and gardens. **Beh** Nests in old beetle tunnels in dead wood, provisioned with small flies. **SS** Other *Crossocerus* species, from which told by microscopic examination [see table *above*].

Trypoxylon 5 spp. | 1 ILL.

Slender Wood-borer Wasp
Trypoxylon attenuatum

Common

BL 7–11 mm. **ID** Very elongate, black. **Hab** Woodland, hedgerows, heathland, parks and gardens. **Beh** Nests in hollow plant stems, provisioned with spiders. **Parasitoids** Blue Cuckoo Wasp (*p. 482*), Golden Cuckoo Wasp (*p. 483*). **SS** Other *Trypoxylon* species, from which told by microscopic examination of distance between eyes, length of abdomen S1 and in ♀ minor clypeus differences.

♀

Rhopalum 3 spp. | 2 ILL.

ABDOMEN red; HIND TIBIAE cream-and-black	Red-bodied Stem Wasp
ABDOMEN black with red tip; HIND TIBIAE whitish, black and reddish-brown	Red-tipped Stem Wasp

The little-known **Fen Stem Wasp** *Rhopalum gracile* [N/I] (BL 5–6 mm) has been recorded from a few wetland sites in East Anglia. The flight period is June to August. Nest sites are yet to be discovered in Britain, although they have been recorded in Common Reed and Lyme Grass stems in locations elsewhere.

Red-bodied Stem Wasp
Rhopalum clavipes

Common

BL 4–6mm. **ID** Elongate, head and thorax black; petiole long, abdomen red. Legs, mandibles and antennae with cream marks. **Hab** Open woodland, hedgerows and gardens. **Beh** Nests in hollow stems, provisioned with barklice. **SS** None.

Red-tipped Stem Wasp
Rhopalum coarctatum

Common

BL 5–7mm. **ID** Elongate, black; petiole long, tip of abdomen red in ♀. Hind tibiae rather swollen, whitish, black and reddish-brown. **Hab** Open woodland, hedgerows and gardens. **Beh** Nests in hollow stems, provisioned with flies and occasionally barklice. **SS** None but superficially like some small black wasps, from which told by body shape and colour.

♀

♀

Astata 1 sp.

Shieldbug Digger Wasp
Astata boops

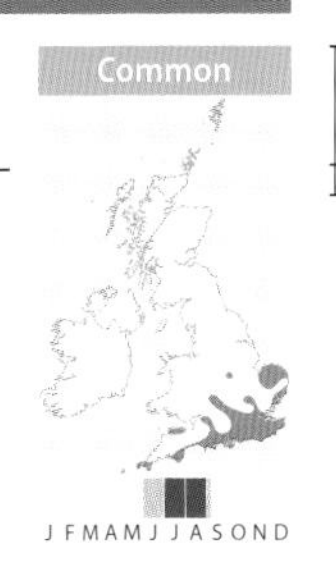

BL 9–13 mm. **ID** Black-and-red, with distinctive large head and eyes. **Hab** Sand dunes, inland heathland, grassland. **Beh** Provisions ground nest with shieldbug nymphs and adults. **Parasitoid** Dull Cuckoo Wasp (*p. 484*). **SS** Other robust-looking red-and-black wasps, from which told by body shape; also head and thorax with whitish hairs.

Harpactus 1 sp.

White-spotted Digger Wasp
Harpactus tumidus

BL 8–10 mm. **ID** Black-and-red, with white scutellum and abdomen with three white spots. **Hab** Heathland, coastal dunes and sandpits.
Beh Nests in sandy soil provisioned with froghoppers. **Cleptoparasite** Small Spurred Digger Wasp (*p. 536*). **SS Small Spurred Digger Wasp**, from which told by scutellum and abdomen colours.

Tachysphex 3 spp. | 1 ILL.

Common Tachysphex
Tachysphex pompiliformis

BL 5–7 mm. **ID** Robust-looking; black, except metanotum upper half red. **Hab** Heathland, grassland. **Beh** Nests in sandy soil, provisioned with grasshopper nymphs. **Parasitoids** Dull Cuckoo Wasp (*p. 484*), Glowing Cuckoo Wasp (*p. 484*). **SS** Superficially like other black-and-red wasps, including pompilids (*p. 503*), as indicated by the scientific name, from which told by body shape and details of forewing cells.

Mimesa 4 spp. | 1 ILL.

Common Mimesa
Mimesa equestris

BL 5–9 mm. **ID** Elongate, black-and-red; petiole long. **Hab** Open areas on mainly light or sandy soils, including heathland. **Beh** Nests in sandy soil, provisioned with bugs. **SS** Other *Mimesa* and *Mimumesa* (*p. 526*) species, from which told by abdomen colour and microscopic examination to determine length of antenna S3, and length/shape of petiole and clypeus (specialist keys needed).

 Guide to Crabronidae *pp.* 524–525

Oxybelus 3 spp.

THORAX + ABDOMEN with silvery hairs; **FEMORA + TIBIAE** ♀ light reddish-brown, ♂ mostly black with cream or yellowish tip		**Silver Spiny Digger Wasp**
THORAX + ABDOMEN without silvery hairs; ♀ **FEMORA** black	**MANDIBLES** black; ♂ **FEMORA** black, tip light reddish-brown, fore femora with white spot beneath	**Common Spiny Digger Wasp**
	MANDIBLES mainly yellow; ♂ **FEMORA** dark brown, fore femora with white beneath	**Pale-jawed Spiny Digger Wasp**

Silver Spiny Digger Wasp
Oxybelus argentatus

BL 5–9 mm. **ID** Black; body with silvery hairs, denser in ♀; abdomen with pale spots, often on all segments, in both sexes. ♂ femora cream or yellowish, except for black two-thirds of mid femora and most of hind femora; ♀ light reddish-brown. Mandibles yellowish with black tip (♀), or often black (♂). **Hab** Sand dunes, inland heathland. **Beh** Fly (Diptera) prey are carried to the nest impaled on the sting. **SS** Pale-jawed Spiny and Common Spiny Digger Wasps [see table *above*].

Common Spiny Digger Wasp
Oxybelus uniglumis

BL 5–8 mm. **ID** Black, abdomen with pale spots. In ♂, femora black with pale reddish-brown tip; fore femora with large white spot beneath. **Hab** Sandy soils; heathland. **Beh** Fly (Diptera) prey are carried to the nest impaled on the sting. **SS** Pale-jawed Spiny and Silver Spiny Digger Wasps [see table *above*].

Common

Pale-jawed Spiny Digger Wasp
Oxybelus mandibularis

BL 5–8 mm. **ID** Black; abdomen segments S2–5 with pale spots. In ♂, femora dark brown; fore femora base white. Mandibles yellow and pale reddish. **Hab** Sand dunes; some inland heathland sites. **Beh** Fly (Diptera) prey are carried to the nest impaled on the sting. **SS** Silver Spiny Digger Wasp, Common Spiny Digger Wasp [see table *above*].

Nationally Scarce

Lindenius 2 spp.

White-lipped Digger Wasp
Lindenius albilabris

Common

BL 5–8 mm. **ID** ♀ Mainly black, including mandibles, antenna base and pronotal collar. Tibiae base yellow. ♂ much yellower, including tibiae and small marks on pronotum (mandibles black). **Hab** Open sandy sites, including heathland and parks. **Beh** Nests in sandy soil, provisioned with small bugs or flies. **Parasitoid** Rugged Cuckoo Wasp (*p. 484*). **SS** Panzer's Digger Wasp, but has all-black mandibles.

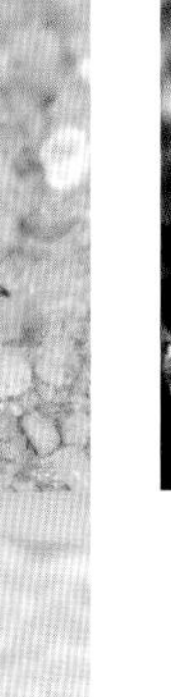

Panzer's Digger Wasp
Lindenius panzeri

Local

BL 5–8 mm. **ID** ♀ Black with yellow on mandibles, antenna base and pronotal collar. Tibiae mainly yellow, hind tibiae with black tip. ♂ similar. **Hab** Open sandy sites, including heathland paths. **Beh** Nests in sandy soil, provisioned with small flies. **SS White-lipped Digger Wasp**, but has yellow on mandibles [see *inset below*].

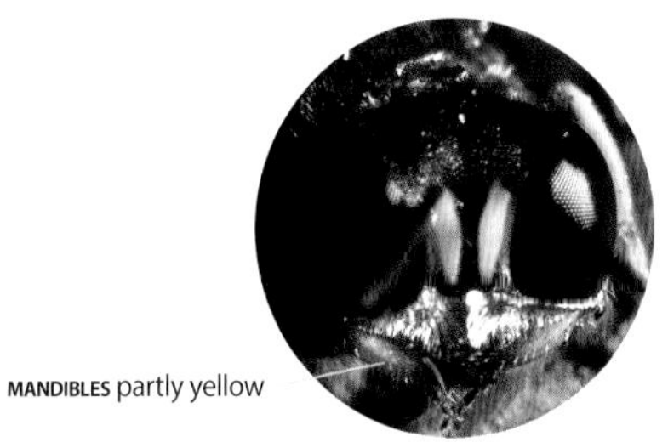

Mellinus 2 spp. | 1 ILL.

Field Digger Wasp
Mellinus arvensis

Common

BL 7–14 mm. **ID** Elongate; abdomen banded black-and-yellow; face yellow. **Hab** Sandy soils; often coastal. **Beh** Nests in sandy soil, provisioned with flies. **SS** Superficially like some other black-and-yellow-banded wasps, but told by body shape. **Orange-legged Digger Wasp** *Mellinus crabroneus* [N/I] (**BL** 6–11 mm) [RDB1] may be extinct in Britain; the abdomen segment marks are white, not yellow.

 Guide to Crabronidae *pp. 524–525*

Cerceris — 6 spp. | 3 ILL.

Species identified by microscopic examination of clypeus and antenna segments, also abdomen markings.

GUIDE TO SELECTED SPECIES

<table>
<tr><td colspan="2">ABDOMEN with 3+ yellow bands, S2 broad with yellow at base, some individuals have yellow marks on other segments; FACE yellow</td><td>Ornate Tailed Digger Wasp</td></tr>
<tr><td rowspan="2">ABDOMEN with 4–5 yellow bands of similar width, S2 base black, only hind part yellow; ♀ FACE with three yellow marks at least</td><td>CLYPEUS black, margin slightly upturned</td><td>Sand Tailed Digger Wasp</td></tr>
<tr><td>CLYPEUS yellow, ♀ with raised plate, ♂ face yellow, clypeus margin with three toothlike processes</td><td>Red-legged Digger Wasp</td></tr>
</table>

Ornate Tailed Digger Wasp
Cerceris rybyensis

BL 6–12 mm. **ID** Black-and-yellow; ♀ abdomen S4 with less yellow than S5 (both often black in ♂), S2 yellow band very broad. Mandibles yellow with black tip. **Hab** Sandy areas, including coastal sites and heathland; also other soils. **Beh** Ground-nester, provisioning the nest with bees. **SS** Other *Cerceris* species [see table *above*].

Sand Tailed Digger Wasp
Cerceris arenaria

BL 8–12 mm. **ID** Black-and-yellow; clypeus black. As typical of the genus, abdomen with 5 yellow bands, each rounded at side. **Hab** Sandy areas including coastal dunes and heathland. **Beh** Ground-nester, provisioning the nest with weevils. **Cleptoparasite** Noble Cuckoo Wasp (*p. 485*). **SS** Other *Cerceris* species [see table *above*].

Red-legged Digger Wasp
Cerceris ruficornis

BL 8–13 mm. **ID** Black and yellow; ♀ clypeus yellow, with raised plate. In ♂ face yellow, clypeus margin with three toothlike processes. Abdomen with 5 yellow bands. **Hab** Heathland. **Beh** Ground-nester, provisioning the nest with weevils. **Cleptoparasite** Niemelä's Cuckoo Wasp (*p. 485*). **SS** Other *Cerceris* species [see table *above*].

Crabro 3 spp.

Slender-bodied Digger Wasp
Crabro cribrarius

BL 10–15 mm. **ID** Large; black, with large yellow bands or spots on abdomen; mesoscutum with longitudinal lines clearly present. ♂ fore tibiae with distinctive large shield, brown with numerous tiny yellow spots. **Hab** Sand dunes, inland heathland. **Beh** Provisions nest with flies. **SS** Other *Crabro* species, from which told by features given; other digger wasps, from which told by fore tibiae shield in ♂ and in general by body form/size and presence of only 1 forewing submarginal cell (2 or 3 in other black-and-yellow Crabronidae).

MESOCUTUM longitudinal lines

FORE TIBIA SHIELD brown with yellow spots

Armed Crabro Digger Wasp
Crabro peltarius

BL 9–13 mm. **ID** Black, with large yellow bands or spots on abdomen, including S1; mesoscutum lacking longitudinal lines. ♂ fore tibiae with large brown and blue shield (can be plain blue). **Hab** Sand dunes, inland heathland. **Beh** Provisions nest with flies. **SS** Other *Crabro* species, from which told by features given; other digger wasps, from which told by same features as for Slender-bodied Digger Wasp *above*.

FORE TIBIA SHIELD blue-and-brown

Scarce Crabro Digger Wasp
Crabro scutellatus

BL 7–13 mm. **ID** Black including thorax; abdomen with cream bands or spots except on S1, which is black; ♂ fore tibiae a large, dark shield, with bold yellow stripes. **Hab** Damp heathland. **Beh** Provisions nest with long-legged flies (Dolichopodidae (*p. 337*)). **SS** Other *Crabro* species, from which told by features given; other digger wasps, from which told by same features as for Slender-bodied Digger Wasp *above*.

FORE TIBIA SHIELD brown with yellow stripes

Argogorytes 2 spp.

Farge's Digger Wasp
Argogorytes fargeii

Nationally Scarce

J F M A M J J A S O N D

BL 10–12 mm. **ID** Large; black, with yellow collar and thorax markings; abdomen with 4 yellow bands. **Hab** Sand dunes, heathland, often nesting in vertical earth banks. **Beh** Nests in sandy or clay soil, provisioned with froghoppers. **Cleptoparasites** Large Spurred Digger Wasp (*p.* 536), Three-banded Digger Wasp (*p.* 536). **SS** **Two-girded Digger Wasp**; **Dark-winged Digger Wasp** (*p.* 537); superficially like other black-and-yellow wasps, from which told by narrower yellow bands on abdomen.

Two-girdled Digger Wasp
Argogorytes mystaceus

Common

BL 10–13 mm. **ID** Large; black, with yellow collar, and thorax markings; abdomen with 2 yellow bands (plus pair of yellow spots on S1); S4 black or with a yellow central mark. **Hab** Open woodland; often seen on spurges. **Beh** Nests in dry banks provisioned with froghoppers. **Cleptoparasite** Large Spurred Digger Wasp (*p.* 536). **SS** **Farge's Digger Wasp**, **Dark-winged Digger Wasp** (*p.* 537), **Large Spurred Digger Wasp**; superficially like other black-and-yellow wasps, from which told by narrower yellow bands on abdomen.

Ectemnius 10 spp. | 4 ILL.

GUIDE TO SELECTED SPECIES

♂ ANTENNAE S3 with a deep notch beneath two teeth; CLYPEUS centre with golden pubescence		**Garden Ectemnius**
ABDOMEN usually with only three yellow bands or spots		**Common Ectemnius**
MESOSCUTUM (behind pronotum) with sideways stripes on front half	ABDOMEN S1 long-haired; ♂ S7 grooved in centre	**Large Ectemnius**
	ABDOMEN S1 short-haired; ♂ S7 not grooved in centre	**Woodland Ectemnius**

Garden Ectemnius

Common

Ectemnius cavifrons

BL 11–16 mm. **ID** One of the larger *Ectemnius* species; black with yellow markings, including on abdomen; ♂ antennae with several teeth, S3 with a deep notch beneath two teeth; centre of clypeus with golden pubescence. **Hab** Woodland, hedgerows, parks and gardens. **Beh** Nests in dead wood, provisioned with hoverflies. **SS** Other *Ectemnius* species, from which told by shape of ♂ antenna segments.

Common Ectemnius

Common

Ectemnius continuus

BL 8–15 mm. **ID** Black; abdomen with only three yellow bands or spots, including S1 and S3. **Hab** Woodland clearings, wetlands and gardens, where umbellifers and dead wood abounds. **Beh** Nests in burrows in dead wood, provisioned with flies. **SS** Other *Ectemnius* species, particularly smaller, scarcer **Bramble Ectemnius** *E. rubicola* [N/I] (**BL** 7–10 mm) that is restricted to S England (requires microscopic examination).

Large Ectemnius

Common

Ectemnius cephalotes

BL 9–17 mm. **ID** One of the larger *Ectemnius* species; black with yellow markings, including on abdomen (S1 with long hairs). Mesoscutum (behind pronotum) with sideways stripes on front half. ♂ abdomen S7 grooved in centre. **Hab** Woodland, parks and gardens. **Beh** Nests in soft dead wood, provisioned with flies. **SS** Other *Ectemnius* species, identification of which requires microscopic examination of thorax for characters referred to above, and use of specialist keys.

Woodland Ectemnius

Common

Ectemnius lituratus

BL 9–15 mm. **ID** Black with yellow markings, including on abdomen (S1 with short hairs). ♂ abdomen S7 grooved in centre. **Hab** Woodland with plenty of umbellifers. **Beh** Nests in dead wood, provisioned with flies. **SS** Other *Ectemnius* species, identification of which requires microscopic examination of thorax for characters referred to above, and use of specialist keys.

Guide to Crabronidae *pp.* 524–52

Nysson — 4 spp.

ABDOMEN S1 red, S2 + S3 with a white spot			Small Spurred Digger Wasp
ABDOMEN black with yellow bands	LEGS pale reddish		Three-banded Digger Wasp
	LEGS blackish	ABDOMEN yellow bands continuous	Large Spurred Digger Wasp
		ABDOMEN yellow bands interrupted	Three-spotted Digger Wasp

Large Spurred Digger Wasp
Nysson spinosus

Local

BL 9–12 mm. **ID** Black-and-yellow; abdomen 3-banded. **Hab** Coastal soft rock cliffs, heathland; coastal dunes and sandpits. **Beh** Favours vertical earth banks for nest sites. **Host** Two-girdled Digger Wasp (*p. 534*), Farge's Digger Wasp (*p. 534*), probably also Dark-winged Digger Wasp. **SS** Other black-and-yellow *Nysson* species [see table *above*].

♀

Three-spotted Digger Wasp
Nysson trimaculatus

Nationally Scarce

BL 6–8 mm. **ID** Rather stocky; black, with 3 pairs of interrupted yellow bands on abdomen. **Hab** Coastal soft rock cliffs, heathland, woodland, parks and gardens. **Beh** Favours vertical earth banks for nest sites. **Host** Broad-banded Digger Wasp, Four-banded Digger Wasp (see *opposite page*), probably also Dark-winged Digger Wasp. **SS** Other black-and-yellow *Nysson* species [see table *above*].

♂

Three-banded Digger Wasp VU
Nysson interruptus

Rare (RDB2)

BL 7–9 mm. **ID** Black-and-yellow; abdomen 3-banded. Legs reddish. **Hab** Coastal soft rock cliffs, heathland, coastal dunes and sandpits. **Beh** Favours vertical earth banks for nest sites. **Host** Farge's Digger Wasp (*p. 534*). **SS** Other black-and-yellow *Nysson* species [see table *above*].

♀

Small Spurred Digger Wasp
Nysson dimidiatus

Nationally Scarce

BL approx. 7 mm. **ID** Black-and-red; abdomen S1 red, S2 and S3 red with two white spots on hind part. **Hab** Heathland; coastal dunes and sandpits. **Beh** Usually seen running over sandy ground. **Host** White-spotted Digger Wasp (*p. 529*), possibly also *Lindenius* spp. (*p. 531*). **SS** White-spotted Digger Wasp, although that species has a bold white spot on scutellum.

♀

Gorytes 2 spp. | 1 ILL.

Broad-banded Digger Wasp *Gorytes laticinctus*

Rare (RDB3)

BL 9–13 mm. **ID** Large; black, with yellow markings on clypeus, collar and thorax; abdomen with 4 yellow bands. **Hab** Often sandy sites including dunes, heathland; also gardens. **Beh** Nests in soil, provisioned with froghoppers. **Cleptoparasite** Three-spotted Digger Wasp. **SS Four-banded Digger Wasp** *Gorytes quadrifasciatus* [N/I] (**BL** 9–11 mm) commoner, clypeus black with little yellow, **Dark-winged Digger Wasp**; superficially like other black-and-yellow wasps, but told by body shape, yellow clypeus and variation in width of yellow bands.

♀

Lestiphorus 1 sp.

Dark-winged Digger Wasp *Lestiphorus bicinctus*

Nationally Scarce

J F M A M J J A S O N D

BL 7–10 mm. **ID** Medium-sized; black, with yellow markings on collar and thorax; abdomen with 3 yellow bands (band on particularly 'waisted' S1 can be a pair of unmerged spots). Forewings dark-smudged. **Hab** Sand dunes, heathland, grassland, woodland, parks and gardens. **Beh** Nests in light soil, provisioned with bugs. **Cleptoparasite** Three-spotted Digger Wasp. **SS Two-girdled Digger Wasp** (*p. 534*); **Broad-banded Digger Wasp**; superficially like other black-and-yellow wasps, but told by forewing smudges and yellow markings, as described above.

♀

Philanthus 1 sp.

Bee-wolf *Philanthus triangulum*

Local (RDB2)

BL 8–17 mm. **ID** Black-and-yellow; clypeus, collar, thorax, most of legs and very broad abdominal bands yellow; ♀ back of head red. **Hab** Sand dunes and heathland. **Beh** Makes a deep burrow, provisioning the nest with Honey Bees (*p. 546*). **SS** None.

♂

♀

Carrying honey bee prey.

FAMILY **Apidae** (Bees) **29 GEN.** (incl. 2†) | **259 spp.** (incl. 18†) | **27 GEN., 88 SPP. ILL.**

Elongate insects often seen visiting flowers for pollen and nectar. Many are solitary, the female excavating and provisioning her own nest burrow. The Honey Bee and bumble bees are social. Cuckoo bees (cleptoparasites) and various insect parasitoids are often associated with specific bee species; there is more to learn about these species, as much of the information available is based on historic observations. **Recent bee books have encouraged more interest in bees, which need care in identification. However, many species, including those shown in this book, have characters that enable identification in the field, rather than requiring microscopic examination.**

The following illustrated guide describes characters that may help with recognition in the field; more detailed examination under a microscope, of wing venation and other features, is important for accurate identification. NOTE: females accumulate pollen while foraging for protein-rich pollen or sugar-rich nectar; the presence of pollen, mainly on their hindlegs, can be striking.

SUBFAMILY MELITTINAE

BL 10–15 mm; **LEGS** hairy with extended pollen brushes on hind tibiae

PANTALOON BEES

Hairy-legged Mining Bee

Dasypoda [Hairy-legged Mining Bee] p. 542

BL 9–11 mm; **BODY** black; **ABDOMEN** glossy with narrow white hair bands. ♂ **FACE** yellow

OIL-COLLECTING BEES

Yellow Loosestrife Bee

Macropis [Yellow Loosestrife Bee] p. 542

BL 10–15 mm; **BODY** dark, hairy (closely resembles *Andrena* (*p. 541*) and *Colletes* (*p. 541*)); **ANTENNAE** tip blunt

BLUNTHORN BEES

Clover Melitta
Melitta leporina (BL 9–12 mm)

Melitta [4 species] p. 542

SUBFAMILY HALICTINAE

FORM small to medium-sized (**BL** 6–16 mm); **ABDOMEN** hair bands on hind margins of segments

END-BANDED FURROW BEES

EN Downland Furrow Bee
Halictus eurygnathus (BL 9–12 mm) (RDB1)

Halictus [6 species (1†)] p. 557

FORM often small (**BL** 4–12 mm); **ABDOMEN** hair bands on base of segments

BASE-BANDED FURROW BEES

Smeathman's Furrow Bee
Lasioglossum smeathmanellum (BL 6–7 mm)

Lasioglossum [32 species] p. 557

FORM small to medium-sized (**BL** 4–12 mm); **ABDOMEN** most species red-and-black

BLOOD BEES

Geoffroy's Blood Bee
Sphecodes geoffrellus (BL 5–8 mm)

Sphecodes [16 species] p. 55

SUBFAMILY APINAE

FORM robust (**BL** 9–17 mm); ♂ **FACE** yellow-marked

FLOWER BEES

♀

Hairy-footed Flower Bee

Anthophora
[5 species]

p. 550

FORM small (**BL** 6–8 mm); metallic blue; ♂ **CLYPEUS** white

SMALL CARPENTER BEES

♂

Little Blue Carpenter Bee

Ceratina
[Little Blue Carpenter Bee]

p. 551

BL 6–11 mm; **BODY** largely hairless; **ABDOMEN** broad, markings colourful

VARIEGATED CUCKOO BEES

♀

Red-thighed Epeolus

Epeolus
[2 species]

p. 551

BL 12–20 mm; **BODY** dark, hairy; **THORAX** often darker than **ABDOMEN**, which is often banded black-and-orange; **LEGS** pollen baskets on hind tibiae

HONEY BEES

♀

Honey Bee

Apis
[Honey Bee]

p. 546

FORM large, robust (**BL** 8–25 mm); **BODY** hairy; **THORAX AND/OR ABDOMEN** often with hair bands, in some tip of abdomen coloured differently. **LEGS** pollen baskets on hind tibiae

BUMBLE BEES

♀

White-tailed Bumble Bee

Bombus
[25 species]

p. 546

FORM robust (**BL** 12–16 mm); **METASTOMA** black with white hair patches forming paired spots (where present)

MOURNING BEES

♀

Common Mourning Bee

Melecta
[2 species (1†)]

p. 551

FORM large and powerful (**BL** 20–28 mm); **BODY** black; **WINGS** metallic blue at certain angles

LARGE CARPENTER BEES

♀

Violet Carpenter Bee

Xylocopa
[Violet Carpenter Bee]

p. 552

FORM robust (**BL** 12–16 mm); ♂ **ANTENNAE** long (shorter in ♀); **CLYPEUS + LABRUM** yellow (black in ♀)

LONG-HORNED BEES

♂

Long-horned Bee

Eucera
[2 species (1†)]

p. 550

FORM small–medium (**BL** 4–15 mm); **BODY** almost hairless, rather wasp-like; **METASTOMA** often bold bands

NOMAD BEES

♀

EN Six-banded Nomad Bee
Nomada sexfasciata (BL 11–15 mm) (RDB1)

Nomada
[34 species]

p. 543

SUBFAMILY MEGACHILINAE

FORM robust (**BL** 9–17 mm); **BODY** black; **ABDOMEN** with paired yellow spots; **CLYPEUS + MANDIBLES** yellow-marked; **FACE** yellow spot above each eye

WOOL-CARDER BEES

Wool-carder Bee

Anthidium [Wool-carder Bee] p. 555

FORM robust (**BL** 7–16 mm); **ABDOMEN** black with whitish hair bands; ♀ with characteristic pointed tip (in ♂, the abdomen tip bears several spines)

SHARP-TAIL BEES

Large Sharp-tail Bee

Coelioxys [6 species] p. 555

BL 5–11 mm; **BODY** black; **ABDOMEN** glossy with narrow white hair bands. ♂ **FACE** yellow

SCISSOR BEES

Sleepy Carpenter Bee

Chelostoma [2 species] p. 55

FORM small, robust (**BL** 4–9 mm); **BODY** black, sharp ridge across top of abdomen S1; ♀ **ABDOMEN** underside with pollen brush

RESIN BEES

Large-headed Resin Bee

Heriades [2 species] p. 554

FORM small to medium-sized, robust (**BL** 5–11 mm); **BODY** mainly black; no pollen-collecting hairs

DARK BEES

Little Dark Bee

Stelis [4 species] p. 554

FORM medium-sized to large, robus (**BL** 6–18 mm); **BEHAVIOUR** ♀s cut sections of leaf and petal, and carry these back to the nest

LEAF-CUTTER AND MUD BEES

Patchwork Leaf-cutter Bee

Megachile [9 species (2†)] p. 55

A ♀ Gold-fringed Mason Bee using an empty snail shell as her nest site.

FORM medium-sized to large, robust (**BL** 6–14 mm); **BODY** some sport bright or metallic colours

MASON BEES

Orange-vented Mason Bee *Osmia leaiana* (BL 5–7 mm)

Osmia [13 species] p. 552

FORM robust (**BL** 6–10 mm) (rathe similar to *Osmia*); **THORAX** pollen brush beneath abdomen cream

LESSER MASON BEES

Welted Mason Bee

Hoplitis [3 species (1†)] p. 55

SUBFAMILY COLLETINAE

FORM medium-sized (**BL** 7–16 mm); **THORAX** brown and hairy; **ABDOMEN** most species strongly banded

PLASTERER BEES

♂

Hairy-saddled Colletes

Colletes
[9 species]

p. 558

FORM small (**BL** 4–8 mm); **BODY** black; **FACE** most with yellow or white spots (useful in identification)

YELLOW-FACE BEES

♀

Little Yellow-face Bee *Hylaeus pictipes* (BL 4–5 mm) (Nat. Scarce)

Hylaeus
[12 species (1†)]

p. 561

Ivy Bee *Colletes hederae* in burrow.

SUBFAMILY ANDRENINAE

FORM small to large (**BL** 5–17 mm); **BODY** variably hairy (colours of some species help with identification, but many have lookalikes); **ANTENNAE** tip pointed

MINING BEES

♀

Buffish Mining Bee

Andrena
[68 species (6†)]

p. 562

FORM small to medium-sized (**BL** 7–12 mm); **BODY** black, sparsely hairy; **HEAD** broad

SHAGGY BEES

♂

Small Shaggy Bee

Panurgus
[2 species]

p. 569

♀ Grey-haired Mining Bees in and next to nest burrows in soil.

MELITTINAE

Dasypoda (Pantaloon bees) 1 sp.

Hairy-legged Mining Bee
Dasypoda hirtipes

Nationally Scarce

BL 10–15 mm. **ID** Thorax brown; abdomen black, segments with narrow whitish hair bands. ♀ thorax with white-haired sides and expanded orange pollen brushes on hind tibiae, giving rise to the alternative name of 'pantaloon bees', although hairy-legged is an apt description, which applies to both sexes. ♂ brown when fresh, becoming grey with age.
Hab Heathland, sand dunes and pits. **SS** None.

Macropis (Oil-collecting bees) 1 sp.

Yellow Loosestrife Bee
Macropis europaea

Nationally Scarce

BL 9–11 mm. **ID** Body black, abdomen S3–4 with narrow white hair bands; ♂ with yellow face.
Hab Wetlands, including in gardens.
Beh Likely to be around Yellow Loosestrife, which is needed for pollen and floral oils; the latter used during nest construction.
SS None.

Melitta (Blunthorn bees) 4 spp. | 2 ILL. (see also *p. 538*)

Gold-tailed Melitta
Melitta haemorrhoidalis

Local

BL 10–13 mm. **ID** Body black, abdomen with narrow pale hair bands and orange-haired tip; hairs on thorax brown, some black. **Hab** Acid and chalk grassland, heathland, woodland, occasionally gardens, around flowering Harebell. **SS** Other *Melitta* species, from which told by abdomen S5 (♀) or S6 (♂) orange-haired, not blackish, plus unique choice of forage plants.

APINAE

Nomada (Nomad bees) — 34 spp. | 10 ILL. (see also *p. 539*)

GUIDE TO SELECTED SPECIES

Features		Species	Page
THORAX dark; ABDOMEN reddish with coloured spots			
THORAX black; ABDOMEN reddish with yellow lateral spots on S2–3		Fabricius' Nomad Bee	*p. 543*
THORAX red-marked	ABDOMEN extensively reddish, usually with yellow spots on S2–3	Little Nomad Bee	*p. 545*
	ABDOMEN S1–2 with red, otherwise white-spotted	Tormentil Nomad Bee	*p. 543*
THORAX dark, with or without reddish markings; ABDOMEN black with yellow bands and/or large spots			
ABDOMEN S1 with red band, other segments with extensive yellow, often as band, combined with narrower black or red bands		Flavous Nomad Bee	*p. 545*
ABDOMEN S1 with red band, other segments with extensive yellow, often as band (but not S2–3), combined with narrower black bands		Panzer's Nomad Bee	*p. 545*
ABDOMEN S1 with broad red and narrow black bands, other segments with extensive yellow and large black bands, expanded in centre		Painted Nomad Bee	*p. 544*
♀ ABDOMEN S2 with unbroken yellow band		Gooden's Nomad Bee	*p. 544*
♀ ABDOMEN S2 with yellow band that has narrow gap in centre		Marsham's Nomad Bee	*p. 544*
ABDOMEN S1 + S2 with yellow spots, not bands; SCUTELLUM with single yellow mark		Black-horned Nomad Bee	*p. 544*

ormentil Nomad Bee

lomada roberjeotiana

Rare (RDB3)

L 7–8 mm. **ID** Body black: thorax with yellow collar and spots, also red spots; abdomen S1–2 with red, otherwise white-spotted. **Hab** Heathlands and moorlands. **Host** Tormentil Mining Bee (*p. 566*). **S** None.

♂

Fabricius' Nomad Bee

Nomada fabriciana

Common

BL 6–11 mm. **ID** Body black: thorax black; abdomen reddish with yellow lateral spots on S2–3. ♀ antenna S4–7 + S12 reddish-orange. **Hab** Wide range of open areas and woodland. **Hosts** Gwynne's Mining Bee *Andrena bicolor* [N/I] and other *Andrena* species (*p. 562*). **SS** Other *Nomada* species [see table *above*].

♂ ♀

♀ ANTENNAE with reddish-orange segments

Marsham's Nomad Bee
Nomada marshamella

Common

BL 8–13 mm. **ID** Body black: thorax yellow-spotted (♀ scutellum with 2 yellow spots); abdomen segments black-and-yellow (S1 with a pair of yellow spots; S2 and in some individuals S3 has a narrow gap in centre of yellow band). Antennae and legs orange. ♂ lower part of face yellow. **Hab** Wide range, including gardens. **Hosts** Chocolate Mining Bee (*p. 564*) and other *Andrena* species (*p. 562*). **SS** Other *Nomada* species [see table on *p. 543*].

J F M A M J J A S O N D

Gooden's Nomad Bee
Nomada goodeniana

EN Common

BL 9–13 mm. **ID** Body black: thorax yellow-spotted and abdomen segments bold black-and-yellow (♀ abdomen S2–5 with complete yellow band). Antennae and legs orange. ♂ face yellow. **Hab** Grasslands to woodlands and urban sites. **Hosts** *Andrena* species (*p. 562*). **SS** Other *Nomada* species [see table on *p. 543*].

Painted Nomad Bee
Nomada fucata

Common

BL 8–12 mm. **ID** Body black: thorax with yellow spots; abdomen S1 with broad red and narrow black bands, other segments with extensive yellow and large black bands, expanded in centre. ♀ antennae, lower part of face and most of legs orange; in ♂ eyes green and lower part of face yellow. **Hab** Open woodlands, chalk grasslands, soft rock cliffs, roadside banks and gardens. **Dist** Has spread since the 1980s, along with the host. **Host** Yellow-legged Mining Bee (*p. 563*). **SS** Other *Nomada* species [see table on *p. 543*].

J F M A M J J A S O N D

Black-horned Nomad Bee
Nomada rufipes

Common

BL 7–10 mm. **ID** Robust-looking. Body black: thorax yellow-marked, notably scutellum with single spot; abdomen S2–3 with large yellow spots, progressing to yellow bands on S4–5. Another ♀ form (fairly common) has red, not black, abdomen. ♀ lower part of face and antenna basal segments red; ♂ antennae black; lower part of face yellow. **Hab** Heathlands. **Hosts** Heather Mining Bee *Andrena fuscipes* VU [N/I] and other *Andrena* species (*p. 562*). **SS** Other *Nomada* species [see table on *p. 543*].

Panzer's Nomad Bee NT
Nomada panzeri

BL 7–12 mm. **ID** Body black: thorax red-marked; abdomen S1 with red band, other segments with extensive yellow, often as band (but not S2–3), combined with narrower black bands; in ♀ the yellow can be much reduced, leaving an almost all-red body, but S2 with large yellow spots and S3 with tiny yellow lateral spots. ♀ antennae orange; ♂ partly black. ♂ face yellow. **Hab** Woodlands. **Hosts** Bilberry Mining Bee *Andrena lapponica* [N/I] and other *Andrena* species (*p.* 562). **SS** Other *Nomada* species, in particular **Flavous Nomad Bee** (♂s practically identical) [see table on *p.* 543].

Flavous Nomad Bee
Nomada flava

BL 7–12 mm. **ID** Body black: thorax red-marked; abdomen S1 with red band, other segments with extensive yellow, often as band, combined with narrower black or red bands; ♀ antennae orange; ♂ partly black. ♂ face yellow. **Hab** Wide range of open areas such as grassland, also woodland and gardens. **Hosts** Chocolate Mining Bee (*p.* 564) and other *Andrena* species (*p.* 562). **SS** Other *Nomada* species, in particular **Panzer's Nomad Bee** (♂s practically identical) [see table on *p.* 543].

Little Nomad Bee
Nomada flavoguttata

BL 5–8 mm. **ID** Body black: thorax red-marked and abdomen extensively reddish, usually with yellow spots on S2–3. **Hab** Wide ranging. **Hosts** Common Mini-miner (*p.* 567) and other 'mini-miners'; rather small, black *Andrena* species (*p.* 562). **SS** Other *Nomada* species [see table on 543].

Apis (Honey bees) — 1 sp.

Honey Bee *Apis mellifera*

BL 12–20 mm. **ID** Thorax usually brown, darker than abdomen in many individuals but same colour in some; may also be a pale orange or brown, or almost entirely black. Abdomen often banded black and orange but can be darker. ♂ plain brown; queen has long abdomen, so forewings are well short of tip. **Hab** Grassland, heathland, woodland, parks and gardens. **Beh** Realistically, queens are only likely to be encountered in bee hives by bee-keepers using protective clothing; neither queens nor males visit flowers. **Parasitoids** include a tiny fly, but the most important is *Varroa destructor*, a destructive mite that is a serious pest, as it can destroy colonies. **SS** None, but superficially similar to some *Andrena* species (*p.* 562), from which told by broadened hindlegs, the presence of a pollen basket (♀); eyes are also hairy (not bare).

Bombus (Bumble bees) — 25 spp. | 12 ILL.

GUIDE TO SELECTED SPECIES

Features		Species	Page
Commoner species, very likely to be seen in gardens, parks and grassland			
BODY + THORAX orange; ABDOMEN with some patches of black		Common Carder Bumble Bee	*p.* 54
ABDOMEN tip red or orange	THORAX + ABDOMEN mainly black	Large Red-tailed Bumble Bee	*p.* 54
	THORAX + ABDOMEN S2 with one yellow band	Early Bumble Bee	*p.* 54
ABDOMEN tip white	THORAX with two yellow bands	Small Garden Bumble Bee	*p.* 54
	THORAX with one yellow band	White-tailed Bumble Bee	*p.* 54
ABDOMEN tip with some buff (queen and male; less distinctive in worker, with only a slight hint of buff on margin); THORAX with one yellow band, slightly darker than in White-tailed Bumble Bee		Buff-tailed Bumble Bee	*p.* 54
Other selected less common species			
THORAX ginger or brown		Tree Bumble Bee	*p.* 54
THORAX straw-coloured with dark band; ABDOMEN similar with orange tip; BEHAVIOUR high pitched buzz in flight		Shrill Carder Bumble Bee	*p.* 54
THORAX black with yellow bands; BEHAVIOUR lacks high pitched buzz in flight	ABDOMEN bright orange	Mountain Bumble Bee	*p.* 54
	ABDOMEN yellow-and-black with white tip	Heath Bumble Bee	*p.* 54
	ABDOMEN black-and-yellow with white tip (being a cuckoo bumble bee, ♀ lacks pollen baskets on hind tibiae)	Vestal Cuckoo Bumble Bee	*p.* 54
THORAX black; ABDOMEN black with red or orange tip		Hill Cuckoo Bumble Bee	*p.* 54

Large Red-tailed Bumble Bee
Bombus lapidarius NT

Common

BL 10–21 mm. **ID** Black, except for red or orange tip to abdomen (in ♀); ♂ thorax has yellow bands. **Hab** Grassland, parks and gardens. **Social parasite** Hill Cuckoo Bumble Bee (*p. 549*). **SS Hill Cuckoo Bumble Bee** [see table *opposite*]; **Red-tailed Carder Bumble Bee** *Bombus ruderarius* VU [N/I] (**BL** 10–20 mm), which has a much shorter abdomen.

♀ queen

Mountain Bumble Bee
Bombus monticola

Local

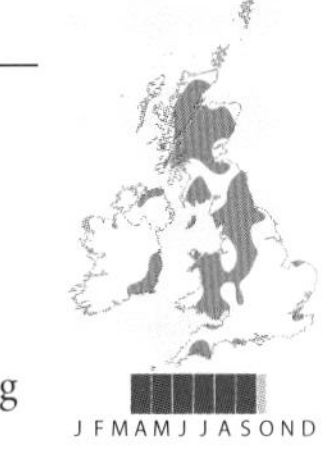

BL 11–30 mm. **ID** Thorax black with two yellow bands; abdomen bright orange from S2. **Hab** Heathland, moorlands and grassland. **Beh** Uses different pollen and nectar sources during the seasons; often Bilberry in spring, various flowers later, including Bell Heather. **Social parasite** Four-coloured Cuckoo Bumble Bee *Bombus sylvestris* [N/I]. **SS** Superficially various other *Bombus* species, particularly **Early Bumble Bee** and **Heath Bumble Bee** (*p. 548*) [see table *opposite*].

Early Bumble Bee
Bombus pratorum

Common

BL 8–17 mm. **ID** Thorax black with one yellow band; abdomen with yellow band on S2 and an orange tip. Yellow more extensive in ♂, including on face. **Hab** Grassland, woodland, parks and gardens. **Social parasites** Four-coloured Cuckoo Bumble Bee *Bombus sylvestris* [N/I], Large Velvet Ant (*p. 500*), parasitic wasps. **SS** None.

Shrill Carder Bumble Bee
Bombus sylvarum EN

Nationally Scarce

BL 9–16 mm. **ID** Thorax straw-coloured with dark band; abdomen similar with orange tip. **Hab** Grassland, brownfield sites, dunes. **Dist** Declining and now rather restricted. **Beh** High-pitched buzz in flight. **Parasitoid** Thick-headed flies (see *p. 338*). **SS White-tailed Bumble Bee** (*p. 548*) [see table *opposite*].

White-tailed Bumble Bee
Bombus lucorum aggregate

Common

J F M A M J J A S O N D

This species group, which can only be identified to species with certainty by molecular studies, comprises **White-tailed Bumble Bee** *Bombus lucorum*, **Northern White-tailed Bumble Bee** *Bombus magnus* DD and **Cryptic Bumble Bee** *Bombus cryptarum* DD.

BL 10–25 mm. **ID** Thorax with one yellow band; abdomen with broad yellow band at base and a white tip. The thorax of some ♂s may have a second yellow band. **Hab** Grassland, parks and gardens. **Social parasite** Gypsy Cuckoo Bumble Bee *Bombus bohemicus* NT [N/I]. **SS** **Buff-tailed Bumble Bee** [see table on *p. 546*].

♀ worker

Heath Bumble Bee
Bombus jonellus

Local

J F M A M J J A S O N D

BL 8–17 mm. **ID** Thorax with two yellow bands; abdomen with broad yellow band at base and a white tip; ♂ face with yellow. **Hab** Heathland, moorland, grassland, sand dunes and coastal cliffs. **Social parasites** Four-coloured Cuckoo Bumble Bee *Bombus sylvestris* [N/I], Large Velvet Ant (*p. 500*). **SS** Other *Bombus* species, particularly **Small Garden Bumble Bee** [see table on *p. 546*].

♂

Buff-tailed Bumble Bee
Bombus terrestris

Common

J F M A M J J A S O N D

BL 10–25 mm. **ID** Thorax with one yellow band; abdomen with yellow band on S2; S6 whitish-buff in queen and ♂. **Hab** Grassland, woodland, parks and gardens. **Social parasite** Vestal Cuckoo Bumble Bee. **SS** **White-tailed Bumble Bee** [see table on *p. 546*].

♀ queen

Small Garden Bumble Bee
Bombus hortorum

Common

J F M A M J J A S O N D

BL 9–25 mm. **ID** Thorax with two yellow bands; abdomen with yellow band at base and a white tip. Face long. **Hab** Open woodland, heathland, moorland, grassland, parks and gardens. **Social parasites** Barbut's Cuckoo Bumble Bee *Bombus barbutellus* EN [N/I], thick-headed flies (see *p. 338*). **SS** *Bombus* species, particularly **Heath Bumble Bee** [see table on *p. 546*].

♂

Vestal Cuckoo Bumble Bee
Bombus vestalis

Common

BL 15–22 mm. **ID** Thorax with one dark yellow band; abdomen black at base, yellow band across middle, white tip. **Hab** Grassland, woodland, parks, gardens. **Host** Buff-tailed Bumble Bee. **SS** Gypsy Cuckoo Bumble Bee *Bombus bohemicus* NT [N/I] (♀) (**BL** 13–21 mm), which has smaller yellow patches on side of abdomen.

Hill Cuckoo Bumble Bee
Bombus rupestris

EN Nationally Scarce

BL 15–24 mm. **ID** ♀ all-black with red or orange tail, wings brownish-black; ♂ similar but has yellow bands on thorax and less darkened wings. **Hab** Grassland and gardens. **Host** Large Red-tailed Bumble Bee (*p. 547*). **SS** Large Red-tailed Bumble Bee, the ♀ of which does not have dark wings [see also table on *p. 546*].

Common Carder Bumble Bee
Bombus pascuorum

Common

BL 8–18 mm. **ID** Pale to dark ginger-brown; abdomen with patches of black hair. **Hab** Grassland, woodland, parks and gardens. **Social parasite** Field Cuckoo Bee *Bombus campestris* [N/I]. **SS** Large Carder Bumble Bee *Bombus muscorum* NT [N/I] (**BL** 15–20 mm) and Brown-banded Carder Bee *Bombus humilis* [N/I] (**BL** 10–19 mm), both of which lack black hair on abdomen.

Tree Bumble Bee
Bombus hypnorum

Common

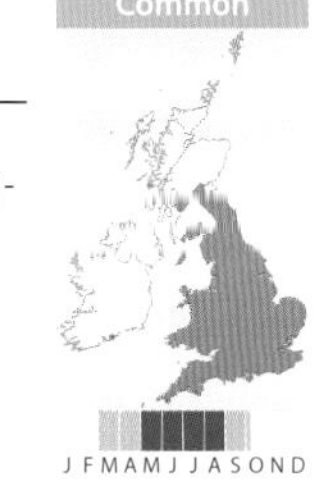

BL 10–19 mm. **ID** Thorax brownish, abdomen black; some partially brown-haired, always with white tip. **Hab** Open woodland, heathland, grassland and gardens. **Dist** First recorded in Britain in 2001 and has since spread rapidly. **Beh** Sometimes nests in bird boxes in gardens. **Parasitoid** Thick-headed flies (see *p. 338*). **SS** None.

Eucera (Long-horned bees) 2 spp. (1†)

Long-horned Bee
Eucera longicornis

Nationally Scarce

BL 12–16 mm. **ID** Large, stocky, brown; ♂ face yellow, antennae very long, black. **Hab** Grassland, coastal cliffs and open woodland, with plentiful legumes. **Dist** Has declined considerably since 1980. **Cleptoparasite** Six-banded Nomad Bee *Nomada sexfasciata* EN (*p.* 539), which is believed to be restricted to one site in Devon. **SS** None, although the ♀ superficially resembles ♂s of *Anthophora* species in the field, from which told by body colour, including lack of a yellow face.

Anthophora (Flower bees) 5 spp. | 1 ILL.

Hairy-footed Flower Bee
Anthophora plumipes

Common

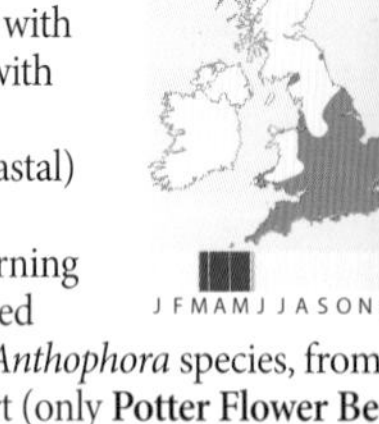

BL 14–17 mm. **ID** Large, stocky; ♀ black, rather bumble bee-like, hindlegs with orange pollen brush. ♂ brown, with yellow face; antennae black. **Hab** Grassland, cliffs (mainly coastal) with plentiful legumes. **Cleptoparasite** Common Mourning Bee. **Parasitoid** Flame-shouldered Blister Beetle (*p.* 288). **SS** Other *Anthophora* species, from which told by body colour in part (only **Potter Flower Bee** *Anthophora retusa* EN [N/I] (**BL** 11–16 mm) (RDB1) has a black ♀; microscopic examination of spurs at tip of hind tibiae needed to confirm identification – orange in Potter Flower Bee, black in Hairy-footed Flower Bee).

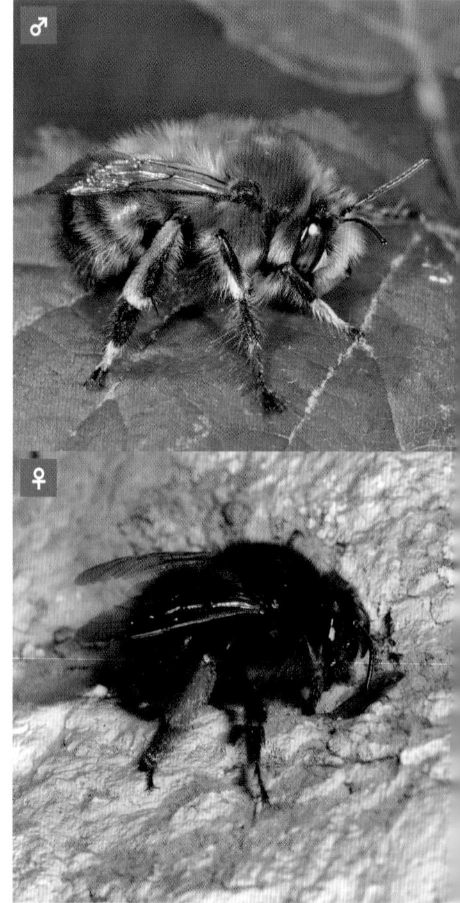

Melecta (Mourning bees) 2 spp. (1†)

Common Mourning Bee
Melecta albifrons

Common

J F M A M J J A S O N D

BL 13–16 mm. **ID** Large, stocky; black with front of thorax grey and abdomen with white lateral spots (striking in fresh adults). An all-back, melanic form also occurs and is not uncommon. **Hab** Open woodland, cliffs, sand pits and gardens. **Host** Hairy-footed Flower Bee and probably the Potter Flower Bee *Anthophora retusa* EN [N/I]. **SS** *Anthophora* species, from which ♀ is easily recognized by leg colour (all-black) and presence of white spots on metanotum.

♀

Epeolus (Variegated cuckoo bees) 2 spp. | 1 ILL.

Red-thighed Epeolus
Epeolus cruciger

BL 6–10 mm. **ID** Robust-looking, non-hairy, black-and-cream bee; legs reddish, ♂ femora black. **Hab** Heathland, moorlands, sand dunes. **Hosts** Girdled Colletes (*p. 559*) and Margined Colletes (*p. 560*), possibly other *Colletes* species (*p. 558*). **SS** Black-thighed Epeolus *Epeolus variegatus* [N/I] (**BL** 6–11 mm), the ♀ of which has dark femora.

Ceratina (Small carpenter bees) 1 sp.

Little Blue Carpenter Bee
Ceratina cyanea

BL 6–8 mm. **ID** Body dark metallic bluish or blue-green, ♂ lower part of clypeus and labrum white. **Hab** Chalk grassland, sand quarries, woodland rides. **Parasitoid** Darwin wasps. **SS** None.

MEGACHILINAE

Xylocopa (Large carpenter bees) 1 sp.

Violet Carpenter Bee
Xylocopa violacea

BL 20–28 mm. **ID** Very large; body and wings dark metallic blue in certain lights, otherwise can appear black. ♂ with antenna S11–12 red. **Hab** Anywhere, including gardens. **SS** None.

Osmia (Mason bees) 13 spp. | 5 ILL. (see also *p. 540*)

GUIDE TO SELECTED SPECIES

THORAX black; ABDOMEN tail red above and below (♀), golden-red (♂)	Two-coloured Mason Bee
THORAX + ABDOMEN golden-haired	Gold-fringed Mason Bee
FORM bulky body, only species with pair of facial 'prongs' below antennae (♀); ANTENNAE much longer than in other spp. (♂)	Red Mason Bee
THORAX with blue or greenish tint; ABDOMEN as thorax (♀), or with a more metallic green sheen (♂)	Blue Mason Bee

Two-coloured Mason Bee *Osmia bicolor*

BL 8–12 mm. **ID** ♀ Black, with bright orange hairs on abdomen; also orange tibiae and tarsi. ♂ brown with paler hairs, long on sides of thorax; abdomen golden-red. **Hab** Calcareous grassland or open calcareous woodland. **Beh** ♀s use empty snail shells as nest sites; these are hidden from view and/or protected against the elements, by covering them with numerous grass stems or parts of dead leaves. **SS** Other *Osmia* species have different body and hair colour

Red Mason Bee *Osmia bicornis*

BL 7–14 mm. **ID** Head black; thorax dark brown; abdomen with orange hair bands. **Hab** Urban areas as well as the wider countryside; often observed on buildings and in gardens. **Beh** Most often nests in crumbling mortar. **Cleptoparasites** Five-spotted Club-horned Wasp (*p. 501*), possibly dark bees (*p. 554*), cuckoo wasps (*p. 481*). **Parasitoids** Parasitic wasps. **SS** Other *Osmia* species, from which told by body colour and antennae length (♂).

Gold-fringed Mason Bee *Osmia aurulenta* NT

BL 7–12 mm. **ID** ♀ body with orange hairs; ♂ brown with black abdomen and orange hair bands. **Hab** Sand dunes, calcareous grassland. **Beh** ♀s use empty snail shells as nest sites. **Cleptoparasite** Five-spotted Club-horned Wasp (*p. 501*). **Parasitoids** Dune Villa (*p. 349*), other flies and parasitic wasps. **SS** Other *Osmia* species, from which told by body colour.

Blue Mason Bee *Osmia caerulescens*

BL 7–11 mm. **ID** Head large. Body with slight blue (♀) or greenish (♂) sheen. ♂ is orange-haired; ♀ has rather fewer hairs with narrow whitish hair bands on abdomen. **Hab** Urban areas as well as the wider countryside, such as woodland clearings; often observed in gardens. **Beh** Nests in decaying wood or crumbling mortar. **Cleptoparasites** Five-spotted Club-horned Wasp (*p. 501*). **SS** Other *Osmia* species, from which told by body colour.

Hoplitis (Lesser mason bees) 3 spp. | 1 ILL.

Welted Mason Bee
Hoplitis claviventris

BL 6–10 mm. **ID** Head large. Body black: thorax with brown hairs; abdomen S1–4 with whitish hair bands. **Hab** Chalk grassland, open woodland and sand dunes. **Cleptoparasite** Spotted Dark Bee *Stelis ornatula* [N/I]. **SS** *Osmia* species (*p.* 552), from which told by pollen hairs all-white in ♀ and flange or ridge on S2 in ♂. **Kirby's Mason Bee** *Hoplitis leucomelana* [N/I] (**BL** 6–9 mm) is doubtfully British, and the larger, more elongate **Viper's Bugloss Mason Bee** *H. adunca* [N/I] (**BL** 8–13 mm), first recorded in London in 2016.

Stelis (Dark bees) 4 spp. | 1 ILL.

Little Dark Bee
Stelis breviuscula

BL 5–7 mm. **ID** Robust-looking. Body black, abdomen with narrow, whitish hair bands. **Hab** Areas with plentiful ragwort flowers, including gardens. **Host** Large-headed Resin Bee. **SS** Other *Stelis* species, particularly **Spotted Dark Bee** *Stelis ornatula* [N/I] (**BL** 6–8 mm), which is larger and only has white spots on abdomen; the other species require microscopic study. Superficially like **Large-headed Resin Bee**, but body darker.

Heriades (Resin bees) 2 spp. | 1 ILL.

Large-headed Resin Bee
Heriades truncorum

BL 5–9 mm. **ID** Robust-looking. Body black, abdomen with narrow, whitish hair bands. ♀ with orange pollen brush beneath abdomen. Face and sides of thorax with brown hairs. **Hab** Grassland, heathland, gardens *etc.* with plentiful ragwort. **Beh** Stem-nesters; cell partitions made from resin, including from pines. **Cleptoparasite** Little Dark Bee. **SS** **Small-headed Resin Bee** *Heriades rubicola* [N/I] (**BL** 4–7 mm) (found few times in Britain, since being discovered in Dorset in 2006); superficially like **Little Dark Bee** or a large-headed *Osmia* species (*p.* 552).

Anthidium (Wool-carder bees) — 1 sp.

Wool-carder Bee
Anthidium manicatum

Common

J F M A M J J A S O N D

BL 9–17 mm. **ID** Robust-looking. Body black, with yellow markings on head, abdomen and legs. **Hab** Areas with plentiful flowers, such as grassland, woodland edge and wetlands, but often also in parks and gardens. **Beh** 'Cards' wool for nest construction from plants such as Lamb's-ear, which is often found in gardens; ♂s are aggressive, defending an area from other ♂s and pollinating insects. **Cleptoparasite** Banded Dark Bee *Stelis punctulatissima* [N/I]. **SS** None.

Coelioxys (Sharp-tail bees) — 6 spp. | 1 ILL.

Large Sharp-tail Bee
Coelioxys conoidea

BL 11–15 mm. **ID** Body black, abdomen with broad, white, partial hair bands. ♀ abdomen with elongated, pointed tip (♂ with spines at tip). **Hab** Mainly coastal sand dunes, landslips, heathland. **Host** Coast Leaf-cutter Bee *Megachile maritima* NT [N/I]. **SS** Other *Coelioxys* species, from which told by microscopic examination of various characters, including shape and structure of abdomen tip.

Megachile (Leaf-cutter and mud bees) — 9 spp. | 1 ILL.

Patchwork Leaf-cutter Bee *Megachile centuncularis* NT

BL 9–12 mm. **ID** Brown, ♀ with orange pollen brush beneath abdomen, reaching to tip (lacking in similar-looking ♂). **Hab** Wide range and likely in many gardens. **Beh** Nests in cavities and cuts leaves from bushes and shrubs for nest-building. **Cleptoparasites** Shiny-vented Sharp-tail Bee *Coelioxys inermis* [N/I], possibly other *Coelioxys* species. **SS** Other *Megachile* species, from which told by diffences in pollen brush in ♀ and by various other minor characters in ♂ that require microscopic examination.

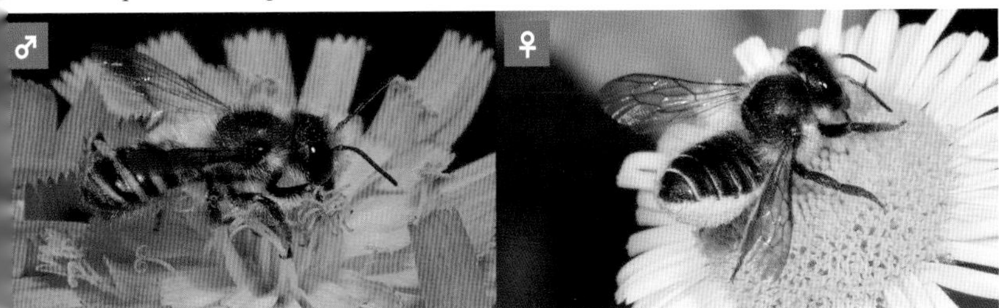

Chelostoma (Scissor bees) 2 spp.

Harebell Carpenter Bee
Chelostoma campanularum
[also known as Small Scissor Bee]

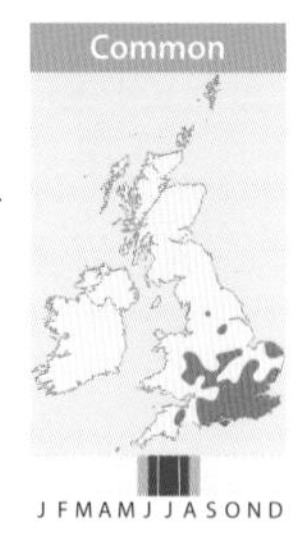

BL 5–7 mm. **ID** Elongate, black. **Hab** Acid and chalk grassland with plentiful bellflowers, also parks and gardens. **Beh** Nesting sites include dead wood and fence posts. **Dist** Also a single record from Scotland in 1977. **SS** Superficially like *Lasioglossum* species, being so small, but differs in wings having 2 submarginal cells, not 3.

Sleepy Carpenter Bee
Chelostoma florisomne
[also known as Large Scissor Bee]

BL 7–11 mm. **ID** Elongate, black; abdomen with narrow whitish hair bands. **Hab** Woodland clearings, hedgerows and grassland, gardens with plentiful buttercups. **Beh** Nesting sites include dead wood and thatching, so thatched roofs are sometimes productive; obtains pollen from buttercups and, at any hint of unfavourable weather, appears to 'fall asleep' i the buttercup flower. **Cleptoparasite** Club-horned Was (*p. 501*). **SS** None.

Sphecodes (Blood bees) 16 spp. | 2 ILL. (see also *p. 538*)

Sandpit Blood Bee DD
Sphecodes pellucidus

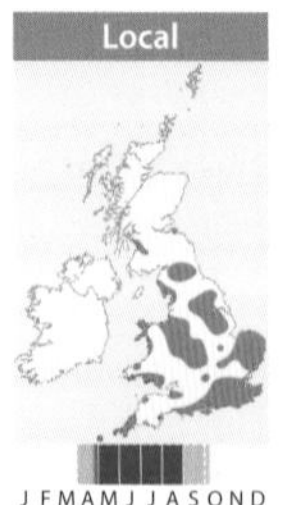

BL 6–10 mm. **ID** Body black; abdomen red except for black final segments. **Hab** Sandy areas. **Host** Sandpit Mining Bee (*p. 568*). **SS** Other *Sphecodes* species, from which told by microscopic examination of various subtle characters on head and fine details of thorax, abdomen segments and genitalia.

Rockford Common in the New Forest, Hampshire. open heathland with sandy areas where blood bees may be found.

Lasioglossum (Base-banded furrow bees) **32 spp. | 3 ILL.** (see also *p. 538*)

Common Furrow Bee
Lasioglossum calceatum

BL 8–11 mm. **ID** Thorax dark brown; ıbdomen black with narrow, whitish ıair bands (♂, abdomen often with :xtensive red). **Hab** Grassland on ·arious soils, gardens. **Cleptoparasites** 3ox-headed Blood Bee *Sphecodes nonilicornis* [N/I]; possibly other *phecodes* species. **SS** Some other *.asioglossum* species, from which told by red coloration ı ♂ (♂ is virtually indistinguishable from **Bloomed 'urrow Bee** *Lasioglossum albipes* [N/I] (**BL** 7–9 mm) and :an be difficult to identify even under a microscope).

♂

Green Furrow Bee
Lasioglossum morio

BL 5–7 mm. **ID** Thorax dark, with metallic green sheen; abdomen darker, ♀ with whitish hair bands. **Hab** Open habitats including parks and gardens. **Cleptoparasites** Dark Blood Bee *Sphecodes niger* [N/I]; probably other *Sphecodes* species. **SS** Other *Lasioglossum* species, from which told (and from other metallic species) by microscopic examination of various subtle characters.

♂

Halictus (End-banded furrow bees) **6 spp. (1†) | 2 ILL.**

Orange-legged Furrow Bee
Halictus rubicundus

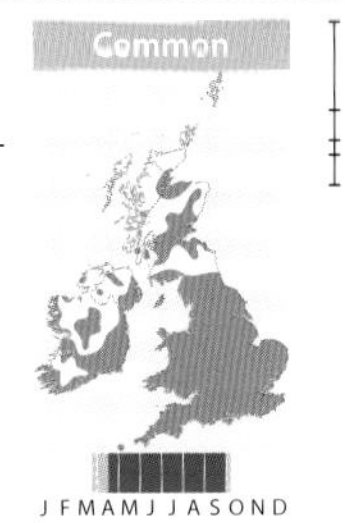

BL 9–11 mm. **ID** Thorax reddish-brown ♀) or brown (♂); abdomen black ·ith narrow, bright white hair bands; ·gs orange. **Hab** Grassland, woodland ·dge, heathland, wherever flowers are ·lentiful. **Cleptoparasites** Dark-·inged Blood Bee *Sphecodes gibbus* CR ·/I], Box-headed Blood Bee *Sphecodes ·onilicornis* [N/I]. **SS** **Orange-footed Furrow Bee** *·asioglossum xanthopus* [N/I] (**BL** 9–12 mm), which has ·uch broader white hair bands on abdomen.

♀

Bronze Furrow Bee NT
Halictus tumulorum

BL 6–8 mm. **ID** Body with bronze-green reflection; abdomen in ♀ with narrow pale hair bands. **Hab** Grassland, gardens. **Cleptoparasites** Bare-saddled Blood Bee *Sphecodes ephippius* [N/I], Geoffroy's Blood Bee (*p. 538*). **SS** Other *Halictus* species, from which told by small size, colour, and microscopic examination of thorax, abdomen hair band margins and ♂ genitalia.

♀

COLLETINAE

Colletes (Plasterer bees) — 9 spp.

Difficult to identify in the field but important features for this genus are size, time of appearance and habitat, as well as colour of thorax and abdomen segments, combined with width of hair bands at margins of these segments. Some characters are unclear when the bees become worn or sun-faded, as the body turns much paler. The bee books provide detailed keys, offering a wide range of characters to check under a microscope, features that are outside the scope of this field guide.

<table>
<tr><td colspan="3">FORM large (BL 12–16 mm); ABDOMEN with indistinct hair bands; FLIGHT spring</td><td>Early Colletes</td><td>p. 558</td></tr>
<tr><td colspan="5">FORM small–medium (BL 7–12 mm); ABDOMEN with distinctive hair bands; FLIGHT summer and/or early autumn</td></tr>
<tr><td colspan="3">ABDOMEN very bright, with broad pale yellow bands; hairs on S1 sides dense; FLIGHT the last Colletes to appear, often not until well into September.</td><td>Ivy Bee</td><td>p. 559</td></tr>
<tr><td colspan="5">ABDOMEN duller, hair bands whitish or with a hint of buff; hairs on S1 sides not extensive</td></tr>
<tr><td rowspan="2">ABDOMEN with narrow hair bands</td><td colspan="2">THORAX hairs brown, central hairs black; DIST very limited in NW England, SW Scotland, Western Isles and SE Ireland</td><td>Northern Colletes</td><td>p. 559</td></tr>
<tr><td colspan="2">THORAX hairs brown; ABDOMEN S1 margin with orange visible; DIST widespread heathland and moorland</td><td>Girdled Colletes</td><td>p. 559</td></tr>
<tr><td rowspan="5">ABDOMEN with fairly broad hair bands</td><td colspan="2">THORAX with sparse, short hairs</td><td>Margined Colletes</td><td>p. 560</td></tr>
<tr><td rowspan="3">THORAX with plentiful orange or brown hairs</td><td>THORAX brown</td><td>Davies' Colletes</td><td>p. 560</td></tr>
<tr><td>♀ THORAX bright orange-brown.</td><td>Hairy-saddled Colletes</td><td>p. 560</td></tr>
<tr><td>♀ ABDOMEN S1 unique in having mostly bare surface; ♀ THORAX reddish-brown</td><td>Bare-saddled Colletes</td><td>p. 560</td></tr>
<tr><td colspan="2">DIST nearly always on saltmarsh margins in S & E England; rarely inland</td><td>Sea Aster Bee</td><td>p. 56</td></tr>
</table>

COLLETES – longest species; hair bands indistinct

Early Colletes

Colletes cunicularius

BL 12–16 mm. **ID** Thorax brown; abdomen black-haired with narrow whitish-brown hair bands.
Hab Open woodland sites.
SS Other *Colletes* species [see table *above*].

HAIR BANDS narrow, not very distinctive

COLLETES – metastoma hair bands broad, bright pale yellow

Ivy Bee *Colletes hederae*

BL 10–16 mm. **ID** Very bright. Thorax brown; broad, pale yellow band on each abdomen segment, sides of S1 densely hairy. **Hab** Coastal cliffs, farmlands, heathland, parks and gardens. **Dist** First recorded in Britain (Dorset) in 2001 and has spread north every year since; further expansion likely. **Beh** Huge populations can develop; the bees are often seen around nesting areas, where 'mating balls' are often seen as the fresh ♀s emerge; also on Ivy flowers, searching for pollen and nectar. **SS** Other *Colletes* species [see table *opposite*].

COLLETES – metastoma hair bands narrow

Girdled Colletes *Colletes succinctus*

BL 7–13 mm. **ID** Thorax hairs brown, some black; ♀ abdomen S1 margin with orange visible. **Hab** Heathland and moorland, where often attracted to Ling and Heather flowers. **Kleptoparasite** Red-thighed Epeolus (p. 551). **Parasitoid** Heath Bee-fly (p. 350). **SS** Other *Colletes* species [see table *opposite*].

Northern Colletes *Colletes floralis*

BL 8–12 mm. **ID** Thorax hairs brown, blacker towards centre; abdomen segments with narrow whitish or greyish hair bands , S1 sparsely punctured and hairy. **Hab** Coastal dunes and machair grassland. **SS** Other *Colletes* species [see table *opposite*].

COLLETES – metastoma hair bands broad

Hairy-saddled Colletes DD
Colletes fodiens

BL 8–11 mm. **ID** ♀ thorax bright orange-brown; abdomen with broad, buff hair bands. **Hab** Coastal dunes, heathland, soft rock cliffs. **Cleptoparasites** Black-thighed Epeolus *Epeolus variegatus* [N/I], possibly Red-thighed Epeolus (*p. 551*). **SS** Other *Colletes* species [see table on *p. 558*].

Bare-saddled Colletes NT
Colletes similis

BL 8–12 mm. **ID** ♀ thorax reddish-brown; abdomen with broad whitish hair bands (S1 mostly bare *i.e.* hairless). **Hab** Coastal cliffs, heathland, open woodland, chalk grassland. **Cleptoparasites** Black-thighed Epeolus *Epeolus variegatus* [N/I] (see *p. 551*). **SS** Other *Colletes* species [see table on *p. 558*].

Davies' Colletes DD
Colletes daviesanus

BL 8–11 mm. **ID** Thorax brown; abdomen with broad buff-white hair bands, S1 sparsely punctured and hairy (similar only to Northern Colletes (*p. 559*)). **Hab** Coastal and inland sites, including parks and gardens on clay soils. **SS** Other *Colletes* species [see table on *p. 558*].

Margined Colletes
Colletes marginatus

BL 7–10 mm. **ID** Smallest *Colletes* species. Thorax brown with sparse, short hairs (as on face); abdomen with broad, whitish hair bands. **Hab** Sand dunes, also Breckland heath inland. **Cleptoparasites** *Epeolus* species (*p. 551*). **SS** Other *Colletes* species [see table on *p. 558*].

Sea Aster Bee
Colletes halophilus

Nationally Scarce

BL 8–12 mm. **ID** Thorax brown; abdomen with broad, whitish hair bands. **Hab** Saltmarsh margins. **Cleptoparasite** Black-thighed Epeolus *Epeolus variegatus* [N/I] (see under Red-thighed Epeolus, *p. 551*). **SS** Other *Colletes* species [see table on *p. 558*].

Hylaeus (Yellow-face bees) 12 spp. (1†) | 3 ILL. (see also *p. 541*)

Common Yellow-face Bee
Hylaeus communis

Common

BL 4–7 mm. **ID** Black; face, thorax, antennae and legs with yellow markings. ♀ facial spots yellow, triangular and hugging eye margins; ♂ face largely yellow, the pattern different from related species, mandibles black [see *inset below*]. **Hab** Open woodland, parks and gardens. **Parasitoid** Javelin Wasp (*p. 478*). **SS** Other *Hylaeus* species, particularly **Reed Yellow-face Bee** *Hylaeus pectoralis* [N/I] (**BL** 6–8 mm), which favours wetlands.

White-jawed Yellow-face Bee
Hylaeus confusus

Common

BL 6–7 mm. **ID** Black; face, thorax, antennae and legs with yellow markings. ♀ facial spots yellow, triangular; ♂ face largely yellow, mandibles whitish-yellow [see *inset below*]. **Hab** Open woodland, heathland and bushy areas. **SS** Other *Hylaeus* species, from which told by microscopic study of various characters; in ♂ the pale mandibles are a key character (black in all except 3 species).

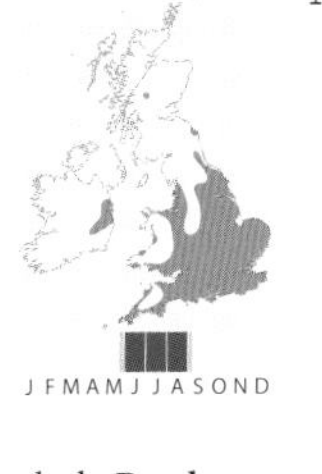

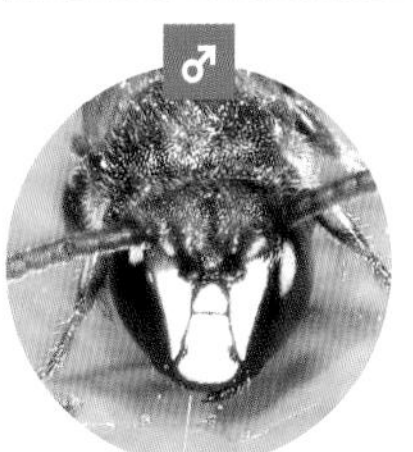

Common Yellow-face Bee
MANDIBLES black

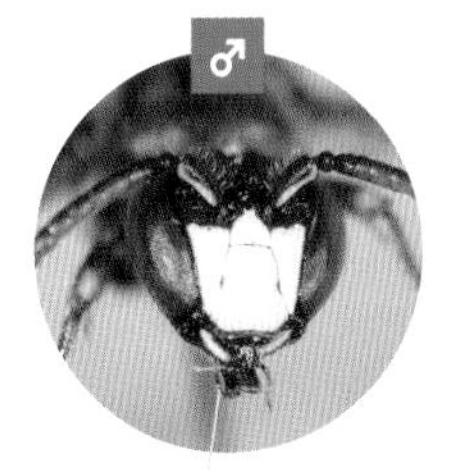

White-jawed Yellow-face Bee
MANDIBLES whitish-yellow

ANDRENINAE

Andrena (Mining bees) — 68 spp. (6†) | 20 ILL.

GUIDE TO SELECTED SPECIES

THORAX distinctly grey-and-black; **ABDOMEN** black	Grey-haired Mining Bee	*p.* 568
THORAX extensively grey-haired	Grey-backed Mining Bee	*p.* 568
THORAX mainly orange-brown haired; ABDOMEN black, with or without pale hair bands		
THORAX reddish-brown (♀) or brown (♂), sides very hairy and paler	Sandpit Mining Bee	*p.* 568
THORAX orange (♀) / brown (♂); **FACE** buff-haired; **HIND TIBIAE** + **TARSI** orange	Clark's Mining Bee	*p.* 565
THORAX dark orange; **HEAD, ABDOMEN** (glossy) + **LEGS** black	Cliff Mining Bee	*p.* 565
FORM small; **ABDOMEN s2–4** hair fringes white (♀) or greyish (♂), only the last unbroken	Short-fringed Mining Bee	*p.* 563
LEGS buff or yellowish	Yellow-legged Mining Bee	*p.* 563
ABDOMEN distinct orange tip	Early Mining Bee	*p.* 562
BODY orange; **HEAD** black; **ANTENNAE** black; **LEGS** black	Tawny Mining Bee	*p.* 564
FORM large; **HEAD** black; **ABDOMEN s1–4** with buff hairs (♀), **s5** with black hairs; **HIND TIBIAE** dark; **POLLEN BRUSH** orange.	Buffish Mining Bee	*p.* 564
FORM large; **THORAX** reddish; **ABDOMEN s1–3** with patches of white hairs	White-haired Mining Bee	*p.* 565
ABDOMEN tip brown-haired, rest hairless except white fringes on **s2–4**	Small Gorse Mining Bee	*p.* 567
BODY rather dark; **ABDOMEN** brown-haired	Chocolate Mining Bee	*p.* 564
ABDOMEN tip orange-haired, ♀ segments with white hair fringes, **s3** fringe broken in centre	Wilke's Mining Bee	*p.* 567
ABDOMEN black or ♀ form red from **s2** with some black spots	Small Scabious Mining Bee	*p.* 567
THORAX brown-haired; ABDOMEN black, with distinct red areas		
ABDOMEN s1–2 with at least some red; **HABITAT** a White Bryony specialist	Bryony Mining Bee	*p.* 563
ABDOMEN s2 with red, black base and tip	Red-girdled Mining Bee	*p.* 566
THORAX black; ABDOMEN black		
BODY black (**ABDOMEN** lacks red): **FORM** small	Common Mini-miner	*p.* 567
BODY black (**ABDOMEN** lacks red): **CLYPEUS** white; **HABITAT** a Tormentil flower specialist	Tormentil Mining Bee	*p.* 566
ABDOMEN all-black or ♀ form red-banded with 1 or 2 black spots; **HABITAT** chalk grassland, on scabious	Large Scabious Mining Bee	*p.* 566

Early Mining Bee *Andrena haemorrhoa*

BL 9–13 mm. **ID** ♀ thorax vivid orange; abdomen black with orange tip. ♂ thorax colour less vivid; face and sides of thorax buff; hind tibiae and tarsi mostly orange. **Hab** Grassland, woodland and gardens. **Cleptoparasite** Fork-jawed Nomad Bee *Nomada ruficornis* [N/I]. **SS** Other *Andrena* species, although ♀ is distinctive in the field [see table *above*

Short-fringed Mining Bee *Andrena dorsata*

BL 7–9 mm. **ID** ♀ thorax with short brown hairs (buff sides and face); abdomen black, S2–4 with white (greyish in ♂) hair fringes; only the last unbroken. **Hab** Heathland, moorland, sand dunes, chalk grassland and gardens. **SS** Other *Andrena* species [see table *opposite*].

Yellow-legged Mining Bee *Andrena flavipes*

BL 9–13 mm. **ID** Thorax brown (yellowish sides and face); abdomen black and mostly not hairy, S2–4 in ♀ with buff hair fringes; hind legs with orange pollen brushes; femora buff. ♂ face black; with buff and black hairs. **Hab** Open woodland, chalk grassland, soft rock cliffs, roadside banks and gardens. **Dist** Range gradually expanding. **Cleptoparasite** Painted Nomad Bee (*p. 544*). **Parasitoids** Dark-edged Bee-fly (*p. 350*), Dotted Bee-fly (*p. 350*) and other flies. **SS** Other *Andrena* species [see table *opposite*].

Bryony Mining Bee *Andrena florea*

BL 10–13 mm. **ID** Thorax brown; abdomen segments black, S1 + S2 with red, the extent of which varies between individuals. **Hab** Areas where White Bryony grows, such as heathland, open woodland, parks and gardens. **SS** Other *Andrena* species [see table *opposite*].

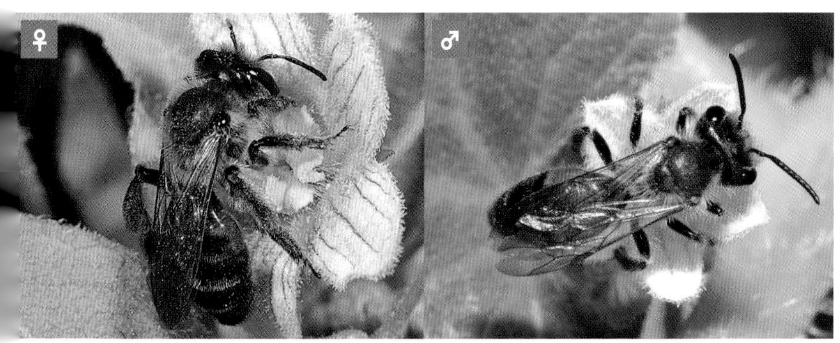

Tawny Mining Bee *Andrena fulva* RE (Ireland)

BL 9–15 mm. **ID** ♀ broad, robust-looking, body orange, contrasting with black head, antennae and legs. ♂ thorax colour less vivid; head with white hairs on lower face and large, black mandibles. **Hab** Grassland, parks and gardens. **Cleptoparasite** Broad-banded Nomad Bee *Nomada signata*. **Parasitoids** Dark-edged Bee-fly (*p. 350*) and *Leucophora* flies (Anthomyiidae (see *p. 343*)). **SS** Other *Andrena* species [see table on *p. 562*], although ♀ is unmistakable due to bright coloration.

Chocolate Mining Bee *Andrena scotica*

BL 10–13 mm. **ID** Thorax and abdomen brown-haired. **Hab** Woodland, grassland, heathland and gardens. **Cleptoparasites** Flavous Nomad Bee (*p. 545*), Marsham's Nomad Bee (*p. 544*). **SS** Other *Andrena* species [see table on *p. 562*].

Buffish Mining Bee *Andrena nigroaenea* VU

BL 10–15 mm. **ID** Large; head black; ♀ thorax brown, abdomen S1–4 with buff hairs, S5 with black hairs; hind tibiae dark, orange pollen brush. ♂ similar, clypeus and sides of face with black hairs. **Hab** Grassland, coastal dunes. **Cleptoparasite** Gooden's Nomad Bee (*p. 544*). **SS** Other *Andrena* species [see table on *p. 562*].

White-haired Mining Bee *Andrena nitida*

BL 10–16 mm. **ID** Large; ♀ thorax hairs reddish; abdomen black, S1–3 with patches of white hairs; also face and femora white-haired. ♂ similar, but face also has black hairs by inner eye margins. **Hab** Grassland, cultivated land and gardens. **Cleptoparasites** Gooden's Nomad Bee (*p. 544*), Marsham's Nomad Bee (*p. 544*). **SS Cliff Mining Bee**, other *Andrena* species [see table on *p. 562*].

Cliff Mining Bee *Andrena thoracica*

BL 9–16 mm. **ID** ♀ thorax dark orange; head black; abdomen glossy black; legs black. **Hab** Mainly coastal soft rock cliffs and adjacent areas. **Cleptoparasite** Gooden's Nomad Bee (*p. 544*) **SS White-haired Mining Bee**, other *Andrena* species [see table on *p. 562*].

Clark's Mining Bee *Andrena clarkella*

BL 9–14 mm. **ID** ♀ thorax dark orange; abdomen black. ♂ brown; face buff-haired. Hind tibiae and tarsi orange. **Hab** Heathland, moorland, sandpits, open woodland. **Cleptoparasite** Early Nomad Bee *Nomada leucophthalma*. **Parasitoid** Dark-edged Bee-fly (*p. 350*). **SS Cliff Mining Bee**; possibly other *Andrena* species [see table on *p. 562*].

Large Scabious Mining Bee *Andrena hattorfiana*

BL 13–17 mm. **ID** Body black; some ♀s with abdomen red-banded, also with one or two black spots; wings dark brown in both sexes, also abdomen tip orange. **Hab** Chalk grassland, usually on Field Scabious or Small Scabious. **Cleptoparasite** Armed Nomad Bee *Nomada armata* [N/I]. **SS** None.

Red-girdled Mining Bee *Andrena labiata*

BL 7–10 mm. **ID** Thorax black; abdomen mainly red, with black base and tip and S2 with black dot on side; ♀ more extensively red. ♂ clypeus white. **Hab** Grassland, woodland rides and churchyards. **Cleptoparasite** Short-spined Nomad Bee *Nomada guttulata* [N/I]. **SS** Superficially similar to *Sphecodes* species (*p.* 556), but those lack black spot on sides of S2.

Tormentil Mining Bee *Andrena tarsata*

BL 6–10 mm. **ID** Black, with glossy thorax and abdomen. Thorax with plentiful silver and some black hairs; abdomen segments with grey hair fringes; hind tibiae and tarsi brown. ♀ mandible 3-toothed. ♂ clypeus white. **Hab** Heathland, moorland, often on Tormentil flowers. **Cleptoparasites** Tormentil Nomad Bee (*p.* 543), **Flat-ridged Nomad Bee** *Nomada obtusifrons* EN [N/I]. **SS** Other *Andrena* species [see table on *p.* 562].

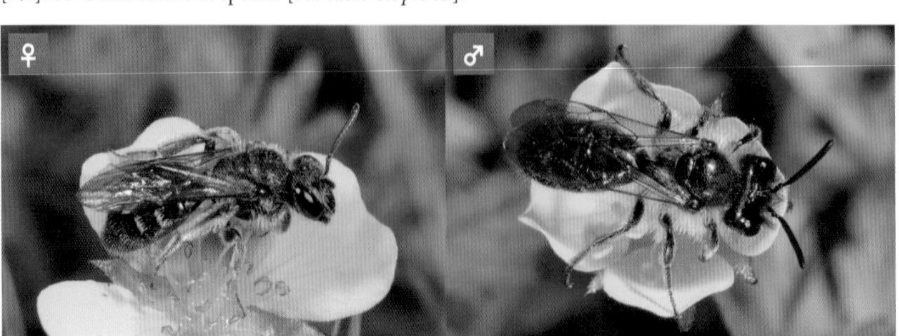

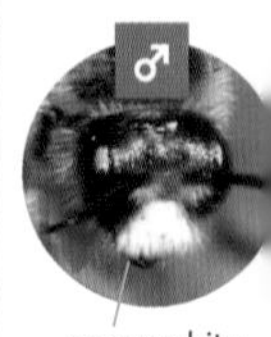

Wilke's Mining Bee DD
Common

Andrena wilkella

BL 8–12 mm. **ID** Thorax dark brown; abdomen black with orange-haired tip, segments with white hair fringes, S3 fringe broken in centre. Hind tibiae and pollen brushes orange. **Hab** Grassland, sand dunes, soft rock cliffs and gardens. **Cleptoparasite** Blunt-jawed Nomad Bee *Nomada striata* EN [N/I]. **SS** Small Gorse Mining Bee, from which told by abdomen tip orange-haired and S3 white hair fringe broken in centre [also see table on *p. 562*].

J F M A M J J A S O N D

♀

Small Scabious Mining Bee CR
Nationally Scarce

Andrena marginata

BL 7–12 mm. **ID** Thorax brown; abdomen in ♀ can range from red from S2 with some black spots, to black (♂ is black, with white or yellow clypeus). **Hab** Chalk grassland, open woodland, heathland and stabilized coastal sand dunes; as name implies, often visits scabious flowers. **Cleptoparasite** Silver-sided Nomad Bee *Nomada argentata* CR [N/I]. **SS** Other *Andrena* species [see table on *p. 562*].

J F M A M J J A S O N D

♀

Small Gorse Mining Bee DD
Local

Andrena ovatula

BL 8–12 mm. **ID** Thorax dark brown; abdomen black with a brown-haired tip, segments hairless except for white hair fringes on S2–4. **Hab** Heathland, grassland and other sites with plentiful gorses. **SS** Wilke's Mining Bee, from which told by abdomen tip brown-haired and S3 white hair fringe continuous; other *Andrena* species [see table on *p. 562*].

J F M A M J J A S O N D

♀

Common Mini-miner
Common

Andrena minutula

BL 5–8 mm. **ID** Black, including face hair in spring generation. **Hab** Open areas including grassland, woodland and gardens. **Cleptoparasite** Little Nomad Bee (*p. 545*)]. **SS** Other *Andrena* species, of which there are several other small to medium-sized black so-called 'mini-miners', from which told by microscopic examination of thorax punctuation and variation in abdomen hair fringes.

J F M A M J J A S O N D

Grey-backed Mining Bee EN
Andrena vaga

Rare (RDB1)

BL 11–16 mm. **ID** ♀ thorax white-haired; abdomen black, glossy, laterally with some white hairs. **Hab** Open habitats with plentiful willows, although little is known about this species. **Dist** Breeding only known at Dungeness, Kent (shingle and sand) and Hampshire. **Cleptoparasite** Lathbury's Nomad Bee *Nomada lathburiana* [N/I]. **SS** Grey-haired Mining Bee, from which told in ♀ by white-haired thorax.

Sandpit Mining Bee NT
Andrena barbilabris

Common

BL 8–12 mm. **ID** ♀ thorax reddish-brown, ♂ thorax brown; sides very hairy, with long, much paler hairs (yellow in ♀, whitish in ♂). Abdomen segments black with narrow white hair fringes. **Hab** Heathland, sand dunes, sandpits, soft rock cliffs. **Cleptoparasite** Sandpit Blood Bee (*p. 556*). **SS** Other *Andrena* species [see table on *p. 562*].

Grey-haired Mining Bee
Andrena cineraria

Local

BL 9–15 mm. **ID** ♀ thorax black-and-white-haired (appears partly grey); abdomen black. ♂ similar but thorax white-haired. Facial hairs white in both sexes. **Hab** Heathland, cliffs, open woodland, chalk grassland, parks and gardens. **Cleptoparasite** Lathbury's Nomad Bee *Nomada lathburiana* [N/I] and probably other nomad bees. **Parasitoids** Dark-edged Bee-fly (*p. 350*), Dotted Bee-fly (*p. 350*). **SS** Grey-backed Mining Bee, from which told by thorax colour (part black in ♀, or base black-haired in ♂).

(Guide to subfamilies and genera pp. 538–541)

Panurgus (Shaggy bees) **2 spp.**

Large Shaggy Bee
Panurgus banksianus

BL 9–12 mm. **ID** Medium-sized to large; the larger of the two *Panurgus* species. Black, sparsely haired; antennae black. **Hab** Heathland, sand dunes, acidic grassland; mainly near coast. **SS** Small Shaggy Bee (smaller).

Small Shaggy Bee
Panurgus calcaratus

BL 7–9 mm. **ID** Medium-sized; the smaller of the two *Panurgus* species. Black, sparsely haired; antennae black with orange segments, conspicuous in ♂, ♀ underside of S9–11 at least, paler. **Hab** Heathland and grasslands. **Beh** Two or more ♀s covered in pollen are often seen gaining access to a single entrance burrow, as this is a communal nest. **SS** Large Shaggy Bee (larger).

♀

♂

FURTHER READING AND USEFUL WEBSITES

Archer, M.E. 2014. ***The Vespoid Wasps (Tiphiidae, Mutillidae, Sapygidae, Scoliidae and Vespidae) of the British Isles***. Handbooks for the identification of British insects 6, part 6. Royal Entomological Society.

Baldock, D.W. 2010. ***Wasps of Surrey***. Surrey Wildlife Trust.

Broad, G.R., Shaw, M.R. & Fitton, M.G. 2018. ***Ichneumonid wasps (Hymenoptera: Ichneumonidae): their classification and biology***. Royal Entomological Society.

Chinery, M. 2011. ***Britain's Plant Galls***. Princeton University Press (Princeton WILD*Guides*).

Else, G.R., Bolton, B. & Broad, G.R. 2016. Checklist of British and Irish Hymenoptera – aculeates (Apoidae, Chrysidoidea and Vespoidea). *Biodiversity Data Journal* 4: e8050.

Else, G.R. & Edwards, M. 2018. ***Handbook of the Bees of the British Isles***. Vols. 1 & 2. The Ray Society.
Major textbook; includes detailed keys to bees in Vol. 1 and the latest, detailed distribution maps.

Falk, S. & Lewington, R. 2015. ***Field Guide to the Bees of Great Britain and Ireland***. British Wildlife Field Guides, Bloomsbury Publishing.

Lacourt, J. 2020. ***Sawflies of Europe. Hymenoptera of Europe 2***. NAP Editions.

Lebas, C., Galkowski, C., Blatrix. R. & Wegnez, P. 2019. ***Ants of Britain and Europe: a photographic guide***. Bloomsbury Publishing.

Skinner, G.J & Allen, G.W. 1996. ***Ants***. Naturalists' Handbooks 24. The Richmond Publishing Co. Ltd. [digital reprint 2013, Pelagic Publishing].

Wiśniowski, B. 2015. ***Cuckoo-wasps (Hymenoptera: Chrysididae of Poland***. Ojców: Ojców National Park.
Useful in Great Britain and Ireland, including detailed close-up images to aid identification.

Yeo, P.F & Corbet, S.A. 1995. ***Solitary Wasps***. 2nd Edition. Naturalists' Handbooks 3. The Richmond Publishing Co. Ltd. [digital reprint 2015, Pelagic Publishing].

www.bwars.com Bees Wasps and Ants Recording Society
Upon joining the Society, members receive a useful Members' Handbook

www.royensoc.co.uk Royal Entomological Society
RES Handbooks Volume 6, parts 1–5 inclusive on various Hymenoptera can be downloaded free

www.sawflies.org.uk The Sawflies (Symphyta) of Britain and Ireland

Conservation status

A summary of the conservation status is included in this book for species that are rare and threatened. Further information can be obtained for the UK at http://jncc.defra.gov.uk/page-3408, from which an up-to-date spreadsheet that includes details of legislation can be downloaded. For Ireland, information is available from www.biodiversityireland.ie/national-standards/irish-red-lists, although few insect orders have yet been assessed.

Although a number of UK species assessments have been published since 2000 (using 2001 threat categories), many insects have, as yet, only been assessed using the old pre-1994 International Union for Conservation of Nature (IUCN) categories; others not at all. The latest IUCN criteria consider the rate of decline in species when assessing the Red List grading, rather than just rarity status, as in the previous system. The codes used for the various IUCN categories for species that have been assessed as Nationally Rare (*i.e.* being threatened with extinction – Red Listed) are:

RE **Regionally Extinct**

CR **Critically Endangered** } Threatened

EN **Endangered** } Threatened

VU **Vulnerable** } Threatened

NT Near Threatened

DD Data Deficient, which effectively means the data are inadequate to assess the status

NOTE: **The Red List categories to which individual species are assigned may change periodically as new assessments are submitted to the IUCN.**

Species that have been assessed but considered not to be at risk are categorized as Least Concern. Since the majority of the species included in this book are common and widespread, and by default Least Concern, these have not been coded; only those species that fall into one of the categories outlined above have been coded.

Separate Red List assessments apply to Ireland and have been made for some groups of insects (www.biodiversityireland.ie/resources/irish-red-lists). For those species that have been categorized as anything other than Least Concern, the Irish Red List code is given, using the same icons as above but with a green border (*e.g.* VU).

For each of the species that is afforded a full account, a colour-coded status box is included above the map. Six broad categories are used, as summarized below, the criteria for which are necessarily rather loose and often reflect the lack of records, even for some common species:

Common	Common and easily found in suitable habitats at the appropriate time of year within the range indicated on the map. This relates to the majority of the insects covered in this book.
Local	Recorded from 101–300 10-km squares in Great Britain since 1980. The species may, however, be common where it is found.
Nationally Scarce	This category is used to denote official conservation status and relates to species recorded from 16–100 10-km squares in Great Britain since 1980/90. This status is still used by some involved with Red Lists. Many of the old assessments are in need of revision and some species should now be classified as rare. Some insect orders and families are poorly recorded and their status has never been assessed.
Rare	Applies to species red-listed since 2000 (*i.e.* those threatened with extinction), but if assessed under the old system is used for those recorded from fewer than 16 10-km squares in Great Britain since 1980. For species categorized under the old Red List system: **RDB1** = **Endangered**; **RDB2** = **Vulnerable**; **RDB3** = Rare; **RDBk** = **Insufficiently known** (this is shown in the red status box). Where species are clearly rare, but not officially Red Listed, "(not Red Listed)" is shown in the red status box.
Introduced; status indicated	Accidentally introduced or imported; may become naturalized.
Immigrant	Has reached Great Britain or Ireland by flying or being transported by the wind (true migrants return to their area of origin and there are very few examples of these).

Legislation

Some, mainly endangered insects are protected by law under the Wildlife and Countryside Act 1981 (as amended), Nature Conservation (Scotland) Act 2004 and Wildlife (Northern Ireland) Order 1985. The few species that are afforded full protection are coded as follows in the relevant species accounts.

● Legally protected in Great Britain

Ⓘ Legally protected in Northern Ireland

In addition, a number of butterflies are at least partially protected, but this information has not coded here. Up to date information on legislation in the UK is given at www.legislation.gov.uk and a spreadsheet is available for download from www.jncc.defra.gov.uk. A useful overall summary is available from https://cdn.buglife.org.uk/2019/07/Policy-and-legislation-summary-final-2014_0.pdf.

Reporting pest species

Sightings of the following pest species should be reported:
Asian Hornet: www.gov.uk/government/publications/asian-hornet-uk-sightings
Oak Processionary: www.gov.uk/government/news/public-urged-to-report-oak-processionary-moth-caterpillar-sightings

Further reading and sources of useful information

This section provides details of general insect identification guides not already listed within the text on each insect order, and suggestions of online and other resources that may help with identification, insect recording schemes and societies.

Further reading

Brock, P.D. 2019. ***A comprehensive guide to insects of Britain & Ireland***. Revised Edition. Pisces Publications.
Covers 2,100 species with over 2,700 photographs, therefore suitable for enthusiasts wishing to see an additional range of species to those covered in this book. Comprehensive enough for many naturalists, who have no need to purchase a range of books on several insect orders.

Barnard, P.C. 2011. ***The Royal Entomological Society book of British insects***. Wiley-Blackwell.
Guide to insect families, with all genera listed and indexed. Includes lists of numerous specialist identification works.

Chinery, M. 2005. ***Collins complete guide to British insects***. HarperCollins Publishers Ltd.

Fowles, A. 1994. ***Invertebrates of Wales: a review of important sites and species***. JNCC.

Lake, S., Liley, D., Still, R. & Swash, A. 2020. ***Britain's Habitats – A field guide to the wildlife habitats of Great Britain and Ireland***. 2nd Edition. Princeton University Press (Princeton WILDGuides).

McCormack, S. & Regan, E. 2014. ***Insects of Ireland***. The Collins Press.
Basic guide to some 120 selected insects, covering several popular groups.

Journal on migrant insects

Atropos www.atropos.info
Publishes Atropos, a useful journal that includes the latest news of migrant insects, articles and notes on identification and occasionally site guides.

Online resources

BioImages – Virtual Field-Guide (UK) www.bioimages.org.uk
Photographs of numerous species.

Field Studies Council www.field-studies-council.org/publications.aspx
AIDGAP guides, fold-out identification charts.

Freshwater Biological Association www.fba.org.uk/
Identification keys for insects associated with freshwater.

InvertebrateIreland Online www.habitas.org.uk/invertebrateireland
Includes checklists covering many of Ireland's insects.

ispot www.ispotnature.org
Help with identification; one of various on-line options.

Nature Conservation Imaging – Jeremy Early's images www.natureconservationimaging.com/Pages/nature_conservation_imaging_pictureindex_invertebrates.htm
Photographs of numerous species.

Nature spot – recording the wildlife of Leicestershire and Rutland www.naturespot.org.uk
Useful basic information and photographs featuring a wide range of insects.

NBN Atlas https://nbnatlas.org
Useful for searchable, indicative distribution maps of British insects. More accurate atlases and versions for some orders/families may, however, be found in various publications, details accessed via websites provided.

Nottinghamshire's online invertebrate resource www.eakringbirds.com/index.htm
Photographs of numerous species, also various downloads.

Royal Entomological Society www.royensoc.co.uk
RES Handbooks Volume 6, parts 1–5 inclusive on various Hymenoptera can be freely downloaded.

Steven Falk's images on Flickr www.flickr.com/photos/63075200@N07/collections
Photographs of selected insect orders (mainly Diptera and Hymenoptera) with identification tips.

Web ID Resources http://insectrambles.blogspot.com/2012/12/web-id-resources.html
A comprehensive list of web ID resources.

Museums

Many naturalists find that checking dead material in museum collections is helpful for identification. The Natural History Museum, London is one of the best places to start including the Angela Marmont Centre for UK Biodiversity. Researchers can arrange in advance to have free access to the UK reference collection, workspace with microscope and imaging facilities and othe support www.nhm.ac.uk/take-part/centre-for-uk-biodiversity.html. There are many regional museums throughout Britain and Ireland with collections of some or all orders, but sadly some o these are closing down due to lack of funding.

Insect recording schemes

National Biodiversity Centre, Ireland www.biodiversityireland.ie
Links to surveys, distribution maps, submitting records in Ireland.

Biological Records Centre www.brc.ac.uk/recording-schemes
Survey schemes for beginners and experienced entomologists.

Insect societies

The following national societies are additional to those listed in the introductions to the various insect orders.

Amateur Entomologists' Society
www.amentsoc.org

The AES, founded in 1935, is a registered charity that aims to promote the study of insects, especially amongst amateurs and young people.

The society publishes periodicals including the general AES *Bulletin*, the more specialist *Entomologist's Record* and a monthly Newsletter, and also a range of entomological books. It has an active Conservation Committee that, among other activities, produces the journal *Invertebrate Conservation News*, which is freely available to members.

In addition to a very popular Annual Exhibition and Trade Fair, the AES arranges museum visits, workshops and field events, as well as an annual Members' Day to coincide with its Annual General Meeting. One of the society's great strengths is that it is always keen to work in partnership with other like-minded organisations.

Younger members receive the *Bug Club Magazine*, to which they are encouraged to contribute (there are prizes!) and there is an annual Young Entomologists' Day that features talks by well-known entomologists as well as by the children themselves. Every other year there is a residential camp for Bug Club members and their families.

British Entomological and Natural History Society www.benhs.org.uk

Buglife – The Invertebrate Conservation Trust www.buglife.org.uk

Royal Entomological Society www.royensoc.co.uk
The RES has produced a Code for Insect Collecting. *Identification guides are published on insects, with many of the older titles that are still useful freely available.*

Selected relevant national organizations

Other national organizations with some involvement in entomological conservation and/or the management of nature reserves include:

Forestry and Land Scotland www.forestryandland.gov.scot

Forestry England www.forestryengland.uk

National Parks & Wildlife Service Ireland www.npws.ie

Natural England www.naturalengland.org.uk

People's Trust for Endangered Species www.ptes.org

RSPB www.rspb.org.uk

The Wildlife Trusts www.wildlifetrusts.org

In addition, there are numerous regional nature, entomological and wildlife groups, which there is no room to list.

Acknowledgements and photo credits

Many people have contributed to the production of this book and my thanks are due to everyone who has been involved. The acknowledgements that follow do not include the names of all the many entomologists who, over the years, have helped me (including staff at the Natural History Museum, London), whether with information, or pointing me towards finding particular species, grateful as I am to them. Instead, I have focussed on those who provided specific help with this book. The text relating to some orders was commented on by specialists and I would particularly like to thank **George Else** (Hymenoptera and generally), **Judith Marshall** (orthopteroid orders), **Maria Justamond** (Hemiptera) and **John Bingham** (Coleoptera) for constructive comments. **Rob Still** and **Andy Swash** of WILD*Guides* offered very considerable support and expertise, and maintained their enthusiasm for the project throughout – Rob's particular skills in designing the book and preparing the illustrations being clear to see. **Rachel** and **Anya Still** also supported production, including the time-consuming process of producing the maps. In addition, thanks are due to **Gill Swash**, **Brian Clews** and **Chris and Jude Gibson** for their contributions to proof reading, and **Steve Holmes** for help with the indexing. I am also very grateful to the **Amateur Entomologists' Society** for their support of this project.

With such a vast subject range, I have relied on the generosity of a number of photographers to ensure the necessary coverage to complete this book. These include some of Britain's and Ireland's foremost field naturalists and/or photographers, and the significant contribution of every one of them is gratefully acknowledged. Each image not taken by me is listed in this section, together with the photographer's name. The following photographers kindly agreed to help in response to specific requests; those marked with an asterisk (*) have 20 or more of their images included, and are due particular thanks:

Ian Andrews, **Phil Attewell**, **Tristan Bantock***, **Peter Barnard**, **George Beccaloni**, **Richard Becker**, **John Bingham***, **Geoff Campbell**, **Richard Comont**, **David Cooke**, **Peter Creed**, **Stuart Crofts**, **Stephen Doggett**, **Robert Dransfield** (InfluentialPoints.com), **Jeremy Early*** (natureconservationimaging.com), **Mike Edwards**, **Peter Eeles**, **Graham Ekins**, **Brian Eversham**, **Steven Falk** (stevenfalk.co.uk), **Christian Fischer**, **Martin Goodey**, **Mark Gurney**, **Jan Hamrsky**, **Paul Harris**, **Frank Hennemann**, **William Harvey**, **Jeff Higgott**, **Peter Hillman**, **Graham J. Holloway**, **Jim Johnson**, **Gus Jones** (bscg.org.uk), **Maria Justamond**, **Malcolm Lee**, **Kate McCusker**, **Gilles San Martin**, **Penny Metal**, **Simon Mitchell**, **Richard Naylor**, **Brian Nelson**, **Erland Refling Nielsen**, **Tim Norriss**, **Geoff Oxford**, **Christophe Quintin**, **Trevor and Dilys Pendleton*** (eakringbirds.com), **Neil Phillips** (uk-wildlife.co.uk), **Maurice Pugh**, **RHS / Rebekah Robinson**, **Tim Ransom**, **Stuart Read**, **Jeremy Richardson**, **Paul Ritchie** (hampshiredragonflies.co.uk/wordpress), **Francisco Rodriguez**, **Daniel Rojas**, **Andrew Salisbury**, **Dave Shute**, **Neils Sloth** (biopix.com), **Dave Smallshire**, **John Smit**, **Robert Still**, **Malcolm Storey**, **Andy and Gill Swash** (worldwildlifeimages.com), **Keith Tailby** (mothshots.com), **David Taylor**, **Gerard Visser** (microcosmos.nl), **John Walters** (johnwalters.co.uk), **Rosie Williams**, **Rosemary Winnall**. **Lena Ward** kindly gave permission to reproduce the late **Jim Grant**'s cicada image.

One of the images is reproduced under the terms of the Creative Commons Attribution-ShareAlike 2.0 Generic license, and another is in the Public Domain. These are indicated by "(CC 2.0)" and (PD) respectively after the photographer's name in the list.

The photographers named *above* occasionally assisted in other ways, along with the following, who helped with discussions on the insect fauna, finding insects, confirming identification from specimens or photographs, or in other ways: **Denise Bingham**, **Gavin Broad**, **Helen Brock**, **Paul Brown**, **Oskar Conle**, **Faye Da Costa**, **Adrian Durkin**, **Colin Easton**, **Florin Feneru**, **Tony Hopkins**, **Ben Price**, **Chris Raper**, **Mick Webb**. I mainly examined reference material in the Natural History Museum, London, including the Angela Marmont Centre for UK Biodiversity, which is a fantastic resource. **Baudewijn Odé** generously allowed use of his Orthoptera song recordings and supplied oscillograms. Any errors in identification and omissions remain mine.

Page 10 **Mayfly** [Stuart Crofts]
Page 11 **Bristletail** [Paul R. Sterry (Nature Photographers Ltd. / Alamy Stock Photo)]; **Silverfish** [H Helene (Alamy Stock Photo)]; **Webspinner** [RHS / Rebekah Robinson]; **Flea** [Richard Naylor]; **Louse** [Gilles San Martin]
Page 12 **Mantid** [Life on white (Alamy Stock Photo)]; **Stick-insect** [PLG (Alamy Stock Photo)]; **Earwig** [H Helene (Alamy Stock Photo)]; **Grasshopper** [Wil Leurs (Agami.nl), **Bush-cricket** [Wil Leurs (Agami.nl)]; **Cockroach** [Kim Taylor (Nature Picture Library / Alamy Stock Photo)]
Page 13 **Click beetle** [Paul R. Sterry (Nature Photographers Ltd. / Alamy Stock Photo)]; **Rove beetle** [Paul R. Sterry (Nature Photographers Ltd. / Alamy Stock Photo)]; **Soldier beetle** [Wil Leurs (Agami.nl)]; **Ground beetle** [Wil Leurs (Agami.nl)]; **Longhorn beetle** [Kim Taylor (Warren Photographic Ltd.)]; **Weevil** [StellaNature (Alamy Stock Photo)]; **Leaf beetle** [Kim Taylor (Warren Photographic Ltd.)]; **Ladybird** [Redmond Durrell (Alamy Stock Photo)]; **Scarab beetle** [Wil Leurs (Agami.nl)]; **beetle mouthparts** [Wil Leurs (Agami.nl)]; **Aphid** [Edward Phillips (Alamy Stock Photo)]; **Froghopper** [Wil Leurs (Agami.nl)]; **Plant bug** [Wil Leurs (Agami.nl)]; **Lace bug** [Wil Leurs (Agami.nl)]; **Shieldbug** [Wil Leurs (Agami.nl)]; **Leatherbug** [Wil Leurs (Agami.nl)]; **Damsel bug** [Wil Leurs (Agami.nl)]
Page 14 **Caddisfly** [Kim Taylor (Warren Photographic Ltd.)]; **Mayfly** [Kim Taylor (Warren Photographic Ltd.)]; **Alderfly** [H Helene (Alamy Stock Photo)]; **Stonefly** [H Helene (Alamy Stock Photo)]; **Snakefly** [H Helene (Alamy Stock Photo)]; **Scorpionfly** (*side*) [MYN / Tim Hunt (Nature Picture Library / Alamy Stock Photo)]; **Scorpionfly** (*above*) [INSADCO Photography (Alamy Stock Photo)]
Page 15 **Lacewing** [Wil Leurs (Agami.nl)]; **Antlion** [Wil Leurs (Agami.nl)]; **Dragonfly** [Tim Hunt (Nature Picture Library / Alamy Stock Photo)]; **Damselfly** [Matt Cole (Nature Picture Library / Alamy Stock Photo)]; **Aphid** [Kim Taylor (Warren Photographic Ltd.)]; **Jumping Plant Louse** [Tristan Bantock]; **Louse** [Richard Naylor]; **Booklouse** [Peter Barnard]; **Stylops** male [John Smit]
Page 16 **Darwin wasp** female [Kim Taylor (Warren Photographic Ltd.)]; **Darwin wasp** male [Wil Leurs (Agami.nl)]; **Nomad Bee** [Wil Leurs (Agami.nl)]; **Ant** wingless [Richard Becker (Alamy Stock Photo)]; **Ant** winged [Domiciano Pablo Romero Franco (Alamy Stock Photo)]; **Cuckoo wasp** [H Helene (Alamy Stock Photo)]; **Mason wasp** [Wil Leurs (Agami.nl)]; **Sawfly** [Alex Hyde (Nature Picture Library / Alamy Stock Photo)]; **Social wasp** [Alex Hyde (Nature Picture Library / Alamy Stock Photo)]; **Furrow bee** [Richard Becker (Alamy Stock Photo)]; **Honey bee** [Wil Leurs (Agami.nl)]; **Bumble bee** (*side*) [Wil Leurs (Agami.nl)]; **Bumble bee** (*above*) [Robert Pickett (Papilio / Alamy Stock Photo)]
Page 17 **Yellow-barred Peat Hoverfly** (*side*) [Wil Leurs (Agami.nl)]; **Median Wasp** (*side*) [Wil Leurs (Agami.nl)]; **Hoverfly** (*top left from above*) [Wil Leurs (Agami.nl)]; **Deerfly** [Wil Leurs (Agami.nl)]; **Bee-fly** [Paul R. Sterry (Nature Photographers Ltd. / Alamy Stock Photo)]; **Cranefly** [Geoff Oxford]; **Hoverfly** (*centre left from above*) [Wil Leurs (Agami.nl)]; **Horsefly** [Kim Taylor (Warren Photographic Ltd.)]; **Fungus gnat** [Keith Burdett (Alamy Stock Photo)]; **Hoverfly** (*bottom left from above*) [Wil Leurs (Agami.nl)]; **Mosquito** [Akil Rolle-Rowan (Alamy Stock Photo)]; **Robberfly** [Thijs de Graaf (Alamy Stock Photo)]
Page 18 **Burnet moth** [Andy and Gill Swash]
Page 22 **Snakefly** [Trevor and Dilys Pendleton]
Page 32 **Hornet Robberfly** [Robert Still]
Page 36 **Chater's Bristletail** and **Southern Bristletail** [Brian Eversham]
Page 37 **Cave Bristletail** [Christophe Quintin]; **Western Sea Bristletail** [Brian Eversham]
Page 38 **Grey Silverfish** [Graham J. Holloway]
Page 39 **Grey Silverfish** [Graham J. Holloway]
Page 40 **Sepia Dun** [Stuart Crofts]
Page 41 **Fishing fly** [David Taylor (CC 2.0)]; **Yellow May Dun** and **Sepia Dun** [Stuart Crofts]
Page 42 **Large Dark Olive** [Stuart Crofts]

Page 43 **Yellow May Dun** and **Large Brook Dun** [Stuart Crofts]
Page 44 **Blue-winged Olive** and **Sepia Dun** [Stuart Crofts]
Page 45 **Drake Mackerel Mayfly** [Peter Creed]
Page 46 **Emperor Dragonfly** [Andy and Gill Swash]
Page 49 **Club-tailed Dragonfly** [Dave Smallshire]
Page 57 **Irish Damselfly** [Geoff Campbell]
Page 58 **Red-eyed Damselfly** female [Erland Refling Nielsen]; **Small Red-eyed Damselfly** male [Andy and Gill Swash], female [Dave Smallshire]
Page 61 **Vagrant Emperor** [Daniel Rojas]; **Green Darner** [Jim Johnson]
Page 62 **Azure Hawker** [Kate McCusker]
Page 63 **Southern Migrant Hawker** [Paul Ritchie]
Page 65 **Common Clubtail** [Dave Smallshire]; **river scene** [Andy and Gill Swash]
Page 66 **Northern Emerald** face [Simon Mitchell]; **Yellow-spotted Emerald** [Dave Smallshire]
Page 67 **Downy Emerald** and **Brilliant Emerald** [Dave Smallshire]; **Northern Emerald** [Andy and Gill Swash]
Page 68 *Leucorrhinia* [Christian Fischer (PD)] ; **Yellow-spotted Whiteface** [Erland Refling Nielsen]; **Wandering Glider** [Dave Smallshire]
Page 69 **Yellow-winged Darter**, **Banded Darter** and **Vagrant Darter** [Dave Smallshire]
Page 71 **Black Darter** male [Andy and Gill Swash]
Page 75 **Northern February Red** [Gus Jones]; **Orange-striped Stonefly** [Rosie Williams]
Page 76 **Willow Fly** (*inset*) [Malcolm Storey]
Page 77 **Northern February Red** [Gus Jones]
Page 78 **Orange-striped Stonefly** [Rosie Williams]
Page 79 **Large Two-spotted Stonefly** head & pronotum [Malcolm Storey]
Page 81 **Ring-legged Earwig** [Francisco Rodriguez]; **Lesser Earwig** [Martin Goodey]
Page 83 **Short-winged Earwig** [Stuart Read]
Page 87 **Greenhouse Camel-cricket** [Graham Ekins]
Page 88 **Large Cone–head** [Dave Shute]
Page 94 **Large Cone–head** (*both*) [Dave Shute]
Page 102 **Greenhouse Camel-cricket** [Graham Ekins]
Page 110 **Lesser Mottled Grasshopper** (*both*) [Brian Nelson]
Page 115 **Wisley Webspinner** (*top right*) [Andrew Salisbury]; (*both*) [RHS / Rebekah Robinson]
Page 118 **Unarmed Stick-insect** eggs [Malcolm Storey]
Page 123 **Thailand Stick-insect** (*main image*) [Frank Hennemann]
Page 129 **Variable Cockroach** (*both*) [George Beccaloni]
Page 132 **Chewing Louse** [Jeremy Richardson]
Page 133 **Head Louse** structure [Gilles San Martin]; **Head Louse** [Richard Naylor]
Page 136 **Cloudy Brown Barklouse** [Brian Eversham]
Page 140 **Blue Shieldbug** [Maria Justamond]
Page 142 **Lesser Water Boatman** and **River Bug** [Brian Eversham]; **Pondweed Bug** [Gerard Visser]
Page 143 **Pygmy Backswimmer** [Gerard Visser]; **Sphagnum Bug** [Brian Nelson]
Page 146 **Beet Leaf Bug** [Tristan Bantock]; **Bed Bug** [Richard Naylor]
Page 151 **Petite Shieldbug** [Jeff Higgott]
Page 153 **Rambur's Pied Shieldbug** [Maria Justamond]
Page 154 **Lesser-streaked Shieldbug** (*top image*) [Stuart Read]
Page 156 **Blue Shieldbug** nymphs [Maria Justamond]; **Elongate Shieldbug** [Paul Harris]
Page 158 **Scarlet Shieldbug** [Maria Justamond]
Page 159 **Blue Shieldbug** [Maria Justamond]
Page 168 **Cryptic Leatherbug** [Tristan Bantock]
Page 170 **Knapweed Rhopalid** [Tristan Bantock]
Page 172 **Plane Groundbug** [Penny Metal]
Page 175 **White-spotted Groundbug** [Tristan Bantock]
Page 177 **Straw Groundbug** [Tristan Bantock]
Page 182 **Spiny Groundbug** [Tim Ransom]
Page 183 **Red-winged Groundbug** and **Beet Leaf Bug** [Tristan Bantock]
Page 184 **Spined Stiltbug** and **Straw Stiltbug** [Tristan Bantock]
Page 185 **Common Stiltbug** [Tristan Bantock]
Page 187 **Moss Lacebug** [John Bingham]; **Gorse Lacebug** [Tristan Bantock]
Page 190 **Bed Bug** eggs, feeding, bites, comparison with **Bat Bug** [Richard Naylor]
Page 203 **Glasswort Plant Bug** [Tristan Bantock]
Page 207 **Water Stick-insect** [Brian Eversham]
Page 208 **Common Water Boatman** [Brian Eversham]

Page 209 **Common Backswimmer** [Brian Nelson]
Page 210 **New Forest Cicada** [Jim Grant]; **Chalk Planthopper** [Tristan Bantock]
Page 211 **New Forest Cicada** [Jim Grant]
Page 217 **Lime Leafhopper** [John Bingham]
Page 220 **Broadened Planthopper** [Penny Metal]; **Rey's Planthopper** [Tristan Bantock]
Page 222 **Chalk Planthopper** and **Jumping Plant Lice** [Tristan Bantock]; **Soft Scales** [Richard Comont]
Page 224 **Dark Green Nettle Aphid** [Robert Dransfield]; **Willow-umbellifer Aphid** and **Tansy Aphid** [Robert Dransfield]
Page 225 **Woolly Apple Aphid** [John Bingham]; **Rose Aphid** [Robert Dransfield]
Page 226 **Ash Psyllid** [Tristan Bantock]
Page 227 **Horse-chestnut Scale Insect** [Richard Comont]; **Woolly Currant Scale Insect** [Richard Comont]
Page 232 **European Antlion** [David Cooke]
Page 233 **Orange-headed Lacewing** (*main image*) [Jeremy Early]
Page 234 **Dash Lacewing** [Brian Eversham]
Page 235 **Hook-winged Lacewing** [Gilles San Martin]
Page 239 **Common Alderfly** end of abdomen [Malcolm Storey]
Page 243 **Common Pool Helophorus** [Trevor and Dilys Pendleton]
Page 244 **Common Marsh Beetle** [Trevor and Dilys Pendleton]
Page 245 **Orange-patched Fungus Beetle** [Trevor and Dilys Pendleton]
Page 246 **Common Throscid** [Trevor and Dilys Pendleton]
Page 248 **King-Alfred's Beetle** [Tristan Bantock]
Page 249 **Lycoperd's Fungus Beetle**, **Common Shining Flower Beetle** and **Orange-patched Flower Beetle** [Trevor and Dilys Pendleton]
Page 250 **Common Cerylonid** and **Brown Mould Beetle** [Trevor and Dilys Pendleton]
Page 251 **Anchor Beetle** [John Bingham]
Page 252 **Red-snouted Bark Beetle** [Trevor and Dilys Pendleton]
Page 253 **Aspen leaf-miner Beetle** [Tristan Bantock]
Page 254 **Leaf-rolling Weevil** [John Bingham]
Page 257 **Great Diving Beetle** male [Neil Phillips]
Page 258 **Large-grooved Diving Beetle** [Brian Nelson]; **Supertramp** [Trevor and Dilys Pendleton]; **Spangled Diving Beetle** [Dave Shute]; **Lesser Spangled Diving Beetle** [Jan Hamrsky]
Page 259 **Lesser Diving Beetle** and **Semi-black Diving Beetle** [Neil Phillips]; **Scarce Lesser Diving Beetle** [Neils Sloth (Biopix.com)]
Page 266 **Copper Peacock** and **Green-socks Peacock** [John Bingham]
Page 268 **Necklace Ground Beetle** [John Walters]
Page 270 **Golden-dimpled Ground Beetle** [John Walters]
Page 271 **Black-headed Bullatus** [Trevor and Dilys Pendleton]; **2-spotted Ground Beetle** [Tristan Bantock]
Page 272 **Common Sun Beetle** [Trevor and Dilys Pendleton]
Page 273 **Müller's Ground Beetle** [Trevor and Dilys Pendleton]
Page 278 **Hairy Rove Beetle** and **Beautiful Philonthus** [Trevor and Dilys Pendleton]
Page 289 **Short-necked Oil Beetle**, **Rugged Oil Beetle** and **Mediterranean Oil Beetle** [John Walters]
Page 290 **Red-thorax Click Beetle** [Mark Gurney]
Page 291 **Pectinate Click Beetle** [Ian Andrews]; **Violet Click Beetle** [William Harvey]
Page 293 **Common Brown Agriotes** [Trevor and Dilys Pendleton]
Page 294 **Ponel's Click Beetle** [Trevor and Dilys Pendleton]
Page 297 **Blood-red Longhorn Beetle** [Gus Jones]
Page 300 **Ribbed Pine Borer** [Gus Jones]
Page 302 **Ladder-marked Longhorn Beetle** [Trevor and Dilys Pendleton]
Page 303 **White-banded Longhorn Beetle** [John Bingham]
Page 307 **Black Spruce Longhorn Beetle** and **Larch Longhorn Beetle** [John Bingham]
Page 310 **Scarce 7-spot Ladybird** (*inset*) [Phil Attewell]; **Hieroglyphic Ladybird** [Stuart Read]
Page 311 **13-spot Ladybird** [Brian Nelson]
Page 313 **Ant-nest Ladybird** [Tristan Bantock]
Page 314 **Red-headed Rhyzobius** and **Angle-spotted Scymnus** [John Bingham]; **Red Marsh Ladybird** and **Pine Scymnus** [Trevor and Dilys Pendleton];
Page 321 **Six-spotted Pot Beetle** [John Bingham]
Page 325 **Common Leaf Weevil** [Trevor and Dilys Pendleton]
Page 326 **Small Elm Bark Beetle** and **bark patterning** [Trevor and Dilys Pendleton]
Page 329 **Small Snakefly** [Rosemary Winnall]; **Scarce Snakefly** [Tristan Bantock]
Page 336 **Four-barred Major** [John Bingham]
Page 337 **Yellow Flat-footed Fly** and a **big-headed fly** [Peter Creed]

Page 338 **Mantis Fly** [Steven Falk]
Page 339 **Patched Diastatid** and **Orange Lauxaniid** [Steven Falk]; **Woodland Megamerinid** [Jeremy Early]
Page 342 **Orange-legged Lesser Housefly** [Steven Falk]
Page 344 **Giant Sabre Comb-horn** [John Bingham]
Page 350 **Western Bee-fly** [John Bingham]
Page 366 **Scarce Scorpionfly** male genital capsule [Ian Andrews]
Page 367 **Scarce Scorpionfly** [Ian Andrews]
Page 371 **Land Caddis** (*both*) [John Bingham]
Page 375 **Grouse Wing** [Jeremy Early]
Page 377 **Common Stylops** mating (*S. ater*) [John Smit], male [John Bingham]
Page 378 **Cat Flea** [Richard Naylor]
Page 379 **Human Flea** [Stephen Doggett]; **Hen Flea** [Richard Naylor]
Page 380 **Green-veined White** [Robert Still]
Page 381 **Cinnabar** (*bottom image*) [John Bingham]
Page 393 **Scarce Tortoiseshell** [Maurice Pugh]
Page 400 **Wall** mating pair [Andy and Gill Swash]
Page 402 **Purple Hairstreak** male [Peter Eeles]
Page 413 **Wood Tiger** [John Walters]
Page 423 **Ghost Moth** male [Tristan Bantock]
Page 426 **Transparent Burnet** [Penny Metal]
Page 429 **Common Lutestring** [Tristan Bantock]
Page 439 **Treble-bar** [Tim Norriss]
Page 443 **Oak Processionary** [Jeremy Early]
Page 449 **Wood Tiger** [John Walters]
Page 462 **Burren Green** [Keith Tailby]
Page 463 **Scarce Silver-lines** [Maurice Pugh]
Page 468 **Solomon's Seal Sawfly** [Peter Creed]
Page 470 **Honeysuckle Sawfly** and **Birch Sawfly** [John Bingham];
Birch Sawfly yellow form [Maria Justamond]
Page 473 **White-tipped/Confused Darwin Wasp** male [Malcolm Storey];
Yellow-striped Darwin Wasp male [Peter Hillman]
Page 475 **Holly Blue Darwin Wasp** [Jeremy Early]; **Oak Eggar Darwin Wasp** [Malcolm Lee]
Page 476 **Black-and-yellow Chalcid Wasp** [Penny Metal]
Page 477 **A gall wasp *Andricus* sp.** [Brian Eversham]
Page 478 **Scelionid** [Maria Justamond]
Page 480 **Collared Dryinid** [Jeremy Early]
Page 483 **Punctured Cuckoo Wasp** [Penny Metal]; **Violet Cuckoo Wasp** [Jeremy Early]
Page 487 **Narrow-bodied Cuckoo Wasp** [Jeremy Early]
Page 488 **Part-golden Cuckoo Wasp** [Mike Edwards]
Page 491 **Shining Guest Ant** [John Bingham]
Page 497 **Shining Guest Ant** [John Bingham]
Page 498 **Common Red Ant** antenna and **Sabulet's Red Ant** antenna [Phil Attewell]
Page 499 **Sabulet's Red Ant** [Richard Becker]
Page 502 **Large Tiphia** male [Jeremy Early]
Page 505 **White-spurred Spider Wasp** and **Blood Spider Wasp** [Jeremy Early]
Page 507 **Yellow-faced Spider Wasp** [Jeremy Early]
Page 508 ***Polistes* paper wasp** [Jeremy Early]
Page 509 **Little Mason Wasp** and **Four-banded Mason Wasp** [Jeremy Early]
Page 511 **Asian Hornet** [Tim Ransom]
Page 515 **European Paper Wasp** [Jeremy Early]
Page 517 **Garden Mason Wasp** [Steven Falk]
Page 520 **Little Mason Wasp** [Jeremy Early]
Page 526 **Thrips Wasp** and **Dahlbom's Digger Wasp** [Jeremy Early]
Page 528 **Slender Wood-borer Wasp** and **Red-bodied Stem Wasp** [Jeremy Early]
Page 533 **Armed Crabro Digger Wasp** leg (*inset*) and **Scarce Crabro Digger Wasp** leg (*inset*) [Jeremy Early]
Page 536 **Large Spurred Digger Wasp** [Steven Falk]
Page 539 **Long-horned Bee** male [John Bingham]
Page 550 **Long-horned Bee** male [John Bingham]

Index

This index includes the English and scientific (*in italics*) names of all the insects in this book.

Bold black text highlights species that are the subject of a main species account.
Semibold text is used for species that are illustrated but not the subject of a full account.
Regular text is used for other species that are referred to in the book but not illustrated, and for English names of genera; regular numbers indicate pages where other key information may be found.
BOLD CAPITAL text is used for orders or higher taxonomic levels.
REGULAR CAPITAL text is used for subdivisions of orders.
BOLD SMALL CAPITAL text is used for families and REGULAR SMALL CAPITAL text for subfamilies.
Italicized numbers indicate photos of early life stages (larvae or nymphs).

A

B

C

D

E

F

G

H

N

O

T

X

Y

Z